AF835344

HERBICIDE-RESISTANT CROPS

Agricultural, Environmental, Economic, Regulatory, and Technical Aspects

Edited by
Stephen O. Duke

CRC Press
Taylor & Francis Group
Boca Raton London New York

CRC Press is an imprint of the
Taylor & Francis Group, an **informa** business

First published 1996 by CRC Press
Taylor & Francis Group
6000 Broken Sound Parkway NW, Suite 300
Boca Raton, FL 33487-2742

Reissued 2018 by CRC Press

© 1996 by CRC Press, Inc.
CRC Press is an imprint of Taylor & Francis Group, an Informa business

No claim to original U.S. Government works

This book contains information obtained from authentic and highly regarded sources. Reasonable efforts have been made to publish reliable data and information, but the author and publisher cannot assume responsibility for the validity of all materials or the consequences of their use. The authors and publishers have attempted to trace the copyright holders of all material reproduced in this publication and apologize to copyright holders if permission to publish in this form has not been obtained. If any copyright material has not been acknowledged please write and let us know so we may rectify in any future reprint.

Except as permitted under U.S. Copyright Law, no part of this book may be reprinted, reproduced, transmitted, or utilized in any form by any electronic, mechanical, or other means, now known or hereafter invented, including photocopying, microfilming, and recording, or in any information storage or retrieval system, without written permission from the publishers.

For permission to photocopy or use material electronically from this work, please access www.copyright.com (http://www.copyright.com/) or contact the Copyright Clearance Center, Inc. (CCC), 222 Rosewood Drive, Danvers, MA 01923, 978-750-8400. CCC is a not-for-profit organization that provides licenses and registration for a variety of users. For organizations that have been granted a photocopy license by the CCC, a separate system of payment has been arranged.

Trademark Notice: Product or corporate names may be trademarks or registered trademarks, and are used only for identification and explanation without intent to infringe.

Catalog record is available from the Library of Congress

A Library of Congress record exists under LC control number: 95043474

Publisher's Note
The publisher has gone to great lengths to ensure the quality of this reprint but points out that some imperfections in the original copies may be apparent.

Disclaimer
The publisher has made every effort to trace copyright holders and welcomes correspondence from those they have been unable to contact.

ISBN 13: 978-1-315-89409-6 (hbk)
ISBN 13: 978-1-351-07319-6 (ebk)

Visit the Taylor & Francis Web site at http://www.taylorandfrancis.com and the
CRC Press Web site at http://www.crcpress.com

PREFACE

I have had an interest in herbicide-resistant crops since the idea of using biotechnology to produce such crops began to be discussed around 1980. My interest intensified in 1990, when I was asked to chair a panel to write a report on this topic for the Council for Agricultural Science and Technology. Since then, I have co-authored two more reviews that deal exclusively with the topic and have attended various workshops, symposia, seminars, and panels on herbicide-resistant crops all over the United States. In 1994, I visited with biotechnologists, regulatory officials, and scientists assessing risk who are dealing with herbicide-resistant crops in Denmark. In 1995, I attended a workshop on herbicide-resistant crops in Australia and the First International Conference on Herbicide-Resistant Crops and Weeds in Spain. I have watched the scientific and regulatory interest in herbicide-resistant crops become intense throughout most of the developed world.

The amount of literature on herbicide-resistant crops has grown from a trickle fifteen years ago to a flood. Yet, no book that deals exclusively with this topic has been produced. This book was conceived to fill the void. Moreover, my intention was to produce a book that would cover most, if not all, significant aspects of the topic, from the technology used to produce these crops to various potential impacts of their use, as perceived by authors with different points of view. This book should be a good reference for regulatory officials, biotechnologists, economists, agriculturists, and environmentalists with an interest in herbicide-resistant crops.

Herbicide-resistant crops have been a matter of hot debate over the past decade, a time when they have had no significant impact on agriculture. At the time of the printing of this book, the first significant impacts of these products of biotechnology will begin to be felt in the United States. Over the next several years comparing the predictions and conclusions of the various authors in this book with the actual use, performance, and impact of herbicide-resistant crops should be an interesting endeavor.

I thank the authors who contributed their time, effort, and expertise to this volume. All of the authors have had a deep involvement in herbicide-resistant crops, and most of them have published substantially in the areas that they cover. Many of the chapters in this book are the most up-to-date and detailed coverage of the specific topic available.

As with most edited books, the final list of chapters and authors differs somewhat from the one with which I started. It was difficult to find authorities in some areas who could produce a chapter in a reasonable period. I feel very fortunate to have accomplished this, despite having some of my initial authors drop out after some time. This put the new authors under considerable pressure to generate manuscripts in a limited time period. I especially thank those authors who came to my rescue to keep the book well balanced in terms of context.

Stephen O. Duke

THE EDITOR

Stephen O. Duke, Ph.D., is Director of the Southern Weed Science Laboratory, Agricultural Research Service, United States Department of Agriculture at Stoneville, Mississippi.

Dr. Duke received a B.S. in biology from Henderson State College (Arkadelphia, Arkansas) in 1966, a M.S. in botany from the University of Arkansas (Fayetteville) in 1968, and a Ph.D. in botany from Duke University (Durham, North Carolina) in 1975.

He is a member of numerous scientific societies and organizations in which he has held many offices and assignments and has won numerous awards. These include: the American Association for the Advancement of Science (Fellow); the American Society for Plant Physiologists (President of the Southern Section); the Weed Science Society of America (President-Elect, Fellow, Scientist of the Year), the Southern Weed Science Society (President, Scientist of the Year), the American Society for Photobiology, the American Chemical Society (Vice Chair of Natural Products Subdivision), and the Council for Agricultural Science and Technology (Board of Directors, Chair of Editorial Committee). Among other awards, he has won the USDA Edminster Award, the USDA-ARS Mid-South Area Scientist of the Year Award, and the Ciba-Geigy Agricultural Recognition Award.

Dr. Duke has published more than 100 refereed research journal articles and more than 50 invited reviews. He is co-author of *Physiology of Herbicide Action* and editor of four other volumes. He has presented well over 100 invited lectures at national and international meetings and has organized numerous symposia for many different organizations. His current research interests include the use of natural products as herbicides and the interaction of biological weed control agents with pesticides.

CONTRIBUTORS

Greg Baldwin, M.S.
Stoneville Pedigreed Seed Company
Stoneville, Mississippi

Gerard G. Barry, Ph.D.
The Agriculture Group
Monsanto
St. Louis, Missouri

Newell F. Bascomb, Ph.D.
American Cyanamid Company
Princeton, New Jersey

Peter Böger, Ph.D.
Department of Plant Physiology
University of Konstanz
Konstanz, Germany

Margaret R. Bolick, Ph.D.
Nebraska State Museum
University of Nebraska
Lincoln, Nebraska

Orvin C. Burnside, Ph.D.
Department of Agronomy and Plant Genetics
University of Minnesota
St. Paul, Minnesota

Harold D. Coble, Ph.D.
Crop Science Department
North Carolina State University
Raleigh, North Carolina

Bruce Coulombe, Ph.D.
Stoneville Pedigreed Seed Company
Stoneville, Mississippi

Xavier Delannay, Ph.D.
The Agriculture Group
Monsanto
St. Louis, Missouri

Michael J. Donnelly-Vanderloo
Department of Crop Science
University of Guelph
Guelph, Ontario, Canada

Stephen O. Duke, Ph.D.
Agricultural Research Service
United States Department Agriculture
Stoneville, Mississippi

Donald N. Duvick, Ph.D.
Department of Agronomy
Iowa State University
Ames, Iowa

William E. Dyer, Ph.D.
Department of Plant, Soil, and Environmental Sciences
Montana State University
Bozeman, Montana

Charlotte V. Eberlein, Ph.D.
Department of Plant, Soil, and Entomological Sciences
University of Idaho
Aberdeen, Idaho

David E. Eichholtz
The Agriculture Group
Monsanto
St. Louis, Missouri

Graeme Evans, Ph.D.
Bureau of Resource Sciences
Department of Primary Industries and Energy
Canberra, Australia

Robert T. Fraley, Ph.D.
The Agriculture Group
Monsanto
St. Louis, Missouri

Roy L. Fuchs, Ph.D.
The Agriculture Group
Monsanto
St. Louis, Missouri

Rebecca J. Goldburg, Ph.D.
Environmental Defense Fund
New York, New York

Jonathan B. Gressel, Ph.D.
Department of Plant Genetics
Weizmann Institute of Sciences
Rehovot, Israel

J. Christopher Hall, Ph.D.
Department of Environmental Biology
University of Guelph
Guelph, Ontario, Canada

F. Dan Hess, Ph.D.
Sandoz Agro, Inc.
Palo Alto, California

D. Douglas Hopkins, J.D.
Environmental Defense Fund
New York, New York

Catherine M. Houck, Ph.D.
Department of Molecular Biology
Calgene, Inc.
Davis, California

David J. Hume, Ph.D.
Department of Crop Science
University of Guelph
Guelph, Ontario, Canada

Kathleen H. Keeler, Ph.D.
School of Biological Sciences
University of Nebraska
Lincoln, Nebraska

Kyle E. Keller, Ph.D.
United States Department of Agriculture
Agricultural Research Service
National Soil Tilth Laboratory
Ames, Iowa

John A. Kiser, M.S.
Calgene, Inc.
Davis, California

Ganesh M. Kishore, Ph.D.
The Agriculture Group
Monsanto
St. Louis, Missouri

David Last, Ph.D.
C.S.I.R.O.
Division of Plant Industry
Canberra, Australia

Danny Llewellyn, Ph.D.
C.S.I.R.O.
Division of Plant Industry
Canberra, Australia

Carol Mallory-Smith, Ph.D.
Department of Crop and Soil Science
Oregon State University
Corvallis, Oregon

C. Jeffry Mauvais, Ph.D.
Stine-Haskell Research Center
DuPont Agricultural Products
Wilmington, Delaware

George D. McLean, Ph.D.
Department of Primary Industries and Energy
Bureau of Resource Sciences
Canberra, Australia

Norihiko Misawa, Ph.D.
Kirwin Brewery Company
Central Laboratories for Key Technology
Yokohama, Japan

David W. Monks, Ph.D.
Horticultural Science Department
North Carolina State University
Raleigh, North Carolina

Thomas B. Moorman, Ph.D.
United States Department of Agriculture
Agricultural Research Service
National Soil Tilth Laboratory
Ames, Iowa

Stephen R. Padgette, Ph.D.
The Agriculture Group
Monsanto
St. Louis, Missouri

Diane B. Re
Department of Regulatory Affairs
Monsanto
St. Louis, Missouri

Stephen G. Rogers, Ph.D.
Department of Regulatory Affairs
Monsanto
St. Louis, Missouri

Leonard L. Saari, Ph.D.
Stine-Haskell Research Center
DuPont Agricultural Products
Newark, Delaware

Gerhard Sandmann, Ph.D.
Biosynthesis Group
Department of Botany
J. W. Goethe University
Frankfurt, Germany

Frank S. Serdy, Ph.D.
Department of Regulatory Affairs
Monsanto
St. Louis, Missouri

Dale L. Shaner, Ph.D.
American Cyanamid Co.
Princeton, New Jersey

Timothy D. Sherman, Ph.D.
Department of Biological Sciences
University of Southern Alabama
Mobile, Alabama

Wendy Smith, Ph.D.
American Cyanamid Co.
Princeton, New Jersey

David A. Somers, Ph.D.
Department of Agronomy and Plant Genetics
University of Minnesota
St. Paul, Minnesota

David M. Stalker, Ph.D.
Calgene Inc.
Davis, California

Terry B. Stone
Department of Regulatory Affairs
Monsanto
St. Louis, Missouri

Loren W. Tauer, Ph.D.
Department of Agricultural, Resource, and Managerial Economics
Cornell University
Ithaca, New York

Donald C. Thill, Ph.D.
Department of Plant, Soil, and Entomological Sciences
University of Idaho
Moscow, Idaho

Charles E. Turner, Ph.D.
United States Department of Agriculture
Agricultural Research Service
Western Regional Research Center
Albany, California

Indra K. Vasil, Ph.D.
Department of Horticultural Sciences
University of Florida
Gainesville, Florida

Kevin C. Vaughn, Ph.D.
United States Department of Agriculture
Agriculture Research Service
Stoneville, Mississippi

Stephen Waters, Ph.D.
Monsanto
Europe S.A.
Brussels, Belgium

John W. Wilcut, Ph.D.
Department of Crop Science
North Carolina State University
Raleigh, North Carolina

Alan C. York, Ph.D.
Department of Crop Science
North Carolina State University
Raleigh, North Carolina

CONTENTS

Part II Effects on Weed Control

Part III The Economics

Part IV Environmental Concerns

Part V Regulation

Part VI Risk Assessment

HERBICIDE-RESISTANT CROPS

Agricultural, Environmental, Economic, Regulatory, and Technical Aspects

Chapter 1

HERBICIDE-RESISTANT CROPS — BACKGROUND AND PERSPECTIVES

Stephen O. Duke

CONTENTS

1. INTRODUCTION

Biotechnology may alter our future in many ways. One is by altering crop species to provide new traits that have been impossible or very slow to produce by conventional breeding. Imparting herbicide resistance to normally herbicide-susceptible crops to produce herbicide-resistant crops (HRCs) has been the most intensively exploited area of plant biotechnology. However, this work has been the focus of the ire of various environmental and other advocacy groups.

The purpose of this chapter is to provide the reader with some background and perspective before reading the other chapters of this book. This information is provided with the caveat that I am certain to have biases. My background is not that of a biotechnologist, but that of a plant scientist, most of whose career has dealt with weeds and weed management topics. I have, however, been a member of environmentalist groups and consider myself to be very concerned about environmental

1-56670-045-0/96/$0.00+$.50
© 1996 by CRC Press Inc.

issues. In trying to be dispassionate about this topic, I hope to avoid the hyperbole generated by some members of both sides of this issue. I have tried to generate an impartial treatment of the topic, based on the best scientific information available. Nevertheless, please consider that this chapter could have been quite different if penned by a scientist with another background.

The topic of HRCs has been reviewed previously and commented on by authors with diverse backgrounds and divergent points of view (e.g., Bright 1992; Claus 1992; Comstock 1989, 1990; Dekker and Comstock 1992; Dekker and Duke 1995; Duke et al. 1991; Dyer et al. 1993; Goldburg et al. 1991; Gressel 1993; Hoyle 1993; Krimsky and Wrubel 1993; Miller 1991; Rüdelsheim 1993).

2. THE STATUS OF WEED MANAGEMENT AND ITS FUTURE

Weeds are major agricultural pests that can devastate a crop if not properly controlled or managed. Until World War II, weeds were managed primarily by cultural methods, with heavy emphasis on tillage and/or hand hoeing. These methods were extremely labor intensive and contributed to soil erosion. Over the past 50 years, such methods have been largely replaced by the use of chemical herbicides. This technology has contributed to increased yields, with a dramatically reduced farm population. Herbicides are used on virtually 100% of the acreage of the major agronomic crops in developed countries.

To date, herbicides have been largely tailored to be used with particular crops, rather than the crops being bred to tolerate the herbicide. In other words, selective herbicides have been marketed to which the crop is naturally resistant. Natural resistance is often incomplete, and sublethal phytotoxicity to the crops has been a problem that farmers have learned to endure. Nonselective herbicides can only be used when the crop is not growing or by using specialized and often inadequate application technology to avoid contact with the crop.

The technology of herbicides has been enormously successful from the standpoint of the farmer (Duke 1994). Weeds can be managed quickly with one or more applications of a herbicide, and the farmer and the farmer's family can then pursue other problems and interests. In industrialized nations herbicides account for 60 to 70% of pesticides sold for agricultural use, by volume (e.g., Kirshner 1994). In undeveloped countries herbicides account for a much smaller percentage of total pesticide use, and a much higher proportion of the population is required for farming (Klassen 1995).

Yet concern is growing about the contamination of food, surface and ground waters, and the air by herbicides. Part of the concern is fueled by the improving detection technology that allows for the identification of xenobiotics at diminishingly low concentrations. However, little doubt exists that xenobiotics, including herbicides, accumulate in compartments of the environment in which they degrade very slowly, such as in groundwater. Furthermore, the toxicological implications of complex mixtures of pesticides and their degradation products in food, air, and water are not well understood. Thus, much of the public sector funding of research on weed management is focused on reducing or eliminating reliance on synthetic chemical herbicides.

Nevertheless, public sector funding for weed research has lagged far behind that focused on insect and plant pathogen pests (Duke 1992). Despite the fact that approximately two thirds of agricultural pesticides used in the U.S. are herbicides, <10% of public research dollars for pest management research in the U.S. is spent on weed science. The reasons for this were reviewed recently (Duke 1992; Gianessi and Darby 1993). As a partial result of weed management research having a relatively low priority for public research funding, there are few promising economical and effective replacements for herbicides in agronomic crops in the near future (Duke et al. 1993; Gianessi 1993a, b).

For example, the same or fewer biocontrol products commercially are available for weed control now than a decade ago (Leathers et al. 1993), despite a plethora of patented potential products (Watson 1993). The reasons that these products have not been successful were reviewed by Zorner et al. (1993). Such products are not likely to replace herbicides in the near future. Tillage operations for weed control such as ridge tillage are not compatible with the movement to reduce or eliminate tillage operations. Reduced and no tillage farming is usually dependent on postemergence herbicides, however, allelopathic cover crops can reduce this dependency (Smeda 1993).

Although there do not appear to be economical replacements for herbicides in agronomic crops in the near future, evidence exists that the amount of herbicide can be reduced significantly. In Scandinavia reductions in the amount of herbicide used and the number of herbicide applications were simply mandated by law. The reduction goals for the amount of herbicides used were met largely by switching from low unit to high unit activity products (Bellinder et al. 1994). This legislative approach does not take into account the relative toxicities, potential for environmental pollution, and other properties of the products used. In the U.S. much public sector research promises to greatly reduce the amount of herbicide used without eliminating herbicides.

This research can be divided into three categories: (1) improvements in application technology; (2) advanced formulation technology; and (3) the use of guidelines and expert systems to prescribe minimal amounts of herbicides for the desired effect.

Improved application technology has the potential to reduce significantly the volume of herbicides used over the next decade. Herbicides are usually sprayed over an entire area with the assumption that the weeds will be contacted, along with everything else. Electronic advances allow the detection of foliage and the direction of spray only onto detected foliage. Such spray systems are already commercially available in the U.S. and Australia (Houtsma 1994). More advanced image analysis detection systems that will differentiate between weeds and crops are under development. With the current rate in advances in electronics and computers, spray systems that will identify a weed species and direct the proper amount of the proper herbicide to kill that species should be available within a decade. The amount of herbicide applied can also be reduced through the use of geographic information system/global positioning system technologies that map fields for precise herbicide applications based on spatial variations in soil types and other factors. Such systems will greatly reduce the amount of herbicide needed for adequate weed management. However, this technology could result in the loss of some herbicides from the marketplace because of lost sales volume.

Improved herbicide formulations, combined with appropriate application technology, can improve the efficacy of herbicides so that less herbicide is needed for the same result. Formulation in oils or invert emulsions have reduced the effective dose of some herbicides by 30 to 70% (e.g., McWhorter et al. 1992). Specialized application devices, chemical industry resistance, and potential acute toxicity problems when spraying highly concentrated formulations have impeded this technology.

Using reduced herbicide rates and minimizing herbicide use by decision aid computer programs are also reducing herbicide use. Herbicides are costly, yet farmers often use them in a prophylactic manner, following the recommendations of herbicide manufacturers. In some cases the recommended rates are more than double what is required for adequate weed management. The manufacturer must ensure that its recommended rates are adequate for any contingency. Farmers are learning to use reduced herbicide rates (e.g., DeFelice et al. 1989) and to use them at the proper rates only when computer-run expert systems predict that spraying herbicides is cost effective for the observed weed species and populations (e.g., Rydahl and Thonke 1993; Stigliani and Resina 1993).

This public sector research, along with the focus of the herbicide industry on high unit activity herbicides that are effective at ounces per acre or grams per hectare, is leading to significant reductions in the amount of herbicide used per unit area to manage weeds in agronomic crops. Nevertheless, herbicides remain the most cost effective and reliable method of managing weeds and apparently will remain so into the 21st century.

3. THE HERBICIDE INDUSTRY — ITS CURRENT STATUS

During the past 15 years, the number of companies involved in herbicide discovery, development, and marketing has undergone considerable attrition, especially in the U.S., in which more than half of the companies in this business a decade ago are no longer involved (Klassen 1995). A general downward trend can also be found in the number of new pesticide active ingredients introduced annually (Klassen 1995). The attrition is due to multiple factors, including greatly increased costs of discovering a product that will compete well with existing products (a move from new markets to replacement markets); much higher costs of meeting regulatory requirements for registration of new and old products; uncertainty about the future regulatory environment; predictions of movement toward reduced herbicide use due to new methods and technologies; and public and internal pressure for large companies, in which agrochemicals represent a relatively small proportion of business, to divest themselves of their pesticide business. The general feeling among those in this industry is that attrition will continue, especially in the U.S. In Europe the attrition rate has been slower, and in Japan little or no attrition has occurred. Japan, Western Europe, and the U.S. account for virtually all of the herbicide discovery and development effort in the world.

In the developed world the herbicide market is essentially a replacement market. Virtually all acreage of agronomic crops are treated at least once with a herbicide, and farmers of agronomic crops generally have choices of several efficacious herbicides. In the U.S. the herbicide choices are very wide for corn (maize) and soybeans because these two crops represent the largest acreage (markets) for herbicides. The cost of herbicide registration generally precludes the development of herbicides for minor crops. Even if there is not an economic problem, there are more potential problems with horticultural crops, such as higher consumer complaint rates with unprocessed fruits and vegetables, and sometimes a substantial addition to the allowable daily intake for that herbicide. Thus, the herbicide industry is focused primarily on the very competitive markets for corn, soybean, rice, wheat, small grains, and cotton herbicides.

Considering the cost of discovery, development, and registration of new products, more effort is being made to expand the use of existing products. Working against this strategy are the attrition of older products caused by the re-registration process and the efforts in the public sector to reduce herbicide use. One of the strategies to increase the use of old herbicides is to expand the crops that the product can be used with by generating HRCs. Still, many in the herbicide industry view this strategy as one that will be commercially successful with only a few herbicides in major crops. For some of the same reasons that there is little interest in producing new herbicides for minor crops, there is little interest in making minor crops resistant to existing herbicides by biotechnology.

4. THE AGRICULTURAL BIOTECHNOLOGY INDUSTRY — ITS CURRENT STATUS

The plant agricultural biotechnology industry can be divided roughly into that conducted by large companies and corporations, that done by seed companies, and that being done by small venture-capital companies. There are numerous examples of large, diversified companies contracting biotechnology projects with seed or biotechnology companies that specialize in biotechnology. Seed companies have been bought by both small biotechnology companies and large corporations involved in plant biotechnology, because selling seed is the ultimate way to get the biotechnology to the marketplace.

The U.S. has led the way in agricultural biotechnology, followed by Europe, Japan, and the rest of the world, in that order. Efforts to create HRCs also generally follow this pattern. The reasons for U.S. leadership in these efforts are not entirely due to our leadership in this technology. In Europe the incentive to create transgenic crops is much lower than in the U.S. because there is even less need to increase crop production than in the U.S. and the "green" movement of Europe is much more opposed to introduction of transgenic organisms than most environmentalists in the U S. (see Chapter 23). Furthermore, there seems to be more skepticism in Europe about such technology paying off financially to the producing company. The reasons for the apparent lack of activity in this area in Japan are unclear.

Hundreds of millions of dollars have been spent in the creation and development of transgenic crops. In addition to traits that impart herbicide resistance, work is progressing on improving agronomic characteristics, food quality, storage characteristics, and resistance to insects and plant pathogens. Herbicide resistance is the only trait that is tied to the sale of another product. Little of the money spent on biotechnology has been returned by sales of improved crop seeds; however, the industry is poised to release many transgenic products in the next decade. The regulatory and economic fate of HRCs will strongly influence the business of plant biotechnology.

5. PUBLIC SECTOR RESEARCH AND HERBICIDE-RESISTANT CROPS

The amount of public sector research on HRCs in the U.S. has almost ground to a standstill. After publication of *The Bitter Harvest* (Goldburg et al. 1990), in which tax-supported research in this area was strongly criticized, efforts were made to prohibit such federally funded research through legislation. Although the statute has not been enacted at the time of this writing, federal agencies sensitized to the negative political aspect of such research phased out the small amount of research in this area. There was considerable disagreement as to how much research of this type was being funded by the U.S. Department of Agriculture (Goldburg et al. 1990; Duke et al. 1991). At this time, to my knowledge, there is no research to create HRCs for commercial purposes being funded by the U.S. Government. Some have argued that in some cases public funding to create and develop HRCs in which the herbicide industry has no interest may benefit the public (Gressel 1993).

6. HERBICIDE-RESISTANT CROPS — THEIR FIT

Virtually all HRCs being commercially developed are major agronomic crops for which many herbicide products already exist and for which more are in the development stage. Thus, the herbicide for which the HRC has been designed must offer an advantage over existing products. For universal adoption of an HRC for a particular crop, the herbicide for which the HRC has been engineered must offer an overwhelming advantage. There is skepticism within the public sector of agricultural research that such an HRC will emerge. The most likely successes are considered by many to be HRCs for nonselective herbicides such as glyphosate or glufosinate or HRCs that lie in an intermediate zone between truly major crops such as rice, soybeans, and corn and minor crops. The best example is cotton, on which several herbicides are used, but for which the availability of suitable postemergence herbicides is very limited.

If HRC minor crops were available, they would reduce the cost of weed management significantly. Relatively few herbicides are available for these crops for the reasons mentioned above. Thus, in some crops (e.g., lettuce) farmers must expend several hundred dollars per hectare for weed management by cultivation (manual and mechanical) and/or expensive and environmentally questionable plastic mulches. Nevertheless, because the cost of herbicide registration and the potential risks are so high, herbicide manufacturers have little interest in marketing herbicides for minor crops, even if they solve many problems for the farmer.

Crops resistant to postemergence herbicides are a potential solution to parasitic weeds such as witchweed or dodder. These weeds are not overwhelming problems in the U.S., but they reduce agricultural output in Africa substantially (Akobundu 1992), and contribute substantially to famine in parts of sub-Saharan Africa. Such parasitic weeds cannot be removed by cultivation and nonselective herbicides will kill the crop as well as the weed.

Preemergence herbicides and tillage are commonly used together to preempt potential weed problems. Preemergence herbicides are used as prophylactic treatments at relatively high rates. Some of these herbicides are those most often found to contaminate groundwater. There is growing pressure to reduce tillage and the use of preemergence herbicides. Crops made resistant to good nonselective, postemergence herbicides will lend themselves to these trends. Such a change in weed management methods should reduce soil erosion and herbicide movement to groundwater.

The evolving resistance of weeds to herbicides is a growing problem (LeBaron 1991), although the extent of the problem is not as great as that of insecticide resistance. Herbicide-resistant crops could be used to manage herbicide resistance, in that they will give the farmer the option of a different herbicide with which to rotate other herbicides. However, HRCs could increase the selection pressure for herbicide resistance if used unwisely (e.g., extended use, year after year).

There are several potential fits for HRCs in agriculture, and in some cases the fit will offer advantages that ensure their adoption. However, the HRCs must compete with existing technologies, and there is no guarantee that they will always compete well. Herbicide-resistant crops will offer one more tool for farmers in their war with weeds, but they are not likely to revolutionize weed management technology.

7. HERBICIDE-RESISTANT CROPS — THEIR ECONOMICS

Anything that allows for more efficiency in production of a product, whether industrial or agricultural, reduces the cost of production, ultimately providing a competitive (hence economic) advantage to those who use the new method or tool. Herbicide-resistant crops will be no different. Agriculture has become much more efficient in developed countries and in many third-world countries during the last half of this century. Without these gains, the world would be much closer to the Malthusian scenario predicted by some to occur near the end of this century (Ehrlich 1968).

In many developing or third-world countries improvements in agricultural productivity has certainly prevented or postponed mass starvation. In developed countries it has reduced dramatically the percentage of the average worker's salary that goes to feed themselves and their family. It has also meant the demise of subsistence farmers and many small farms in developed countries. The profit from most small farms is no longer sufficient to support a farm family, no matter how productive the farm. This has resulted in a great reduction in the number of farmers and a corresponding deterioration of rural infrastructure. Those who produce our food now generally manage much larger farms than the average farmer of 50 years ago. Present-day farmers have more scientific, technical, and business training than farmers of half a century ago. Many of these individuals prefer to be called by the vague terms "producer" or "grower", suggesting that they do not care for the term "farmer". The small farmers that are left generally have other sources of income to supplement their farm profits. These trends are the result of the interplay of technology and economics. Will HRCs perpetuate these trends?

The only analyses of the economics of HRCs suggests that this technology will have similar effects as any of the past agricultural technological innovations (Hayenga et al. 1992; Tauer and Love 1989). Economics dictate that farmers (growers, producers) will adopt available technology that will improve their profits. In some, but not all, cases HRCs will have such an effect. Tauer and Love (1989) point out that those farmers who use such technology first will reap some economic benefit, while those who do not will lose. However, when universally adopted, there will be no significant benefit to the farmer. On a worldwide scale, farmers in countries which prohibit HRCs that improve efficiency will be at an economic disadvantage to those in countries that allow the technology, provided they are not protected from international competition by government policies.

Herbicide manufacturers, biotechnology companies, and seed companies involved in the development of HRCs have obvious potential economic benefits. However, not all seed companies are enamored with this technology, but they must be involved in order to compete, in the event that the technology is very successful (Duvick 1992).

As with other technology that has reduced the cost of agricultural production, the consumer will be the principal beneficiary, provided the technology has no hidden costs, such as environmental or toxicological problems. Some have argued that the hidden costs of modern agriculture with its technological fixes are unacceptable. This argument has reached a crescendo when focused on HRCs.

8. HERBICIDE-RESISTANT CROPS — ENVIRONMENTAL AND FOOD SAFETY RISKS

Concern is increasing over the contamination of the environment and food supply with synthetic chemicals (e.g., Cockburn 1995; Long 1994; van Straalen et al. 1995). Agrochemicals represent a large amount of these materials that have been introduced deliberately into the environment every year for several decades. Most pesticides, by volume, have been herbicides. Herbicides and their breakdown products can be found in soil, groundwater, surface water, and food. The potential occurrence of particular herbicides in each of these components can vary between herbicides markedly, depending on many factors. Most concern is focused on a few herbicides used in certain environments.

Herbicides and their breakdown products that can be identified as synthetic all have environmental half-lives which vary with their location in the environment. Thus, contamination with any of these materials is not permanent. Furthermore, the average use of a herbicide active ingredient is less than 20 years, so the time period of accumulation of any particular compound is limited, unlike that of many generic industrial xenobiotics. Little is known of how these factors translate into the dynamics of herbicide and herbicide breakdown product accumulation and loss from and movement between different environmental compartments. If half-lives are sufficiently long in some compartments, the compartment will eventually accumulate a complex mixture of herbicides and their residues. If any of these compounds are at some point found to be of toxicological or environmental concern at the levels found in that compartment (e.g., groundwater), a serious and difficult problem arises.

Most of the toxicity concerns are with toxicity to animals. There is evidence from North America (Duke et al. 1991) and Japan (Kanazawa 1991) that there are no long-term phytotoxicity problems with proper use of herbicides, for three reasons: (1) the relatively short half-life of most herbicides in the environment in which plants grow (usually <1 month); (2) movement of herbicides into other parts of the environment by volatilization or movement to ground- and surface water; and (3) rapid inactivation of some herbicides (e.g., glyphosate, paraquat) by binding to soil particles. Herbicide degradation products generally have very little phytotoxicity (Duke et al. 1991).

Adoption of a HRC will increase the use of the herbicide to which the crop has been made resistant. The toxicological profile of that herbicide could be more or less desirable than the herbicide(s) that it replaces. Many factors must be considered in evaluating whether the HRC reduces or improves food safety. Another aspect of food safety is whether the genetic alteration of the crops causes the crop to manufacture a potentially toxic compound. Exotic genes could cause unpredictable pleiotrophic effects related to food safety (Olempska-Beer et al. 1993). However, even conventional plant breeding can generate toxic products in crops.

There is some concern that the HRC itself could become an environmental hazard if the new gene(s) causes the crop to become "weedy". This is considered highly unlikely. However, there is a more legitimate concern that under the selection pressure of the herbicide, introgression between the crop and wild relatives could produce new biota that might disrupt ecological balances (Dale et al. 1993; Keeler and Turner 1991). Studies have suggested cause for concern with some crops (e.g., Boudry et al. 1993) and little risk with others (Crawley et al. 1993). However, for decades many agronomic crops have been treated over vast areas with herbicides to which they are naturally resistant and there are no reported cases of weeds becoming resistant to these herbicides by introgression or of new weedy hybrids occurring due to this selection pressure. No experiment to test for the potential of introgression to occur can be this definitive.

9. HERBICIDE-RESISTANT CROPS — WHO DO THEY BENEFIT? WHO DO THEY HARM?

New technology usually has both proponents and opponents because of perceived beneficial and adverse effects. Perceptions can deviate substantially from reality. Furthermore, short-term effects can often differ dramatically from long-term results. Table 1 lists some of the perceived benefits and risks of various proponents and opponents of HRCs. Whether there are any substantive risks or beneficial effects of HRCs will depend largely on which HRCs are adopted. Certainly, there is the potential to increase risk with this technology. For example, crops can be made resistant to the very poisonous herbicide paraquat by transgenic methods (Perl et al. 1993). However, the induction of such HRCs is exceedingly unlikely because of regulatory constraints and public pressure. Most of the herbicides to which HRCs intended for introduction have been generated are of less concern than the herbicides that they will replace (Duke et al. 1991).

One perceived reason for the relative paucity of public funding for weed management research is that herbicides have been highly successful from an agricultural viewpoint (Duke 1992; Gianessi and Darby 1993). Nevertheless, the cost of weed management and the economic losses due to incomplete weed control represent a major portion of the costs of pests and their management (Bridges 1992). If HRCs improve weed management and reduce the costs of weeds, there will probably be less pressure to find nonchemical weed management alternatives, continuing the pattern of poor funding of public sector research in this area. Environmentalist support for increased funding of research on alternative weed management technology is probably the only mechanism by which this pattern will be broken.

10. SUMMARY

On the eve of their adoption, the controversy over HRCs swirls throughout the agricultural, environmental, legislative, regulatory, and biotechnological communities. There is debate over their potential utility, their environmental impact, and their effects on agrarian society. This technology poses a rather large target because it involves both the use of pesticides and crops that are sometimes

TABLE 1 Perceived Benefits and Risks of HRCs by Opponents or Proponents of This Technology

Risk/Benefit	Opponents	Proponents
Water contamination	Increased	Decreased
Herbicide use	Increased	Decreased
Food contamination	Increased	Decreased
Accelerated loss of small farms	Increased	No effect
Cost of farm production	Increased	Decreased
Consumer costs	Little effect	Decreased
Herbicide resistance	Increased	Decreased or no effect

transgenic. Regulatory agencies are entrusted with protecting the public from unacceptable problems with both pesticides and transgenic organisms. These agencies are under opposing pressures from environmental groups and agribusiness. Whether HRCs have benefited society or harmed it should be more clear at the end of this decade.

REFERENCES

Akobundu, I.O. 1992. Weeds in human affairs in sub-Saharan Africa. *Weed Technol.* 5: 680.

Bellinder, R.R., Gummesson, G., and Karlson, C. 1994. Percentage-driven government mandates for pesticide reduction: the Swedish model. *Weed Technol.* 8: 350.

Boudry, P., Mörchen, M., Saumitou-Laprade, P., Vernet, Ph., and Van Dijk, H. 1993. The origin and evolution of weed beets: consequences for the breeding and release of herbicide-resistant transgenic sugar beets. *Theor. Appl. Genet.* 87: 471.

Bridges, D.C. (Ed.). 1992. *Crop Losses Due to Weeds in the United States - 1992*, Weed Science Society of America, Champaign, IL.

Bright, S.W.J. 1992. Herbicide-resistant crops. In *Biosynthesis and Molecular Regulation of Amino Acids in Plants.* Ed. by Singh, B.K., Flores, H.E., and Shannon, J.C., American Society of Plant Physiologists, Rockville, MD, p. 184.

Claus, J.S. 1992. Integrating herbicide-resistant crops into discovery research. In *Pest Management in Soybeans.* Elsevier, London, p. 356.

Cockburn, A. 1995. Health risk assessment — critical issues for improved decision making. In *Options 2000 - Eighth International Congress of Pesticide Chemistry.* Ed. by Ragsdale, N.N., Kearney, P.C., and Plimmer, J.R., American Chemical Society, Washington, D.C., p. 404.

Comstock, G. 1989. Genetically engineered herbicide resistance. I. *J. Agric. Ethics* 2: 263.

Comstock, G. 1990. Genetically engineered herbicide resistance. II. *J. Agric. Ethics* 3: 114.

Crawley, M.J., Hails, R.S., Rees, M., Kohn, D., and Buxton, J. 1993. Ecology of transgenic oilseed rape in natural habitats. *Nature* 363: 620.

Dale, P.J., Irwin, J.A., and Scheffler, J.A. 1993. The experimental and commercial release of transgenic crop plants. *Plant Breeding* 111: 1.

DeFelice, M.S., Brown, W.B., Aldrich, R.J., Sims, B.D., Judy, D.T., and Guethle, D.R. 1989. Weed control in soybeans (*Glycine max*) with reduced rates of postemergence herbicides. *Weed Sci.* 37: 365.

Dekker, J. and Comstock, G. 1992. Alternative conceptions and models of sustainability. XXXI. Ethical and environmental considerations in the release of herbicide resistant crops. *Agric. Human Values* 9(No. 3): 31.

Dekker, J. and Duke, S.O. 1995. Herbicide-resistant field crops. *Adv. Agron.* 54: 69.

Duke, S.O. 1992. Weed science — the need and the reality. *Phytoparasitica* 20: 183.

Duke, S.O. 1994. Weed science. In *Encyclopedia of Agricultural Science, Vol. 4.* Academic Press, New York, p. 485.

Duke, S.O., Holt, J.S., Hess, F.D., and Christy, A.L. 1991. *Herbicide-Resistant Crops.* Comments from CAST, no. 1991-1, Council for Agricultural Science and Technology, Ames, IA.

Duke, S.O., Menn, J.J., and Plimmer, J.R. 1993. Challenges of pest control with enhanced toxicological and environmental safety. *ACS Symp. Ser.* 524: 1.

Duke, S.O., Moorman, T.B., and Bryson, C.B. 1991. Phytotoxic effects of pesticide degradation products. *Am. Chem. Soc. Symp. Ser.* 459: 188.

Duvick, D.N. 1992. Concerns of seed company officials with herbicide-tolerant cultivars. *Weed Technol.* 6: 640.

Dyer, W.E., Hess, F.D., Holt, J.S., and Duke, S.O. 1993. Potential benefits and risks of herbicide-resistant crops produced by biotechnology. *Hortic. Rev.* 15: 367.

Erhlich, P.R. 1968. *The Population Bomb.* Ballantine Books, New York.

Gianessi, L.P. 1993a. Chemicals seen vital to weed, pest control. *Forum Appl. Res. Public Policy 8: Fall,* 87.

Gianessi, L.P. 1993b. The quixotic quest for chemical-free farming. *Issues Sci. Technol. 10: Fall,* 29.

Gianessi, L.P. and Darby, C.P. 1993. The Potential Roles of Herbicide-Resistant Crops in the United States. Paper presented at Fourth National Conference on Pesticides, November 1 to 3, 1993.

Goldburg, R., Rissler, J., Shand, H., and Hassebrook, C. 1990. *Biotechnology's Bitter Harvest.* Environmental Defense Fund, Biotechnology Working Group, New York.

Gressel, J. 1993. Advances in achieving the needs for biotechnologically-derived herbicide resistant crops. *Plant Breeding Rev.* 11: 155.

Hayenga, M., Thompson, L.C., Chase, C., and Kaaria, S. 1992. Economic and environmental implications of herbicide-tolerant corn and processing tomatoes. *J. Soil Water Conserv.* 47: 411.

Hoyle, R. 1993. Herbicide-resistant crops are no conspiracy. *Bio/Technology* 11: 783.

Houtsma, J. 1994. Visionary sprayers outsmart weeds. *Farm Ind. News* 27(4): 4.

Kanazawa, J. 1991. Herbicide residues in soil in long-term application — long effects of herbicides in paddy rice. *Shocucho* 24: 388.

Keeler, K.H. and Turner, C.E. 1991. Management of transgenic plants in the environment, in *Risk Assessment in Genetic Engineering*. Ed. by Levin, M.A. and Strauss, H.S., McGraw-Hill, New York, p. 189.

Kirschner, E.M. 1994. Agricultural chemical producers rebound from floods of 1993. *Chem. Eng. News* 72(40): 13.

Klassen, W. 1995. World food security up to 2010 and the global pesticide situation. In *Options 2000 - Eighth International Congress of Pesticide Chemistry*. Ed. by Ragsdale, N.N., Kearney, P.C., and Plimmer, J.R., American Chemical Society, Washington, D.C., p. 1.

Krimsky, S. and Wrubel, R. 1993. *Agricultural Biotechnology: An Environmental Outlook*. Tufts University, Medford, MA.

Leathers, T.D., Gupta, S.C., and Alexander, N.J. 1993. Mycopesticides: status, challenges and potential. *J. Ind. Microbiol.* 12: 69.

LeBaron, H.M. 1991. Distribution and seriousness of herbicide-resistant weed infestations worldwide. In *Herbicide Resistance in Weeds and Crops*. Ed. by Caseley, J.C., Cussans, G.W., and Atkin, R.K., Butterworth-Heinemann Ltd., Oxford, Chap. 2.

Long, J. 1994. Report says herbicides foul drinking water. *Chem. Eng. News* 72(43): 8.

McWhorter, C.G., Barrentine, W.L., and Hanks, J.E. 1992. Postemergence grass control with herbicides applied at ULV in paraffinic oil. *Weed Technol.* 6: 262.

Miller, M. 1991. The promise of biotechnology: developing pesticide- and herbicide-resistant crops. *J. Environ. Health* 54(2): 13.

Olempska-Beer, Z.S., Kuznesof, P.M., DiNovi, M., and Smith, M.J. 1993. Plant biotechnology and food safety. *Food Technol.* December: p. 66.

Perl, A., Perl-Treves, R., Galili, S., Aviv, D., Shalgi, E., Malkin, S., and Galun, E. 1993. Enhanced oxidative-stress defense in transgenic potato expressing tomato Cu, Zn superoxide dismutases. *Theor. Appl. Genet.* 85: 568.

Rüdelsheim, P. 1993. Engineering crops for tolerance to specific herbicides: a valid alternative. *Brighton Crop Protect. Conf.* 1: 265.

Rydahl, P. and Thonke, K.E. 1993. PC-plant protection: optimizing chemical weed control. *Bull. OEPP/EPPO Bull.* 23: 589.

Smeda, R.J. 1993. Potential of spring-seeded sorghum-Sudangrass for weed management in soybean. *Proc. South Weed Sci. Soc.* 46: 57.

Stigliani, L. and Resina, C. 1993. SELOMA: expert system for weed management in herbicide-intensive crops. *Weed Technol.* 7: 550.

Tauer, L.W. and Love, J. 1989. The potential economic impact of herbicide-resistant corn in the U.S.A. *J. Prod. Agric.* 2: 202.

van Straalen, N.M., van Rijn, J.P., and van Gestel, C.A.M. 1995. Ecotoxicological risk assessment of pesticides in soil. In *Options 2000 - Eighth International Congress of Pesticide Chemistry*. Ed. by Ragsdale, N.N., Kearney, P.C., and Plimmer, J.R., American Chemical Society, Washington, D.C. p. 418.

Watson, A.K. (Ed.) 1993. *Biological Control of Weeds Handbook*. Weed Science Society of America, Champaign, IL.

Zorner, P.S., Evans, S.L., and Savage, S.D. 1993. Perspectives on providing a realistic technical foundation for the commercialization of bioherbicides. *ACS Symp. Ser.* 524: 79.

PART I

THE TECHNOLOGY

Chapter 2

MECHANISMS OF ACTION AND RESISTANCE TO HERBICIDES

Timothy D. Sherman
Kevin C. Vaughn
Stephen O. Duke

CONTENTS

1-56670-045-0/96/$0.00+$.50
© 1996 by CRC Press Inc.

1. INTRODUCTION

Herbicides are phytotoxins used by humans to kill unwanted vegetation. The compounds can be completely nonselective, killing every species of plant, or very selective, killing only certain plants. Several hundred herbicide active ingredients have been marketed since the dawn of chemical control of weeds in the 1940s. The mechanisms of action of most of these compounds are known (Devine et al. 1993a; Duke 1990; Pilmoor et al. 1995), although there are still many gaps in our knowledge.

Most herbicides are thought to directly affect only one molecular target in the plant, resulting in a cascade of secondary and tertiary effects that eventually kill the plant. This is probably an oversimplification for some herbicides. For example, most photosynthetic inhibitors also affect plant respiration at higher doses (Devine et al. 1993a). Under some circumstances, these less important primary effects may contribute to phytotoxicity. Nevertheless, it is assumed that almost all herbicides have only one important primary site of action. Approximately 15 molecular sites of action of herbicides are known or strongly suspected (Devine et al. 1993a; Duke 1990, 1992; Pilmoor et al. 1995) (Table 1). Only a few of these are molecular sites found also in animals.

Understanding the molecular site(s) of action of herbicides can be crucial in the production of a herbicide-resistant crop (HRC). If the site of action is to be manipulated to produce the resistant crop, this knowledge is essential. Even if other strategies are used to obtain resistance in the crop (limitation of movement to the site of action or metabolic protection), the possibility of other, normally less important sites of action of the herbicide may be germane.

This chapter also deals with physiological and biochemical mechanisms of natural and evolved resistance to herbicides. We do not discuss mechanisms of genetically engineered resistance to

TABLE 1 Known and Proposed Molecular Sites of Action of Commercial Herbicides

Site of Action[a]	Herbicide(s)
Amino acid synthesis	
Acetolactate synthase	Imidazolinones, sulfonylureas, triazolopyrimidines, and others
EPSP synthase	Glyphosate
Glutamine synthetase	Glufosinate (phosphinothricin), bialaphos
Auxin-binding protein	Several, including 2,4-D, triclopyr, dicamba, picloram, and clopyralid
Carotenoid synthesis	
Geranylgeranyl synthase?	Clomazone
Phytoene desaturase	Several, including norflurazon
Cellulose synthesis	Diclobenil, isoxaben
Folate synthesis	
Dihydropteroate synthase	Asulam
Lipid synthesis	
Acetyl CoA carboxylase	Aryloxyphenoxyproprionates (e.g., diclofop) and cyclohexanediones (e.g., sethoxydim)
Microtubule function	
Tubulin	Dinitroanilines and phosphoric amides
Photosynthesis	
D-1 of photosystem II	Many, including bromoxynil, triazines, substituted ureas
Energy diversion from photosystem I	Paraquat and diquat
Protoporphyrinogen oxidase	Many, including oxadiazon, acifluorfen, oxyfluorfen, and lactofen

[a] Italics denotes that the molecular site is also found in mammals.

herbicides in crops, except in a few cases which are not covered in other chapters. In this chapter we use the term "resistance" rather than "tolerance" for all forms of immunity to herbicides. The type of resistance will be provided by modifiers.

Both crops and weeds are naturally resistant to many selective herbicides. The mechanisms of preexisting, natural resistance of weeds and crops to herbicides are often the same for a particular herbicide. Hundreds of weed species throughout the world have become resistant to various herbicides. This resistance has generally taken longer to evolve than the evolution of insecticide resistance due to the relatively long generation time of weeds as compared to insects and the residual herbicide-susceptible biotypes present in a usually long-lived seed bank. Mechanisms of herbicide resistance in weeds are extremely varied (Duke 1993; Vaughn and Duke 1991). The mechanism of evolved resistance is almost always completely different than the mechanism of natural resistance in crops to the same herbicide. This is an important point in that it suggests that acquisition of resistance is not due to the movement of resistance-imparting genes from the crop to weeds.

We organized this chapter by site of action rather than by chemical class. A brief explanation of the sites of action, the physiological implications of each site of action, and the mechanisms of resistance to herbicides that target the site of action are provided. We also attempt to point out any special features of particular sites of action that have implications with respect to HRCs.

2. MECHANISMS OF RESISTANCE

Herbicide resistance can be due to three basic mechanisms: resistance at the site of action, metabolic detoxification, and prevention of the herbicide from reaching the site of action.

The first mechanism is generally due to an alteration in the herbicide target site (generally a protein) that prevents the herbicide from binding to that site and inhibiting the vital process mediated by the target site. An uncommon variation on this mechanism is for the plant to have many more copies of the target site than are found in susceptible plants, so that it takes much more herbicide to effectively block the process for which the target molecule is required.

Metabolic detoxification generally means that the plant can degrade the herbicide faster than the herbicide can cause irreversible damage to the plant. The speed of the enzymatic degradation can vary with plant growth stage, weather, and other factors. Thus, the same herbicide application rate might be phytotoxic to a species under some circumstances and benign under others. The degradation products of herbicides generally have little or no phytotoxicity, except in the case of proherbicides, which must be metabolically activated by the target plant (Duke et al. 1991b). Some herbicides exert their effects through production of other toxic molecules in the plant. Resistance to these herbicides can sometimes be attributed to metabolic detoxification of these toxic compounds.

For a herbicide to be effective, it must reach its molecular target site. The molecular target site might be far removed from the site of herbicide application. For example, the herbicide may need to be translocated from leaves to apical meristems to be effective. Thus, blockage of movement to the site of action can occur at the plant surface (leaf cuticle or root epidermis), just inside the plant (cell wall or plasma membrane), or in vascular tissues in which translocation occurs. Sequestration of the herbicide in a metabolically inactive site is a variation on reducing movement of the herbicide to the site of action. Just as plants sequester phytotoxins that they produce themselves (Duke 1995), herbicides sometimes can be partitioned into metabolically inactive sites. Sequestration has been used to explain the natural resistance of cotton to some herbicides (Duke 1992) and evolved paraquat resistance in some weeds (Preston 1994).

A species, weed or crop, can be resistant to a particular herbicide by one or more of these mechanisms. The level of resistance can range from low susceptibility at recommended field rates of the herbicide to complete resistance, even at doses 100 or more times than the recommended field rate. An example of a very low level of resistance is the evolved resistance of cocklebur (*Xanthium strumarium* L.) to the organic arsenical herbicide MSMA (Nimbal et al. 1995), whereas resistance of most atrazine-resistant weeds is >100-fold the recommended dose rate (Fuerst et al. 1986).

3. AMINO ACID SYNTHESIS INHIBITORS

Plants synthesize all of their essential amino acids. Theoretically, blocking the synthesis of any essential amino acid will eventually kill a plant. There are commercial herbicides that target three enzymes involved in amino acid synthesis: (1) glutamine synthetase (GS), (2) acetolactate synthase (ALS), and (3) enolpyruvylshikimatephosphate synthase (EPSP synthase). We discuss each of these sites separately.

3.1. GLUTAMINE SYNTHETASE

Glutamine synthetase activity is essential for the assimilation of organic nitrogen in the form of ammonia into plants. Lack of GS activity, due either to a genetic lesion or inhibition of GS, in a plant that is reducing nitrite to ammonia can lead to very high ammonia levels. Two commercial herbicides, bialaphos and glufosinate (also known as phosphinothricin, when from a natural source), act through the inhibition of GS. Bialaphos is a fermentation product of *Streptomyces* spp. that has a limited market in Japan. It has no effect on GS as bialaphos, but it is metabolically converted to phosphinothricin by target weeds. Thus, bialaphos is a proherbicide. Phosphinothricin is also produced by some *Streptomyces* sp., however, only the synthetic version of this compound, glufosinate, is sold as a herbicide. Other naturally occurring GS inhibitors are discussed by Duke and Lydon (1993) and Duke et al. (1995).

The accumulation of ammonia caused by GS inhibitors was thought to be the principal cause of phytotoxicity. However, it is now known that herbicidal GS inhibitors cause most of their phytotoxicity by rapidly elevating glyoxylate, an inhibitor of the first enzyme of carbon fixation, RUBP carboxylase (Devine et al. 1993a). This causes a rapid halt in carbon fixation, resulting in a complex series of destructive events due to the inability of the chloroplast to dissipate captured light energy by carbon fixation.

These herbicides have not been used long enough for weeds to develop resistance to them. Glufosinate is a relatively nonselective herbicide, whereas bialaphos has some selectivity based on whether a species can bioactivate this proherbicide. There are some differences in sensitivities of different species to glufosinate, but these differences do not appear to be due to differences in GS (Ridley and McNally 1985). Some crop varieties are resistant to other naturally occurring GS inhibitors that are chemical analogues of glufosinate (Knight et al. 1988), but whether these varieties are also resistant to glufosinate has not been reported.

3.2. ACETOLACTATE SYNTHASE

This is an enzyme common to the synthesis of the three essential branched chain amino acids: leucine, isoleucine, and valine. There are currently three major herbicide chemical classes which target this enzyme: the imidazolinones, triazolopyrimidines, and sulfonylureas (Devine et al. 1993a). Reviews are available on the sulfonylureas (Beyer et al. 1988) and triazolopyrimidines (Subramanian and Gerwick 1989), and an entire book is available on the imidazolinones (Shaner and O'Connor 1991). Several other chemical classes of herbicides that inhibit ALS are under development.

These herbicides are generally highly active and selective, with particular ALS inhibitors being marketed for use with specific crops. This has led to the marketing of a relatively large and growing number of ALS-inhibitor herbicides. These inhibitors are used in most cases for postemergence weed management.

Also known as acetohydroxy acid synthase or AHAS, ALS is the first eznyme in the branched chain amino acid pathway. All of the enzymes of this pathway are found only in the plastid (chloroplast when green). Many compounds are good inhibitors of ALS, apparently binding a vestigial ubiquinone binding site (Schloss et al. 1988). Inhibition of ALS stops growth and then kills the plant relatively slowly. Acetalactate synthase is an excellent molecular target site for herbicides, because only a few grams per hectare of certain ALS inhibitors are needed to kill target weeds.

Natural crop resistance to ALS-inhibiting herbicides is due to rapid metabolic degradation of the herbicide (Beyer et al. 1988; Shaner and Mallipudi 1991). In contrast, weeds that have evolved resistance to ALS inhibitors have almost always evolved an altered ALS that is resistant to the herbicide (Devine et al. 1993a; Schmitzer et al. 1993). Different cross-resistance patterns can occur within a single species, depending on the number and location of mutations (e.g., Hattori et al. 1992). Alterations in at least ten mutation sites can confer resistance without seriously reducing enzyme activity (Mazur and Falco 1989).

Weeds have evolved resistance to ALS-inhibitor herbicides comparatively rapidly, appearing in only 3 to 5 years of selection with sulfonylureas (Thill et al. 1991) and imidazolinones (Schmitzer et al. 1993). In all of these cases resistance is due to herbicide-resistant ALS. These weeds are apparently as fit as the susceptible biotypes. The ALS enzyme efficiency is not significantly affected by most of the mutations that result in resistance. This may be due largely to the fact that the herbicide binding site is different from the active site of the enzyme. From the large number of good inhibitors and the different types of mutations conferring different patterns of cross-resistance to different ALS inhibitors, one may infer that the herbicide-binding site ALS is a very plastic molecular domain.

3.3. ENOLPYRUVYLSHIKIMATEPHOSPHATE SYNTHASE

Glyphosate (*N*-(phosphonomethyl)glycine) is the only commercial herbicide that attacks EPSP synthase, an enzyme of the shikimate pathway. It is an enzyme common to the synthesis of the aromatic amino acids tryrosine, phenylalanine, and tryptophan. In plants these amino acids are essential for critical processes other than protein synthesis, including cell wall formation, defense against pathogens and insects, production of hormones, and production of compounds required in energy transduction such as plastoquinone (Duke 1988).

There is evidence that some of the phytotoxicity caused by glyphosate is due to deregulation of the shikimate pathway due to reduced feedback inhibition, causing greatly increased amounts of carbon to flow into this pathway. This results in two possible detrimental effects: (1) debilitation of other pathways due to loss of carbon skeletons to the shikimate pathway (Killmer et al. 1981) and (2) accumulation of massive, possibly phytotoxic, levels of shikimate and benzoic acids (Becerril et al. 1989; Lydon and Duke 1988).

Secondary products derived from aromatic amino acids are involved in the interactions of plants with plant pathogens and herbivores (insects and others). The reductions in these compounds in glyphosate-treated plants have been demonstrated to reduce their resistance to plant pathogens (Christy et al. 1993; Lydon and Duke 1989, 1993). This process may play an important part in the phytotoxicity of glyphosate. In some studies much lower levels of glyphosate were needed to kill plants in nonsterilized than in sterilized soil, because of the involvement of plant pathogens in the mechanism of action of glyphosate (Rahe et al. 1990).

Another aspect of the mode of action of glyphosate is that the pattern of translocation in plants is much the same as that of sucrose (Duke 1988; Gougler and Geiger 1984). Thus, it moves to metabolic sinks such as meristems, fruits, nodules, and storage tissues. Furthermore, most if not all higher plants metabolically degrade glyphosate slowly, if they degrade it at all. Thus, if a gene(s) that encode a glyphosate-degrading enzyme are not inserted into a glyphosate-resistant crop, the herbicide will accumulate in one or more of these metabolic sinks. There are no published data to indicate whether this could cause unacceptable residue problems and/or inhibition of nodulation or nitrogen fixation, as *Bradyrhizobium japonicum* is not naturally resistant to glyphosate (Moorman et al. 1992).

There has been speculation that glyphosate might have another primary site of action in some plant species. Carbon fixation is markedly reduced within 15 min of spraying sugarbeets with glyphosate (Servaites et al. 1987). Furthermore, rapid effects of glyphosate on the variable fluorescence of chlorophyll in sugarbeets can be detected within 1 h after spraying, and these effects correlate well with phytotoxicity (Madsen et al. 1995). These effects are difficult to explain in terms of inhibition of EPSP synthase. Whether these effects are really important in the mode of action of glyphosate in sugarbeets could be determined with transgenic sugarbeets with a resistant EPSP synthase. Such data are not yet available.

Glyphosate is a completely nonselective herbicide, and no weeds have yet developed resistance to it. Some species are somewhat resistant to glyphosate, probably due to uptake or translocation mechanisms, but no crops have sufficient resistance for glyphosate to be used directly on the crop as a selective herbicide. Genetic diversity in resistance to glyphosate between biotypes of birdsfoot trefoil (Boerboom et al. 1991) indicates that resistance could develop. The poor results in attempting to produce glyphosate-resistant crops by cell selection technology and the complicated procedure used to genetically engineer a successful glyphosate-resistant crop (Dyer et al. 1993; Padgette et al. 1995) would suggest that the evolution of glyphosate-resistant weeds will be a very slow process, and that the level of resistance from field selection will be relatively low. Furthermore, glyphosate selection pressure is short-lived, due to its strong binding to soil components (Duke 1988).

4. MITOTIC DISRUPTERS

So-called mitotic disrupter herbicides include the dinitroaniline, phosphoric amide, and *N*-phenyl carbamate groups as well as the chemically distinct herbicides DCPA, dithiopyr, and pronamide. To our knowledge, no crops are being genetically engineered to be resistant to any of these herbicides, although one patent for the production of trifluralin-resistant crops by transformation with the gene encoding the modified tubulin of goosegrass has been filed (Cronin et al. 1993). Virtually all of these herbicides are used for the control of annual grasses in broadleaf crops or in turfgrass. The dinitroanilines are characteristically utilized by incorporating the herbicide into the top layer (ca. 5 cm) of the soil prior to planting the crop (pre-plant incorporation). Large seeded crop plants such as cotton are able to grow through the zone of incorporation, whereas smaller-seeded annual grasses are not, thus accounting for some of the selectivity of these compounds. Dinitroanilines are very lipophilic (water solubility ~1 mg/l) and crop seeds with large lipid reserves are able to partition or sequester the herbicides in lipid bodies, away from the site of action (Hilton and Christianson 1972). Most of the other mitotic inhibitor herbicides are utilized after planting. Both DCPA and dithiopyr are utilized extensively in crabgrass control in turf. Pronamide provides effective weed control in the very resistant lettuce plant. *N*-Phenyl carbamates are utilized both for pre- and postplanting weed control in a number of crops. Besides herbicidal use, the *N*-phenyl carbamate herbicide CIPC (chlorpropham) is utilized extensively as an antisproutant in stored potatoes.

The most characteristic symptom of all of these mitotic disrupter herbicides is root tip swelling to a club shape, identical to that formed after treatment with the classic mitotic disrupter colchicine (Hess 1987; Vaughn and Lehnen 1991). The affected weeds generally have not only small swollen roots, but the region around the intercalary meristem of grass roots is significantly broadened. Plant death probably results from the lack of a functional root system, leading to a lack of water flow into the aerial portions of the plant, and subsequent consequences of water deficit.

As mentioned above, these mitotic disrupter herbicides cause the same root tip swelling as the classic mitotic disrupter colchicine. Both structural and biochemical investigations indicate that many, if not all, of these mitotic disrupter compounds work in the same manner as colchicine. Structural studies at both the light and electron microscopic levels indicate the same symptomology: arrested prometaphase figures, lobed reformed 4C nuclei, aberrant cell plates, isodiametric cell shape, disorganization of or lack of all microtubule arrays (Hess 1987; Vaughn and Lehnen 1991; Hoffman and Vaughn 1994). All of these phenomena result as consequences of the last of these effects, as microtubules are the cellular structure responsible for both cellular shaping and the movement of chromosomes during mitosis.

Biochemically we know much less about the mode of action of these compounds than for most herbicides. Dinitroanilines, phosphoric amides, and pronamide were shown to inhibit microtubule polymerization from free tubulin subunits *in vitro* (Hess and Bayer 1977; Morejohn and Fosket 1986; Morejohn et al. 1987; Akashi 1988). Trifluralin and oryzalin are also known to bind tubulin. Limited biochemical data are available for the other compounds, although certain of their symptomology indicate that they have a similar biochemical mechanism. The *N*-phenyl carbamates and DCPA are much less effective herbicides than the others described herein, and some of their effects are characteristic of marginal effects of the more effective herbicides (Hoffman and Vaughn 1994). Whereas complete microtubule loss may be observed after dinitroaniline herbicide treatment, the *N*-phenyl carbamates cause only partial loss, leading to multipolar mitosis and aberrant cell plate formation.

Considering the efficacy and widespread use for decades of dinitroaniline herbicides, especially trifluralin and pendimethalin for weed control in cotton, it is not surprising that several cases of evolved herbicide resistance have been reported. The most well-documented cases were of goosegrass (*Eleusine indica*), which was reported by a number of states throughout the U.S. "cotton belt" (Mudge et al. 1984). The most common biotype is resistant to aqueous saturated concentrations of all of the dinitroaniline herbicides, and this biotype exhibits similar levels of cross-resistance to the phosphoric amide herbicides, with slightly lower levels of resistance to dithiopyr (Vaughn et al. 1987). *In vitro* polymerization studies indicate that the tubulin from the R biotype of goosegrass is able to polymerize into microtubules in the presence of oryzalin, whereas the S biotype is not, indicating the difference is related to intrinsic differences in tubulin protein (Vaughn and Vaughan 1990). The R biotype is very sensitive to the microtubule stabilizer taxol, indicating that this biotype has hyperstabilized microtubules; i.e., the microtubules are more likely to stay in the polymerized state and exhibit slower turnover (Vaughn and Vaughan 1991). The inheritance of this mutant appears complex, although an F1 population of R $\times$ S is only slightly less resistant than the R parent, indicating that at least one of the genes for resistance is semidominant (Vaughn, unpublished).

A biotype of green foxtail (*Setaria viridis*) found in Canada was originally reported to be trifluralin resistant (2 to 10X), although subsequent cross-resistance studies indicate that this biotype is much more resistant to DCPA (>100X) and terbutol (~30X), an obsolete carbamate herbicide (Smeda et al. 1993; Smeda and Vaughn 1994). Surprisingly, the biotype is actually more sensitive to CIPC. No differences in α– or ß-tubulin are found, although no further biochemical characterization of this biotype was reported. Genetic studies indicate that the resistance characteristic is a monogenic recessive (Jasieniuk et al. 1994).

Biotypes of several other weed species (johnsongrass, Palmer amaranth) were reported to be dinitroaniline resistant, although other than the reports of the resistance, little data have been accumulated. Mutants of the green alga *Chlamydomonas reinhardii* have been selected for resistance to oryzalin in laboratory experiments (James et al. 1988; Bolduc et al. 1988). Most of these are cross-resistant to amiprophos methyl, although this alga exhibits a fairly high level of natural resistance to all of these herbicides. Genetic investigations indicate that most of the mutations map to the *uni* linkage group, a group of genes involved in flagellar functions. Biochemical investigations have not been reported.

5. CELLULOSE BIOSYNTHESIS INHIBITORS

Although they have a similar mechanism of action, the herbicides dichlobenil and isoxaben are utilized in totally different ways. Dichlobenil is used after germination and after establishment of the crop for total weed control; earlier treatments are generally phytotoxic to the crop. In contrast, isoxaben is a preemergent herbicide that selectively controls a number of dicotyledonous weeds in a variety of crops. Dichlobenil also has been utilized in aquatic weed control. To our knowledge, no crops are being made resistant to either of these herbicides by biotechnology.

Although the specific molecular site of action of these herbicides is not known, both herbicides were shown to inhibit the incorporation of radiolabeled glucose into cellulose (Montezinos and Delmer 1980; Heim et al. 1990). In higher plants neither isolation of the enzymes involved in cellulose biosynthesis nor the demonstration of cellulose biosynthesis *in vitro* has been successful, which makes it difficult to determine the molecular sites of action of these compounds. An azido analog of dichobenil identified a polypeptide of about 14 kDa which may be a subunit of the cellulose biosynthesis complex in higher plants (Delmer et al. 1987). At the structural level treatment with either herbicide results in the appearance of incomplete or reticulate (or totally absent) cell plates (Vaughn, unpublished). Cytokinesis is thus not completed. Because cell walls are required for growth, growth stops. Eventually this results in plant death.

No field resistance has been reported for dichlobenil, and isoxaben is so new that it is unlikely that field resistance would have appeared. Mutants of *Arabidopsis* resistant to both herbicides and cell lines with resistance to dichlobenil have been selected in the laboratory, however. Interestingly, the mutants are not cross-resistant to the other compound, and in the case of isoxaben, two loci are involved in resistance. Both mutants are recessive. The dichlobenil-resistant cell lines produce aberrant cell walls, even when grown in the absence of herbicide, indicating that they may utilize alternative (dichlobenil insensitive) mechanisms for cell wall production. The monocot grass *Agrostis palustris* is naturally resistant to isoxaben. The level of resistance cannot be accounted for by decreased uptake or metabolic degradation (Heim et al. 1993). A major part of the resistance was concluded to be due to decreased binding of isoxaben to a site involved in cellulose biosynthesis.

6. PHOTOOXIDATIVE STRESS GENERATORS

Although some herbicides can cause photooxidative stress indirectly by stopping photosynthesis, there are two fundamental mechanisms by which herbicides kill plants more rapidly through causing lethal photooxidative stress: (1) diversion of photosynthetic energy from photosystem I to reduce molecular oxygen to superoxide radical, a toxic compound, and (2) inhibition of protoporphyrinogen oxidase (Protox) to cause the massive accumulation of the phototoxic compound protoporphyrin IX (Proto IX).

6.1. PHOTOSYSTEM I DIVERTERS

Paraquat and diquat are widely utilized nonselective postemergent herbicides, whereas the related herbicide morfamquat is a selective dicot herbicide for use in cereals (Summers 1980). All of these bipyridilium herbicides bring about rapid cell death; treated plants appear water-soaked and necrotic and die shortly after treatment. Besides their use as herbicides, both diquat and paraquat are utilized as harvest aids by dessicating foliage of potatoes and soybeans.

All bipyridilium herbicides require active photosynthesizing plants for activity, with the non-green portions of the plants much less affected than the photosynthetically competent tissues. Biochemical investigations indicate that bipyridilium herbicides are reduced by the primary acceptor of photosystem I (X). Upon reduction, the radical dication formed reacts with molecular oxygen to produce superoxide, hydroxyl radical, and hydrogen peroxide, which are the agents that cause cellular breakdown. By accepting electrons from X, the bipyridilium herbicides also stop the flow of electrons to NADP, depriving the plant of reducing equivalents for the dark reactions of photosynthesis. Cellular effects are first noted in the thylakoids, with a swelling and ballooning of the thylakoid lumen (Vaughn and Duke 1983). In many respects paraquat damage resembles accelerated tissue senescence.

In addition to a number of nonphotosynthetic mutants that were shown to be resistant to paraquat (Vaughn and Duke 1983), a number of weed species have evolved resistance to paraquat, chiefly species of *Conyza* or *Erigeron* (the fleabanes), despite having functional photosynthesis (Preston 1994). Two theories have been put forth to explain paraquat resistance in these plants: detoxification of the toxic oxygen species and a sequestration/compartmentalization of the paraquat in nonchloroplast sites. It is the latter theory which seems the most reasonable for the majority of weed species. Evidence for this theory is as follows: (1) paraquat-induced fluorescence suppression (indicating paraquat has reached the active site) is found in the S biotype, but not in the R (Fuerst et al. 1985; Vaughn et al. 1989); (2) radiolabelled paraquat movement is restricted in the R biotypes but not the S biotypes (Fuerst et al. 1985; Norman et al. 1994); (3) there is no correlation of enzymes involved in toxic oxygen protection and paraquat resistance (Norman et al. 1993, Vaughn and Fuerst 1985), (4) cross-resistance of paraquat to other herbicides that generate toxic oxygen species, including morfarquat, could not be demonstrated (Vaughn et al. 1989); and (5) the self-regenerating nature of bipyridiliums indicates that toxic oxygen protection would only slow down, but not prevent herbicide damage (reviewed by Fuerst and Vaughn 1990; Preston 1994). A biotype of *Hordeum* was shown to be altered in the uptake of paraquat into cells and this may be related to alteration in polyamine uptake sites. A paraquat-resistant biotype of *Lolium* that exhibits very low levels of resistance does experience an increase of protective enzymes and is cross-resistant to diquat and morfamquat (Preston 1994). The resistance in *Conyza* biotypes is a dominant trait, but the *Hordeum* is recessive (Preston 1994).

Some crops have been made resistant to paraquat by using transgenic methods to introduce genes encoding antioxidant enzymes. However, there is apparently no intent to commercialize these crops. In genetically engineered tobacco an *Escherichia coli* gene encoding glutathione reductase increased resistance to paraquat (Aono et al. 1993), and the gene from pea for copper/zinc chloroplast superoxide dismutase imparted paraquat resistance (Sen Gupta et al. 1993). Genes for tomato superoxide dismutases transferred to potato conferred enhanced tolerance to paraquat (Perl et al. 1993). Although these studies were apparently done for academic reasons, it is possible that such modifications in crops could be used to prevent other types of oxidative stress, such as that caused by ozone or sulfur dioxide (Hérouart et al. 1993).

6.2. PROTOX INHIBITORS

No discussion of herbicides should exclude the Protox inhibitors, because they are one of the most patented groups of herbicides in recent years (Anderson et al. 1994). Acifluorfen, lactofen, oxadiazon, and fomesafen are examples of herbicides with this mode of action that are sold in the U.S.

Protox inhibitors are applied as foliar sprays which cause very rapid cellular collapse and desiccation. Some of these compounds are effective at use rates of only a few grams per hectare. A major problem with Protox inhibitors is that few crops (primarily soybean and rice) are naturally resistant to them. Production of herbicide-resistant crops by biotechnology with which Protox inhibitors could be used would create new opportunities for the many companies that manufacture this class of herbicide.

Protox is the last enzyme in the porphyrin pathway that is common to both heme and chlorophyll synthesis. Protox inhibitors cause massive levels of the enzyme product, but not the substrate protoporphyrinogen IX (Protogen IX), to accumulate through a complex mechanism involving both a herbicide-susceptible chloroplast Protox and a herbicide-resistant extraplastidic Protogen IX-oxidizing enzyme (Duke et al. 1991a, 1994; Jacobs et al. 1991, 1994; Lee et al. 1993, 1994).

Proto IX is a photosensitizing agent, generating highly reactive singlet oxygen in the presence of sunlight. Proto IX is the only photodynamic porphyrin intermediate caused to accumulate to high levels by Protox inhibitors (Becerril and Duke 1989b; Matsumoto and Duke 1990), and the amount of Proto IX accumulated correlates well with herbicidal injury (Becerril and Duke 1989a). Thus, Proto IX, a metabolic intermediate, is the acutely toxic agent causing phytotoxicity.

A wide range exists in the natural resistance of crops and weeds to Protox inhibitors (Sherman et al. 1991; Matsumoto et al. 1994); however, there is no evidence that any weeds have become resistant to these herbicides as the result of selection pressure. There are apparently several mechanisms of resistance, including reduced sensitivity of Protox (Matsumoto et al. 1994; Pornprom et al. 1994), rapid metabolic degradation of the herbicide (Frear et al. 1983), and resistance to singlet oxygen (Matsumoto et al. 1994).

Efforts are being made to produce a plant with a herbicide-resistant chloroplast Protox by either selection with Protox inhibitors (Sato et al. 1994) or introduction and expression of a resistant Protox from another organism. The ultimate practical result of this research would be the production of new crops resistant to Protox inhibitors.

7. PHOTOSYSTEM II INHIBITORS

Photosynthesis is an obvious target for herbicidal attack because the process is critical, yet limited to plants, algae, and certain bacteria, and therefore has lowered potential for animal toxicity. The commercial production of photosynthesis-inhibiting herbicides began in the early to mid-1950s with the introduction of the phenylureas and the *s*-triazines. During the years since, a wide variety of different chemistries have come into use, including a second family of compounds known collectively as the "phenols". Substituted ureas and *s*-triazines make up 62% of commercial products presently available that inhibit Photosystem II (PS II) (Devine et al. 1993a). Although the initial release was intended for weed control in cotton, subsequently released formulations have either weed-specific or general herbicidal effectiveness. These compounds have been very effective in preemergence control of annual weed species in crops and for general postemergence removal of unwanted vegetation.

At the time of their introduction, it was known only that the phenylureas and *s*-triazines affected photosynthesis. They soon were demonstrated to specifically inhibit the Hill reaction of PS II (Wessels and van der Veen 1956). As a result, these herbicides became not only tools for weed control but for plant biochemists and physiologists to begin unravelling the functional aspects of the light reactions of photosynthesis. This widespread experimental usage of these compounds led to the first and certainly most detailed characterization of any herbicide chemistry. It is now well known that these compounds of quite varied chemical structure all bind to the same protein in the thylakoid membrane, known alternately as D-1, the 32-kDa protein, or the *psb* A gene product. Binding of the phenylureas and *s*-triazines occurs at different but largely overlapping areas of the protein (reviewed in Bowyer et al. 1991). Herbicide binding blocks electron flow between the two photosystems and effectively shuts down photosynthetic output of reducing equivalents utilized in carbon fixation. Secondarily, ß-carotene levels decrease, and chloroplasts show visible damage, presumably because of the lost protection against oxidative damage that is imparted by these carotenoids. The phenol-type PS II inhibitors show similar herbicidal effect and are believed to interact with the same regions of the D-1 but in a still different but overlapping site, as they can displace the urea/triazine family members in competition experiments (Trebst and Draber 1986; Oettmeier et al. 1987).

A wide range of resistance to PS II inhibitors in crop species is mediated through decreased uptake and translocation as well as metabolic detoxification of the herbicides by a variety of mechanisms (reviewed in Ashton and Crafts 1981; Hatzios and Penner 1982). Mechanisms utilized vary by species, with some possessing a single means, while others such as maize gain a high level of resistance through the use of hydroxylation, dealkylation, and conjugation to glutathione to inactivate these herbicides.

The vast majority of weed species exhibiting resistance to PS II inhibitors have been shown to possess an additional mechanism that is not found in any crop plants. Through years of selective pressure under field conditions, this resistance mechanism has appeared independently in a variety of different weed species. The first triazine-resistant mutants found in the field were of *Senecio* (Ryan 1970). This mutation was determined to be a single, maternally inherited gene (Souza Machado et al. 1978). Through the use of this and other naturally occuring and experimentally generated mutants, extensive characterization of the binding site for these herbicides has been conducted, the results of which are briefly summarized below.

The herbicide binding protein is composed of a series of five membrane-spanning helices and intermediate loops that interconnect them (Trebst et al. 1990). It is part of a protein complex along with another almost identical protein known as D-2. These two plastoquinone-binding proteins are essentially mirror images of one another, with the exception of the quinone-binding properties. In D-2 the quinone (Q_a) is stably bound; in D-1, however, binding of quinone (Q_b) is reversible, thus providing for competition with PS II herbicides (Tischer and Strotmann 1977). The herbicide (quinone) -binding sites of D-1 is formed by two of the spanning helices and one interconnecting segment. All of the cases of resistance mediated through herbicide binding are due to amino acid substitutions in this region of the protein. The only naturally occuring mutation demonstrated in higher plants is a serine to glycine change at position number 264, which confers 1000-fold resistance to triazines, but little or no resistance to the substituted ureas or phenolics. This has been shown for *Amaranthus hybridus, Phalaris paradoxa, Brassica napus, Senescio vulgaris,* and *Solanum nigrum* (Hirschberg et al. 1987). A number of other mutations have been generated in the laboratory by selection with sublethal doses of a variety of PS II inhibitors in tobacco, potato, green algae (*Chlamydomonas*), cyanobacteria (*Anacystis*), and photosynthetic bacteria (*Synechococcus, Rhodopseudomonas,* and *Rhodobacter*) (Smeda et al. 1993; Trebst 1991). Strangely, in algae and cyanobacteria the mutation at position 264 shows heightened resistance to both triazines and ureas (10- to 100-fold). The binding of members of the phenol family must certainly involve other sites on D-1, because mutations at position 264 have no effect or actually cause increased sensitivity to this family. Mutations in the green alga *Chlamydomonas* at positions 219 and 251, however, allow for slight resistance (2- to 50-fold) to some phenols, but are without effect with other PS II inhibitors (Trebst 1991).

Mechanisms of resistance utilized by naturally resistant crop species have evolved in biotypes of three weed species. *Abutilon theophrasti* biotypes have appeared that utilize glutathione conjugation as a means of detoxification (Anderson and Gronwald 1987, 1991; Gronwald et al. 1989). In Europe and Australia biotypes of *Alopecurus myosuroides* (Moss 1992) and *Lolium rigidum* (Burnet et al. 1991, 1992), respectively, have developed resistance to PS II inhibitors as well as a variety of other unrelated herbicides. The mechanism in both cases is through detoxification by cytochrome P-450 monooxygenases.

The incidence of resistance to PS II inhibitor herbicides has increased steadily since first appearing in the field in 1968. Recent surveys indicate that 58 species (comprising thousands of independently evolved biotypes) show resistance to the triazines in 22 countries worldwide, making this group by far the most prevalent type of resistance among weeds (Holt et al. 1993; Moss and Rubin 1993). Resistance to the substituted ureas is much less common, appearing in 7 species at over 50 independent sites in 3 countries. Resistance to members of the phenol family has not been reported.

8. LIPID BIOSYNTHESIS INHIBITORS

Several different classes of herbicides have been reported to interfere with some aspect of lipid biosynthesis. These include the aryloxyphenoxypropionates, cyclohexanediones, pyridazinones, thiocarbamates, and chloroacetamides. Of these, the site of action was identified definitively only for the first two groups that share the same target enzyme. In the pyridazinones and thiocarbamates target enzymes were inferred from experimental evidence, but were not identified positively. The chloroacetamides, although first described almost 40 years ago, still remain an enigma, having been associated with the inhibition of biosynthesis of lipids, proteins, gibberellins, and phenylpropanoids and of histone acetylation. We will not discuss thiocarbamates, as there is little likelihood that crops will be made resistant to these older, more environmentally suspect herbicides. Collectively, the lipid biosynthesis inhibitors were reviewed by Gronwald (1991), Harwood et al. (1989), and Hoppe (1989). We address these groups in the order listed above. This order represents not only the level of knowledge about sites of action but also the number of occurrences of resistance in weed populations, in decreasing order in both cases.

The aryloxyphenoxypropionates and cyclohexanediones, commonly known as the "fops" and "dims", respectively, are relatively recent additions to the list of herbicides used for the control of grasses. Both are used for the postemergence control of annual and perennial grasses in dicots and nongraminaceous monocots, and thus are called selective graminicides. They are applied foliarly and show rapid translocation to meristematic regions of the targeted weeds. Symptomology is similar between these two chemically dissimilar groups of herbicides, with chlorosis and growth cessation followed by necrosis at the meristems and apices as consequences of treatment (Duke and Kenyon 1988; Shimabukuro 1990).

These herbicides were shown in a number of studies to block the uptake of ^{14}C-acetate into lipids in grasses and not dicots, but to have no effect upon the uptake of radiolabeled amino acids or nucleotides (Burton et al. 1987; Hoppe 1981; Kobek et al. 1988a, b; Stoltenberg et al. 198; Walker et al.,1988). Experiments with ^{14}C-acetate, ^{14}C-pyruvate, and ^{14}C-malonyl-coenzyme A (CoA) in the presence of haloxyfop and sethoxydim showed decreased incorporation of the first two labeled precursors, but no effect on incorporation of the third (Burton et al. 1987). From this work and subsequent studies in other laboratories, it was determined that a molecular site of action for both of these groups of herbicides is the lipid biosynthetic enzyme acetyl-CoA carboxylase (ACCase) (Focke and Lichtenthaler 1987; Kobek et al. 1988a, b; Rendina and Felts 1988; Secor and Cséke 1988). This biotin-containing enzyme is involved in the ATP-dependent carboxylation of acetyl-CoA to form malonyl-CoA, and is essential for the formation and elongation of fatty acids (Harwood 1988). Without fatty acids, many essential cellular processes are halted.

It has been suggested that the fops and dims also interfere with lipid metabolism at the linoleic acid (C18:2) to linolenic acid (C18:3) conversion, tipping the balance toward linolenic acid, as is seen normally in cold-adapted plants (Banas et al. 1990, 1993a). This is the opposite of the effect of the pyridazinones (see below). In fact, the pyridazinones are antagonistic to the effects of haloxyfop and alloxydim in wheat (Banas et al. 1993a). The reversal of haloxyfop and alloxydim effects could also be caused by other antioxidants (Banas et al. 1993b), thus implying that part of the mechanism of these herbicides involves increases in polyunsaturated lipids and the subsequent oxidation of those lipids.

An additional site of action unrelated to lipid biosynthesis was suggested by a variety of laboratories (Cobb and Barnwell 1989; DiTomaso et al. 1991; Häusler et al. 1991; Lucas et al. 1984; Shimabukuro and Hoffer 1992; Shimabukuro et al. 1982; 1989). It was shown that the aryloxyphenoxypropionate, diclofop, had antagonistic activity to the auxinic herbicides, both in the field and in the laboratory (Shimabukuro et al. 1982; Cobb and Barnwell 1989). This is apparently not true of the cyclohexanediones, however. This anti-auxin effect of the aryloxyphenoxypropionates is thought to be mediated by a protonophore action of these herbicides in grasses (Lucas et al. 1984; Shimabukuro et al. 1989), thus depolarizing the membrane. This allows the hydrogen ion gradient induced by the auxinic herbicides to dissipate. As this mechanism does not involve lipid biosynthesis, it is not addressed further here. This mechanism is reviewed elsewhere (Shimabukuro 1990).

The intrinsic resistance of dicots to these herbicides is mediated by a less sensitive ACCase (Burton et al. 1989; Hoppe and Zacher 1985; Kobek et al. 1988a, b; Rendina and Felts 1988). Although grasses are target species of this group of herbicides, not all are affected equally, and sensitivity differences can occur between varieties or even within a genus. Wheat is resistant to diclofop by virtue of an enhanced metabolic detoxification (Donald and Shimabukuro 1980; Duke and Kenyon 1988; Secor and Cséke 1988), but is sensitive to other members of this chemical class and to the cyclohexanediones (Rendina and Felts 1988). Although maize is normally affected like other grasses, it has been possible to select for resistance in cell culture. This resistance is the result of two different mechanisms — increased gene expression for ACCase (Parker et al. 1990b) and a herbicide-resistant ACCase (Parker et al. 1990a). Here, the selective agent was sethoxydim, but resistance to haloxyfop occurred simultaneously. There are examples of noncrop species that show resistance to both classes of the herbicides as well. Red fescue (*Festuca rubra* L.) is resistant to haloxyfop (Stoltenberg et al. 1989; Secor and Cséke 1988; Owen 1991), sethoxydim (Stoltenberg et al. 1989; Hosaka et al. 1987), and tralkoxydim (Secor and Cséke 1988; Owen 1991), and blue fescues (*F. ovina* var. *glauca* and *F. amethystina* L.) are resistant to fluazifop and sethoxydim (Catanzaro et al. 1993). However, another species within the same genus, white tall fescue (*F. arundinacea* Schreb.) is sensitive to all three. In the three cases within the fescues resistance could be explained by resistant ACCase.

Evolved resistance in the field has been reported in a number of species in a variety of locations, including *Alopecurus myosuriodes* (Moss 1990), *Avena sterilis* and *A. fatua* (Mansooji et al. 1992; Devine et al. 1992; Heap et al. 1993), *Lolium multiflorum* (Stanger and Appleby 1989), *L. rigidum* (Heap and Knight 1986, 1990), *Setaria viridis* (Marles et al. 1993), and *Sorghum halepense* (L.) Pers. (Barrentine et al. 1992). In all of these cases resistance seems to be mediated by more than one mechanism within the resistant populations. *Alopecurus myosuriodes* was originally noted to be resistant to the phenyl urea, chlorotoluron, but also showed cross-resistance to ten other herbicide classes, including the ACCase inhibitors (Moss and Cussans 1991; Chauvel 1992). Resistance was demonstrated for diclofop-methyl, fluazifop-butyl, fenoxyprop-ethyl, but not for quizalofop-ethyl. Among the cyclohexanediones, tralkoxydim and alloxydim-sodium were shown to be ineffective, yet these plants were susceptible to sethoxydim. In at least one case enhanced metabolism of the herbicide (diclofop) is responsible for resistance in this species (Kemp et al. 1990). One biotype of *Avena fatua* isolated from Canada shows resistance to both fops and dims, but shows no decrease in sensitivity at the enzyme level and no enhanced metabolic detoxification abilities (Devine et al. 1992). It was subsequently shown that the only apparent difference between the susceptible and resistant biotypes is the ability to regenerate transmembrane potentials that are rapidly lost upon exposure to diclofop (Devine et al. 1993b). Four new biotypes of *A. fatua* were identified in Canada that show cross-resistance to the fops and dims, but show no resistance to herbicides with other modes of action (Heap et al. 1993). No mechanism for the resistance was addressed in this work. Diclofop-methyl-resistant Italian ryegrass (*Lolium multiflorum*), found in Oregon, was shown to possess a resistant ACCase (Betts et al. 1992). Within the many biotypes of cross-resistant *L. rigidum* found in Australia, cross-resistance patterns vary extensively, with some biotypes more or less sensitive to fops than dims or showing equal levels of resistance to both classes of herbicides (Heap and Knight 1990). Resistance has been demonstrated at the enzyme level in some, but in others it seems that a mechanism involving membrane repolarization, similar to that seen in *Avena,* may be responsible (Häusler et al. 1991; Holtum et al. 1991; Shimabukuro and Hoffer 1992). In the case of *Setaria* resistance to a wide variety of fops and dims was determined to result from a resistant ACCase (Marles et al. 1993). From these observations, it is obvious that acquired resistance to the aryloxyphenoxypropionates and cyclohexanediones will continue to be a problem in weed populations worldwide.

The pyridazinones, characterized by norflurazon, BASF 13-338, SAN 6706, and pyrazon, form a chemically similar group of herbicides. Subtle differences in substituent chemistry cause them to affect different cell processes. In addition to the inhibition of fatty acid biosynthesis, they were reported to inhibit photosynthesis in a manner similar to the triazines and phenyl ureas, and to block carotenoid biosynthesis (St. John and Hilton 1976; Eder 1979) (see below). Inhibition of PS II is thought to be important *in vivo* only for pyrazon (Fedtke 1982). This mechanism probably is irrelevant, however, because the pyridazinones are used for preemergence control of both broadleaf and monocot weed species, and result in bleached seedlings in which no photosynthesis occurs. The reason for this bleached appearance is related to the ability of a number of these herbicides to inhibit the carotenoid biosynthetic enzyme phytoene desaturase (Mayer et al. 1989; Sandmann et al. 1984; Linden et al. 1990). Blockage of carotenoid biosynthesis may well be the most important plant process effected by most of these herbicides, but there is also strong evidence that some members of the group have a role in reducing the degree of desaturation of certain lipids (Duke 1985; Harwood et al. 1989). BASF 13-338 (SAN 9785) and SAN 6706 are two such herbicides. Both were demonstrated to effect the conversion of linoleic acid (C18:2) to linolinic acid (C18:3) mediated by linoleic acid desaturase (St. John 1976; St. John et al. 1979). These are prominent galactolipids in the membrane of plastids, with the ratio of the two determining membrane fluidity as a function of temperature. Higher ratios of C18:3 to C18:2 maintain fluidity at lower temperatures and are found in cold-adapted plastid membranes (St. John et al. 1979; Wang et al., 1987). Herbicidal blockage of this conversion is unlikely to be seriously deleterious to the plant, and thus explains the low effectiveness of BASF 13-338. Other herbicides within this group, such as SAN 6706, block both carotenoid biosynthesis and lipid desaturation (Eder 1979; Harwood et al. 1989; Khan et al. 1979). This compound was shown to be metabolized to norflurazon in some species (Strang and Rogers 1974). It then appears that the pyridazinones as a group act as inhibitors of a variety of chloroplastic desaturases.

9. CAROTENOID SYNTHESIS INHIBITORS

Carotenoids are essential to the protection of green plants from photooxidative processes. The chlorophyll of plants that lack these energy-absorbing compounds because of a genetic lesion or chemical treatment is rapidly lost in the light, resulting in an albino plant. Thus, inhibitors of carotenoid synthesis cause albinism, which is lethal. Apparently, several molecular sites of action of commercial herbicides inhibit carotenoid biosynthesis; however, only one of these was clearly demonstrated, phytoene desaturase. Therefore, we break this discussion into phytoene desaturase inhibitors and other sites in the carotenoid pathway.

9.1. PHYTOENE DESATURASE

Only two phytoene desaturase inhibitors, fluridone and norflurazon, are commercially available in the U.S., and of these, only norflurazon is used on agronomic crops. Diflufenican, another phytoene desaturase inhibitor, is being developed for use in wheat and barley. Norflurazon is generally used as a preemergence, soil-applied herbicide. Several important crops are not tolerant to it and there is sometimes phytotoxicity to cotton. Furthermore, due to its persistence in soil, phytotoxicity can sometimes be experienced the next year in rotated crops. A major problem with phytoene desaturase inhibitors is their lack of selectivity. Thus, the advent of crops with resistance to these herbicides at the phytoene desaturase level would be especially useful.

There is a practical level of resistance in crop species such as cotton, red beet, soybeans, and cranberries. Resistance is conferred by reduced absorption and translocation (Strang and Rogers 1974; Stephenson and Ries 1967) or metabolism (Ries et al. 1968; Yaklich et al. 1974). Metabolic degradation of norflurazon in cotton is not faster than in some susceptible weeds, and there is evidence that sequestration of norflurazon in the lysigenous glands that cover the hypocotyl and cotyledons of cotton imparts tolerance to cotton (Duke 1992). Glandless cotton varieties are more susceptible to norflurazon than those with glands (Stegink and Vaughn 1988).

No cases of evolved resistance to these herbicides have been reported as a result of selection by phytoene desaturase inhibitors. The only documented case of resistance in weed species is to the PS II-inhibiting pyridazinone, pyrazon. Pyrazon is not used as a herbicide in the U.S. In both *Amaranthus hybridus* (Arntzen et al. 1982) and *Chenopodium album* (Gressel et al. 1982) triazine-resistant biotypes show cross-resistance to pyrazon. A resistant mutation was created in the cyanobacteria, *Synechococcus* PCC 7942, by mutagenesis with ethyl methanesulfonate (Linden et al. 1990) and *Synechocystis* sp. PCC 6803 (Martinez-Férez and Viogue 1992). The *Synechococcus* mutant showed decreased sensitivity to norflurazon as well as decreased phytoene desaturase activity and lowered carotenoid content. The gene responsible for resistance was cloned subsequently and shown to be a phytoene desaturase containing a single point mutation that induces an amino acid substitution of a glycine for a valine (Chamovitz et al. 1991). The *Synechocystis* mutant was generated by continuous selection on norflurazon-containing medium. This single nucleotide mutation converts an arginine to a cysteine. The gene from *Synechocystis* is 82 and 61% homologous to the phytoene desaturase genes from *Synechococcus* and soybean, respectively (Martinez-Férez and Viogue 1992). Because of the sequence similarity between these species, the mutants isolated in cyanobacteria may be useful in planning site-directed mutagenesis in crop species to obtain norflurazon-resistance.

9.2. OTHER SITES OF THE CAROTENOID PATHWAY

Clomazone is a relatively new selective herbicide that apparently inhibits a site of the carotenoid pathway earlier than phytoene desaturase (Duke et al. 1985). Some data suggest that it may inhibit the conversion of farnasyl-PP to geranylgeranyl-PP (Duke et al. 1991c). This site of action would explain its apparent inhibition of the synthesis of products of the terpenoid pathway other than carotenoids, such as phytol and gibberellic acid (Duke et al. 1985; Vencill et al., 1989). This compound may be a proherbicide, in that this herbicide does not inhibit any of the enzymes of the terpenoid pathway tested *in vitro* (Croteau 1992; Scott et al. 1994). Its mechanism of selectivity is not clear; however, there is some evidence that some crops are resistant due to rapid metabolic inactivation (ElNaggar et al. 1992). Cotton can be protected from clomazone by organophosphate insecticides (Jordan et al. 1993). The biochemical mechanism of this effect is unknown.

10. AUXIN FUNCTION

The phenoxycarboxylic acids (PCAs) such as 2,4-D and 2,4,5-T are among the oldest synthetic herbicides. These compounds are used primarily for postemergence management of dicot weeds in grass crops, pastures, and lawns. The butryric acid derivative of 2,4-D, 2,4-DB is a proherbicide that is metabolized to 2,4-D in sensitive species. Some legumes that are sensitive to 2,4-D are tolerant to 2,4-DB because they lack the ß-oxidation activity required to activate it. Phenoxycarboxylic acids have been off patent for many years and are very effective and inexpensive.

Sensitive crops and ornamentals are often injured by 2,4-D spray drift from applications to tolerant crops or turf. Introduction of 2,4-D-resistant crops could eliminate damage to these crops from such a source, the purported or implied objective of some research to generate 2,4-D-resistant cotton (Bayley et al. 1992; Lyon et al. 1993). This herbicide would be extremely useful and cost effective as a postemergence herbicide in cotton.

Despite many years of study, the actual molecular site of action of this family of herbicides is unknown. The PCAs are sometimes called hormone-type herbicides because they mimic in many ways lethal doses of the plant hormone indoleacetic acid (IAA), also known as auxin (Devine et al. 1993). Both IAA and the PCAs are thought to act by influencing plasma membrane properties to act at a molecular site in the plasma membrane.

Tolerance of grasses to these herbicides appears to be related to rapid irreversible metabolic conversion to nontoxic products, whereas in dicotyledonous species the herbicide is often found in the form of reversible conjugates (Devine et al. 1993). Several weed species have evolved resistance to 2,4-D (LeBaron 1991). Where studied, the resistance appears to be due to accelerated metabolic detoxification.

11. MULTIPLE RESISTANCE

Although the first cases of herbicide resistance were all cases of single resistance to a single herbicide, with cross-resistance only to other herbicides at the same binding site, weed biotypes have appeared more recently that have resistance to a number of herbicides with diverse mechanisms of action. These are termed multiple resistant biotypes.

The most well-documented cases of multiple resistance are noted in the *Lolium* biotypes found in Australia, analyzed by Powles and colleagues (Hall et al. 1994). Resistance was first noted to the lipid biosynthesis-inhibiting herbicides such as diclofop and sethoxydim. The binding site of these herbicides is the enzyme acetyl-CoA carboxylase, but no alterations in the sensitivity of this enzyme to these herbicides were found. Treatment of the biotype with inhibitors of cytochrome P-450 ameliorated the resistance to a certain extent for some of these herbicides, but for other herbicides there was no amelioration by these same compounds, indicating that more than a simple acceleration in metabolism had occurred in these biotypes. For example, diclofop resistance did not correlate with the ability to metabolize diclofop nor changes in the active site, but did correlate with the resistance to membrane depolarization. Moreover, there were many combinations of cross-resistance patterns in the individual biotypes, which indicate that there may even be multiple mechanisms for the resistance by a number of means. By interbreeding of the survivors, a given new population may have at its disposal a variety of biochemical/physiological options for coping with xenobiotic stress.

12. CONSPECTUS

Several major questions are still facing weed scientists/plant physiologists in the area of herbicide mode of action. Considering that alachlor and metoalachlor are two of our most commonly utilized herbicides (probably used on >80% of the acreage of corn in the Corn Belt), that we still have no idea as to how these herbicides work is astounding. With the cellulose biosynthesis inhibitors and the mitotic disrupter herbicides, the molecular site of action has not been identified. The selection of mutants resistant to these herbicides and the identification and cloning of the genes for resistance may aid us in our investigation of these recalcitrant modes of action.

Resistance at the site of action has occurred most rapidly to those herbicides that have (1) a single site of action, (2) a high persistence in the environment, (3) a high level of efficacy in causing plant death, (4) a high rate of mutation at the active site, and (5) a less debilitating effect by alteration of this herbicidal site. The rapid occurrence of herbicide resistance to the sulfonylureas and imidazolinones is a perfect example of this scenario for the development of resistance. The development of other mechanisms of resistance, such as compartmentalization or metabolism, is more difficult to predict because in many cases these cellular factors involved in herbicide movement/sequestration are less well-characterized than those relating to the site of action. The development of multiple resistance mechanisms is particularly frightening, especially because there are fewer new herbicides being released and many of these newer herbicides have sites of action or mechanisms of metabolism in common with older herbicides to which resistance has already developed. Staying one step ahead of the evolution of resistance will be a full-time job for the weed scientist of the future.

REFERENCES

Akashi, T. 1988. Effects of propyzamide on tobacco cell microtubules *in vivo* and *in vitro*. *Plant Cell Physiol.* 29: 1053-1062.
Anderson, R.J., Norris, A.E., and Hess, F.D. 1994. Synthetic chemicals that act through the porphyrin pathway: a survey. *ACS Symp. Ser.* 559: 18-33.

Anderson, R.N. and Gronwald, J.W. 1987. Noncytoplasmic inheritance of atrazine tolerance in velvetleaf (*Abutilon theophrasti*). *Weed Sci.* 35: 496-498.

Anderson, R.N. and Gronwald, J.W. 1991. Atrazine resistance in velvetleaf (*Abutilon theophrasti*) biotype due to enhanced glutathione S-transferase activity. *Plant Physiol.* 96: 104-109.

Aono, M., Kubo, A., Saji, H., Tanaka, K., and Kondo, N. 1993. Enhance tolerance to photooxidative stress of transgenic *Nicotiana tabacum* with high chloroplastic glutathione reductase activity. *Plant Cell Physiol.* 34: 129-135.

Arntzen, C.J., Pfister, K., and Steinback, K.E. 1982. The mechanism of chloroplast triazine resistance: alterations in the site of herbicide action. In *Herbicide Resistance in Plants.* Ed. by LeBaron, H.M. and Gressel, J., Wiley-Interscience, New York. pp. 185-214.

Ashton, F.M. and Crafts, A.S. 1981. *Mode of Action of Herbicides.* John Wiley & Sons, New York.

Banas, A., Johansson, I., Stenlid, G., and Stymne, S. 1990. The effect of haloxyfop-ethoxyethyl on lipid metabolism in oat and wheat shoots. *Swedish J. Agric. Res.* 20: 97-104.

Banas, A., Johansson, I., Stenlid, G., and Stymne, S. 1993a. The effect of haloxyfop and alloxydim on growth and fatty acid composition of wheat roots. *Swedish J. Agric. Res.* 23: 55-65.

Banas, A., Johansson, I., Stenlid, G., and Stymne, S. 1993b. Free radical scavengers and inhibitors of lipoxygenases as antagonists against the herbicides haloxyfop and alloxydim. *Swedish J. Agric. Res.* 23: 67-75.

Barrentine, W.L., Snipes, C.E., and Smeda, R.J. 1992. Herbicide resistance confirmed in johnsongrass biotypes. *Miss. Agric. Forest. Exp. Stn. Res. Rep.* 17: No. 5.

Bayley, C., Trolinder, N., Ray, C., Morgan, M., Quisenberry, J.E., and Ow, D.W. 1992. Engineering 2,4-D resistance into cotton. *Theor. Appl. Genet.* 83: 645-649.

Becerril, J.M. and Duke, S.O. 1989a. Protoporphyrin IX content correlates with activity of photobleaching herbicides. *Plant Physiol.* 90: 1175-1181.

Becerril, J.M. and Duke, S.O. 1989b. Acifluorfen effects on intermediates of chlorophyll synthesis in green cucumber cotyledon tissues. *Pestic. Biochem. Physiol.* 35: 119-126.

Becerril, J.M., Duke, S.O., and Lydon, J. 1989. Glyphosate effects on shikimate pathway products in leaves and flowers of velvetleaf (*Abutilon theophrasti* Medic.). *Phytochemistry* 28: 695-699.

Betts, K.J., Ehlke, N.J., Wyse, D.L., Gronwald, J.W., and Somers, D.A. 1992. Mechanism of inheritance of diclofop resistance in Italian ryegrass (*Lolium multiflorum*). *Weed Sci.* 40: 184-189.

Beyer, E.M., Duffy, M.J., Hay, J.V., and Schlueter, D.D. 1988. Sulfonylureas. In *Herbicides—Chemistry, Degradation and Mode of Action Vol. III.* Ed. by Kearney, P.C. and Kaufman, D.D. Marcel Dekker, New York, pp. 117-189.

Boerboom, J., Ehlke, N.J., Wyse, D.L., and Somers, D.A. 1991. Recurrent selection for glyphosate tolerance in birdsfoot trefoil. *Crop Sci.* 31: 1124-1129.

Bolduc, C., Lee, V.D., and Huang, B. 1988. ß-Tubulin mutants of the unicellular green alga *Chlamydomonas reinhardtii.* *Proc. Natl. Acad. Sci. U.S.A.* 85: 131-135.

Bowyer, J.R., Camilleri, P., and Vermaas, W.F.J. 1991. Photosystem II and its interaction with herbicides. In *Herbicides: Topics in Photosynthesis.* Ed. by Baker, N.R. and Percival, M.P. Elsevier, Amsterdam, pp. 27-86.

Burnet, M.W.M., Hildebrand, O.B., Holtum, J.A.M., and Powles, S.B. 1991. Amitrole, triazine, substituted urea, and metribuzin resistance in a biotype of rigid ryegrass (*Lolium rigidum*). *Weed Sci.* 39: 317-323.

Burnet, M.W.M., Loveys, B.R., Holtum, J.A.M., and Powles, S.B. 1992. Increased detoxification is a mechanism of simazine resistance in annual ryegrass (*Lolium rigidum*). *Pestic. Biochem. Physiol.* 46: 207-218.

Burton, J.D., Gronwald, J.W., Somers, D.A., Connelly, J.A., Gengenbach, B.G., and Wyse, D.L. 1987. Inhibition of plant acetyl-coenzyme A carboxylase by the herbicides sethoxydim and haloxyfop. *Biochem. Biophys. Res. Commun.* 148: 1039-1044.

Burton, J.D., Gronwald, J.W., Somers, D.A., Gengenbach, B.G., and Wyse, D.L. 1989. Inhibition of corn acetyl-CoA carboxylase by cyclohexanedione and aryloxyphenoxypropionate herbicides. *Pestic. Biochem. Physiol.* 34: 76-85.

Catanzaro, C.J., Burton, J.D., and Skroch, W.A. 1993. Graminicide resistance of acetyl-CoA carboxylase from ornamental grasses. *Pestic. Biochem. Physiol.* 45: 147-153.

Chamovitz, D., Pecker, I., and Hirschberg, J. 1991. The molecular basis of resistance to the herbicide norflurazon. *Plant Mol. Biol.* 16: 967-974.

Chauvel, B. 1992. Cross resistance to herbicides in a weed (*Alopecurus myosuroides* Hudson). *C. R. Seances Soc. Biol. Filial.* 186: 287-292.

Christy, A.L., Herbst, K.A., Kostka, S.J., Mullen, J.P., and Carlson, P.S. 1993. Synergizing weed biocontrol agents with chemical herbicides. *ACS Symp. Ser.* 524: 87-100.

Cobb, A.H. and Barnwell, P. 1989. Anti-auxin activity of graminicides. In *Brighton Crop Protection Conf. — Weeds,* 1989: 183-190.

Cronin, K.E., Hussey, P.J., Ray, J.A., Waldin, T.R., and Ellis, J.R. 1993. Herbicide-Resistant Plants Expressing a Modified Tubulin Gene. PCT Int. Appl. WO 93 24,637, December 9.

Croteau, R. 1992. Clomazone does not inhibit the conversion of isopentenyl pyrophosphate to geranyl, farnesyl, or geranylgeranyl pyrophosphate in vitro. *Plant Physiol.* 85: 469-473.

Delmer, D.P., Read, S.M., and Cooper, G. 1987. Identification of a receptor protein in cotton fibers for the herbicide 2,6-dichlorobenzonitrile. *Plant Physiol.* 84: 415-420.

Devine, M.D., Duke, S.O., and Fedtke, C. 1993a. *Physiology of Herbicide Action.* Prentice Hall, Englewood Cliffs, NJ.

Devine, M.D., Hall, J.C., Romano, M.L., Marles, M.A.S., Thomson, L.W., and Shimabukuro, R.H. 1993b. Diclofop and fenoxaprop resistance in wild oat is associated with an altered effect on the plasma membrane electrogenic potential. *Pestic. Biochem. Physiol.* 45: 167-177.

Devine, M.D., MacIsaac, S.A., Romano, M.L., and Hall, J.C. 1992. Investigation of diclofop-resistance in two biotypes of *Avena fatua. Pestic. Biochem. Physiol.* 42: 88-96.

DiTomaso, J.M., Brown, P.H., Stowe, A.E., Linscott, D.L., and Kochian, L.V. 1991. Effects of diclofop and diclofop-methyl on membrane potentials in roots of intact oat, maize, and pea seedlings. *Plant Physiol.* 95: 1063-1069.

Donald, W.W. and Shimabukuro, R.H. 1980. Selectivity of diclofop-methyl between wheat and wild oat: growth and herbicide metabolism. *Physiol. Plant.* 49: 459-464.

Duke, S.O. 1985. Effects of herbicides on nonphotosynthetic biosynthetic processes. In *Herbicide Physiology*, Vol. 2. Ed. by Duke, S.O. CRC Press, Boca Raton, FL, pp. 92-112.

Duke, S.O. 1988. Glyphosate. In *Herbicides — Chemistry, Degradation and Mode of Action Vol. III.* Ed. by Kearney, P.C. and Kaufman, D.D. Marcel Dekker, New York, pp. 1-70.

Duke, S.O. 1990. Overview of herbicide mechanisms of action. *Environ. Health Perspect.* 87: 263-271.

Duke, S.O. 1992. Modes of action of herbicides used in cotton. In *Weeds of Cotton: Characterization and Control.* Ed. by McWhorter, C.G. and Abernathy, J.L. The Cotton Foundation, Memphis, TN, pp. 403-437.

Duke, S.O. 1993. Mechanisms for resistance of weeds to herbicides. *Proc. Beltwide Cotton Conf. 1993.* 3: 1509-1511.

Duke, S.O. 1995. Commentary: glandular trichomes — a focal point of chemical and structural interactions. *Int. J. Plant Sci.* 155: 617-620.

Duke, S.O., Abbas, H.K., Amagasa, T., and Tanaka, T. 1995. Phytotoxins of microbial origin with the potential for use as herbicides. In *Natural Products and Their Potential in Agriculture.* Critical Reviews in Applied Chemistry Series. Ed. by Copping, L.G. in press.

Duke, S.O., Becerril, J.M., Lydon, J., Matsumoto, H., and Sherman, T.D. 1991a. Protoporphorinogen oxidase-inhibiting herbicides. *Weed Sci.* 39: 465-473.

Duke, S.O., Kenyon, W.H., and Paul, R.N. 1985. FMC 57020 effects on chloroplast development in pitted morningglory (*Ipomoea lacunosa*) cotyledons. *Weed Sci.* 33: 786-794.

Duke, S.O. and Kenyon, W.H. 1988. Polycyclic alkanoic acids. In *Herbicides — Chemistry, Degradation and Mode of Action, Vol. III.* Ed. by Kearney, P.C. and Kaufman, D.D. Marcel Dekker, New York, pp. 71-116.

Duke, S.O., Lee, H.J., Nandihalli, U.B., and Duke, M.V. 1994. Why protoporphyrinogen oxidase is probably the optimal herbicide site in the porphyrin pathyway. *ACS Symp. Ser.* in press.

Duke, S.O. and Lydon, J. 1993 Natural phytotoxins as herbicides. *ACS Symp. Ser.* 524: 110-124.

Duke, S.O., Lydon, J., and Paul, R.N. 1989. Oxadiazon activity is similar to that of *p*-nitro-diphenyl ether herbicides. *Weed Sci.* 37: 152-160.

Duke, S.O., Moorman, T.B., and Bryson, C.B. 1991b. Phytotoxic effects of pesticide degradation products. *ACS Symp. Ser.* 459: 188-204.

Duke, S.O., Paul, R.N., Becerril, J.M., and Schmidt, J.H. 1991c. Clomazone causes accumulation of sesquiterpenoids in cotton (*Gossypium hirsutum* L.). *Weed Sci.* 39: 339-346.

Dyer, W.E., Hess, F.D., Holt, J.S., and Duke, S.O. 1993. Potential benefits and risks of herbicide-resistant crops produced by biotechnology. *Hortic. Rev.* 15: 367-408.

Eder, F.A. 1979. Pyridazinones, their influence on the biosynthesis of carotenoids and the metabolism of lipids in plants (survey of literature). *Z. Naturforsch.* 34c: 1052-1054.

ElNaggar, S.F., Creekmore, R.W., Schocken, M.J., Rosen, R.T., and Robinson, R.A. 1992. Metabolism of clomazone herbicide in soybean. *J. Agric. Food Chem.* 40: 880-883.

Fedtke, C. 1982. *Biochemistry and Physiology of Herbicide Action.* Springer-Verlag, Berlin.

Focke, M. and Lichtenthaler, H.K. 1987. Inhibition of the acetyl-CoA carboxylase of barley chloroplasts by cycloxydim and sethoxydim. *Z. Naturforsch.* 42c: 1361-1363.

Frear, D.S., Swanson, H.R., and Mansager, E.R. 1983. Acifluorfen metabolism in soybean: diphenylether bond cleavage and the formation of homoglutathione, cysteine, and glucose conjugates. *Pestic. Biochem. Physiol.* 20: 299-310.

Fuerst, E.P., Arntzen, C.J., Pfister, K., and Penner, D. 1986. Herbicide cross-resistance in triazine resistant biotypes of four species. *Weed Sci.* 34: 344-353.

Fuerst, E.P., Nakatani, H.Y., Dodge, A.D., Penner, D., and Arntzen, C.J. 1985. Paraquat resistance in *Conyza. Plant Physiol.* 77: 984-989.

Fuerst, E.P. and Vaughn, K.C. 1990. Mechanisms of paraquat resistance. *Weed Technol.* 4: 150-156.

Gougler, J.A. and Geiger, D.A. 1984. Carbon partitioning and herbicide transport in glyphosate-treated sugarbeet (*Beta vulgaris*). *Weed Sci.* 32: 546-551.

Gressel, J., Ammon, H.U., Fogelfors, H., Gasquez, J., Kay, Q.O.N., and Kees, H. 1982. Discovery and distribution of herbicide-resistant weeds outside North America. In *Herbicide Resistance in Plants.* Ed. by LeBaron, H.M. and Gressel, J. Wiley-Interscience, New York, pp. 32-55.

Gronwald, J.W. 1991. Lipid biosynthesis inhibitors. *Weed Sci.* 39: 435-449.

Gronwald, J.W., Anderson, R.N., and Yee. C. 1989. Atrazine resistance in velvetleaf (*Abutilon theophrasti*) due to enhanced atrazine detoxification. *Pestic. Biochem. Physiol.* 34: 149-163.

Hall, L.M., Holtum, J.A.M., and Powles, S.B. 1994. Mechanisms responsible for cross resistance and multiple resistance. In *Herbicide Resistance in Plants.* Ed. by Powles, S.B. and Holtum, J.A.M. Lewis Publishers, Boca Raton, FL, pp. 243-261.

Harwood, J.L. 1988. Fatty acid metabolism. *Annu. Rev. Plant Physiol. Molec. Biol.* 39: 101-138.

Harwood, J.L., Ridley, S.M., and Walker, K.A. 1989. Herbicides inhibiting lipid synthesis. In *Herbicides and Plant Metabolism.* Ed. by Dodge, A.D. Cambridge University Press, New York, pp. 73-96.

Hattori, J., Rutledge, R., Labbé, H., Brown, D., Sunohara, G., and Miki, B. 1992. Multiple resistance to sulfonylureas and imidazolinones conferred by an acetohydroxyacid synthase gene with separate mutations for selective resistance. *Mol. Gen. Genet.* 232: 167-173.

Hatzios, K.K. and Penner, D. 1982. *Metabolism of Herbicides in Higher Plants.* Burgess Publishing, Minneapolis.

Häusler, R.E., Holtum, J.A.M., and Powles, S.B. 1991. Cross-resistance to herbicides in annual ryegrass (*Lolium rigidum*). IV. Correlation between membrane effects and resistance to graminicides. *Plant Physiol.* 97: 1035-1043.

Heap, I. and Knight, R. 1986. The occurrence of herbicide cross-resistance in a population of annual ryegrass, *Lolium rigidum*, resistant to diclofop-methyl. *Aust. J. Agric. Res.* 37: 149-156.

Heap, I. and Knight, R. 1990. Variations in herbicide cross-resistance among populations of annual ryegrass (*Lolium rigidum*) resistant to diclofop-methyl. *Aust. J. Agric. Res.* 41: 121-128.

Heap, I.M., Murray, B.G., Loeppky, H.A., and Morrison, I.N. 1993. Resistance to aryloxyphenoxypropionate and cyclohexandione herbicides in wild oat (*Avena fatua*). *Weed Sci.* 41: 232-238.

Heim, D.R., Bjelk, L.A., James, J., Schneegurt, M.A, and Larrinua, I.M. 1993. Mechanism of isoxaben tolerance in *Agrostis palustris* var. Penncross. *J. Exp. Bot.* 44: 1185-1189.

Heim, D.R., Skomp, J.R., Tschabold, E.E., and Larrinua, I.M. 1990. Isoxaben inhibits the synthesis of acid insoluble cell wall materials in *Arabidopsis thaliana. Plant Physiol.* 93: 695-700.

Hérouaart, D., Bowler, C., Willekens, H., Van Camp, W., Slooten, L., Van Montagu, M., and Inzé, D. 1993. Genetic engineering of oxidative stress resistance in higher plants. *Philos. Trans. R. Soc. Lond. B* 342: 235-240.

Hess, F.D. 1987. Herbicide effects on the cell cycle of meristematic plant cells. *Rev. Weed Sci.* 3: 183-203.

Hess, F.D. and Bayer, D.E. 1977. Binding of the herbicide trifluralin to *Chlamydomonas* flagellar tubulin. *J. Cell Sci.* 24: 351-360.

Hilton, J.L. and Christiansen, M.N. 1972. Lipid contribution to selective action of trifluralin. *Weed Sci.* 20: 351-360.

Hirschberg, J., Yehuda, A.B., Pecker, I., and Ohad, N. 1987. Mutations resistant to photosystem II herbicides. In *Plant Molecular Biology.* Ed. by von Wettstein, D. and Chua, N.-H. Plenum Press, New York, pp. 357-366.

Hoffman, J.C. and Vaughn, K.C. 1994. Mitotic disrupter herbicides act by a single mechanism but vary in efficacy. *Protoplasma* 179: 16-25.

Holt, J.S., Powles, S.B., and Holtum, J.A.M. 1993. Mechanisms and agronomic aspects of herbicide resistance. *Annu. Rev. Plant Physiol. Molec. Biol.* 44: 203-229.

Holtum, J.A.M., Matthews, J.M., Häusler, R.E., Liljegren, D.R., and Powles, S.B. 1991. Cross-resistance to herbicides in annual ryegrass (*Lolium rigidum*). III. On the mechanism of resistance to diclofop-methyl. *Plant Physiol.* 97: 1026-1034.

Hoppe, H.H. 1981. Effect of diclofop-methyl on protein, nucleic acid, and lipid biosynthesis in tips of radicles from *Zea mays* L. *Z. Pflanzenphysiol.* 102: 189-197.

Hoppe, H.H. 1989. Fatty acid biosynthesis — a target site of herbicide action. In *Target Sites for Herbicide Action.* Ed. by Böger, P. and Sandmann, G. CRC Press, Boca Raton, FL, pp. 65-83.

Hoppe, H.H. and Zacher, H. 1985. Inhibition of fatty acid biosynthesis in isolated bean and maize chloroplasts by herbicidal phenoxyphenoxypropionic acid derivatives and structurally related compounds. *Pestic. Biochem. Physiol.* 24: 298-305.

Hosaka, H., Inaba, H., Satoh, A., and Tanoue, T. 1987. Selectivity mechanisms of sethoxydim between red fescue (*Festuca rubra*) and tall white fesuce (*Festuca arundinacea*). *Weed Res.* 32: 255-262.

Jacobs, J.M., Jacobs, N.J., Sherman, T.D., and Duke, S.O. 1991. Effect of diphenyl ether herbicides on oxidation of protoporphyrinogen to protoporphyrin in organellar and plasma membrane-enriched fractions of barley. *Plant Physiol.* 97: 197-203.

Jacobs, J.M., Wehner, J.M., and Jacobs, N.J. 1994. Porphyrin stability in plant supernatant fractions: implication for the action of porphyrinogenic herbicides. *Pestic. Biochem. Physiol.* 50: 23-30.

James, S.W., Ranum, L.P.W., Silflow, C.D., and Lefebvre, P.A. 1988. Mutants resistant to anti-microtubule herbicides map to a locus on the *uni* linkage group in *Chlamydomonas reinhardtii. Genetics* 118: 141-147.

Jasieniuk, M., Brûlé-Babel, A.L., and Morrison, I.N. 1994. Inheritance of trifluralin resistance in green foxtail (*Setaria viridis*). *Weed Sci.* 42: 123-127.

Jordan, D.L., York, A.C., McClelland, M.R., and Frans, R.E. 1993. Clomazone as a component in cotton (*Gossypium hirsutum*) herbicide programs. *Weed Technol.* 7: 201-211.

Kemp, M.S., Moss, S.R., and Thomas, T.H. 1990. Herbicide resistance in *Alopecurus myosuroides*. In *Managing Resistance to Agrochemicals From Fundamental Research to Practical Strategies*. Ed. by Green, M.B., LeBaron, H.M., and Moberg, W.K., American Chemical Society, Washington, D.C., pp. 376-393.

Khan, M., Lem, N.W., Chandorkar, K.R., and Williams, J.P. 1979. Effects of substituted pyridazinones (San 6706, San 9774, San 9785) on glycerolipids and their associated fatty acids in the leaves of *Vicia faba* and *Hordeum vulgare*. *Plant Physiol.* 64: 300-305.

Killmer, J., Widholm, J., and Slife, F. 1981. Reversal of glyphosate inhibition of carrot cell culture growth by glycolytic intermediates and organic amino acids. *Plant Physiol.* 68: 1299-1302.

Kobek, K., Focke, M., and Lichtenthaler, H.K. 1988a. Fatty-acid biosynthesis and acetyl-CoA carboxylase as a target of diclofop, fenoxaprop and other aryloxy-phenoxy-propionic acid herbicides. *Z. Naturforsch.* 43c: 47-54.

Kobek, K., Focke, M., Lichtenthaler, H.K., Retzlaff, G., and Würzer, B. 1988b. Inhibition of fatty acid biosynthesis in isolated chloroplasts by cycloxydim and other cyclohexane-1,3-diones. *Physiol. Plant.* 72: 492-498.

Knight, T.J. Bush, D.R., and Langston-Unkefer, P.J. 1988. Oats tolerant of *Pseudomonas syringae* pv. *tabaci* contain tabtoxinine-β-lactam-insensitive leaf glutamine synthetases. *Plant Physiol.* 88: 333-339.

LeBaron, H.M. 1991. Distribution and seriousness of herbicide-resistant weed infestations worldwide. In *Herbicide Resistance in Weeds and Crops*. Ed. by Caseley, J.C., Cussans, G.W.,and Atkin, R.K., Butterworth-Heinemann, Oxford, pp. 27-43.

Lee, H.J., Duke, M.V., and Duke, S.O. 1993. Cellular localization of protoporphyrinogen-oxidizing activities of etiolated barley (*Hordeum vulgare* L.) leaves. Relationship to mechanism of action of protoporphyrinogen oxidase inhibiting herbicides. *Plant Physiol.* 102: 881-889.

Lee, H.J. and Duke, S.O. 1994. Protoporphyrinogen IX-oxidizing activities involved in the mode of action of peroxidizing herbicides. *J. Agric. Food Chem.* 42: 2610-2618.

Linden, H., Sandmann, G., Chamovitz, D., Hirschberg, J., and Böger, P. 1990. Biochemical characterization of *Synechococcus* mutants selected against the bleaching herbicide Norflurazon. *Pestic. Biochem. Physiol.* 36: 46-51.

Lucas, W.J., Cousins, Y.L., Llewellyn, D.J., and Dennis, E.S. 1984. Perturbation of *Chara* plasmalemma transport function by 2(4(2(2′-dichlorophenoxy)phenoxy)propionic acid. *Plant Physiol.* 74: 61-66.

Lydon, J. and Duke, S.O. 1988a. Glyphosate induction of elevated levels of hydroxybenzoic acids in higher plants. *J. Agric. Food Chem.* 36: 813-818.

Lydon, J. and Duke, S.O. 1989. Pesticide effects on secondary metabolism of higher plants. *Pestic. Sci.* 25: 361-373.

Lydon, J. and Duke, S.O. 1993. The role of pesticides on host allelopathy and their effects on allelopathic compounds. In *Pesticide Interactions in Crop Production, Beneficial and Deleterious Effects*. Ed. by Altman, J. CRC Press, Boca Raton, FL, pp. 37-56.

Lyon, B.R., Cousins, Y.L., Llewellyn, D.J., and Dennis, E.S. 1993. Cotton plants transformed with a bacterial degradation gene are protected from accidental spray drift damage by the herbicide 2,4-dichlorophenoxyacetic acid. *Transgenic Res.* 2, 162-169.

Madsen, K.H., Heitholt, J. J., Duke, S.O., Smeda, R.J., and Streibig, J. 1995. Photosynthetic parameters in glyphosate-treated sugarbeets (*Beta vulgaris* L.). *Weed Res.* in press.

Mansooji, A.M., Maneechote, C., Holtum, J.A.M., and Powles, S.B. 1992. The spectrum and mechanism of herbicide resistance in wild oats (*Avena sterilis* ssp. *ludoviciana*, biotype SAS 1). *Proc. 1st Int. Weed Control Cong.* Weed Science Society of Australia, Victoria, pp. 304-306.

Marles, M.A.S., Devine, M.D., and Hall, J.C. 1993. Herbicide resistance in *Setaria viridis* conferred by a less sensitive form of acetyl coenzyme A carboxylase. *Pestic. Biochem. Physiol.* 46: 7-14.

Martinez-Férez, I.M. and Viogue, A. 1992. Nucleotide sequence of the phytoene desaturase gene from *Synochocystis* sp. PCC 6803 and characterization of a new mutation which confers resistance to the herbicide norflurazon. *Plant Molec. Biol.* 18: 981-983.

Matsumoto, H. and Duke, S.O. 1990. Acifluorfen-methyl effects on porphyrin synthesis in intact *Lemna pausicostata* Hegelm. 6746 plants. *J. Agric. Food Chem.* 38: 2066-2071.

Matsumoto, H., Lee, J.J., and Ishizuka, K. 1994. Variation in crop response to protoporphyrinogen oxidase inhibitors. *ACS Symp. Ser.* 559: 120-132.

Mayer, M.P., Bartlett, D.L., Beyer, P., and Kleinig, H. 1989. The *in vitro* mode of action of bleaching herbicides on the desaturation of 15-*cis*-phytoene and *cis*-d-carotene in isolated daffodil chromoplasts. *Pestic. Biochem. Physiol.* 34: 111-117.

Mazur, B.J. and Falco, S.C. 1989. The development of herbicide-resistant crops. *Annu. Rev. Plant Physiol.* 40: 441-470.

Montezinos, D. and Delmer, D.P. 1990. Characterization of inhibitors of cellulose synthesis in cotton fibers. *Planta* 148: 305-311.

Moorman, T.B., Becerril, J.M., Lydon, J., and Duke, S.O. 1992. Production of hydroxybenzoic acids by *Bradyrhizobium japonicum* strains after treatment with glyphosate. *J. Agric. Food Chem.* 40: 289-293.

Morejohn, L.C., Bureau, T.C., Molé-Bajer, J., Bajer, A.S., and Fosket, D.E. 1987. Oryzalin, a dintroaniline herbicide, binds to plant tubulin and inhibits microtubule polymerization in vitro. *Planta* 172: 252-264.

Morejohn, L.C. and Fosket, D.E. 1986. Tubulins from plant, animals and protists. In *Cell and Molecular Biology of the Cytoskeleton*. Ed. by Sharp, J.W. Plenum Press, New York, pp. 257.

Moss, S.R. 1990. Herbicide cross-resistance in slender foxtail (*Alopecurus myosuroides*). *Weed Sci.* 38: 492-496.

Moss, S.R. 1992. Herbicide resistance in the weed *Alopecurus myosuroides* (black-grass): the current situation. In *Achievements and Developments in Combating Pesticide Resistance.* Ed. by Denholm, I., Devonshire, A.L., and Holloman, D.W. Elsevier Applied Science Publishers, London, pp. 75-87.

Moss, S.R. and Cussans, G.W. 1991. The development of herbicide-resistant populations of *Alopecurus myosuroides* (Black-Grass) in England. In *Herbicide Resistance in Weeds and Crops.* Ed. by Caseley, J.C., Cussans, G.W., and Atkins, R.K. Butterworth-Heinemann, Oxford, pp. 45-55.

Moss, S.R. and Rubin, B. 1993. Herbicide-resistant weeds: a worldwide perspective. *J. Agric. Sci.* 120: 141-148.

Mudge, L.C., Gossett, B.J., and Murphy, T.R. 1984. Resistance of goosegrass (*Eleusine indica*) to dinitroaniline herbicides. *Weed Sci.* 32: 591-594.

Nimbal, C.I., Shaw, D.R., Duke, S.O., and Byrd, J.D., Jr. 1995. Response of MSMA-resistant and wild-type common cocklebur biotypes to cotton herbicides and cross-resistance to arsenicals and membrane disruptors. *Weed Technol.* in press.

Norman, M.A., Fuerst, E.P., Smeda, R.J., and Vaughn, K.C. 1993. Evaluation of paraquat resistance mechanisms in *Conyza. Pestic. Biochem. Physiol.* 46: 236-249.

Norman, M.A., Smeda, R.J., Vaughn, K.C., and Fuerst, E.P. 1994. Differential movement of paraquat in resistant and sensitive biotypes of *Conyza. Pestic. Biochem. Physiol.* 50: 31-42.

Oettmeier, W., Kude, C., and Soll, H. 1987. Phenolic herbicides and their methylethers: binding characteristics and inhibition of photosynthetic electron transport and photophosphorylation. *Pestic. Biochem. Physiol.* 27: 79-87.

Owen, W.J. 1991. Differential inhibition of plant acetyl CoA carboxylase — the biochemical basis for the selectivity of the aryloxy-phenoxypropanoate and cyclohexanedione herbicides. In *Herbicide Resistance in Weeds and Crops.* Ed. by Caseley, J.C., Cussans, G.W., and Atkins, R.K. Butterworth-Heinemann, Oxford, pp. 199-211.

Padgette, S.R., Re, D., Barry, G.F., Eichholtz, D.E., Delannay, X., Fuchs, R.L., Kishore, G.M., and Fraley, R.T. 1995. New weed control opportunities: development of soybeans with a Roundup Ready™ gene. In *Herbicide-Resistant Crops: Agricultural, Economic, Environmental, Regulatory, and Technological Aspects.* Ed. by Duke, S.O. CRC Press, Boca Raton, FL, this volume.

Parker, W.B., Marshall, L.C., Burton, J.D., Somers, D.A., Wyse, D.L., Gronwald, J.W., and Gengenbach, B.G. 1990a. Dominant mutations causing alterations in acetyl-coenzyme A carboxylase confer tolerance to cyclohexanedione and aryloxyphenoxypropionate herbicides in maize. *Proc. Natl. Acad. Sci. U.S.A.* 87: 7175-7179.

Parker, W.B., Somers, D.A., Wyse, D.L., Keith, R.A., Burton, J.D., Gronwald, J.W., and Gengenbach, B.G. 1990b. Selection and characterization of sethoxydim-tolerant maize tissue cultures. *Plant Physiol.* 92: 1220-1225.

Perl, A., Perl-Treves, R., Galili, S., Aviv, D., Shalgi, E., Malkin, S., and Galun, E. 1993. Enhanced oxidative-stress defense in transgenic potato expressing tomato Cu, Zn superoxide dismutases. *Theor. Appl. Genet.* 85: 568-576.

Pilmoor, J.B., Lindell, S.D., Briggs, G.G., and Wright, K. 1995. In *Eighth International Congress of Pesticide Chemistry, Options 2000.* Ed. by Ragsdale, N.N., Kearney, P.C., and Plimmer, J.R. American Chemical Society, Washington, D.C., pp. 292-303.

Pornprom, T., Matsumoto, H., Usui, K., and Ishizuka, K. 1994. Characterization of oxyfluorfen tolerance in selected soybean cell line. *Pestic. Biochem. Physiol.* 50: 107-114.

Preston, C. 1994. Resistance to photosystem I disrupting herbicides. In *Herbicide Resistance in Plants.* Ed. by Powles, S.B. and Holtum, J.A.M. Lewis Publishers, Boca Raton, FL, pp. 61-82.

Rahe, J.E., Lévesque, CA., and Johal, G.S. 1990. Synergistic role of soil fungi in the herbicidal efficacy of glyphosate. *ACS Symp. Ser.* 439: 260-275.

Rendina, A.R. and Felts, J.M. 1988. Cyclohexanedione herbicides are selective and potent inhibitors of acetyl-CoA carboxylase from grasses. *Plant Physiol.* 86: 983-986.

Ridley, S.M. and McNally, S.F. 1985. Effects of phosphinothricin on the isozymes of glutamine synthetase isolated from plant species which exhibit varying degrees of susceptibility to the herbicide. *Plant Sci.* 39: 31-36.

Ries, S.K., Zabik, M.J., Stephenson, G.R., and Chen, T.M. 1968. *N*-Glucosyl metabolite of pyrazon in red beets. *Weed Sci.* 16: 40-41.

Ryan, G.F. 1970. Resistance of common groundsel to simazine and atrazine. *Weed Sci.* 28: 614-616.

Sandmann, G., Clark, I.E., Bramely, P.M., and Böger, P. 1984. Inhibition of phytoene desaturase — the mode of action of certain bleaching herbicides. *Z. Naturforsch.* 39c: 443-449.

Sato, R., Yamamoto, M., Shibata, H., Oshio, H., Harris, E.H., Gillham, N.W., and Boynton, J.E. 1994. Characterization of a Protox mutant of *Chlamydomonas reinhardtii* resistant to Protox inhibitors. *ACS Symp. Ser.* 559: 91-104.

Schloss, J.V., Ciskanik, L.M., and Van Dyk, D.E. 1988. Origin of the herbicide binding site of acetolactate synthase. *Nature* 331: 360-362.

Schmitzer, P.R., Eilers, R.J., and Céske, C. 1993. Lack of cross-resistance of imazaquin-resistant *Xanthium strumarium* acetolactate synthase to flumetsulam and chlorimuron. *Plant Physiol.* 103: 281-283.

Scott, J.E., Weston, L.A., Chappell, J., and Hanley, K. 1994. Effects of clomazone on IPP isomerase and prenyl transferase activities in cell suspension cultures and cotyledons of solanaceous species. *Weed Sci.* 42: 509-516.

Secor, J. and Cséke, C. 1988. Inhibition of acetyl-CoA carboxylase activity by haloxyfop and tralkoxydim. *Plant Physiol.* 86: 10-12.

Sen Gupta, A., Heinen, J.L., Holaday, A.S., Burke, J.J., and Allen, R.D. 1993. Increased resistance to oxidative stress in transgenic plants that overexpress chloroplastic Cu/Zn superoxide dismutase. *Proc. Natl. Acad. Sci. U.S.A.* 90: 1629-1633.

Servaites, J.C., Tucci, M.A., and Geiger, D.R. 1987. Glyphosate effects on carbon assimilation, ribulose bisphosphate carboxylase activity, and metabolite levels in sugar beet leaves. *Plant Physiol.* 85: 370-374.

Shimabukuro, M.A., Shimabukuro, R.H., and Walsh, W.C. 1982. The antagonism of IAA-induced hydrogen ion extrusion and coleoptile growth by diclofop-methyl. *Plant Physiol.* 56: 444-452.

Shimabukuro, R.H. 1990. Selectivity and mode of action of the postemergence herbicide diclofop-methyl. *Plant Growth Reg. Soc. Am.* 18: 37-54.

Shimabukuro, R.H. and Hoffer, B.L. 1992. Effect of diclofop on the membrane potentials of herbicide-resistant and -susceptible annual rye grass root tips. *Plant Physiol.* 98: 1415-1422.

Shimabukuro, R.H., Walsh, W.C., and Wright, J.P. 1989. Effect of diclofop-methyl and 2,4-D of transmembrane proton gradient: a mechanism for their antagonistic interaction. *Plant Physiol.* 77: 107-114.

Shaner, D.L. and Mallipudi, N.M. 1991. Mechanisms of selectivity of the imidazolinones. In *The Imidazolinone Herbicides*. Ed. by Shaner, D.L. and O'Connor, S.L. CRC Press, Boca Raton, FL, pp. 91-102.

Shaner, D.L. and O'Connor, S.L. (Eds.) 1991. *The Imidazolinone Herbicides*. CRC Press, Boca Raton, FL.

Sherman, T.D., Becerril, J.M., Matsumoto, H., Duke, M.V., Jacobs, J.M., Jacobs, N.J., and Duke, S.O. 1991. Physiological basis for differential sensitivities of plant species to protoporphyrinogen oxidase-inhibiting herbicides. *Plant Physiol.* 97: 280-287.

Smeda, R.J., Hasegawa, P.M., Goldsbrough, P.B., Singh, N.K., and Weller, S.C. 1993. A serine-to-threonine substitution in the triazine herbicide-binding protein in potato cells results in atrazine resistance without impairing productivity. *Plant Physiol.* 103: 911-917.

Smeda, R.J. and Vaughn, K.C. 1994. Resistance to dinitroaniline herbicides. In *Herbicide Resistance in Plants*. Ed. by Powles, S.B. and Holtum, J.A.M. Lewis Publishers, Boca Raton, FL, pp. 215-228.

Smeda, R.J., Vaughn, K.C., and Morrison, I.N. 1992. A novel pattern of herbicide cross-resistance in a trifluralin-resistant biotype of green foxtail [*Setaria viridis* (L.) Beauv.]. *Pestic. Biochem. Physiol.* 42: 227-241.

Souza-Machado, V., Bandeen, J.D., Stephenson, G.R., and Lavigne, P. 1978. Uniparental inheritance of chloroplast atrazine tolerance in *Brassica campestris*. *Can. J. Plant Sci.* 58: 977-981.

St. John, J.B. 1976. Manipulation of galactolipid fatty acid composition with substituted pyridazinones. *Plant Physiol.* 57: 38-40.

St. John, J.B., Christiansen, M.N., Ashworth, E.N., and Gentner, W.A. 1979. Effect of BASF 13-338, a substituted pyridazinone, on linolenic acid levels and winterhardiness of cereals. *Crop Sci.* 19: 65-69.

St. John, J.B. and Hilton, J.L. 1976. Structure versus activity of substituted pyridazinones as related to mechanism of action. *Weed Sci.* 24: 579-582.

Stanger, C.E. and Appleby, A.P. 1989. Italian ryegrass (*Lolium multiflorum*) accessions tolerant to diclofop. *Weed Sci.* 37: 350-352.

Stegink, S.J. and Vaughn, K.C. 1988. Norflurazon (SAN-9789) reduces abscisic acid levels in cotton seedlings: a glandless isoline is more sensitive than its glanded counterpart. *Pestic. Biochem. Physiol.* 31: 269-275.

Stephenson, G.R. and Ries, S.K. 1967. The movement and metabolism of pyrazon in tolerant and susceptible species. *Weed Res.* 7: 51-60.

Stoltenberg, D.E., Gronwald, J.W., Wyse, D.L., Burton, J.D., Somers, D.A., and Gengenbach, B.G. 1989. Effect of sethoxydim and haloxyfop on acetyl-coenzyme A carboxylase activity in *Festuca* species. *Weed Sci.* 37: 512-516.

Strang, R.H. and Rogers., R.L. 1974. Behavior and fate of two phenylpyridazinone herbicides in cotton, corn, and soybean. *J. Food Agric. Chem.* 22: 1119-1125.

Subramanian, M.V. and Gerwick, B.C. 1989. Inhibition of acetolactate synthase by triazolopyrimidines. A review of recent developments. *ACS Symp. Ser. 389: 277-288.*

Summers, L.A. 1980. *The Bipyridinium Herbicides*. Academic Press, New York.

Thill, D.C., Mallory-Smith, C.A., Saari, L.L., Cotterman, J.C., Primiani, M.M., and Saladini, J.L. 1991. Sulfonylurea herbicide resistant weeds: discovery, distribution, biology, mechanism, and management. In *Herbicide Resistance in Crops and Weeds*. Ed. by Caseley, J.C., Cussans, G.W., and Atkin, R.K., Butterworth-Heinemann, Oxford, pp. 115-128.

Tischer, W. and Strotmann, H. 1977. Relationship between inhibitor binding by chloroplasts and inhibtion of photosynthetic electron transport. *Biochim. Biophys. Acta* 460: 113-125.

Trebst, A. 1991. The molecular basis of resistance of photosystem II herbicides. In *Herbicide Resistance in Weeds and Crops*. Ed. by Caseley, J.C., Cussans, G.W., and Atkins, R.K. Butterworth-Heinemann, Oxford, pp. 145-164.

Trebst, A., Depka, B., and Kipper, M. 1990. The topology of the reaction center polypeptides of photosystem II. *Curr. Res. Photosyn.* 1: 217-222.

Trebst, A. and Draber, W. 1986. Inhibitors of photosystem II and the topology of the herbicide and Q_B binding polypeptide in the thylakoid membrane. *Photosyn. Res.* 10: 381-392.

Vaughn, K.C. and Duke, S.O. 1983. *In situ* localization of the sites of paraquat action. *Plant Cell Environ.* 6: 13-20.

Vaughn, K.C. and Duke, S.O. 1991. Mechanisms of resistance to herbicides. In *Chemistry of Plant Protection, Vol. 7, Herbicide Resistance — Brassinosteroids, Gibberellins, Plant Growth Regulators,* Ed. by Ebing, W. Springer-Verlag, Berlin, pp. 141-169.

Vaughn, K.C. and Fuerst, E.P. 1985. Structural and physiological studies of paraquat-resistant *Conyza. Pestic. Biochem. Physiol.* 24: 86-94.

Vaughn, K.C. and Lehnen, L.P. 1991. Mitotic disrupter herbicides. *Weed Sci.* 39: 450-457.

Vaughn, K.C., Marks, M.D., and Weeks, D.P. 1987. A dinitroaniline-resistant mutant of *Eleucine indica* exhibits cross-resistance and supersensitivity to antimicrotubule herbicides and drugs. *Plant Physiol.* 83: 956-964.

Vaughn, K.C. and Vaughan, M.A. 1990. Structural and biochemical characterization of dinitroaniline-resistant *Eleusine. ACS Symp. Ser.* 421: 364-375.

Vaughn, K.C. and Vaughan, M.A. 1991. Dinitroaniline resistance in *Eleucine indica* may be due to hyperstabilized microtubules. In *Herbicide Resistance in Weeds and Crops.* Ed. by Caseley, J.C., Cussans, G.W., and Atkin, R.K. Butterworth-Heinemann, Oxford, pp. 177-186.

Vaughn, K.C., Vaughan, M.A., and Camilleri, P. 1989. Lack of cross-resistance of paraquat-resistant hairy fleabane (*Conyza bonariensis*) to other toxic oxygen generators indicates enzymatic protection is not the resistance mechanism. *Weed Sci.* 37: 5-11.

Vencill, W.K., Hatzios, K.K., and Wilson, H.P. 1989. Growth and physiological responses of normal, dwarf and albino corn (*Zea mays*) to clomazone treatments. *Pestic. Biochem. Physiol.* 35: 81-88.

Walker, K.A., Ridley, S.M., and Harwood, J.L. 1988. Effects of the selective herbicide fluazifop on fatty acid synthesis in pea (*Pisum sativa*) and barley (*Hordeum vulgare*). *Biochem. J.* 254: 811-817.

Wang, X., Hildebrand, D.F., Norman, H.D., Dahmer, M.L., St. John, J.B., and Collins, G.B. 1987. Reduction of linolenate content in soybean cultures by a substituted pyridazinone. *Phytochemistry* 26: 955-960.

Wessels, J.S.C. and van der Veen, R. 1956. The action of some derivatives of phenylurethan and 3-phenyl-1,1-dimethylurea on the Hill reaction. *Biochim. Biophys. Acta* 19: 548-549.

Yaklich, R.W., Karczmarczyk, S.J., and Devlin, R.M. 1974. Metabolism of ^{14}C-SAN-6706 and ^{14}C-norflurazon in cranberry plants. *Weed Sci.* 22: 595-599.

Chapter 3

TECHNIQUES FOR PRODUCING HERBICIDE-RESISTANT CROPS

William E. Dyer

CONTENTS

1-56670-045-0/96/$0.00+$.50
© 1996 by CRC Press Inc.

1. INTRODUCTION

Herbicide-resistant crops (HRCs) represent one of the first and most widely publicized applications of plant biotechnology. Most HR cultivars were created through the stable integration of a foreign gene using recently developed techniques of molecular biology and plant transformation. This approach takes advantage of the potential ability of every plant cell to be regenerated into a whole plant after receiving a foreign gene. Although some species were intransigent, advances in transformation and regeneration methodology are occurring at a rapid pace, making transformation the method of choice for most species. However, it is important to note that HRCs also were developed using traditional plant breeding methods as well as hybridization with conspecific HR relatives. This chapter discusses the methods used to create HRCs, and describes some more recently developed techniques that may have application in future HRC research. The technology and development of HRCs were reviewed by Botterman and Leemans (1988), Dyer et al. (1993); Gressel (1993); Mazur and Falco (1989); Mullineaux (1992); Oxtoby and Hughes (1989); Quinn (1990); Schulz et al. (1990); and Stalker (1991).

2. SELECTION METHODS

The most straightforward way to obtain HRCs is through selecting from the existing germplasm. Plant species possess a wide variation for resistance to different herbicides, and even within a crop species, individual cultivars have demonstrated substantial differences in tolerance. Isolation of resistant individuals or populations is usually combined with some type of recurrent selection to enhance resistance levels and stabilize the phenotype.

2.1. WHOLE PLANT SELECTION

Screening germplasm collections often identifies variability among lines in resistance levels to a particular herbicide. For example, resistance to difenzoquat varies about two- to three-fold within some wheat cultivars (Snape et al. 1991); soybean species and cultivars likewise show differential response to metribuzin (Fedtke 1991) and other herbicides (Hartwig 1987; Loux et al. 1987). Graminicide resistance varies among ornamental grasses, including several fescues (Catanzaro et al. 1993). Thus, variability in herbicide resistance may be exploited and perhaps enhanced through recurrent selection and traditional crossing programs to create new crop cultivars. Such an approach was used successfully for perennial pasture and turf grasses in the U.K. (Johnston and Faulkner 1991). Several cultivars of perennial ryegrass and other species have been selected with useful levels of resistance to dalapon, paraquat, or glyphosate. However, this approach has not been widely embraced by plant breeders, perhaps because of the relatively low priority traditionally placed on traits such as herbicide resistance. Also, a lack of sufficient variability in resistance levels within some species may have hindered breeding efforts (e.g., Foy et al. 1988; Kibite and Harker 1991).

2.1.1. Seed Selection

Mutagenesis may be combined with selection to enhance the likelihood of recovering HR individuals from large populations. Several methods of mutagenesis such as chemicals and X-rays have been used to induce point mutations and deletions, respectively, in genes of interest. A specific method of mutagenesis may be chosen to maximize the chances of obtaining a gene alteration or knockout and thus the desired phenotype. This approach was used successfully to select for HR individuals in several species. Mass selection of chemically mutagenized soybean seed was used to select a sulfonylurea-resistant line from about 400,000 seeds germinated in the presence of chlorsulfuron (Sebastian et al. 1989). The resistant line was backcrossed into commercial lines and is being field tested in anticipation of commercial release. Chemical mutagenesis of rapeseed (*Brassica napus* L. var. *napus*) seed was likewise used to select sulfonylurea-resistant lines (Tonnemaker et al. 1992). Similar approaches have been used to select a fluazifop-resistant barley line (E.P. Fuerst, personal communication) and imidazolinone-resistant wheat (W.A. Smith and K.E. Newhouse, personal communication).

2.1.2. Pollen Selection

Mutagenesis of pollen may also be used to recover a desired phenotype. This approach was used, particularly in maize to obtain a number of biologically interesting mutants (Huang 1992). The advantage of this methodology is that it is easy to mutagenize a large population of haploid pollen grains and, after fertilization, obtain many individuals for screening. Although use of the technique has not been reported in the development of HRCs, it remains a valuable approach for generating novel phenotypes.

2.2. CELL CULTURE SELECTION

When plant cells are cultured as callus or cells in suspension, numerous changes occur in their genetic composition, including gene rearrangements, altered ploidy levels, chromosome alterations, and others. This process, termed somaclonal variation, was put forth in the early 1980s as a rich source of variability that could be exploited to obtain novel and improved crop cultivars (e.g., Chaleff 1983; Scowcroft and Larkin 1988). While some success has been achieved, the number of useful mutants recovered generally has not lived up to expectations, most likely because somaclonal variants with a desired mutation often also possess numerous undesirable alterations. For example, selection for glyphosate resistance in long-term tobacco cell cultures resulted in the recovery of stably resistant lines, but regenerated plants were partially sterile and otherwise agronomically impaired (Dyer et al. 1988). In several cases resistance was shown to be due to the amplification of genes encoding the herbicide target enzyme (Goldsbrough et al. 1990; Tischer et al. 1986), a resistance mechanism which is probably a result of the tissue culturing process itself. Presence of this "genetic baggage" may necessitate the use of substantial backcrossing programs to restore the mutant to agronomic usefulness.

There are numerous examples of selecting HR mutant lines using protoplast, microspore (Swanson et al. 1988), callus (Chaleff and Ray 1984; Dyer et al. 1988; Tuberosa and Lucchese 1990), and especially cell suspension cultures (Nafziger et al. 1984; Parker et al. 1990; Singer and McDaniel 1985). Use of plant tissue culture for this purpose was the subject of several reviews (Chaleff 1988; Hughes 1983; Maliga et al. 1987; Smeda and Weller 1991). One of the best-characterized examples involved the exposure of embryogenic maize callus cultures to imidazolinone herbicides and the subsequent isolation of resistant sectors (Anderson and Georgeson 1989). Several herbicide-resistant lines were regenerated carrying either of two unlinked, semidominant alleles (Newhouse et al. 1991) conferring varying degrees of resistance. The lines were backcrossed to agronomically useful inbred lines, and 14 imidazolinone-resistant hybrid maize cultivars were released for commercial use in early 1993 (R. Chaleff, personal communication). Other examples include selection for chlorsulfuron resistance in sugarbeet cell cultures and regeneration of resistant plants (Hart et al. 1992). Also, resistant birdsfoot trefoil (*Lotus corniculatus*) plants were recovered after selection of calli, shoots, and whole plants using the sulfonylurea herbicide thiameturon methyl (Pofelis et al. 1992).

3. HYBRIDIZATION

Naturally occurring resistance in weed biotypes has been exploited by transferring the resistance trait into crop plants using traditional plant breeding methods. The atrazine-resistant weed

Brassica campestris was hybridized with several *Brassica* crops of commercial importance, including Chinese cabbage, canola, and rutabaga, and led to the release of 'Triton' atrazine-resistant canola (Beversdorf et al. 1980). Although there is about a 20% yield penalty in this cultivar due to the type of chloroplast mutation conferring resistance, it is grown in situations in which high weed pressure would otherwise preclude the use of susceptible cultivars. A similar approach was used to transfer atrazine resistance from the weed *Setaria viridis* into the crop *Setaria italica* (Darmency and Pernes 1989) and sulfonylurea resistance from prickly lettuce (*Lactuca serriola*) to domestic lettuce (*Lactuca sativa*) (Mallory-Smith et al. 1993). Because this approach requires the availability of a compatible HR weed (or other plant) and crop pair, its applications may be limited.

4. GENE TRANSFER

Most HRCs created to date were developed by introducing a foreign gene (transgene; transgenic plants) that confers resistance (Dyer et al. 1993). The gene may encode a herbicide-resistant form of an endogenous enzyme or a novel enzyme that alters and thereby inactivates the herbicide. Although a number of techniques were used to transform plant cells with a foreign gene (see below), some aspects are commonly practiced in conjunction with all methods. In particular, confirmation of stable transgene incorporation into the host plant genome during the early stages of plant regeneration is usually desirable. DNA hybridization (Southern blotting) (Sambrook et al. 1989), the polymerase chain reaction (PCR) (Mullis and Faloona 1987), and assaying enzyme activity from the transgene are commonly used for this purpose. Survival of regenerating plants in the presence of the herbicide in question provides strong but not unequivocal evidence for successful transformation. General reviews of gene transfer methodology are available (Gasser and Fraley 1989; Jenes et al. 1993; Lindsey 1992; Potrykus 1990a; Wettstein 1993), including information specifically related to transformation research in trees (Jouanin et al. 1993), crops (Dale et al. 1993), temperate woody crops (Schuerman and Dandekar 1993), and tropical and subtropical crops (Gardner 1993).

A primary prerequisite for plant transformation research is the use of a selectable marker gene, so named because it confers the ability to survive in the presence of a normally toxic compound. Only transgenic cells and regenerating plants are thus selected for and maintained. The first selectable marker widely used in plant tranformation work was a bacterial Tn5 gene encoding neomycin phosphotransferase (NPTII) which confers resistance to the aminoglycoside antibiotics kanamycin, neomycin, and G-418. However, many important crop plants including wheat are not very sensitive to these antibiotics, so they do not provide stringent selection after transformation (Vasil 1987). Other work showed that these compounds interfere with callus formation and regeneration in several species (e.g., Catlin 1990; Orlikowska et al. 1993). To circumvent such problems, other selectable marker genes were proposed for plants, including genes conferring resistance to hygromycin (Waldron et al. 1985) and chloramphenicol (Herrera-Estrella et al. 1983). Bacterial genes for aspartate kinase and dihydrodipicolinate synthase may have application as plant-selectable markers by conferring resistance to normally toxic amino acid analogues (Perl et al. 1993). Several of the herbicide resistance genes described below were proposed as efficient selectable markers for plants in order to recover transgenic plants containing tandemly inserted but otherwise unrelated genes of interest. For example, the *bar* gene was used as a selectable marker to recover transgenic plants from regenerable rice protoplasts (Rathore et al. 1993). Likewise, the *dehl* gene from *Pseudomonas putida* encoding a 2,2-dichloropropionic acid (dalapon) dehalogenase (Buchanan-Wollaston et al. 1992) and the *sul*I gene encoding a sulfonamide (asulam) -resistant dihydropteroate synthase (Guerineau et al. 1990) were used to confer herbicide resistance in tobacco and may have utility as selectable markers in other species.

4.1. *AGROBACTERIUM TUMEFACIENS*-MEDIATED TRANSFORMATION

Agrobacterium tumefaciens is a Gram-negative soil bacterium that causes tumorous growths, or crown galls, on a wide range of dicotyledonous plants and gymnosperms (reviewed in Kado 1991). The basis for this disease is the transfer of a segment of DNA from a large Ti plasmid contained in the bacterium to wounded plant cells. The transferred segment of DNA contains genes for phytohormone production which lead to the crown gall phenotype and genes for biosynthesis of amino acid derivatives catabolized by the *Agrobacterium*. In most cases this transferred segment of DNA is stably incorporated into the host plant genome and is inherited in a normal Mendelian fashion. This natural gene transfer system has been exploited for use in creating transgenic plants by replacing the tumor-inducing genes in the Ti plasmid with desired gene(s) of interest to create "disarmed" vectors (reviewed in Hookyaas 1989) (Figure 1). Ti plasmid-derived cloning vectors were designed for replication in both *A. tumefaciens* and *Escherichia coli* bacterial strains, thus facilitating the cloning steps in *E. coli* and subsequent transfer of the completed vector to *A. tumefaciens* (An et al. 1985; Bevan 1984). Recently developed vectors include selectable marker genes for herbicide resistance and conveniently assayed reporter genes (Becker et al. 1992).

Gene transfer using Ti plasmid-derived cloning vectors in *A. tumefaciens* is probably the most widely used method of creating transgenic plants. It has been most successfully applied to transformation research with *Nicotiana* and *Petunia* spp. (Horsch et al. 1985) as well as many other agronomically important dicotyledonous crops (reviewed in Lindsey 1992). However, a major impediment to the universal application of this technique is that most monocotyledonous species are not suitable hosts for *A. tumefaciens*, especially the agronomically important graminaceous monocots such as wheat, rice, and maize. Although DNA can be introduced into monocot cells by *A. tumefaciens* (Grimsley et al. 1987), these species apparently lack the proper wound response and infected cells do not regenerate (Potrykus 1990b). Thus, alternative strategies were developed for cereal species as discussed in the next section.

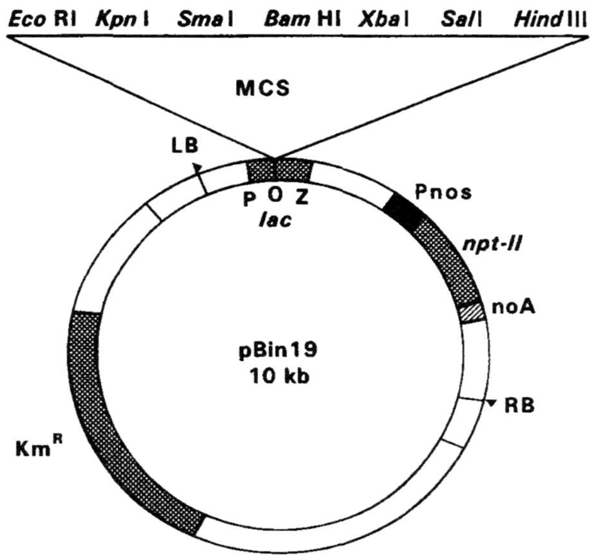

FIGURE 1 Structure of the binary Ti-based vector pBin19 used in *Agrobacterium tumefaciens*-mediated transformation. LB, RB = left and right border sequences; *lac* = α-complementary region of *lac* operon; Pnos = nopaline synthase gene promoter; *npt-II* = neomycin phosphotransferase-II coding sequence from Tn*5*; noA = nopaline synthase gene polyadenylation site; MCS = multiple cloning site.

A. tumefaciens-mediated transformation methods were developed for whole plant explants, cultured cells, and protoplasts. Transformation of explants in combination with direct shoot regeneration has proven to be an efficient technique for amenable species (Horsch et al. 1985; Orlikowska et al. 1993) because it avoids undesirable somaclonal variation associated with plants regenerated from cultured cells or calli. Factors affecting success using *A. tumefaciens*-mediated transformation include choice of bacterial strain, presence of phytohormones, presence of inducing agents such as acetosyringone, and tissue culture conditions before and after transformation (reviewed in Kado 1991).

Most initial research on introducing herbicide resistance into plants used *A. tumefaciens*-mediated transformation. A gene encoding the wild type 5-enolpyruvylshikimate 3-phosphate (EPSP) synthase, the target enzyme of glyphosate inhibition, was placed under the control of a high-level expression promoter and transformed into *Petunia* cells in a Ti plasmid-based construction; the resulting transgenic cells and regenerated plants showed resistance to glyphosate (Shah et al. 1986). Glyphosate resistance was also achieved by using *A. tumefaciens*-mediated transformation to introduce a mutant *Salmonella typhimurium* (Comai et al. 1985) or *E. coli* (Della-Cioppa et al. 1987) *aroA* gene into tobacco or petunia cells, respectively (reviewed in Dyer 1994). Similarly, a mutant *Arabidopsis thaliana* gene encoding an insensitive acetolactate synthase (ALS), the target enzyme of sulfonylurea and imidazolinone inhibition, was transferred using *A. tumefaciens* into tobacco (Haughn et al. 1988), sugarbeet (D'Halluin et al. 1992), canola (Miki et al. 1990), and chicory (*Cichium intybus* L.) (Vermeulen et al. 1992) to obtain transgenic resistant plants.

The first example of introducing herbicide resistance by means of a detoxifying enzyme was carried out using *A. tumefaciens*-mediated transformation. Glufosinate (a synthetic version of the natural product phosphinothricin) is a herbicide that inhibits glutamine synthetase, causing rapid ammonia accumulation and eventual membrane breakdown and cell death (Tachibana et al. 1986). *Streptomyces* spp. that synthesize bialaphos, the tripeptide precursor of phosphinothricin, contain the *bar* gene which encodes an acetyl transferase able to inactivate the herbicide (Thompson et al. 1987). Glufosinate-resistant tobacco, potato, and tomato plants were obtained from cells transformed with the *bar* gene (DeBlock et al. 1987). Similarly, metabolically based resistance to bromoxynil was obtained in transgenic tomato and potato plants expressing the *bxn* gene, which encodes a bromoxynil-specific nitrilase (Stalker et al. 1988). Other research using *A. tumefaciens*-mediated transformation with the *bar* gene has created glufosinate-resistant sugarbeet (D'Halluin et al. 1992) and safflower (*Carthamus tinctorius*) (Stringfellow et al. 1993). Transformation with the *tfdA* gene, encoding a 2,4-D monoxygenase, led to recovery of 2,4-D-resistant tobacco (Lyon et al. 1989; Streber and Wilmitzer 1989) and cotton plants (Bayley et al. 1992).

4.2. PARTICLE BOMBARDMENT

As discussed above, the lack of success using *A. tumefaciens*-mediated transformation in monocots as well as other considerations led to the development of alternative methods of plant transformation. As these methods were investigated it became increasingly clear that DNA entry into plant cells and nuclei may be facilitated by many types of physical and chemical treatment (Van Wert and Saunders 1992). The foremost technique of direct DNA transfer involves the delivery of foreign genes through bombardment with a "biolistic" (biological ballistic) device. The process basically consists of high-velocity acceleration of microprojectiles (about 1-μm diameter tungsten or gold particles) coated with DNA through the plant cell wall and plasma membrane into the nucleus. Particle bombardment was first described by Klein et al. (1987) and is the subject of recent reviews (Christou 1992; Klein et al. 1992; Sanford et al. 1993). The original design used gunpowder to accelerate the microprojectiles while newer models are helium driven (Figure 2).

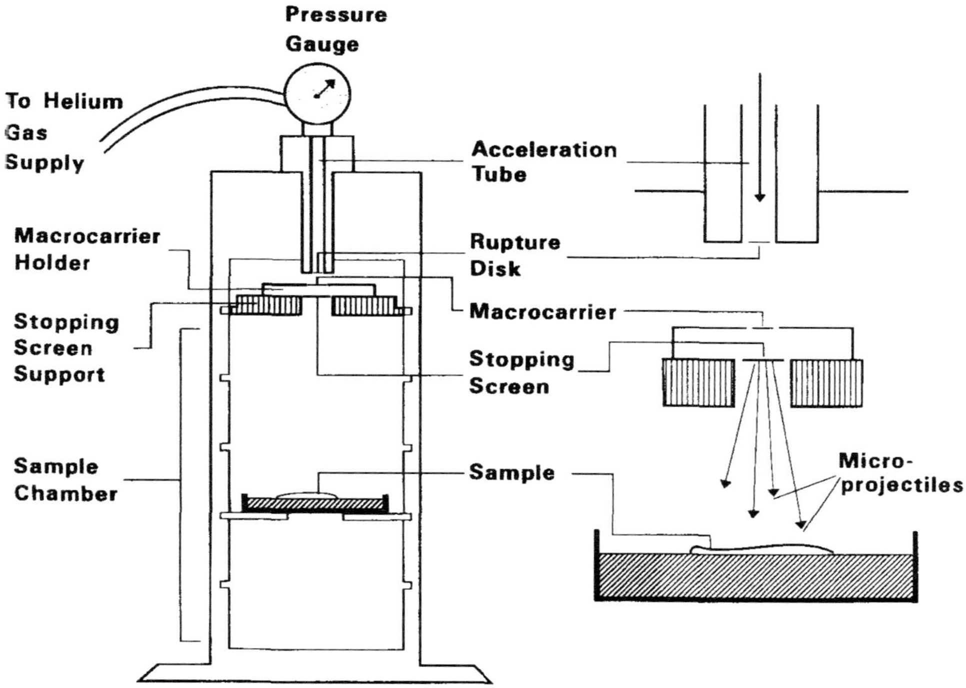

FIGURE 2 Schematic diagram of typical particle bombardment device construction. (Adapted from the BIOLISTIC Particle Delivery System, Model PDS-1000/He, DuPont/BioRad.)

Particle bombardment has shown broad utility and appears to be effective for all plant species tested thus far. Essentially any tissue or cell type can be used as a target for bombardment, thus shifting the research emphasis away from DNA transfer methodology toward identifying potentially regenerable cells or tissues in the species of interest. The technique is versatile, practically simple, and should continue to provide a valuable alternative for plant transformation, especially in cereals. The only potential disadvantage discussed so far is that regenerated plants are often chimeric (partially transgenic) due to random bombardment of a small number of cells within a multiple cell system (Jenes et al. 1993).

One of the early successful applications of particle bombardment was the transformation and regeneration of fertile maize plants, a species notoriously recalcitrant to these manipulations. Cells from embryogenic suspension cultures were bombarded with the *bar* gene and transgenic plants with heritable glufosinate resistance were recovered (Gordon-Kamm et al. 1990). Maize transformation has since been accomplished in many laboratories and is becoming routine. Similar methodology was used to obtain stable glufosinate-resistant sugarcane (*Saccharum* spp.) calli (Chowdhury and Vasil 1992) and rice plants (Cao et al. 1992) after bombarding embryogenic cell suspension cultures with the *bar* gene. Glufosinate-resistant oat plants were likewise obtained from bombarding embryogenic callus (Somers et al. 1992). Regeneration of fertile glufosinate-resistant wheat plants was obtained from bombarded embryogenic calli using a system that required the identification and maintenance of a specific regenerable callus type (Vasil et al. 1992). A potentially more universal technique was demonstrated by bombarding calli from immature wheat embryos with the *bar* gene and regenerating glufosinate-resistant fertile wheat plants (Weeks et al. 1993). Most recently, particle bombardment of several barley tissues was successfully used to produce fertile herbicide-resistant transgenic plants (Wan and Lemaux 1994). A related technology used electrical discharge to accelerate DNA-coated particles into immature rice embryos to obtain resistant plants (Christou et al. 1991).

4.3. PROTOPLAST TRANSFORMATION

Significant improvements in plant regeneration from protoplasts (isolated cells with cell walls removed) were achieved in the last decade. The advantages of using protoplasts over intact cells include more efficient DNA transfer across the plasma membrane without interference from the cell wall, and the fact that chimeric plants are not regenerated from transformed protoplasts. The primary drawback to this approach is that it is limited to cultivars from which fertile transgenic plants can be regenerated from protoplasts. Nevertheless, it has become the method of choice for some crops such as rice (Peng et al. 1992). Most monocot systems use embryogenic cell suspension cultures to obtain regenerable protoplasts for transformation research (Vasil 1987).

4.3.1. Polyethylene Glycol-Mediated Transformation

Various physical and chemical treatments have been used to stimulate DNA uptake by protoplasts. Polyethylene glycol (PEG) acts to increase the permeability of plant cell membranes and is probably the most widely used method for direct DNA uptake. Glufosinate-resistant tall fescue (*Festuca arundinacea* Schreb.) plants were obtained by PEG-facilitated import of the *bar* gene into protoplasts isolated from embryogenic cell suspension cultures (Wang et al. 1992). Likewise, PEG-mediated transformation of the *bar* gene into Indica rice protoplasts yielded transgenic glufosinate-resistant plants (Datta et al. 1992).

4.3.2. Electroporation

Electroporation uses short electrical pulses of high field strength to facilitate DNA uptake by increasing the permeability of cell membranes. Gene transfer efficiency and protoplast viability are influenced by the amplitude and duration of the electric pulse as well as the electroporation medium. The method has been successfully used for rice (Shimamoto et al. 1989) and maize (Fromm et al. 1986) protoplasts. Stable glufosinate-resistant calli were obtained from sugarcane protoplasts electroporated with the *bar* gene (Chowdhury and Vasil 1992).

4.3.3. Sonication

Sonication has recently been proposed as an alternative to electroporation for direct DNA uptake into protoplasts. Although the method was successful for sugarbeet (Joersbo and Brunstedt 1990) and tobacco (Zhang et al. 1991) protoplasts, its applicability to other plant species is unknown.

4.3.4. Microinjection

Microinjection is an extremely precise method for delivering DNA into cells, and is the method of choice for transformation of animal cells. The technique involves the use of micromanipulators and microcapillaries to hold cells and inject DNA, respectively, into the nucleus. Although the method was demonstrated in plant protoplasts (Reich et al. 1986) and zygotic proembryos of several species (Potrykus 1990b), it has not been widely adopted, probably because of the considerable time and skill required.

4.3.5. Other Techniques for DNA Transfer

Zhou et al. (1983) injected total DNA from one cotton variety (donor) into immature embryos of another variety (recipient) and observed morphological or other characteristics of the donor variety in the recipient offspring. Likewise, DNA containing a selectable marker gene was injected into immature rye (*Secale cereale*) tillers to obtain putative transgenic seedlings (De la Peña et al. 1987). However, these results have not been reproduced in other laboratories. Silicon carbide fibers were used to physically penetrate cultured cell walls and facilitate DNA entry (Kaeppler et al. 1992). Other proposed methods for DNA transfer include the pollen tube pathway (Luo and Wu 1988), DNA electroporation into intact leaf tissues (Dekeyser et al. 1990), and dry embryo incubation in DNA solutions (Töpfer et al. 1989).

5. RELATED TECHNIQUES OF PLANT MOLECULAR BIOLOGY

5.1. ANTISENSE RNA

Elimination of a specific plant trait by inhibition of gene expression may be desirable under certain conditions. For example, plant breeders may wish to eliminate natural mechanisms of herbicide resistance in order to preferentially remove certain progeny in breeding programs. This can be achieved by *in vivo* production of a messenger RNA (mRNA) molecule synthesized from the noncoding (antisense) strand of the gene conferring resistance. The antisense mRNA molecule interferes with translation of the endogenous mRNA through one or more mechanisms. The system has been shown to function in plants (Sandler et al. 1988) and received much publicity for its use by Calgene to create the slow-ripening (Flavr Savr™) tomato (Beck and Ulrich 1993). Antisense RNA technology may provide useful advantages in future HRC research.

5.2. TRANSPOSON TAGGING

Transposable elements (transposons) are mobile genetic elements that can be used to clone genes by virtue of the unstable (mutable) phenotype created upon insertion into genes regulating an identifiable phenotype (Gierl and Saedler 1992). For example, an unknown gene conferring herbicide resistance could be inactivated by transposon insertion and simultaneously "tagged" by presence of the known transposon DNA sequences. The *Ac/Ds* system has been successfully used to clone several loci (Federoff et al. 1984), but its use is restricted to maize and *Antirrhinum majus* (snapdragon) because these are the only two species that possess well-characterized active transposable element systems. More recently, a three-component transposition system based on the *En/Spm* transposable element family was characterized and shows promise for use in heterologous host plant species (Cardon et al. 1993).

5.3. *LOXP*-CRE DIRECTED EXCISION

In some cases it may be desirable to not only control expression of a transgene during the transgenic plant's life cycle, but also to manipulate the gene in subsequent generations. In particular, commercial considerations may dictate expression of a trait such as herbicide resistance in purchased seed only, but not in the progeny. The *loxP*-Cre site-specific recombination system directs recombination by the Cre protein, which recognizes and interacts with two 34-bp *loxP* sites to direct excision, integration, or inversion of adjoining DNA sequences (Odell et al. 1990). This system was shown to function in several plant species and was used to disable expression of a mutant *ALS* selectable marker gene while the linked nonselectable trait was retained (Russell et al. 1992). To be used in HRC production, the gene conferring resistance would be flanked by *loxP* sites prior to insertion. Exposure to the Cre protein, either in the F_1 hybrid or through retransformation, would thus disable expression of the original transgene for herbicide resistance. Although its use has not been reported in HRC research, this system may prove to be of value in transgenic crop development and protection.

5.4 SUBTRACTION HYBRIDIZATION

Cloning unknown genes conferring traits such as herbicide resistance may be accomplished if gene expression is differentially regulated in resistant vs. susceptible types (Dyer 1991). An effective technique designed for this purpose is known as subtraction hybridization, in which complementary DNAs (cDNAs) common to both resistant and susceptible conditions would be "subtracted" out of a mixture, leaving behind the unique resistance-related cDNAs (Duguid et al. 1988). cDNAs thus isolated can then be used to create a library, from which specific genes associated with the resistance trait can be obtained, tested, and characterized.

5.5 DIFFERENTIAL DISPLAY

A recently developed and technically straightforward technique with the same goal of isolating genes based on differential expression was described by Liang and Pardee (1992). Briefly, cDNA populations are synthesized from subsets of mRNAs extracted from resistant and susceptible types and subjected to PCR in the presence of ^{35}S-dATP. The products are visualized side by side on DNA sequencing gels, on which an increase in the relative signal strength of a band from resistant mRNA indicates the presence of a differentially expressed gene potentially involved in the resistance phenotype. PCR products of interest may then be eluted from the gel and subcloned for further characterization such as sequencing and gene expression analysis. This approach was used successfully to isolate cDNAs associated with the regulation of seed dormancy in *Avena fatua* (Cranston et al. 1994; Johnson et al. 1994), and has potential for investigating unknown mechanisms of herbicide action and resistance.

6. CONCLUSIONS

Herbicide-resistant plants have been obtained for a large number of species and HRCs are being commercialized for many agronomically important crops (Table 1). Rapid strides in plant tissue culture, regeneration, and transformation methodology have made these developments possible and even routine. Further research in fundamental plant biology and gene expression coupled with these advances may promote the development of crop cultivars with additional agronomically useful traits.

TABLE 1 Herbicides, Plant Species, Methods Used, and References for Herbicide-Resistant Plant Development

Herbicide	Plant Species	Method	Ref.
Asulam	Tobacco	AT	Guerineau et al. 1990
Atrazine	Canola	H	Beversdorf et al. 1980
	Setaria italica	H	Darmency and Pernes 1989
Bromoxynil	Tobacco	AT	Stalker et al. 1988
	Cotton	AT	Duke et al. 1991
Dalapon	Tobacco	AT	Buchanan-Wollaston et al. 1992
	Perennial ryegrass	S	Johnston and Faulkner 1991
Glufosinate	Alfalfa, canola	AT	D'Halluin et al. 1990
	Barley	PB	Wan and Lemaux 1994
	Maize	PB	Gordon-Kamm et al. 1990
	Oat	PB	Somers et al. 1992
	Poplar	AT	Devillard 1992
	Rice	PB	Cao et al. 1992
	Rice	P	Datta et al. 1992
	Safflower	AT	Stringfellow et al. 1993
	Sugarbeet	AT	Botterman and Leemans 1988
	Sugarcane	PB, P	Chowdhury and Vasil 1992
	Sweet potato, tomato	AT	DeGreef et al. 1989
	Tall fescue	P	Wang et al. 1992
	Tobacco	AT	DeBlock et al. 1987
	Wheat	PB	Vasil et al. 1992; Weeks et al. 1993
Glyphosate	Cotton, flax	AT	Waters 1991
	Perennial ryegrass	S	Johnston and Faulkner 1991
	Petunia	AT	Della Cioppa et al. 1987; Shah et al. 1986
	Poplar	AT	Fillatti et al. 1988
	Soybean	AT	Hinchee et al. 1988
	Tobacco	AT	Comai et al. 1985
Imidazolinones	Canola	C	Swanson et al. 1988

TABLE 1 (continued) Herbicides, Plant Species, Methods Used, and References for Herbicide-Resistant Plant Development

Herbicide	Plant Species	Method	Ref.
	Maize	C	Anderson and Georgeson 1989
Sulfonylureas	Birdsfoot trefoil	C	Pofelis et al. 1992
	Canola	AT	Miki et al. 1990
		C	Swanson et al. 1989
		S	Tonnemaker et al. 1992
	Chicory	AT	Vermeulen et al. 1992
	Lettuce	H	Mallory-Smith et al. 1993
	Soybean	S	Sebastian et al. 1989
	Sugarbeet	AT	D'Halluin et al. 1992
		C	Hart et al. 1993
	Tobacco	AT	Haughn et al. 1988
2,4-D	Cotton	AT	Bayley et al. 1992
	Tobacco	AT	Lyon et al. 1989; Streber and Willmitzer 1989

AT = gene transfer by *Agrobacterium tumefaciens*; C = cell selection; H = hybridization; P = direct DNA uptake by protoplasts; PB = gene transfer by particle bombardment; S = whole plant or seed selection.

ACKNOWLEDGMENT

Work in the author's laboratory is supported by the National Science Foundation (OSR-9350546), USDA/NRICGP (92-37303-7614), and the Montana Agricultural Experiment Station.

REFERENCES

An, G., Watson, B.D., Stachel, S., Gordon, M.P., and Nester, E.W. 1985. New cloning vehicles for transformation of higher plants. *EMBO J.*, 4: 277.

Anderson, P.A. and Georgeson, M. 1989. Herbicide-tolerant mutants of corn. *Genome* 31: 994.

Bayley, C., Trolinder, N., Ray, C., Morgan, M., Quisenberry, J.E., and Ow, D.W. 1992. Engineering 2,4-D resistance into cotton. *Theor. Appl. Genet.* 83: 645.

Beck, C.I. and Ulrich, T. 1993. Biotechnology in the food industry. *Bio/Technology* 11: 895.

Becker, D., Kemper, E., Schell, J., and Masterson, R. 1992. New plant binary vectors with selectable markers located proximal to the left T-DNA border. *Plant Mol. Biol.* 20: 1195.

Bevan, M. 1984. Binary *Agrobacterium* vectors for plant transformation. *Nucl. Acids Res.* 12: 8711.

Beversdorf, W.D., Weiss-Lerman, J., Erickson, L.R., and Souza-Machado, V. 1980. Transfer of cytoplasmically-inherited triazine resistance from bird's rape to cultivated *Brassica campestris* and *Brassica napus*. *Can. J. Genet. Cytol.* 22: 167.

Botterman, J. and Leemans, J. 1988. Engineering herbicide resistance into plants. *Trends Genet.* 4: 219.

Buchanan-Wollaston, V., Snape, A., and Cannon, F. 1992. A plant selectable marker gene based on the detoxification of the herbicide dalapon. *Plant Cell Rep.* 11: 627.

Cardon, G.H., Frey, M., Saedler, H., and Gierl, A. 1993. Definition and characterization of an artificial *En/Spm*-based transposon tagging system in transgenic tobacco. *Plant Mol. Biol.* 23: 157.

Cao, J., Duan, X., McElroy, D., and Wu, R. 1992. Regeneration of herbicide resistant transgenic rice plants following microprojectile-mediated transformation of suspension culture cells. *Plant Cell Rep.* 11: 586.

Catanzaro, C.J., Skroch, W.A., and Burton, J.D. 1993. Resistance of selected ornamental grasses to graminicides. *Weed Technol.* 7: 326.

Catlin, D.W. 1990. The effect of antibiotics on the inhibition of callus induction and plant regeneration from cotyledons of sugarbeet (*Beta vulgaris* L.). *Plant Cell Rep.* 9: 285.

Chaleff, R.S. 1983. Isolation of agronomically useful mutants from plant cell cultures. *Science* 219: 676.

Chaleff, R.S. 1988. Herbicide-resistant plants from cultured cells. In *Applications of Plant Cell and Tissue Culture*. Ciba Foundation Symp. 137, Wiley, New York, p. 3.

Chaleff, R.S. and Ray, T.B. 1984. Herbicide-resistant mutants from tobacco cell cultures. *Science* 223: 1148.

Chowdhury, M.K.U. and Vasil, I.K. 1992. Stably transformed herbicide resistant callus of sugarcane via microprojectile bombardment of cell suspension cultures and electroporation of protoplasts. *Plant Cell Rep.* 11: 494.

Christou, P., Ford, T.L., and Kofron, M. 1991. Production of transgenic rice (*Oryza sativa* L.) plants from agronomically important indica and japonica varieties via electric discharge particle acceleration of exogenous DNA into immature zygotic embryos. *Bio/Technology* 9: 957.

Christou, P. 1992. Genetic transformation of crop plants using microprojectile bombardment. *Plant J.* 2: 275.

Comai, L., Facciotti, D., Hiatt, W.R., Thompson, G., Rose, R.E., and Stalker, D.M. 1985. Expression in plants of a mutant *aroA* gene from *Salmonella typhimurium* confers tolerance to glyphosate. *Nature* 317: 741.

Cranston, H.J., Johnson, R.R., and Dyer, W.E. 1994. Differential display: a new technique for isolating genes based on mRNA abundance. *Proc. West. Soc. Weed Sci.* 47: 27.

D'Halluin, K., Bossut, M., Bonne, E., Mazur, B., Leemans, J., and Botterman, J. 1992. Transformation of sugarbeet (*Beta vulgaris* L.) and evaluation of herbicide resistance in transgenic plants. *Bio/Technology*, 10: 309.

Dale, P.J., Irwin, J.A., and Scheffler, J.A. 1993. The experimental and commercial release of transgenic crop plants. *Plant Breeding* 111: 1.

Darmency, H. and Pernes, J. 1989. Agronomic performance of a triazine resistant foxtail millet (*Setaria italica* [L.] Beauv.). *Weed Res.* 29: 147.

Datta, S.K., Datta, K., Soltanifar, N., Donn, G., and Potrykus, I. 1992. Herbicide-resistant Indica rice plants from IRRI breeding line IR72 after PEG-mediated transformation of protoplasts. *Plant Mol. Biol.* 20: 619.

DeBlock, M., Botterman, J., Vandewiele, M., Dockyx, J., Thoen, C., Gossele, V., Movva, N.R., Thompson, C., Van Montagu, M., and Leemans, J. 1987. Engineering herbicide resistance in plants by expression of a detoxifying enzyme. *EMBO J.* 6: 2513.

DeGreef, W., Delon, R., DeBlock, M., Leemans, J., and Botterman, J. 1989. Evaluation of herbicide resistance in transgenic crops under field conditions. *Bio/Technology* 7: 61.

Dekeyser, R.A., Claes, B., DeRycke, R.M.U., Habets, M.E., Van Montagu, M.C., and Caplan, A.B. 1990. Transient gene expression in intact and organized rice tissues. *Plant Cell* 2: 591.

De la Peña, A., Lörz, H., and Schell, J. 1987. Transgenic rye plants obtained by injecting DNA into young floral tillers. *Nature* 325: 274.

Della-Cioppa, G., Bauer, S.C., Taylor, M.L., Rochester, D.E., Klein, B.K., Shah, D.M., Fraley, R.T., and Kishore, G.M. 1987. Targeting a herbicide-resistant enzyme from *Escherichia coli* to chloroplasts of higher plants. *Bio/Technology* 5: 579.

Devillard, C. 1992. Transformation in vitro du tremble (*Populus tremula x Populus alba*) par *Agrobacterium rhizogenes* et régénération de plantes tolérantes au basta. *C. R. Acad. Sci. Paris* 314(Ser III): 291.

Duguid, J.R., Rohwer, R.G., and Seed, B. 1988. Isolation of cDNAs of scrapie-modulated RNAs by subtractive hybridization of a cDNA library. *Proc. Natl. Acad. Sci. U.S.A.* 85: 5738.

Duke, S.O., Christy, A.L., Hess, F.D., and Holt, J.S. 1991. Herbicide-Resistant Crops. Comments from CAST 1991-1, Council for Agricultural Science and Technology, Ames, IA.

Dyer, W.E. 1991. Applications of molecular biology in weed science, *Weed Sci.* 39: 482.

Dyer, W.E. 1994. Resistance to glyphosate. In *Herbicide Resistance in Plants: Biology and Biochemistry.* Ed. by Powles, S. and Holtum, J. Lewis Publishers, Chelsea, MI, p. 229.

Dyer, W.E., Hess, F.D., Holt, J.S., and Duke, S.O. 1993. Potential benefits and risks of herbicide-resistant crops produced by biotechnology. *Hortic. Rev.* 15: 367.

Dyer, W.E., Weller, S.C., Bressan, R.A., and Herrmann, K.M. 1988. Glyphosate tolerance in tobacco (*Nicotiana tabacum* L.). *Plant Physiol.* 88: 661.

Federoff, N., Furtek, D., and Nelson, O.E. 1984. Cloning of the *bronze* locus in maize by a simple and generizable procedure using the transposable controlling element *Ac. Proc. Natl. Acad. Sci. U.S.A.* 81: 3825.

Fedtke, C. 1991. Deamination of metribuzin in tolerant and susceptible soybean (*Glycine max*) cultivars. *Pestic. Sci.* 31: 175.

Fillatti, J.J., Haissig, B., McCown, B., Comai, L., and Riemenschneider, D. 1988. Development of glyphosate-tolerant *Populus* plants through expression of a mutant *aroA* gene from *Salmonella typhimurium.* In *Genetic Manipulation of Woody Plants.* Ed. by Hanover, J.W. and Keathley, D.E. Plenum Press, New York, p. 243.

Foy, C.L., Jacobsohn, R., and Jain, R. 1988. Screening of *Lycopersicon* spp. for glyphosate and/or *Orobanche aegyptiaca* Pers. resistance. *Weed Res.* 28: 383.

Fromm, M.E., Taylor, L.P., and Walbot, V. 1986. Stable transformation of maize after gene transfer by electroporation. *Nature* 319: 791.

Gardner, R.C. 1993. Gene transfer into tropical and subtropical crops. *Sci. Hortic.* 55: 65.

Gasser, C.S. and Fraley, R.T. 1989. Genetically engineering plants for crop improvement. *Science* 244: 1293.

Gierl, A. and Saedler, H. 1992. Plant transposable elements and gene tagging. *Plant Mol. Biol.* 19: 39.

Goldsbrough, P.B., Hatch, E.M., Huang, B., Kosinski, W.G., Dyer, W.E., Herrmann, K.M. and Weller, S.C. 1990. Gene amplification in glyphosate tolerant tobacco cells. *Plant Sci.* 72: 53.

Gordon-Kamm, W.J., Spencer, T.M., Mangano, M.L., Adams, T.R., Daines, R.J., Start, W.G., O'Brien, J.V., Chambers, S.A., Adams, W.R., Willetts, N.G., Rice, T.B., Mackey, C.J., Krueger, R.W., Kausch, A.P., and Lemaux, P.G. 1990. Transformation of maize cells and regeneration of fertile transgenic plants. *Plant Cell* 2: 603.

Gressel, J. 1993. Advances in achieving the needs for biotechnologically-derived (B-D) herbicide-resistant crops. *Plant Breed. Rev.* 11: 155.

Grimsley, N.H., Hohn, T., Davies, J.W., and Hohn, B. 1987. *Agrobacterium*-mediated delivery of infectious maize streak virus into maize plants. *Nature* 325: 177.

Guerineau, F., Brooks, L., Meadows, J., Robinson, C., and Mullineaux, P. 1990. Sulfonamide resistance gene for plant transformation. *Plant Mol. Biol.* 15: 127.

Hart, S.E., Saunders, J.W., and Penner, D. 1992. Chlorsulfuron-resistant sugarbeet: cross-resistance and physiological basis of resistance. *Weed Sci.* 40: 378.

Hartwig, E.A. 1987. Identification and utilization of variation in herbicide tolerance in soybean (*Glycine max*) breeding. *Weed Sci.* 35(Suppl. 1): 4.

Haughn, G.W., Smith, J., Mazur, B., and Somerville, C. 1988. Transformation with a mutant *Arabidopsis* acetolactate synthase gene renders tobacco resistant to sulfonylurea herbicides. *Mol. Gen. Genet.* 211: 266.

Herrera-Estrella, L., DeBlock, M., Van Montagu, M., and Schell, J. 1983. Chimeric genes as dominant selectable markers in plant cells. *EMBO J.* 2: 987.

Hinchee, M.A.W., Conner-Ward, D.V., Newell, C.A., McDonnell, R.E., Sato, S.J., Gasser, C.S., Fischoff, D.A., Re, D.B., Fraley, R.T., and Horsch, R.B. 1988. Production of transgenic soybean plants using *Agrobacterium*-mediated DNA transfer. *Bio/Technology* 6: 915.

Hocykaas, P.J.J. 1989. Transformation of plant cells via *Agrobacterium. Plant Mol. Biol.* 13: 327.

Horsch, R.B., Fry, J.E., Hoffmann, N.L., Eichholtz, D., Rogers, S.G., and Fraley, R.T. 1985. A simple and general method for transferring genes into plants. *Science* 227: 1229.

Huang, B. 1992. Genetic manipulation of microspores and microspore-derived embryos. *In Vitro Cell. Dev. Biol. Plant* 28P: 53.

Hughes, K. 1983. Selection for herbicide resistance. In *Handbook of Plant Cell Culture, Vol. 1*. Ed. by Evans, D.A., Sharp, W.R., Ammirato, P.V., and Yamada, Y. Macmillan, New York, p. 442.

Jenes, B., Moore, H., Cao, J., Zhang, W., and Wu, R. 1993. Techniques for gene transfer. In *Transgenic Plants, Vol. 1, Engineering and Utilization*. Ed. by Kung, S. and Wu, R. Academic Press, San Diego, CA, chap. 4.

Joersbo, M. and Brunstedt, J. 1990. Direct gene transfer to plant protoplasts by mild sonication. *Plant Cell Rep.* 9: 207.

Johnson, R.R., Cranston, H.J., Chaverra, M.E., and Dyer, W.E. 1995. Screening for differentially expressed genes in dormant and nondormant *A. fatua* embryos. *Plant Mol. Biol.* (In press).

Johnston, D.T. and Faulkner, J.S. 1991. Herbicide resistance in the graminaceae — a plant breeder's view. In *Herbicide Resistance in Weeds and Crops*. Ed. by Caseley, J.C., Cussans, G.W., and Atkin, R.K. Butterworth-Heinemann, Oxford, p. 319.

Jouanin, L., Brasileiro, A.C.M., Leple, J.C., Pilate, G., and Cornu, D. 1993. Genetic transformation — a short review of methods and their applications, results and perspectives for forest trees. *Ann. Sci. Forest.* 50: 325.

Kado, C.I. 1991. Molecular mechanisms of crown gall tumorigenesis. *Crit. Rev. Plant Sci.* 10: 1.

Kaeppler, H.F., Somers, D.A., Rines, H.W., and Cockburn, A.F. 1992. Silicon carbide fiber-mediated stable transformation of plant cells. *Theor. Appl. Genet.* 84: 560.

Kibite, S. and Harker, S.N. 1991. Evaluation of oat germplasm for resistance to diclofop-methyl. *Can. J. Plant Sci.* 71: 491.

Klein, T.M., Wolf, E.D., Wu, R., and Sandford, J.C. 1987. High velocity microprojectiles for delivery of nucleic acids into living cells. *Nature* 327: 70.

Klein, T.M., Arentzen, R., Lewis, P.A., and Fitzpatrick-McElligot, S. 1992. Transformation of microbes, plants and animals by particle bombardment. *Bio/Technology* 10: 286.

Liang, P. and Pardee, A.B. 1992. Differential display of eukaryotic messenger RNA by means of the polymerase chain reaction. *Science* 257: 967.

Lindsey, K. 1992. Genetic manipulation of crop plants. *J. Biotechnol.* 26: 1.

Loux, M.M., Liebl, R.A., and Hymowitz, T. 1987. Examination of wild perennial *Glycine* species for glyphosate tolerance. *Soybean Genet. Newslett.* 14: 268.

Luo, Z. and Wu, R. 1988. A simple method for the transformation of rice via the pollen-tube pathway. *Plant Mol. Biol. Rep.* 6: 165.

Lyon, B.R., Llewellyn, D.J., Huppatz, J.L., Dennis, E.S., and Peacock, W.J. 1989. Expression of a bacterial gene in transgenic tobacco plants confers resistance to the herbicide 2,4-dichlorophenoxyacetic acid. *Plant Mol. Biol.* 13: 533.

Maliga, P., Fejes, E., Steinback, K., and Menczel, L. 1987. Cell culture approaches for obtaining herbicide-resistant chloroplasts in crop plants. *ACS Symp. Ser.* 334: 115.

Mallory-Smith, C.A., Thill, D.C., and Dial, M.J. 1993. ID-BR1: sulfonylurea herbicide resistant lettuce germplasm. *HortScience* 28: 63.

Mazur, B.J. and Falco, S.C. 1989. The development of herbicide resistant crops. *Annu. Rev. Plant Physiol. Plant Mol. Biol.* 40: 441.

Miki, B.L., Labbé, H., Hattori, J., Ouellet, T., Gabard, J., Sunohara, G., Charest, P.J., and Iyer, V.N. 1990. Transformation of *Brassica napus* canola cultivars with *Arabidopsis thaliana* acetohydroxyacid synthase genes and analysis of herbicide resistance. *Theor. Appl. Genet.* 80: 449.

Mullineaux, P.M. 1992. Genetically engineered plants for herbicide resistance. In *Plant Genetic Manipulation for Crop Protection*. Biotechnology in Agriculture Series Vol. 7. Ed. by Gatehouse, A.M.R., Hilder, V.A., and Boulder, D. chap. 4.

Mullis, K.B. and Faloona, F.A. 1987. Specific synthesis of DNA *in vitro* via a polymerase catalyzed chain reaction. *Methods Enzymol.* 255: 335.

Nafziger, E.D., Widholm, J.M., Steinrücken, H.C., and Killmer, J.L. 1984. Selection and characterization of a carrot cell line tolerant to glyphosate. *Plant Physiol.* 76: 571.

Newhouse, K., Singh, B., Shaner, D., and Stidham, M. 1991. Mutations in corn (*Zea mays* L.) conferring resistance to imidazolinone herbicides. *Theor. Appl. Genet.* 83: 65.

Odell, J., Caimi, P., Sauer, B., and Russell, S. 1990. Site-directed recombination in the genome of tobacco. *Mol. Gen. Genet.* 223: 369.

Orlikowska, T., Cranston, H.C., and Dyer, W.E. 1994. Factors influencing *Agrobacterium tumefaciens*-mediated transformation and regeneration of the safflower cv. 'Centennial'. *Plant Cell Tissue Org. Cult.* 40: 85.

Oxtoby, E. and Hughes, M.A. 1989. Breeding for herbicide resistance using molecular and cell techniques. *Euphytica* 40: 173.

Parker, W.B., Somers, D.A., Wyse, D.L., Keith, R.A., Burton, J.D., Gronwald, J.W., and Gengenbach, B.G. 1990. Selection and characterization of sethoxydim-tolerant maize tissue cultures. *Plant Physiol.* 92: 1220.

Peng, J., Kononowicz, H. and Hodges, T.K. 1992. Transgenic indica rice plants. *Theor. Appl. Genet.* 83: 855.

Perl, A., Galili, S., Shaul, O., Ben-Tzvi, I., and Galili, G. 1993. Bacterial dihydrodipicolinate synthase and desensitized aspartate kinase: two novel selectable markers for plant transformation. *Bio/Technology* 11: 715.

Pofelis, S., Le, H., and Grant, W.F. 1992. The development of sulfonylurea herbicide-resistant birdsfoot trefoil (*Lotus corniculatus*) plants from in vitro selection. *Theor. Appl. Genet.* 83: 480.

Potrykus, I. 1990a. Gene transfer methods for plants and cell cultures. In *1990 Bioactive Compounds from Plants. Ciba Foundation Symp.* 154, Wiley, Chichester, U.K., p. 198.

Potrykus, I. 1990b. Gene transfer to cereals: an assessment. *Bio/Technology* 6: 530.

Quinn, J.P. 1990. Evolving strategies for the genetic engineering of herbicide resistance in plants. *Biotech. Adv.* 8: 321.

Rathore, K.S., Chowdhury, V.K., and Hodges, T.K. 1993. Use of bar as a selectable marker gene and for the production of herbicide-resistant rice plants from protoplasts. *Plant Mol. Biol.* 21: 871.

Reich, T.J., Iyer, V.N., and Miki, B.L. 1986. Efficient transformation of alfalfa protoplasts by the intranuclear microinjection of Ti plasmids. *Bio/Technology* 4: 1001.

Russell, S.H., Hoopes, J.L., and Odell, J.T. 1992. Directed excision of a transgene from the plant genome. *Mol. Gen. Genet.* 234: 49.

Sambrook, J., Fritsch, E.F., and Maniatis, T. 1989. *Molecular Cloning — A Laboratory Manual*, 2nd ed., Cold Spring Harbor Laboratory Press, Cold Spring Harbor, New York.

Sandler, S.J., Stayton, M., Townsend, J.A., Ralston, M.L., Bedbrook, J.R., and Dunsmuir, P. 1988. Inhibition of gene expression in transformed plants by antisense RNA. *Plant Mol. Biol.* 11: 301.

Sanford, J.C., Smith, F.D., and Russell, J.A. 1993. Optimizing the Biolistic process for different biological applications. *Methods Enzymol.* 217: 483.

Schuerman, P.L. and Dandekar, A.M. 1993. Transformation of temperate woody crops — progress and potentials. *Sci. Hortic.* 55: 101.

Schulz, A., Wengenmayer, F., and Goodman, H.M. 1990. Genetic engineering of herbicide resistance in higher plants. *Crit. Rev. Plant Sci.* 9: 1.

Scowcroft, W.R., and Larkin, P.J. 1988. Somaclonal variation. In *Applications of Plant Cell and Tissue Culture*. Wiley, Chichester, U.K., p. 21.

Sebastian, S.A., Fader, G.M., Ulrich, J.F., Forney, D.R., and Chaleff, R.S. 1989. Semidominant soybean mutation for resistance to sulfonylurea herbicides. *Crop Sci.* 29: 1403.

Shah, D.M., Horsch, R.B., Klee, H.J., Kishore, G.M., Winter, J.A., Tumer, N.E., Hironaka, C.M., Sanders, P.R., Gasser, C.S., Aykent, S., Siegel, N.R., Rogers, S.G., and Fraley, R.T. 1986. Engineering herbicide tolerance in transgenic plants. *Science* 233: 478.

Shimamoto, K., Terada, R., Izawa, T., and Fujimoto, H. 1989. Fertile transgenic rice plants regenerated from transformed protoplasts. *Nature* 338: 274.

Singer, S.R. and McDaniel, C.N. 1985. Selection of glyphosate-tolerant tobacco calli and the expression of this tolerance in regenerated plants. *Plant Physiol.* 78: 411.

Smeda, R.J. and Weller, S.C. 1991. Plant cell and tissue culture techniques for weed science research. *Weed Sci.* 39: 497.

Somers, D.A., Rines, H.W., Gu, W., Kaeppler, H.F., and Bushnell, W.R. 1992. Fertile, transgenic oat plants. *Bio/Technology* 10: 1589.

Snape, J.W., Leckie, D., Parker, B.B., and Nevo, E. 1991. The genetical analysis and exploitation of differential responses to herbicides in crop species. In *Herbicide Resistance in Weeds and Crops*. Ed. by Caseley, J.C., Cussans, G.W., and Atkin, R.K. Butterworth-Heinemann, Oxford, p. 305.

Stalker, D.M. 1991. Developing herbicide resistance in crops by gene transfer technology. *Plant Biotechnol.* 1: 82.

Stalker, D.M., McBride, K.E., and Malyj, L.D. 1988. Herbicide resistance in transgenic plants expressing a bacterial detoxification gene. *Science* 242: 419.

Streber, W.R., and Willmitzer, L. 1989. Transgenic tobacco plants expressing a bacterial detoxifying enzyme are resistant to 2,4-D. *Bio/Technology* 7: 811.

Stringfellow, C.B., Orlikowska, T., and Dyer, W.E. 1993. Transformation and regeneration of glufosinate-resistant safflower (*Carthamus tinctorius* L.) plants. *Weed Sci. Soc. Am. Abstr.* 33: 197.

Swanson, E.B., Coumans, M.P., Brown, G.L., Patel, J.D., and Beversdorf, W.D. 1988. The characterization of herbicide tolerant plants in *Brassica napus* L. after in vitro selection of microspores and protoplasts. *Plant Cell Rep.* 7: 83.

Tachibana, K., Watanabe, T., Sekizawa, T., and Takemutsu, T. 1986. Action mechanism of bialaphos. II. Accumulation of ammonia in plants treated with bialaphos. *J. Pestic. Sci.* 11: 33.

Thompson, C., Movva, N., Tizard, R., Crameri, R., Davies, J., Lauwereys, M., and Botterman, J. 1987. Characterization of the herbicide-resistance gene *bar* from *Streptomyces hygroscopicus*. *EMBO J.* 6: 2519.

Tischer, E., DasSarma, S., and Goodman, H.M. 1986. Nucleotide sequence of an alfalfa glutamine synthetase gene. *Mol. Gen. Genet.* 203: 221.

Tonnemaker, K.A., Auld, D.L., Thill, D.C., Mallory-Smith, C.A., and Erickson, D.A. 1992. Development of sulfonylurea-resistant rapeseed using chemical mutagenesis. *Crop Sci.* 32: 1387.

Töpfer, R., Gronenborn, B., Schell, J., and Steinbiss, H.H. 1989. Uptake and transient expression of chimeric genes in seed-derived embryos. *Plant Cell* 1: 133.

Tuberosa, R. and Lucchese, C. 1990. Selection of maize cell lines tolerant to the nonselective herbicide Basta. *Chim. Oggi* 8(12): 43.

Van Wert, S.L. and Saunders, J.A. 1992. Electrofusion and electroporation of plants. *Plant Physiol.* 99: 365.

Vasil, I.K. 1987. Developing cell and tissue culture systems for the improvement of cereals and grass crops. *J. Plant Physiol.* 128: 193.

Vasil, V., Castillo, A., Fromm, M., and Vasil, I. 1992. Herbicide resistant fertile transgenic wheat plants obtained by microprojectile bombardment of regenerable embryogenic callus. *Bio/Technology* 10: 667.

Vermeulen, A., Vaucheret, H., Pautot, V., and Chupeau, Y. 1992. *Agrobacterium* mediated transfer of a mutant *Arabidopsis* acetolactate synthase gene confers resistance to chlorsulfuron in chicory (*Cichorium intybus* L.). *Plant Cell Rep.* 11: 243.

Waldron, C., Murphy, E.B., Roberts, J.L., Gustafson, G.D., Armour, S.L., and Malcolm, S.K. 1985. Resistance to hygromycin B. A new marker for plant transformation studies. *Plant Mol. Biol.* 5: 103.

Wan, Y.C. and Lemaux, P.G. 1994. Generation of large numbers of independently transformed fertile barley plants. *Plant Physiol.* 104: 37.

Wang, Z., Takamizo, T., Iglesias, V.A., Osusky, M., Nagel, J., Potrykus, I., and Spangenberg, G. 1992. Transgenic plants of tall fescue (*Festuca arundinacea* Schreb.) obtained by direct gene transfer to protoplasts. *Bio/Technology* 10: 691.

Waters, S. 1991. Glyphosate-tolerant crops for the future: development, risks and benefits. *Brighton Crop Prot. Conf., Weeds-1991* 1: 165.

Weeks, J.T., Anderson, O.D., and Blechl, A.E. 1993. Rapid production of multiple independent line of fertile transgenic wheat *(Triticum aestivum). Plant Physiol.* 102: 1077.

von Wettstein, D. 1993. Genetic engineering and plant breeding, especially cereals. *Food Rev. Int.* 9: 411.

Zhang, W., McElroy, D., and Wu, R. 1991. Analysis of rice *Act1* 5′ region activity in transgenic rice plants. *Plant Cell* 3: 1155.

Zhou, G.Y., Weng, J., Zeng, Y., Huang, J., and Qian, S. 1983. Introduction of exogenous DNA into cotton embryos. *Methods Enzymol.* 101: 433.

Chapter 4

NEW WEED CONTROL OPPORTUNITIES: DEVELOPMENT OF SOYBEANS WITH A ROUNDUP READY™ GENE

Stephen R. Padgette
Diane B. Re
Gerard F. Barry
David E. Eichholtz
Xavier Delannay
Roy L. Fuchs
Ganesh M. Kishore
Robert T. Fraley

CONTENTS

1-56670-045-0/96/$0.00+$.50
© 1996 by CRC Press Inc.

1. INTRODUCTION

United States consumers continue to spend less of their disposable income on food every year (11.4% in 1992) (Dunham 1993). Two factors contribute to this low percentage of income spent on food - increasing personal incomes and improvements in food production efficiency. Improvements in food production efficiency arise from fertilizer use, farm equipment and farm management advances, plant breeding, and pesticide use. Further improvements in agriculture will be required if food production efficiency is to keep pace with the ever-rising world population. Higher productivity agriculture will allow for reduced usage of marginal lands, marshes, and other environmental habitats.

Control of weeds with herbicides has been critical in food production because weeds compete with crops for water, nutrients, and soil. In addition, weeds can harbor insect and disease pests, and noxious weeds and weed seeds can greatly undermine crop quality. For these reasons, close to 100% of the U.S corn and soybean acres are treated with herbicides (Gianessi et al. 1991). Farmers consider several factors before choosing a weed management system using herbicides: weed-control spectrum, lack of crop injury (selectivity), cost, environmental characteristics, and fit with modern conservation trends such as reduced tillage. Several classes of herbicides are effective for broad-spectrum weed control, but many of the herbicides lack selectivity and either kill crop plants or they significantly injure crops at the application rates required for effective weed control.

There are several reasons why new weed control systems are needed. First, improved weed control and greater flexibility in herbicide application mean better production systems. Second, the use of herbicides with alternate modes-of-action in a given weed management system offers advantages in extending the lives of current programs, such as in weed resistance management.

Finally, new weed control systems featuring herbicides with beneficial environmental characteristics are needed to meet current and future rigorous demands on active ingredients.

Several options exist today for developing new weed control systems. First, the catalog of registered compounds can be expanded through the development of new formulations, mixtures, and safeners, for example. Alternately, new herbicides can be discovered and developed. For new herbicides, extensive testing must be performed to evaluate not only weed control spectrum, but crop safety, toxicology profile, and environmental characteristics. These studies required for new herbicide development are both time and cost intensive. Nonchemical approaches to weed control are also being investigated; these approaches include cover crops, intense cultivation, and biological control methods.

Another approach to new weed control systems is the development of herbicide-tolerant crops (HTCS*). Crops that are tolerant to herbicides are not new. Herbicide development programs have always assessed the crop safety of a given herbicide in a weed control system, so that no crop yield reduction would result from herbicide application. In fact, crop safety has been a focus of chemical herbicide discovery and plant breeding for the last 30 years, and thousands of chemical analogues have been screened to allow identification of herbicides that control target weeds and are safe to crops. However, it is now possible, using the modern techniques of biotechnology such as selection in tissue culture and genetic modification by transgenic methods, to build herbicide selectivity into the crop. This approach thus expands the utility of proven, previously nonselective, broad-spectrum herbicides.

The isopropylamine salt of glyphosate (*N*-[phosphonomethyl]glycine) (Figure 1), the active ingredient in the foliar-applied, postemergent herbicide Roundup® (a registered trademark of Monsanto Company) (Baird 1971; Malik et al. 1989), is one nonselective, broad-spectrum herbicide which currently cannot be applied "over-the-top" to crops due to severe plant injury. Glyphosate is the world's most popular herbicide. This is primarily due to its excellent weed control capabilities and its well-known, very favorable environmental and safety characteristics (Malik et al. 1989). The extension of the use of glyphosate to allow in-season application in major crops would provide new weed control options for farmers. Glyphosate is highly effective against the majority of annual and perennial grasses and broad-leaved weeds. Glyphosate's positive environmental features include rapid soil binding (resistance to leaching) and biodegradation (which decreases persistence), as well as extremely low toxicity to mammals, birds, and fish (Malik et al. 1989). Glyphosate is classified by the EPA as Category E (evidence of noncarcinogenicity for humans) (57 FR 8739). Importantly,

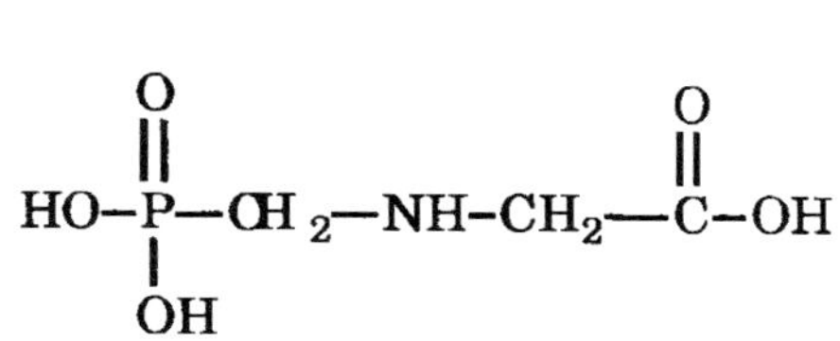

FIGURE 1 Glyphosate structure.

* The abbreviations used are ai, active ingredient; ALS, acetolactate synthase; AMPA, aminomethyl phosphonic acid; cDNA, complementary DNA; cm, centimeters; CP4 EPSPS, EPSPS from *Agrobacterium* sp. strain CP4; CTP, chloroplast transit peptide; EMS, ethylmethanesulfonate; ENU, ethylnitroso urea; EPA, Environmental Protection Agency; EPSP, 5-enolpyruvylshikimate-3-phosphate; EPSPS, 5-enolpyruvylshikimate-3-phosphate synthase; FDA, Food and Drug Administration; GTS, glyphosate-tolerant soybeans; GUS, β-glucuronidase; GOX, glyphosate oxidoreductase; ha, hectare; HA, hydroxylamine; HTC, herbicide-tolerant crops; P_i, inorganic phosphate; NG, nitrosoguanidine; PEP, phosphoenolpyruvate; S3P, shikimate-3-phosphate; USDA, United States Department of Agriculture; UWGCG, University of Wisconsin Genetics Computer Group; wt, wild-type.

the use of Roundup® herbicide has not resulted in the occurrence of even one case of spontaneous weed resistance to glyphosate in almost 20 years of widespread use. This benefit of glyphosate relative to other classes of herbicides should not be underemphasized.

The use of crops containing Roundup Ready™ genes (Monsanto's trademark for its genes to confer tolerance to Roundup® herbicide) will enable farmers to utilize glyphosate for effective control of weed pests and to take advantage of this herbicide's beneficial features. Crops with these genes can positively impact current agronomic practices by:

1. offering the farmer a new wide-spectrum weed control option
2. allowing the use of an environmentally sound herbicide
3. providing a new herbicidal mode of action for in-crop weed control (which has not resulted in weed resistance development in almost 20 years of use, based on the extensive use history of glyphosate)
4. increasing the ability to treat weeds on an "as needed" basis (flexibility of use)
5. offering less dependence on herbicides used before planting
6. providing an excellent fit with no-till systems, which results in increased soil moisture, while reducing soil erosion and fuel use
7. providing cost-effective weed control, not only because Roundup® herbicide may be less expensive than most current options, but because the total number of herbicides used may be reduced, compared to the farmer's current weed management program.

2. DEVELOPMENT OF ROUNDUP READY™ GENES — OVERVIEW

The development of crops tolerant to glyphosate has been ongoing since the early 1980s. Three basic mechanisms have been employed to confer tolerance: (1) overproduction of 5-enolpyruvyl-shikimate-3-phosphate synthase EPSPS (E.C. 2.5.1.19) (Shah et al. 1986); (2) introduction of an EPSPS with decreased affinity for glyphosate (Barry et al. 1992; Comai et al. 1983; Hinchee et al. 1993), and (3) introduction of a glyphosate degradation gene (Barry et al. 1992). Overexpression of a glyphosate-sensitive EPSPS conferred *in planta* glyphosate tolerance, but the tolerance was not sufficient to withstand commercial glyphosate application rates (Shah et al. 1986). In order to obtain higher levels of tolerance, other mechanisms were necessary. Introduction of a herbicide-insensitive EPSPS into crops by genetic modification techniques was successful in conferring glyphosate tolerance to plants. Because the molecular site of action ("target site") of the herbicide glyphosate is the enzyme EPSPS (Steinrucken et al. 1980), the identification of glyphosate-tolerant EPSPS enzymes has facilitated the development of glyphosate-tolerant crops. Upon glyphosate treatment of crops expressing certain glyphosate-tolerant EPSPSs, the plants remain unaffected because the continued action of the glyphosate-tolerant EPSPS enzyme meets the plant's need for aromatic amino acids. This is in contrast to the plant death or severe yield reduction observed upon glyphosate treatment of non-glyphosate-tolerant plants.

Tolerance to herbicides also may be imparted by metabolic inactivation of the active ingredient. Glyphosate is readily metabolized in some environments, especially soil, and although microbes capable of this degradation have been known for some time, the details of the mechanisms, the enzymes, and the genes involved have not been well characterized. Glyphosate may be metabolized by two routes: one pathway leads to sarcosine and inorganic phosphate as the primary products (the C–P lyase pathway) (Barry et al. 1992), and the other, the aminomethyl phosphonic acid AMPA pathway, leads to AMPA and glyoxylate. Advances in the study of both of these pathways has been hampered by the lack of demonstration of cell-free enzymatic activity. However, as discussed below, the bacterial gene which encodes the enzyme catalyzing glyphosate degradation to AMPA, GOX, has now been cloned and utilized in conferring glyphosate tolerance *in planta*.

One focus of this chapter is on the genes utilized to confer glyphosate tolerance to plants. Soybeans with a Roundup Ready™ gene (also called GTS) are discussed in detail as a case study of the development of a glyphosate-tolerant crop and corresponding safety assessment programs. Early research in the area of glyphosate tolerance has been reviewed by Gasser (1989); Hinchee et al. (1993); Padgette et al. (1989); Mazur (1989); and Kishore et al. (1988a, b).

3. EFFORTS TO SELECT GLYPHOSATE TOLERANCE AT THE WHOLE PLANT LEVEL

3.1. *ARABIDOPSIS THALIANA*

Although high-level glyphosate tolerance in crop plants has been obtained using genetic modification methods (see below), selection on glyphosate as a potential method to generate glyphosate-tolerant plants has also been investigated. Mutagenesis of *A. thaliana* seeds and herbicide selection was previously utilized to generate *Arabidopsis* mutants resistant to sulfonylurea herbicides (Haughn et al. 1986) and imidazolinones (Haughn et al. 1990). Seeds from a population of plants which had been heavily mutagenized with the alkylating agent EMS were screened by selecting on ALS inhibitor herbicides. In a typical experiment Somerville and co-workers recovered three mutants conferring very high levels of resistance to the imidazolinone herbicide imazapyr by screening 150,000 M2 seeds derived from a population of 25,000 M1 seeds (Haughn et al. 1990). It was subsequently established that at least one of the mutations was in the gene encoding ALS, the target enzyme for sulfonylurea and imidazolinone herbicides. In similar experiments using the sulfonylurea herbicide chlorsulfuron, they obtained three highly resistant mutants by screening 300,000 M2 plants (Haughn et al. 1986). The basis for the resistance in one case was a single nucleotide change in the gene encoding ALS (Haughn et al. 1988). Because the mutagenesis protocol employed increases the mutation frequency approximately 1000 to 10,000 times over the spontaneous rate, these results also can be used to estimate the frequency with which the same mutations will spontaneously arise in a wild population of plants. It has been argued that this protocol effectively tests all possible single nucleotide changes (Haughn et al. 1987).

Similar experiments were performed by Somerville and co-workers with glyphosate (C. Somerville, personal communication). In those experiments, 1 million EMS-mutagenized seeds were germinated on sterile minimal medium containing glyphosate at ten times the dose required to kill the plant. No survivors were identified. Next, the dose was reduced to five times the glyphosate dose required to kill seedlings, and another 1 million heavily mutagenized seeds were screened from the same population used to select sulfonylurea resistance. Once again, no survivors were identified. These results suggested strongly that single nucleotide changes which will convert a glyphosate-sensitive plant to a glyphosate-tolerant plant are extremely rare, if they exist at all. This conclusion is supported by the lack of the development of glyphosate-resistant weeds, even in systems receiving heavy usage of glyphosate. The lack of success in selecting glyphosate tolerance at the whole plant level in the EMS experiments (and others performed at Monsanto) suggested that transgenic methods would be effectively the only route to developing commercially acceptable glyphosate-tolerant crops.

3.2. TURFGRASS

Johnston and co-workers described the recurrent selection of fescue turfgrass species tolerant to glyphosate (Johnston et al. 1989), which is the only known example of *in planta* selection to yield reasonably high glyphosate tolerance at the whole plant level. These workers planted 30,000 seeds each of Manior slender creeping red fescue and Waldina hard fescue in trays, and applied glyphosate at the two-leaf stage at a dose capable of achieving 95 to 99% mortality in each variety. The survivors, approximately 800 per variety, were repotted and after 8 weeks they were resprayed with glyphosate to achieve 85% mortality. Approximately 120 survivors were allowed to interpollinate, yielding S1 seed. Additional cycles of selection were carried out (up to fourth generation), using more and more severe selection pressure with higher and higher rates of glyphosate. The highest level glyphosate tolerance (given as EC_{50} values) was obtained in the S4 generation plants, 0.87 kg ai/ha for the RF3 selection and 0.71 kg ai/ha for the HF2 selection (relative to 0.37 and 0.13 kg ai/ha in the respective S1 generations). It was reported that the increase in tolerance to glyphosate with successive cycles of selection was much slower with perennial ryegrass (Johnston et al. 1991), with the most advanced line five-fold more tolerant to glyphosate than the control after 11 cycles of recurrent selection. The mechanism of tolerance of these selections is unknown, but the requirement for repeated, tedious selection with increasing but sublethal doses of glyphosate is not suggestive of a spontaneous single mutation yielding a highly glyphosate-tolerant EPSPS.

4. EPSPS: THE TARGET OF GLYPHOSATE ACTION

4.1. EPSPS REACTION AND GLYPHOSATE INHIBITION

EPSPS catalyzes the reversible reaction of S3P and PEP to produce EPSP and P_i, as shown in Figure 2. EPSPS is an enzyme of the shikimate pathway of aromatic amino acid biosynthesis, and is present in plants, bacteria, and fungi, but not animals. In plants EPSPS is localized in the chloroplasts or plastids (della-Cioppa et al. 1986). Extensive work has been performed on the kinetics and mechanism of *Escherichia coli* EPSPS, and the tetrahedral reaction intermediate shown in Figure 2 has been isolated and conclusively identified (Anderson et al. 1988a). That research was reviewed by Anderson et al. (1990).

Glyphosate is a specific inhibitor of EPSPS, and is not known to inhibit any other PEP-dependent enzymatic reactions (Steinrucken et al. 1984a). For EPSPS, glyphosate is a reversible competitive inhibitor with respect to PEP, and it is a noncompetitive inhibitor with respect to S3P (Boocock et al. 1983). Stopped-flow and equilibrium fluorescence studies showed that glyphosate does not bind significantly to the free EPSPS enzyme, and that binding is ordered with S3P binding first, followed by glyphosate binding to form an EPSPS·S3P·glyphosate complex (Anderson et al. 1988b). Recent work modified that model, indicating that the EPSPS reaction is actually random, with synergistic binding occurring between substrates, products, and inhibitors (Gruys et al. 1992, 1993). The random character of the reaction does not, however, change the fact that in the forward

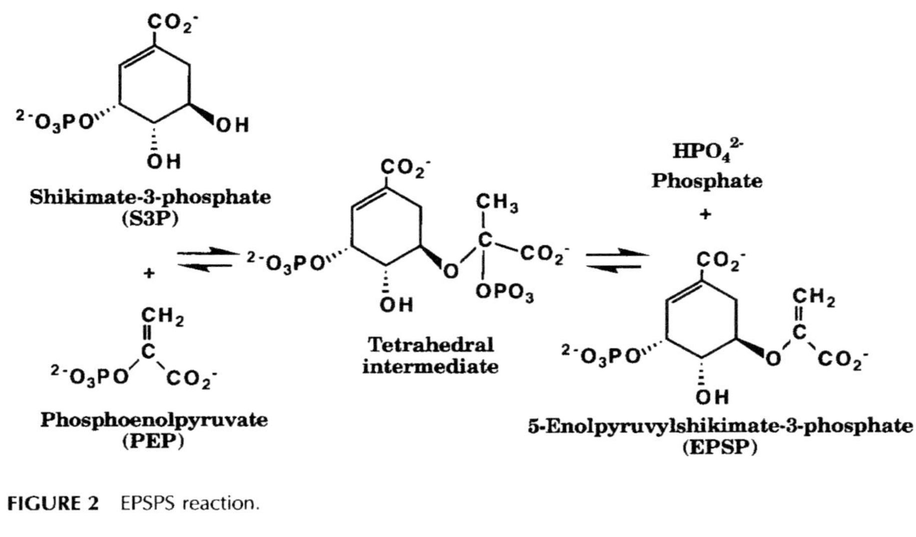

FIGURE 2 EPSPS reaction.

reaction direction, glyphosate binds to EPSPS as the EPSPS·S3P·glyphosate complex, and that this "dead-end" complex prevents the EPSPS reaction from proceeding to form products. Chemical modification data also supports the existence of the EPSPS·S3P·glyphosate complex, because only in the presence of S3P could glyphosate protect EPSPS from inactivation by arginine- and cysteine-specific reagents (Padgette et al. 1988a, b). Thus, glyphosate treatment of plants results in EPSPS inhibition (by the formation of the E·S3P·glyphosate complex), causing the plants to starve for EPSP and the metabolic products derived from EPSP, such as aromatic amino acids. Inhibition of EPSPS also leads to deregulation of the shikimate pathway, as well as accumulation of very high levels of shikimic acid (Devine et al. 1993).

4.2. EPSPS PROTEINS AND GENES

EPSPS has been purified from bacterial, fungal, yeast, and plant sources. These sources include *E. coli* (Duncan et al. 1984b), *Klebsiella pneumoniae* (Steinrucken et al. 1984b), pea (Mousdale et al. 1984), petunia (Steinrucken et al. 1986; Padgette et al. 1987), and sorghum (Ream et al. 1988). The bacterial and plant EPSPSs contain a single polypeptide chain of molecular mass of approximately 44 to 48 kDa. The EPSPSs from *Aspergillus nidulans* (Hawkins et al. 1991) and *Saccharomyces cerevisiae* (Duncan et al. 1987) are part of a pentafunctional polypeptide which catalyzes four other reactions of the shikimate pathway. EPSPS genes have been cloned from numerous sources and have provided a large number of deduced amino acid sequences that have been aligned to identify conserved residues. These conserved residues, shown in Figure 3 (compared to *E. coli* EPSPS) as the "consensus" EPSPS sequence, presumably include the critical residues necessary for

```
  E. coli  .......... .......... .......... ......MESL TLQPIARVDG
Consensus  ---------- ---------- ---------- ---------- ----------

  E. coli  TINLPGSKTV SNRALLLAAL AHGKTVLTNL LDSDDVRHML NALTAL.GVS
Consensus  ----PG-K-- --R------L --G------- L---D----- ----------

  E. coli  YTLSADRTRC EIIGNGGPLH AE....GALE LFLGNAGTAM RPLAAALCL.
Consensus  ---------- ---------- ---------- L--GN--T-- R---------

  E. coli  .....GSNDI VLTGEPRMKE RPIGHLVDAL RLGGAKITYL EQENYPPLRL
Consensus  ---------- ---G------ RP-------L ---------- ----------

  E. coli  QG..GFTG.. ...GNVDVDG SVSSQFLTAL LMTAP.LA.. .PEDTVIRIK
Consensus  ---------- ---------- --S-Q----- L--------- ----------

  E. coli  GDLVSKPYID ITLNLMKTFG VE..IEN.QH YQQFVVKGGQ SYQSPGTYLV
Consensus  ---------- ---------- ---------- ---------- ----------

  E. coli  EGDASSASYF LAAAA.IKGG TVKVTGIGRN SMQGDIRFA. DVLEKMGATI
Consensus  --D-S----- ---------- ---------- ---------- -----MG---

  E. coli  ....CWGDDY I........S CTRGELNAI. DMDMNHIPDA AMTIATAALF
Consensus  ---------- ---------- ---------- --------D- ----------

  E. coli  AKG....... TTRLRNIYNW RVKETDRLFA MATELRKVGA EV.EEGHDYI
Consensus  ---------- ---------- -V----R--- ---------- ----------

  E. coli  RITP..PEKL NFA....... EIATYNDHRM AMCFSLVAL. .........S
Consensus  ---------- ---------- --------R- ---------- ----------

  E. coli  DTPVTILDPK CTAKTFPDYF EQLARISQAA ........
Consensus  ---------- ------P--- ---------- --------
```

FIGURE 3 Consensus sequence of EPSPSs compared to that of *E. coli* EPSPS. The sequence comparisons were performed using the software from UWGCG, Version 7 (Genetics Computer Group 1991). Twenty-three EPSPS amino acid sequences were aligned, including plant, bacterial (Gram positive and negative), and fungal sources.

substrate binding and catalysis at the active site. Several residues which have been identified chemically as at (or near) the active site are conserved, such as Lys^{22} (postulated to be involved in PEP binding) (Huynh et al. 1988); Arg^{27} (involved in S3P binding) (Padgette et al. 1988b); and Arg^{124} (postulated to be involved in PEP binding) (Huynh et al. 1988) (*E. coli* numbering). These residues are highlighted in bold in Figure 3. It is clear, based on the relatively few number of absolutely conserved residues in Figure 3, that there is a wide diversity of amino acid sequences in EPSPSs from different sources.

The wealth of primary sequence information on EPSPS has been augmented considerably by the solution of the 3-dimensional X-ray structure of *E. coli* EPSPS (Stallings et al. 1991) (see Color Plate 1*). EPSPS has a two-lobed structure, and as discussed below in the context of glyphosate binding, the key active site residues listed above are present at the inner surface of the "upper" domain. This arrangement suggests that upon glyphosate and S3P binding, EPSPS may undergo a transition to a more "closed" structure, representative of the "dead-end" complex discussed above. Despite repeated attempts, however, the E·S3P·glyphosate complex has not been crystallized and the structure of the enzyme complex with glyphosate remains unsolved. The structure of the EPSPS·S3P binary complex of *E. coli* EPSPS has been solved and indicates that Arg^{28} of *E. coli* EPSPS binds to the carboxylate of S3P (Stallings et al. unpublished data). This conclusion has been supported by studies involving site-directed mutagenesis at this site·

Extensive research has been devoted to developing new inhibitors of EPSPS. Alberg and co-workers have had considerable success in synthesizing inhibitors based on the tetrahedral intermediate structure shown in Figure 2, resulting in inhibitors with nanomolar-level K_i constants (Alberg et al. 1988, 1992). Other work has combined extensive physical characterization of EPSPS with inhibitor design efforts (Sikorski et al. 1990). Nevertheless, despite years of extensive inhibitor design research, glyphosate remains the only EPSPS inhibitor currently available for weed control use.

5. GLYPHOSATE-TOLERANT EPSPSs

5.1. SALMONELLA TYPHIMURIUM P101S VARIANT

Comai and co-workers at Calgene reported the isolation of glyphosate-tolerant mutants of *Salmonella typhimurium* (Comai et al. 1983). One of the mutations was reported to map in the *aroA* (EPSPS) locus, and the EPSPS activity of this mutant was also reported to be tolerant to inhibition by glyphosate. Additional studies indicated that the glyphosate-tolerant *S. typhimurium* EPSPS contained a single amino acid change of Pro to Ser at position 101 of the polypeptide sequence (Stalker et al. 1985). The *S. typhimurium* mutant *AroA* gene was introduced into tobacco cells via *Agrobacterium*-mediated transformation (Comai et al. 1985); transformation of tomato with this mutant EPSPS also was reported (Fillatti et al. 1987). A glyphosate-tolerant phenotype was reported in the regenerated plants. Note that the *S. typhimurium* mutant *AroA* gene used in this work lacked a chloroplast transit peptide and the enzyme was thus localized in the cytosol. As mentioned above, reactions of the shikimate pathway occur primarily in chloroplasts, and the source of S3P substrate for this cytosolic EPSPS mutant is unknown. Current information on the commercial development of crops expressing the *S. typhimurium* P101S EPSPS is scarce.

5.2. WILD-TYPE PETUNIA EPSPS

Early experiments with *E. coli*, which is sensitive to glyphosate, demonstrated that overexpression of the *E. coli* EPSPS conferred glyphosate tolerance *in vivo* (Rogers et al. 1983). This approach was tested in plants by expressing in petunia the cDNA encoding petunia EPSPS (Gasser et al. 1988),

* Color plates follow page 84.

with its CTP (Shah et al. 1986). Those results showed that overproduction of EPSPS in plants (20- to 40-fold) could also confer tolerance to glyphosate. However, additional experiments with genetically modified crops overexpressing the wild-type glyphosate-sensitive petunia EPSPS indicated that commercially acceptable levels of glyphosate tolerance could not be practically achieved using such glyphosate-sensitive EPSPS proteins.

5.3. *E. COLI* G96A VARIANT AND HOMOLOGS

In 1986 we reported the isolation of a variant EPSPS which was highly tolerant to inhibition by glyphosate. This variant had been isolated following selection of *E. coli* B for growth in the presence of the herbicide (Kishore et al. 1986). This *E. coli* strain (SM-1) had an EPSPS specific activity comparable to the wild-type *E. coli*, but was resistant to inhibition by glyphosate. The *AroA* gene encoding the EPSPS from SM-1 was cloned and shown to contain a single amino acid substitution relative to the wild-type *E. coli* B EPSPS (Ala for Gly at position 96). Kinetic studies indicated that the $appK_i$ for glyphosate for the G96A variant was extremely high, 4 m*M*, as compared to 0.5 μ*M* for the wild-type E. coli enzyme. The variant enzyme thus had an 8000-fold decrease in glyphosate sensitivity. The $appK_m$ for S3P was slightly elevated (~4-fold). However, the $appK_m$(PEP) was increased significantly in the G96A variant, from 17 to 220 μ*M*. Efforts to identify EPSPS enzymes resistant to glyphosate and that exhibit nonelevated $appK_m$ (PEP) constants are discussed in more detail below. Note that Sost et al. (1990) identified a glyphosate-tolerant *Klebsiella pneumoniae* EPSPS which also has a substitution of glycine to alanine at the homologous residue described above for the SM-1 *E. coli* variant.

As shown in Table 1, initial alignment of the amino acid sequences of numerous EPSPSs revealed that the *E. coli* Gly^{96} residue is located in a conserved region of EPSPS, with a consensus sequence $L^{90}XLGNA\underline{G}^{96}TAXRXL^{102}$ (X represents nonconserved amino acids). Consideration of a broader range of EPSPSs, however, indicates that the so-called "conserved region" actually has absolute homology only at the G, T, and R residues in the following sequence: GXXXTXXR

TABLE 1 Sequence Comparison of the Conserved Region of Several EPSPSs Around Gly^{96} of *E. coli* EPSPS

EPSPS Source	Sequence					Ref.
E. coli	L^{90}	F	$LGNA\underline{G}^{96}$	$TAMRP^{101}$	L^{102}	Duncan et al. 1984a
S. typhimurium	L^{90}	F	$LGNA\underline{G}^{96}$	$TAMRP^{101}$	L^{102}	Stalker et al. 1985
Petunia	L^{95}	F	$LGNA\underline{G}^{101}$	$TAMRP^{106}$	L^{107}	Gasser et al. 1988
Tomato	L^{95}	F	$LGNA\underline{G}^{101}$	$TAMRP^{106}$	L^{107}	Gasser et al. 1988
Arabidopsis	L^{95}	Y	$LGNA\underline{G}^{101}$	$TAMRP^{106}$	L^{107}	Klee et al. 1987
B. napus	L^{95}	Y	$LGNA\underline{G}^{101}$	$TAMRP^{106}$	L^{107}	[a]
Soybean	L^{98}	F	$LGNA\underline{G}^{104}$	$TAMRP^{109}$	L^{110}	[a]
Maize	L^{95}	F	$LGNA\underline{G}^{101}$	$TAMRP^{106}$	L^{107}	[a]
A. nidulans[b]	L	Y	$LGNA\underline{G}$	TASRF	L	Charles et al. 1986
S. cerevisiae[b]	L	Y	$LGNA\underline{G}$	TASRF	L	Powell et al. 1992
B. pertussis	L^{90}	F	$LGNA\underline{G}^{96}$	$TAFRP^{101}$	L^{102}	Maskell et al.1988
Consensus[c] 1[d]	L	X	$LGNA\underline{G}$	TAXRX	L	
Consensus 2[e]	X	X	XGXXX	TXXRX	X	

Note: The underlined glycine residues correspond to Gly^{96} of *E. coli* EPSPS (modified from Padgette et al. 1991).

[a] Monsanto unpublished results.
[b] Part of the arom multienzyme complex.
[c] For the consensus sequence, "X" represents nonconserved amino acids.
[d] Taken from O'Connell et al. 1993.
[e] Including information from O'Connell et al. 1993 and Garbe et al. 1990.

(O'Connell et al. 1993; Garbe et al. 1990). The bold italic "X" in this sequence corresponds to the *E. coli* Gly^{96} position described above, which would be a completely conserved residue except for the naturally occurring EPSPS from *Agrobacterium* sp. strain CP4, which is discussed in detail below. It has been shown, however, that this region is part of a conserved EPSPS active site by introducing the equivalent Gly to Ala change in five additional EPSPS enzymes (petunia, tomato, soybean, maize, and *Arabidopsis*) (Padgette et al. 1991). All of the wild-type EPSPSs tested were sensitive to inhibition by glyphosate, while the Ala-substituted variants were extremely resistant to inhibition by glyphosate. Thus, the substitution corresponding to G96A affects the glyphosate sensitivities of both *E. coli* and plant EPSPSs to similar degrees — more than a 500-fold increase in glyphosate IC_{50} (glyphosate concentration causing 50% inhibition of EPSPS activity) is obtained in all cases. The residue corresponding to Ala^{96} of *E. coli* EPSPS therefore interferes with glyphosate binding to EPSPS by a mechanism common to the *E. coli* and plant EPSPSs tested.

In order to better quantitate the effect of the G101A substitution in petunia EPSPS (G96A in the *E. coli* sequence), steady-state kinetic constants were determined for purified wild-type and G101A petunia EPSPSs (Padgette et al. 1991). For the G101A variant, the $appK_i$(glyphosate) vs. PEP was found to be 2.0 ± 0.2 m*M*, which is 5000 times greater than that of the wild-type enzyme; the inhibition remained competitive vs. PEP. The $appK_m$(PEP) was elevated from 5.0 ± 1.3 μ*M* in the wild-type enzyme to 210 ± 20 μ*M* in the G101A variant, and the $appK_m$(phosphate) was elevated from 0.59 ± 0.09 m*M* for the wild-type enzyme to 6.7 ± 1.2 m*M* for the G101A variant. In contrast to the decreases in apparent binding seen for glyphosate, PEP, and P_i in the G101A variant, the $appK_m$ values for both S3P and EPSP were essentially unchanged between the wild-type and G101A EPSPSs. Taken together, these kinetic data are consistent with a perturbation of a binding interaction common for glyphosate and PEP (and potentially P_i), but not S3P or EPSP. If one views the $appK_i$(glyphosate)/$appK_m$(PEP) as a measure of the selectivity for PEP over glyphosate binding, the large increase in its value from the petunia wild-type EPSPS ($K_i/K_m = 0.08$) to the G101A variant ($K_i/K_m = 9.5$) indicates that the introduction of the alanine residue selectively destabilizes the interaction of the enzyme with glyphosate as compared to PEP. In addition to its glyphosate insensitivity, the catalytic efficiency for G101A EPSPS also is altered. The $k_{cat}/appK_m$(PEP) values for the wild-type and G101A EPSPSs were 7.2 and 0.1 s/μ*M*, respectively, indicating that the G101A variant is about 72 times less efficient than the wild-type enzyme. However, the relative catalytic efficiency in the presence of glyphosate, $k_{cat} \cdot appK_i$(glyphosate)/ $appK_m$(PEP), indicates that the G101A variant is about 69 times more efficient than the wild-type enzyme in the presence of glyphosate.

The G101A petunia EPSPS variant and its homologs have diminished glyphosate and PEP binding to the enzyme, albeit to different degrees. If this perturbation was a direct result of the increased steric bulk of the Ala^{101} methyl group as compared to the Gly^{101} hydrogen of the wild-type enzyme, one would predict that increasing the size of the side chain of the amino acid residue at this position would affect both glyphosate and PEP binding to EPSPS. Therefore, the G101S petunia EPSPS variant was constructed (Padgette et al. 1991). Using *E. coli* extracts containing the G101S EPSPS, we were unable to detect any EPSPS activity. This result suggested that the G101S substitution may have significantly affected the binding of PEP to the enzyme. If this were the case, S3P binding should remain intact, although the enzyme is catalytically inactive. Indeed, equilibrium dialysis measurements demonstrated that S3P binding remains unaltered in the G101S variant. Based on fluorescence emission and equilibrium dialysis analyses, however, the G101S·S3P complex was unable to bind glyphosate. The results obtained for the G101S variant, in combination with the increased $appK_m$(PEP) observed for the G101A variant, indicate that the Gly^{101} region of petunia EPSPS (and the analogous region in other EPSPSs) is at the binding sites of glyphosate and PEP, and, by inference, the PEP and glyphosate of EPSPS either overlap or are extremely close to one another.

Transgenic tobacco, soybean, canola, and tomato were generated expressing several different EPSPSs with the substitution corresponding to G96A in *E. coli* EPSPS. These plants demonstrated much higher levels of glyphosate tolerance than plants expressing the wild-type petunia EPSPS. However, field tests performed in the late 1980s indicated that yield reductions were evident in the plants after treatment with approximately two times the anticipated weed control use rate of glyphosate. Because yield reductions upon glyphosate treatment are unacceptable commercially for glyphosate-tolerant crops, these results indicated that improved approaches would be necessary to deliver commercially acceptable levels of tolerance. One approach was the identification of new glyphosate-tolerant EPSPS enzymes with appK$_m$(PEP) constants near that of wild-type plant EPSPS (discussed below).

5.4. RATIONALE FOR REDUCING THE APPK$_M$(PEP) IN GLYPHOSATE-TOLERANT EPSPSs

The rationale for developing glyphosate-tolerant EPSPSs with reduced appK$_m$(PEP) constants for glyphosate-tolerant crop production is based on the fact that the decreased interactions with the PEP substrate will reduce the EPSPS reaction rate at low PEP levels in the chloroplasts. This is because the endogenous EPSPS will be inhibited by glyphosate (upon glyphosate treatment) and the plant will have to rely on the glyphosate-tolerant EPSPS for EPSP synthesis. The only way to increase the amount of EPSP formed by an EPSPS with an elevated appK$_m$(PEP) in this case would be to increase the expression level of the introduced EPSPS in the plant. This is illustrated for the petunia G101A variant in Figure 4A, utilizing the kinetic constants previously reported (Padgette et al. 1991). For this variant, due to the increased appK$_m$(PEP) and the reduced V_{max}, relative to the wild-type EPSPS, it can be seen that the enzyme would have to be overexpressed 43-fold relative to endogenous EPSPS to achieve the same EPSP production under conditions of 20 μ*M* PEP and 500 μ*M* glyphosate. It is also evident that as the PEP level goes higher (above the appK$_m$[PEP]), less overexpression is required. Because it is not feasible in most cases to achieve expression levels *in planta* of introduced proteins to the levels predicted to be required at low PEP concentrations (Figure 4A), the only real solution to ensure sufficient EPSP production at low PEP and high glyphosate levels is to utilize glyphosate-tolerant EPSPSs with much lower appK$_m$(PEP) constants.

5.5. SECOND-SITE VARIANTS OF G101A PETUNIA EPSPS

In order to attempt to identify glyphosate-tolerant EPSPSs with appK$_m$(PEP) constants lower than that of the G101A variant, a random mutagenesis program was employed at Monsanto in which the petunia EPSPS cDNA was mutagenized with EMS, ENU, NG, or HA in *E. coli* (Ruff et al. 1991). These chemical mutagenesis treatments were performed with actively growing cultures of *E. coli* harboring plasmids containing a cloned petunia EPSPS coding sequence that had been engineered for *E. coli* expression. Both the wt and the G101A coding sequences were utilized.

Plasmid DNA was purified from the *E. coli* host and retransformed back into *E. coli*, and glyphosate-resistant colonies were selected on minimal medium containing glyphosate (1° screen). The wt EPSPS coding sequence was expressed from a moderately strong constitutive promoter and selections were performed on 20 m*M* glyphosate. The G101A variants were expressed from a very weak constitutive promoter and selections were performed on 5 or 10 m*M* glyphosate. The glyphosate-resistant colonies from the 1° screen were pooled and grown in a nonselective medium to enrich for the variants. The plasmid DNA was again isolated and the petunia EPSPS coding sequence was excised and cloned into a new vector backbone. *E. coli* was then transformed with the ligation mixture and glyphosate-resistant colonies were selected on minimal medium containing glyphosate (2° screen). This 2° screen eliminated any possible substitutions in the original plasmid, outside of the EPSPS coding region, that might influence or mimic the glyphosate-resistant phenotype, such as promoter strength or plasmid copy number substitutions. These recloned coding

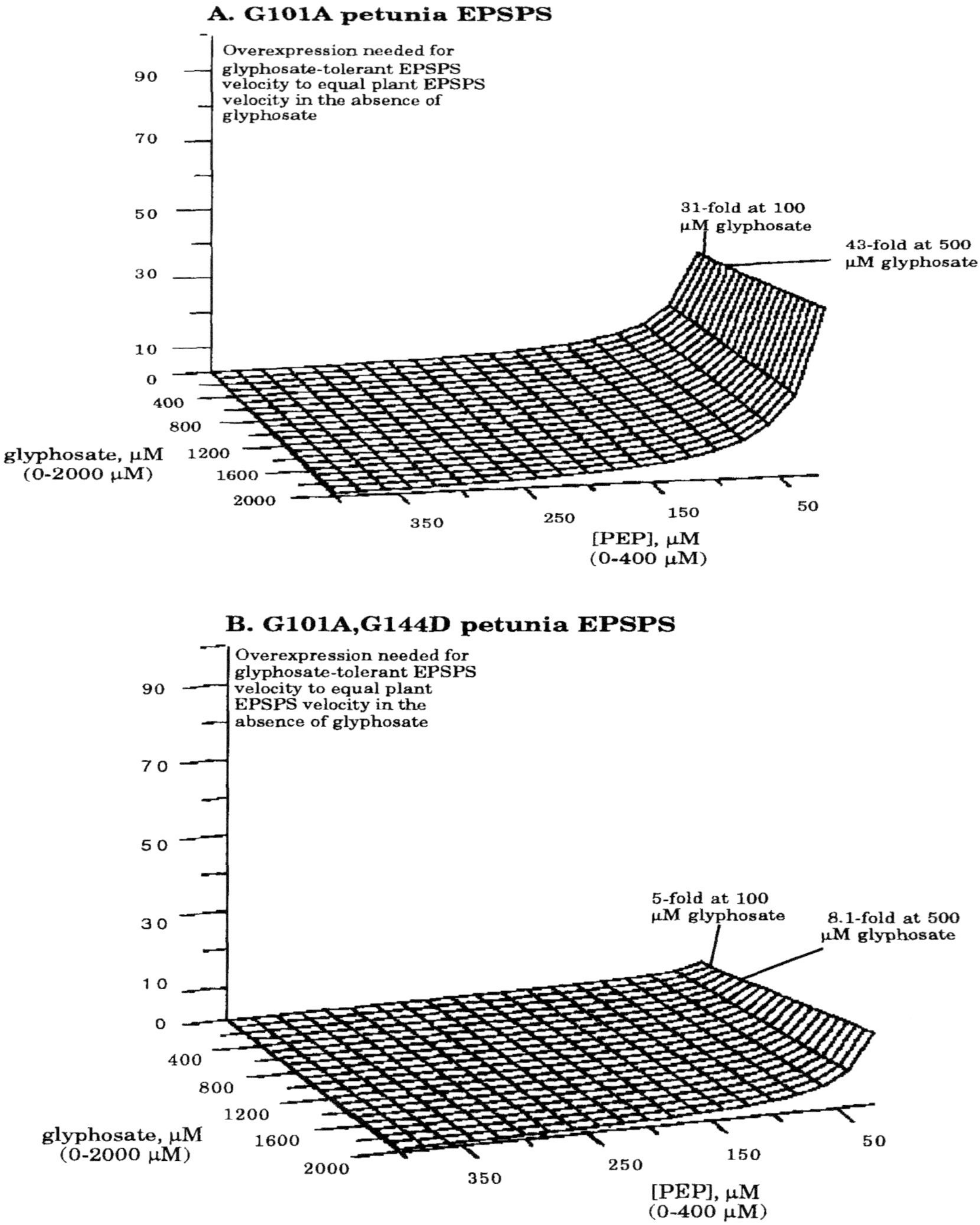

FIGURE 4 Kinetic modeling of glyphosate-tolerant EPSPSs. The plots indicate the influence of different PEP and glyphosate concentrations on the EPSPS reaction rate, for EPSPSs with different appK_i(glyphosate) and appK_m(PEP) constants. The *y* axis represents the fold-overexpression *in planta* needed for the introduced glyphosate-tolerant EPSPS relative to the endogenous plant EPSPS, in order for the glyphosate-tolerant EPSPS reaction rate to equal the endogenous EPSPS reaction rate without glyphosate present under the conditions displayed. At low PEP and high glyphosate concentrations, the glyphosate-tolerant EPSPS reaction rates are lower, causing more expression to be required to get similar reaction rates. (A) G101A petunia EPSPS. The appK_i(glyphosate) is 2.0 m*M* and appK_m(PEP) is 210 μ*M*. (B) G101A, G144D petunia EPSPS. The appK_i(glyphosate) is 0.33 m*M* and appK_m(PEP) is 47 μ*M*.

sequence variants were then screened for effects on growth rates in liquid cultures containing various amounts of glyphosate (3° screen). Variants in which the cell growth was virtually unaffected by high concentrations of glyphosate were then further characterized by enzyme assay and kinetic characterization.

Both EMS and ENU mutagenesis of the wt coding sequence produced 1168 Glp^R (glyphosate resistant) colonies of 9,942,414 selected in the 1° screen. These colonies were pooled, the plasmid DNA was isolated, and the EPSPS coding sequences were subcloned and reselected on glyphosate. This 2° screen advanced 2446 Glp^R coding sequence variants out of 29,920 transformants selected. A total of 250 of the coding sequence variants were then characterized further in liquid screen, and 15 of these were advanced for kinetic characterization. One single site variant isolated, T102I (petunia EPSPS numbering), had an $appK_m$(PEP) of 32 μM, and an $appK_i$(glyphosate) of 21 μM. However, even after extensive mutagenesis of the petunia EPSPs cDNA, no glyphosate-tolerant single-site variants were obtained with $appK_i$(glyphosate) values over 50 μM. These were not pursued for *in planta* expression because at high glyphosate levels, the predicted overexpression was higher than desired (using analyses similar to those in Figure 4).

EMS, ENU, NG, and HA mutagenesis of the G101A backbone resulted in 14,945 Glp^R colonies in the 1° screen out of 116,588,600 transformants selected on glyphosate. The 2° screen advanced 3038 Glp^R coding sequence variants out of 166,200 transformants selected. A total of 647 coding sequence variants were characterized in liquid culture assays, and 39 of those were advanced for kinetic characterization. In addition, glyphosate-resistant coding sequence variants from the NG and HA mutagenized GA backbones were also selected on glyphosate, thereby bypassing the 1° screen. This direct selection resulted in 132 Glp^R coding sequence variants out of 583,000 transformants selected on glyphosate. A total of 120 directly selected variants were then characterized in liquid cultures, and 36 of them were advanced for kinetic characterization.

Utilizing these schemes, second-site variants of the petunia EPSPS G101A variant were identified that had reduced $appK_m$(PEP) constants relative to the G101A enzyme. Three of these double variant petunia EPSPS enzymes were the G101A,G144D EPSPS; the G101A,P167 EPSPS; and the G101A,A192T EPSPS. As shown in Table 2, the $appK_m$(PEP) constants of these three enzymes were indeed reduced from that of the G101A EPSPS progenitor (210 μM), exhibiting values of 47, 35, and 54 μM, respectively. However, as shown in Table 2, the $appK_i$(glyphosate) was also reduced similarly, to 328 to 670 μM, relative to 2000 μM in the G101A progenitor. Therefore, the glyphosate/PEP selectivity of the active site, measured as the $appK_i$(glyphosate)/$appK_m$(PEP) ratio, did not change significantly for the double variants (12.4 for the G101A,A192T EPSPS vs. 9.5 in the progenitor G101A enzyme). A significant improvement of *in planta* EPSPS flux is expected under conditions of low PEP and medium glyphosate due to the reduction in $appK_m$(PEP) for these reduced $appK_m$(PEP) glyphosate-tolerant EPSPSs (Figure 4B). For instance, at 20 μM PEP and 500 μM glyphosate, the G101A,G144D enzyme is predicted to require 8.1-fold overexpression (relative to endogenous EPSPS) to maintain the wt EPSP production as compared to 43-fold overexpression for the G101A progenitor. Therefore, the double variants isolated were predicted to

TABLE 2 Kinetic Characteristics of Several EPSPSs

EPSPS Source	$appK_m$(PEP) (μM)	$appK_i$(glyphosate) (μM)	$appK_i$(glyphosate)/ appKm(PEP)	Ref.
E. coli	8	0.16	0.02	Anderson et al. 1988b,1990
Petunia	5	0.4	0.08	Padgette et al. 1987
Petunia G101A	210	2000	9.5	Padgette et al. 1991
Petunia G101A,G144D	47	330	7.0	Ruff et al. 1991
Petunia G101A,P167S	35	238	6.8	Ruff et al. 1991
Petunia G101A,A192T	54	670	12.4	Monsanto unpublished results
Anabaena variabilis	34	350	10.3	Powell et al.1992

perform better *in planta* than G101A when PEP levels are low, as long as glyphosate levels are not extremely high (because the double variants have a reduced $appK_i$(glyphosate) relative to G101A EPSPS (Figure 4B).

5.6. LOCATION OF GLYPHOSATE-TOLERANCE SUBSTITUTIONS IN THE *E. COLI* EPSPS STRUCTURE

Utilizing the *E. coli* EPSPS X-ray structure (Color Plate 2*), we can locate the amino acid residues corresponding to the second-site substitutions discussed above relative to the active site of the enzyme. The Gly^{96} residue and the so-called "conserved region" are labeled, as are several key active site residues discussed above (Arg^{27}, Lys^{22}, Arg^{100}, and Arg^{124}). The presence of the Gly^{96} residue in this region, and thus the glyphosate binding site, allows one to conclude that glyphosate binds at or very close to the active site of EPSPS. This is in contrast to herbicides that inhibit ALS, which have been shown to bind at a regulatory site distinct from the active site (Subramanian et al. 1991). It is clear, however, that glyphosate does not bind in a manner identical to that of PEP. If this were the case, PEP binding would be reduced comparably to glyphosate binding for the variant EPSPSs described above, relative to the wt EPSPSs. In all of the plant EPSPS variants studied, a decrease in glyphosate binding (measured as an increase in $appK_i$[glyphosate]) always leads to an loss of PEP binding (measured as an increase in $appK_m$[PEP]). There may be an overlap between the PEP and glyphosate sites (perhaps sharing residue[s] interacting with the phosphate/phosphonate moieties), or the two sites may be adjacent to one another and both proximal to the Gly^{96} residue. The exact mode of glyphosate and PEP binding and the extent of overlap of their binding sites will not be completely understood until additional X-ray crystallography results are obtained for EPSPS complexes containing glyphosate. Whatever the relationship between the two sites turns out to be, the mutagenesis data demonstrate that generation of glyphosate-tolerant plant EPSPSs with unperturbed PEP binding interactions would be extremely unlikely.

It is difficult to fully explain the effects of the $appK_m$(PEP) and $appK_i$(glyphosate) lowering effects of the second-site substitutions. The three *E. coli* residues corresponding to the sites of the G144D, P167S, and A192T second-site substitutions in petunia EPSPS (positions 137, 158, and 183 of *E. coli* EPSPS) are spatially located far from the active site and the Gly^{96} residue (Plate 2). However, all three are located on adjacent strands of the protein in the upper domain, and these strands all "feed" into the active site. Clearly, no direct steric interaction could impact binding from that distance, so the effects seen with the second-site substitutions must be subtle and perhaps transferred through the protein strands to the PEP/glyphosate site.

An important point can be made from the extensive research undertaken to generate glyphosate-tolerant EPSPSs with lowered $appK_m$(PEP) constants: Out of the millions of transformants generated by the mutagenesis schemes described above, not a single glyphosate-resistant variant was obtained that had a near-normal $appK_m$(PEP) ($<20\ \mu M$). The petunia EPSPS mutagenesis was, however, more successful in generating glyphosate-resistant EPSPSs than the whole-plant *Arabidopsis* mutagenesis described above. This may be due not only to the increased number of mutagenesis "events" generated, but also because resistant EPSPSs generated *in planta* may not be sufficiently robust to supply aromatic amino acid demand due to the increased $appK_m$(PEP) that accompanies glyphosate tolerance in all glyphosate-tolerant plant EPSPSs identified to date. This observation may contribute to the lack of glyphosate-resistant weeds in nature.

Several of the double variant petunia EPSPS enzymes were introduced into plants (with the CTP), and evaluated for glyphosate tolerance. Modest but measurable improvements in tolerance relative to plants expressing the glycine to alanine variants were observed in canola and soybean, but the average level of tolerance obtained was still less than desirable. Therefore, new approaches were initiated for identifying EPSPSs with low $appK_m$(PEP) and high $appK_i$(glyphosate).

---

* Color plates follow page 84.

5.7. NATURALLY OCCURRING GLYPHOSATE-TOLERANT EPSPSs

It was reported that EPSPSs from a number of bacteria exhibit tolerance to glyphosate (Schulz et al. 1985). Therefore, bacterial cultures from diverse sources were collected and analyzed for their EPSPS appK_m(PEP) and appK_i(glyphosate). The target kinetic parameters for this screen was an EPSPS with an appK_m(PEP) less than ~15 μM and an appK_i(glyphosate)/appK_m(PEP) ratio of >100. One type of bacteria represented heavily in the screen was glyphosate-degrading bacteria (D. Re, G. Barry, M. Weldon, S. Padgette, and G. Kishore, unpublished results), the rationale being that perhaps organisms that can grow in the presence of glyphosate and degrade the herbicide may also express naturally glyphosate-tolerant EPSPSs.

The EPSPS with the highest tolerance to glyphosate found in the screen was CP4 EPSPS. CP4 EPSPS has extremely high glyphosate tolerance (K_i[glyphosate] = 2.7 mM) and tight binding of PEP (appK_m[PEP] = 12 μM) (Barry et al. 1992). Note again that the appK_m(PEP) and appK_i(glyphosate) for the wt petunia EPSPS are 5 and 0.4 μM, respectively (Padgette et al. 1987). The appK_m(PEP) of the highly glyphosate-tolerant CP4 EPSPS is only approximately twofold greater than the wt plant EPSPS. The constant k_{cat} · appK_i(glyphosate)/appK_m(PEP), which is a measure of the catalytic efficiency of the EPSPS in the presence of glyphosate, is approximately tenfold higher for CP4 EPSPS than for the petunia G101A EPSPS, and results primarily from the 17-fold reduction in appK_m(PEP) for the CP4 enzyme relative to G101A petunia EPSPS. Shown in Figure 5 is a plot of the increases in EPSPS enzyme efficiency in the presence of glyphosate obtained in the Monsanto glyphosate-tolerant EPSPS program. It should be noted that an EPSPS from the cyanobacterium *Anabaena variabilis* was reported to have a K_m(PEP) of 34 μM and a K_i(glyphosate) of 350 μM, giving a K_i/K_m of 10.3 (Powell et al. 1991, 1992) (similar to the kinetic constants of the double variant petunia EPSPSs in Table 2).

Based on the kinetic parameters, and thus the suitability for use in conferring glyphosate tolerance to crops, the gene for CP4 EPSPS was cloned from *Agrobacterium* sp. strain CP4 (S. Padgette, G. Barry, D. Re, D. Eichholtz, M. Weldon, and G. Kishore, manuscript in preparation). A cosmid bank was constructed in *E. coli* using DNA from *Agrobacterium* sp. strain CP4, and cosmids from the resistant isolates were tested for complementation of the *AroA* allele and growth in the presence of inhibitory concentrations of glyphosate. Specific oligonucleotide probes, synthesized based on peptide sequences derived from purified CP4 EPSPS, were utilized in the cloning of the EPSPS from one of the cosmids imparting glyphosate tolerance. Based on the nucleotide sequence, CP4 EPSPS is a 47.6-kDa protein consisting of a single polypeptide of 455 amino acids. The identification of codons in the gene encoding four peptide sequences obtained directly from the

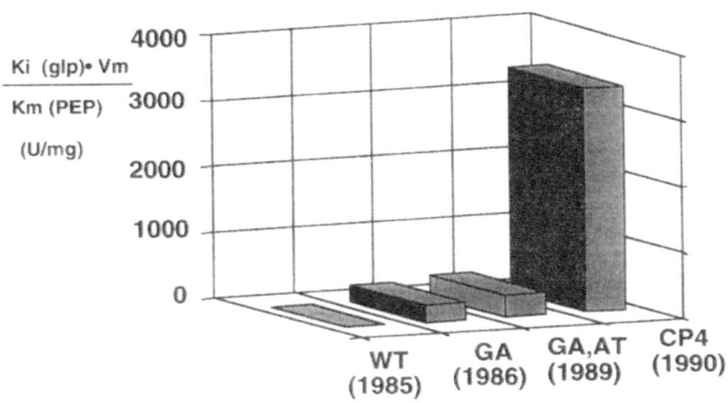

FIGURE 5 *Improvements in efficiencies of glyphosate-tolerant EPSPSs. WT = wild-type petunia EPSPS; GA = G101A petunia EPSPS; GA,AT = G101A, A192T petunia EPSPS; and CP4 = CP4 EPSPS.*

purified enzymatically active CP4 EPSPS conclusively demonstrated that the gene cloned was the EPSPS gene from *Agrobacterium* sp. CP4.

Comparing the deduced amino acid sequences of CP4 EPSPS with EPSPS from soybean, corn, petunia, *E. coli, Bacillus subtilis*, and *Saccharomyces cerevisiae* (Baker's yeast) yields similarities of 51.2, 48.5, 50.1, 52.2, 59.3, and 53.5%, respectively, and identities of 26.0, 24.1, 23.3, 26.0, 41.1, and 29.9%, respectively. The amino acid sequence homology between the CP4 EPSPS and EPSPSs typically present in plants and bacteria (48.5 to 59.3% similar, 23.3 to 41.1% identical) is comparable to the homology between the EPSPSs from, for instance, soybean and *B. subtilis* (55.6% similar, 30.1% identical). This indicates that there is considerable divergence in the EPSPSs that are typically present in foods, and that the divergence of the CP4 EPSPS sequence from typical food EPSPS sequences is of the same order.

Several residues that were identified as important for EPSPS function are also conserved in CP4 EPSPS. The residues include Lys^{28} (which corresponds to Lys^{22} in *E. coli* EPSPS), Arg^{33} (which corresponds to Arg^{27} of *E. coli* EPSPS), and Arg^{128} (which corresponds to Arg^{124} of *E. coli* EPSPS). One striking feature of the CP4 EPSPS primary structure is the presence of an Ala at position 100, which corresponds to Gly^{96} of *E. coli* EPSPS (and Gly^{101} of petunia EPSPS), and was discussed in detail above. The CP4 EPSPS must have subtle changes in the active site residues which sets it apart from the other EPSPSs with alanine at this position, with respect to PEP and glyphosate interactions.

The CP4 EPSPS coding sequence, in preparation for expression in plants and to test for the effectiveness of this EPSPS in imparting glyphosate tolerance, was fused to CTP coding sequences to target the protein to the plastids. The CTPs were derived from the EPSPSs of either Arabidopsis (Klee et al. 1987) or *Petunia* (Shah et al. 1986) and were fused from the normal processing point directly to the fMet of the CP4 EPSPS; this fusion is predicted to liberate the CP4 EPSPS cleanly upon processing in the plastid. Further information on *in planta* results with CP4 EPSPS is given following the discussion of the glyphosate-degrading enzyme GOX.

6. METABOLIC INACTIVATION APPROACH: GLYPHOSATE OXIDOREDUCTASE (GOX)

In addition to focusing on target site modification in the development of crops tolerant to glyphosate, the metabolic inactivation approach was also investigated. Many phosphonates may be metabolized via the C–P lyase pathway, in which the C–P bond is cleaved. Examples of this include the metabolism of AMPA, ethyl- and methylphosphonate to methylamine, methane, and ethane, respectively, in certain *E. coli* strains (Zeleznick et al. 1963; Daughton et al. 1979; Wackett et al. 1987b), and glyphosate to sarcosine and pyrophosphate in *Pseudomonas* sp. PG2982 (Shinabarger et al. 1986; Kishore et al. 1987). Studies of the metabolism of phosphonates were limited to whole-cell analyses due to the lack of demonstration *in vitro* of C–P lyase activity in any organism (McMullan et al. 1991). More recently, genetic and molecular approaches were applied to the study of this metabolic route in *Pseudomonas* PG2982 (G. Barry, K. Fincher, S. Bolten, F. Perlak, M. Weldon, D. Fischhoff, and G. Kishore, unpublished results) and in *E. coli* (Avila et al. 1987; Loo et al. 1987; Wackett et al. 1987a; Chen et al. 1990; Makino et al. 1991). The complexity and number of gene products involved in the C-P lyase operon and the current inability to identify those important for enzyme activity suggest that the use of this pathway to impart glyphosate tolerance would require much more additional information (Barry et al. 1992).

The glyphosate–AMPA pathway appears to be the predominant degradation route for glyphosate in soil (Rueppel et al. 1977; Nomura et al. 1977; Torstensson 1985) and has been identified in a number of Gram-negative and Gram-positive bacteria (Pipke et al. 1988; Jacob et al. 1988). The enzyme activity has not been demonstrated in cell-free extracts. In Monsanto studies of this pathway efficient glyphosate-degrading bacteria were first isolated from a collection of putative glyphosate–AMPA bacteria (Hallas et al. 1988). The *Achromobacter* sp. strain LBAA was chosen for further study (G. Barry, M. Taylor, L. Hallas, and G. Kishore, unpublished results). This microorganism is capable of using glyphosate as a carbon or phosphorus source or as a source for both requirements. Figure 6 shows the two pathways of glyphosate degradation.

A direct approach to the cloning of the gene(s) responsible for the glyphosate–AMPA activity approach was taken by Barry and co-workers (1992) and was based on the ability of certain *E. coli* strains (Mpu$^+$; methylphosphonate utilizing) (Wackett et al. 1987b; Chen et al. 1990; Makino et al. 1991) to utilize AMPA or other phosphonates as phosphorus sources through the action of C–P lyase. A cosmid library of LBAA genomic DNA was screened in an Mpu$^+$ *E. coli* host on the basis of growth on glyphosate as a phosphorus source. The ability to use glyphosate was linked to tolerance to glyphosate. DNA sequencing of subclones revealed a single open reading frame, and the predicted translation product showed little homology to other proteins, except for a region near the N terminus that showed a conserved motif associated with flavin binding in flavoenzymes such as D-amino acid oxidases and sarcosine oxidases (Isogai et al. 1990; Suzuki et al. 1992). The coding region was expressed to high levels in *E. coli* in a number of vectors and the extracts were assayed for glyphosate–AMPA activity. The enzyme, GOX, catalyzes the cleavage of the C–N bond of glyphosate, yielding AMPA and glyoxylate as the reaction products (S. Padgette, M. Taylor, T. Huber, G. Barry, and G. Kishore, unpublished results) (Figure 6). The GOX from LBAA, which utilizes oxygen as a cosubstrate, does not produce hydrogen peroxide as a product of oxygen reduction and appears to be a flavoprotein using FAD as a cofactor.

Glyphosate moves rapidly away from the site of application to the plant, primarily the leaf, and accumulates in the growing meristematic regions (Gougler et al. 1981). A strategy to utilize GOX to impart glyphosate tolerance in plants might require high and constitutive expression in the vegetative tissues and perhaps also in the meristematic tissues. Early attempts at expression of GOX in tobacco resulted in plants expressing the protein only at very low levels. Modification of the coding sequence and ultimately a complete resynthesis of the coding region resulted in increasing GOX expression (G. Barry, K. Kolacz, M. Taylor, S. Padgette, G. Kishore, unpublished results). The expression of GOX imparted glyphosate tolerance to these plants.

The target of glyphosate action in the plant, the EPSPS enzyme, is located within the chloroplast. The effectiveness of GOX in imparting glyphosate tolerance might be improved through directing the GOX protein also to this organelle. Fusions of a CTP derived from that of the SSU1a gene from *Arabidopsis thaliana* (Timko et al. 1988) and GOX were made, tested *in vitro* for import into isolated chloroplasts (della-Cioppa et al. 1987), and introduced into plants. The steady-state level of GOX protein was up to ten-fold higher than that obtained previously with the cytosolic versions; whether this reflects greater stability of the protein within the plastid or a greater overall

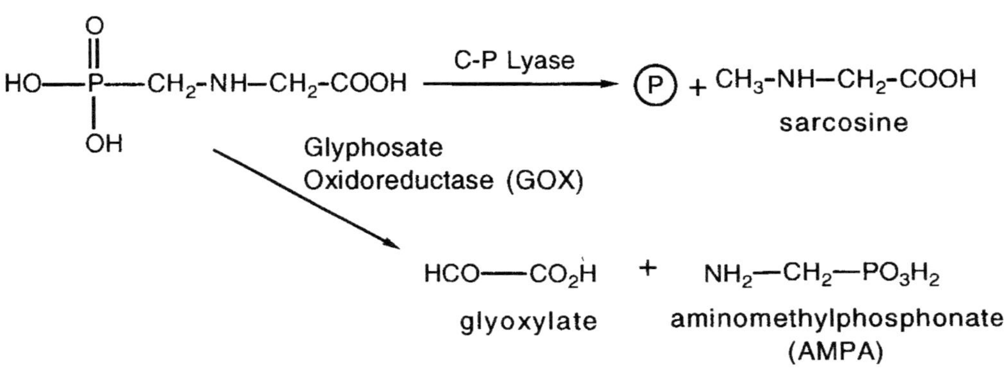

FIGURE 6 Glyphosate degradation pathways.

expression level has not been determined. Tobacco plants expressing the plastid targeted GOX (CTP1-GOX) were sprayed at different rates of glyphosate and were scored for both vegetative and reproductive tolerance. All of the lines showed some level of glyphosate tolerance and in some cases the treatment seemed to have little if any effect, even at the higher spray rates. The reproductive tolerance was also outstanding for many of these lines. The CTP1-GOX fusion was introduced into other crop plants, and under field conditions, GOX provides excellent vegetative and reproductive tolerance (Barry et al. 1992).

7. RECENT RESULTS IN SOYBEANS WITH A ROUNDUP READY™ GENE

7.1. LINE 40-3-2

The lead progenitor soybean line with a Roundup Ready™ gene developed by Monsanto is denoted "40-3-2". The construction and field performance of this soybean line has been described below (Padgette et al. 1995a; Delannay et al. 1995). Soybean line 40-3-2 expresses only the CP4 EPSPS gene product; no metabolic inactivation (GOX) gene is contained in this line.

The CP4 EPSPS gene was engineered for plant expression by fusing the 5′ end of the gene to a CTP sequence derived from petunia EPSPS (Shah et al. 1986; Gasser et al. 1988; della-Cioppa et al. 1986). *In vitro* chloroplast uptake assays verified that the petunia EPSPS CTP indeed delivered CP4 EPSPS to the chloroplasts (M. Taylor, G. Barry, S. Padgette, unpublished results). After the "pre-CP4 EPSPS" protein (containing the CTP amino-terminal extension) reaches the chloroplast or plastid stroma, the CTP is cleaved and degraded, as are the CTPs from other naturally occurring nuclear-encoded chloroplast-targeted proteins (Bartlett et al. 1982). This leaves the "mature" CP4 EPSPS, with no CTP sequences retained, as the CP4 EPSPS species present *in planta*. We have demonstrated, by purification of CP4 EPSPS from seeds of soybeans expressing the CP4 EPSPS gene, that indeed no CTP-derived sequences are present at the amino-terminus of the soybean-produced CP4 EPSPS protein (Harrison et al. 1995).

The plasmid PV-GMGT04, used to transform the parental soybean line to generate line 40-3-2, contains three genes driven by plant promoters: two CP4 EPSPS genes and a gene encoding GUS from *E. coli* (scorable marker). Introduction of DNA into soybean tissue by the particle acceleration method (particle gun) was described by McCabe et al. (1988) and Christou et al. (1988). The shoots that developed from the transformed cells demonstrated the phenotype expected. The expression of the GUS protein was used as evidence of transformation. The positive shoots were then grown to maturity, and the resulting plants were screened for glyphosate tolerance and gene expression.

Line 40-3-2 is an R_1 (first progeny generation) selection from an original R_0 transformant, 40-3. R_1 progenies from 40-3 R_1 seed were grown in a greenhouse during the winter of 1990–1991 and evaluated for glyphosate tolerance by spray test. R_2 seeds from individual R_1 plants were planted in the field during the summer of 1991. Characterization of the R_2 progenies indicated that the original 40-3 R_0 plant received two DNA inserts located at different positions in the genome. The first insert was responsible for the expression of the GUS marker protein. The second insert had a strong expression of the glyphosate tolerance trait, but did not express the GUS protein. The 40-3-2 R_2 progeny that were selected exhibited strong glyphosate tolerance, but no GUS enzyme activity, and thus did not contain the other insert, which had been lost through normal genetic segregation. This was indicated by the fact that none of the progenies from line 40-3-2 expressed GUS, based on leaf GUS enzyme assays. Extensive analyses of line 40-3-2 seed and leaves using a validated GUS enzyme-linked immunosorbent assay confirmed that no GUS protein is expressed in line 40-3-2. Subsequent Southern blot analyses have shown that the DNA insert present in line 40-3-2 contains only a single insert of DNA derived from PV-GMGT04, and that this insert contains a portion of the E35S promoter, the petunia EPSPS CTP, the CP4 EPSPS gene, and a portion of the NOS 3′ terminator (Padgette et al. 1995a).

F_2 progenies of crosses between other soybean lines and line 40-3-2 consistently segregate 3 tolerant to 1 sensitive, indicating that the 40-3-2 insert behaves as a single dominant gene inherited in a Mendelian fashion (Padgette et al. 1995). In addition, the gene encoding CP4 EPSPS in line 40-3-2 is stable. The glyphosate tolerance phenotype and Mendelian transmission has been consistent for every generation of line 40-3-2 soybean tested to date, which spans more than nine generations, including field tests harvested in 1994. GTS line 40-3-2 has been field tested in the U.S., Central and South America, Europe, and Canada since 1991 at approximately 300 locations. Data collected from these trials (including yield, agronomic characteristics, vigor, and disease and insect susceptibility), literature references, and expert opinion letters demonstrate that the GTS line 40-3-2:

1. exhibits no plant pathogenic properties
2. is no more likely to become a weed than the unmodified parental varieties
3. has not revealed any potential to increase the weediness of any other cultivated plant or native wild species
4. does not negatively impact processed agricultural commodities
5. has not revealed any potential to harm other organisms that are beneficial to agriculture

A request was made of USDA/APHIS that line 40-3-2 and any progenies derived from crosses between line 40-3-2 and traditional soybean varieties no longer be considered a regulated article under regulations in 7 CFR part 340 (Re et al. 1993). On May 19, 1994, USDA/APHIS, on the basis of an environmental assessment in response to the petition from Monsanto, reached a finding of no significant impact on the environment from the unconfined, agricultural use of GTS line 40-3-2 and its progeny. Granting of this petition by the USDA, however, did not constitute permission for soybeans with a Roundup Ready™ gene to enter commerce in the U.S.; completion of a food and feed safety consultation with the FDA was still required (see below).

7.2. FIELD TOLERANCE AND AGRONOMIC PERFORMANCE OF LINE 40-3-2

Line 40-3-2 and progeny from crosses between line 40-3-2 and other soybean varieties were yield tested under weed-free conditions in 1992 and 1993. Data from these experiments showed that there is no yield penalty observed upon glyphosate treatment of this line with Roundup® herbicide, even at excessive rates (up to 64 oz/acre) (Delannay et al. 1995). In addition, the window of application is very wide; no yield penalty is seen upon glyphosate applications up to and beyond soybean flowering. This attribute will allow growers great flexibility in using Roundup® herbicide for weed control with soybeans with a Roundup Ready™ gene, in addition to early applications. These cases could be when herbicide treatments are triggered by weed appearance, when weather precludes scheduled treatments, or when "rescue" applications are desired. Additional extensive tests using isopopulations and breeding programs have also verified that there is no yield penalty associated with the CP4 EPSPS DNA insert present in line 40-3-2 (Delannay et al. 1995). Shown in Color Plate 3* are field plots of an experimental soybean line expressing the CP4 EPSPS gene.

7.3. CURRENT WEED CONTROL OPTIONS IN SOYBEANS AND IMPACT OF SOYBEANS WITH A ROUNDUP READY™ GENE

Soybeans are grown primarily in rotation with corn and wheat in both midwest and southern states. In the northern midwest the practice is usually a corn/soybean/corn rotation, while in more

* Color plates follow page 84.

southern states, it is commonly a fall seeded wheat/double cropped soybean/corn rotation. With considerably reduced frequency, soybean rotations may include sorghum, tobacco, and cotton.

The number of herbicide applications currently utilized varies in each soybean production region from one to five treatments per growing season. The actual number is dependent upon weed species, population densities, weather, and production economics. In the Corn Belt region soybean producers generally apply two standard treatments plus one spot spray. No-tillage acreage (increasing in the Corn Belt) commonly receives three treatments. In the Delmarva region (Delaware, Maryland, and Virginia), two to three treatments are common. The number of treatments increases when johnsongrass is present. In the southeast, due to the large number of weed genera present, as many as five treatments may be applied. The weed species include a tough annual grass, Texas panicum, tough annual dicots in the Leguminosae family such as Florida beggarweed and sicklepod, bermudagrass (a perennial), and the perennial yellow and purple nutsedges. In the midsouth area severe infestation of the major weed classes (annual grasses and dicots, perennial grasses and dicots) requires three applications per season at minimum to reduce weed interference to acceptable levels. More commonly, four applications plus spot sprays are needed. This is due to the widespread presence of diverse annual species including crabgrass, broadleaf signalgrass, barnyardgrass, cocklebur, several morningglories, prickly sida, spurred annoda, hemp sesbania, and smartweeds. Additionally, johnsongrass is established throughout the area, itchgrass and wild poinsettia are spreading out of Louisiana, and sicklepod has recently become widely distributed in the area. Each selective herbicide registered for use in soybeans and other crops is specific in mode of action and controls only certain weed species. The large number of weed species present, representing diverse botanical families, requires the application of four to five herbicides in different chemical families for effective weed management in soybean. Glyphosate is active at varying levels on all the above, but has not been used in-season in soybeans due to the lack of crop tolerance.

Several postemergence herbicides are registered for use in soybeans. They are usually applied when the crop is 10 to 25 cm or 25 to 38 cm in height. These herbicides include bentazon, acifluorfen, chlorimuron, fomesafen, fluazifop, imazaquin, imazethapyr, lactofen, sethoxydim, quizalofop, and 2,4-DB. Separate applications of the grass-active and dicot-active products are made to obtain the best performance from each. Tank mixtures of two or more dicot-active products are generally used south of the central Corn Belt of the U.S. in order to obtain the needed control spectrum. Directed sprays of some nonselective products are applied in the southern region in order to safely use these more effective products. The label for many commonly used herbicides include precautions on seeding depth, application timing, stage of crop growth, etc., to reduce the risk of soybean injury. Sometimes restrictions are also placed on planting rotational crops to prevent carryover losses from herbicides that detoxify slowly in the field.

Systems of weed management employed in soybeans vary within and between production regions. No one system is so universal as to be considered dominant. The use of soybeans with glyphosate tolerance will provide an attractive alternative to farmers who wish to have additional options for effective weed control. Glyphosate is highly effective against the majority of annual and perennial grasses and broad-leaved weeds, and will provide benefits to the farmer as discussed in the introduction to this chapter. Currently, farmers are using up to four or five different herbicide families to manage soybean weed problems. One, or possibly two, treatments of glyphosate herbicide will control both annuals and perennials, which would reduce the time, cost (herbicide and application), and number of herbicide treatments per acre. Soybeans with glyphosate tolerance will also allow farmers to plant narrow-row (<25 cm) soybeans. University yield data in the north and south indicates a yield increase from planting narrow-row beans. However, good weed control is generally difficult under narrow-row cultural practices. Soybeans with glyphosate tolerance will help farmers plant narrow-row soybeans more efficiently and take advantage of increased yields and decreased costs. The use of glyphosate over-the-top will allow the farmer to reduce the number of herbicides used in soybean weed control, as well as fit well with emerging trends in agriculture, such as reduced tillage using postemergent herbicides. Conservation tillage not only reduces soil erosion, it increases soil moisture and reduces fuel costs due to fewer trips across the fields for cultivation.

7.4. REGULATORY APPROVAL OF SOYBEANS WITH THE ROUNDUP READY™ GENE

In the U.S. the FDA has oversight relative to the food and feed safety of glyphosate-tolerant soybeans. The EPA has oversight relative to the amending of the Roundup® herbicide label to include use on GTS, and the USDA has oversight relative to the agronomic and ecological aspects of the crop. Currently, Roundup® herbicide is labeled by EPA for preharvest (and preemergent) use on soybeans. More information on the food and feed safety studies performed on GTS is discussed below.

8. SAFETY ASSESSMENT OF SOYBEANS WITH A ROUNDUP READY™ GENE

Thorough safety assessments will be required prior to commercial introduction of GTS. These assessments cover environmental safety and food/feed safety, and, as mentioned previously, Roundup® herbicide also must be approved for foliar application on the crop. A summary of the information obtained related to environmental and food/feed safety of GTS is presented below. No further discussion will be given herein on the herbicide approval because it follows standard requirements for herbicide registration. Approval by the EPA for Roundup® herbicide use on Roundup Ready™ soybeans was granted in May, 1995.

8.1. ENVIRONMENTAL ASPECTS OF SOYBEANS WITH A ROUNDUP READY™ GENE

8.1.1. Weediness Potential of the Line 40-3-2

The introduction of a glyphosate tolerance gene to a soybean cultivar is not expected to increase the "weediness" of the plant. A general consensus of the traits common to many weeds was developed by Baker (1974). They include:

1. germination requirement fulfilled in many environments
2. discontinuous germination and great longevity of seed
3. rapid growth through vegetative phase to flowering
4. continuous seed production for as long as growing conditions permit
5. self-compatibility but not completely autogamous and apomictic
6. when cross-pollinated, unspecialized visitors or wind pollinated
7. high seed output in favorable environment and some seed production in a wide range of environments
8. adaption for short- and long-distance dispersal
9. if perennial, vegetative production or regeneration from fragments and brittleness (so not easily removed from the ground)
10. ability to compete interspecifically by special means (rosette formation and presence of allelochemicals)

Not all weeds have all of these characteristics. Soybean possesses few of the characteristics of plants that are notably successful weeds. It is an annual crop which is considered to be a highly domesticated, well-characterized crop plant that is not persistent in undisturbed environments without human intervention. The soybean cultivar which has been genetically modified is not considered to be a weed, and introduction of the glyphosate tolerance trait into this cultivar has not imparted any new "weedy" characteristics. No increase was noted with the transformed cultivar with respect to the number of seeds produced (yield data), and no changes were noted with respect to the germination characteristics of seeds, final stands, or disease or insect susceptibility. Thus, we concluded that GTSs were equivalent to commercial soybean varieties with respect to weediness potential (Re et al. 1993).

8.1.2. Effects of Glyphosate-Tolerant Soybeans on Nontarget Organisms

Extensive field testing of GTS line 40-3-2 and progeny varieties shows that GTS exhibit no toxicity toward insects, birds, or other species that frequent soybean fields. We have also performed feeding studies using either raw soybeans or soybean meal on cows, rats, chickens, quail, and catfish with no adverse effects noted (see below). EPSPS enzymes are present in plants (including soybeans) and microorganisms, and are therefore ubiquitous in nature and ordinarily present in food and feeds derived from plant and microbial sources. Therefore, no effects on nontarget organisms were expected.

8.1.3. Potential for Outcrossing

The only wild species that could cross with the cultivated soybean are members of the genus *Glycine* (Re et al. 1993). No other genus is closely enough related to soybean to allow for the possibility of outcrossing. The only known perennial species of *Glycine* occur in Australia, the south Pacific Islands, the west central Pacific Islands, mainland China, Papau New Guinea, the Philippines, and Taiwan, where the wild perennial species are endemic. Soybean production in these areas is mostly in China and Australia. There are no known reports of successful natural hybridization between the cultivated soybean and the wild perennial species. Thus, the possibility of gene transfer is nonexistent because hybridization does not occur without *in vitro* seed culture (R. Palmer, USDA, personal communication). Even in those cases, the F_1 plants obtained are generally sterile.

The wild annual species, *G. soja* is found in China, Taiwan, Japan, Korea, and the former USSR. Natural hybridizations between *G. soja* and the cultivated soybean occurs (Kwon 1972). In fact, the semi-wild form, intermediate in many phenotypic traits between *G. max* and *G. soja*, has been recognized as *G. gracilis* (Skvortzov 1927). *G. soja* is not native to North America and occurs only in research plots. There are no reports of its escape or dispersal from research plots. *G. soja* has never been found as a weed or naturalized in the U.S. Thus, the possibility of gene transfer is very low within the U.S.

Hybridization among cultivated soybeans is generally <1%. Insect activity does increase the outcrossing rate, but soybeans are not a preferred plant (Erickson 1984). Male-sterile, female-fertile mutants are used in breeding studies, but it is very unlikely that chance pollination with transformed soybeans would occur. In any case, soybean seeds generally do not survive the winter, and soybean does not establish itself as a volunteer weed in other crops. Even if it did, the plants could easily be controlled by current herbicides which are presently active against soybean.

8.1.4. Likelihood of the Appearance of Glyphosate-Resistant Weeds and Volunteer Soybeans

Several decades ago, herbicide-resistant weeds were virtually unknown. Today there are some 109 herbicide resistant weed biotypes, with over half of them resistant to triazines (LeBaron 1991). Major factors which can contribute to the development of resistant weeds include: a single target site and a specific mode of action, broad spectrum of activity, long residual activity and the capacity to control weeds year-long, and frequent applications without rotation to other herbicides or cultural control practices. Using these criteria and based on current use data, glyphosate is considered to be a herbicide with low risk for weed resistance (Benbrook 1991). Glyphosate has been used for over 20 years in various preplant, directed, spot or postharvest weed management systems with no known reports of weed resistance (Holt et al. 1993). This is most likely due to biological and chemical properties demonstrated by glyphosate and the use patterns of the herbicide. Glyphosate essentially has no residual activity in the soil and is relatively quickly broken down by microorganisms in the soil. Also, there is no other herbicide on the market today that has the same mode of action as glyphosate.

Another factor which may play a role in lack of glyphosate-tolerant weeds is the fact that substitutions which impart high-level glyphosate tolerance to EPSPS are rare, as described in detail above. Even more rare are glyphosate-tolerant EPSPSs which are efficient with respect to PEP binding, as discussed above. In fact, only the CP4 EPSPS comes close to normal EPSPS catalytic efficiency. The data obtained to date indicate that it is extremely difficult to disrupt glyphosate binding without negatively impacting PEP binding. Taking into account the extended history of glyphosate use without development of weed resistance, as well as the *Arabidopsis* and EPSPS gene mutagenesis results, it can be postulated that any mutations in weed EPSPSs which significantly decrease glyphosate binding (leading to glyphosate-resistant weeds) also are deleterious to PEP binding and harmful to the weed. This lack of "fitness" for glyphosate-tolerant weeds resulting from a modified target site, should they occur, could account for the lack of glyphosate-resistant weeds. On the other hand, it could be argued that EPSPS is not the rate-limiting step in aromatic amino acid biosynthesis, and thus an increase in $appK_m$(PEP) may not be "lethal" to the weed. Although this argument is rational, the impact of an elevated $appK_m$(PEP) is highly dependent on the chloroplastic PEP level. If the PEP level is extremely low in the weed (or low at a critical growth stage), and a mutated glyphosate-tolerant EPSPS in the weed had the expected elevated $appK_m$(PEP), then flux through the shikimate pathway will be negatively impacted and a negative impact on the fitness of the weed could be expected. One important point is that glyphosate-tolerant crops will have utility themselves as an effective management tool to reduce the potential for further spread of herbicide resistance among new weed biotypes or in other herbicide families. Glyphosate is already a component of the weed resistance management strategy for herbicides, and the ability of the farmer to utilize Roundup® herbicide in-crop for weed control will provide a new chemical mode of action to utilize in a weed resistance management strategy.

Experts tend to agree that eventually one will see a shift in weed populations due to the use of glyphosate in soybeans; however, this would occur with any new herbicide use. In fact, any significant change in weed management systems will cause a shift in weed species, but usually these shifts cannot be related to a single variable (combination of tillage, rotation, herbicides, etc.).

Finally, soybeans have no innate dormancy, therefore, overwintering is rare. Due to this lack of dormancy (selected for in commercial soybean seed), soybean seeds germinate quickly with adequate temperature and moisture, so all seed that might shatter and/or fall to the ground due to harvest losses eventually will germinate, emerge, and be killed by frost during the fall/early winter of the year that they were produced. Even in the unlikely case that some seed did survive, there are several other methods that can be used to remove GTS. The genes inserted into soybeans to provide glyphosate tolerance, whether CP4 EPSPS (as in the case of line 40-3-2), GOX, or both, confer tolerance only to glyphosate. Thus, any herbicide which currently controls soybeans also controls GTS, except glyphosate. All glyphosate-tolerant soybean field sites have been monitored for 1 year after harvest for volunteer soybeans, and very few, if any, volunteers have been noted.

8.2 FOOD AND FEED SAFETY: SOYBEANS WITH A ROUNDUP READY™ GENE AS A CASE STUDY

8.2.1. Soybean-Based Products and Human/Animal Consumption

In order to design a relevant food/feed safety assessment program for GTS, it was crucial to understand the uses of soybeans. There are three major markets for soybean products: beans, oil, and meal. Soybeans yield about 80% by weight protein-rich meal and 18% oil (Soy Stats Reference Guide 1991). The primary use of soybean meal is as a protein supplement for animal feeds. The use of soybean meal in animal feeds accounts for about 97% of the soybean meal produced annually (Horan 1974). An estimate of the livestock feed distribution of soybean meal in the U.S. in 1990 is poultry, 49%; swine, 28%; dairy, 8%; beef, 8%; and other, 7% (Soy Stats Reference Guide 1992).

During the last several decades, the soybean has been developed as a major source of protein. Soybeans are capable of producing the greatest amount of protein per unit of land of any major plant or animal source used as food by people today (Considine et al. 1982). The amino acid profile of soy protein is unusually well rounded for a plant protein (Erdman 1988). Especially important is the soybean's high content of essential amino acids, particularly lysine, leucine, and isoleucine (Coppock 1974). Extensive literature information exists on the nutritional value of soy protein (Liener 1972; Wilke et al. 1979; Wolf 1983). Soybeans naturally contain certain factors, such as trypsin inhibitors, which may act as antinutrients if the soybeans are not properly heated during preparation (Rackis 1974). This is the main reason that virtually 100% of soybean protein products are heated prior to human and most animal consumption.

As mentioned above, in terms of domestic usage of the U.S. soybean crop, only about 3% of the total protein derived from soybean is used in human food (Horan 1974; Mounts et al. 1987). Foods containing soybean protein include bakery products, confections, meat products, textured foods, and nutritional supplements (Waggle et al. 1979). The soybean is the highest natural source of dietary fiber, and soy hulls are processed into fiber bran breads, cereals, and snacks. Soybean oil is extensively used in the food industry in products such as cooking and salad oils, salad dressings, shortening, and oleo margarine. Soybean oil is currently the major edible oil used in the U.S. (Mounts 1988).

8.2.2. Safety Studies

Based on the extensive use of soybeans in feed and food described above, it is clear that the safe use of soybeans has been well established. Thus, the characteristics of soybean per se, or the specific progenitor line utilized for transformation, do not warrant special analytical or toxicological tests. Typically, soybean breeders make genetic crosses to generate new cultivars with enhanced commercial value, and they evaluate new varieties primarily based on yield as well as protein and oil content. However, the FDA Food Policy (U.S. Food and Drug Administration 1992) has recommended that additional key compositional components of genetically modified plant varieties be assessed prior to commercial introduction. We have performed extensive analytical studies to compare the compositional quality of soybeans from line 40-3-2 to the parental control soybean variety (Padgette et al. 1995b).

Over 1400 individual analyses demonstrate that line 40-3-2 soybeans are not materially different from commercially available soybeans in essential nutrients or antinutrients. To assess as thoroughly as possible whether a material difference might exist, we have conducted numerous compositional studies and accessed existing relevant data and information. Upon quantitatively and qualitatively evaluating all of the data available, we have been able to ensure that in all but a few instances, for every parameter there are no statistically significant differences between the 40-3-2 line of soybeans and its parental variety. In those few instances in which a difference was noted and in the cases in which the available data did not permit statistical analysis, we have been able to establish that the values or effects are well within established ranges (or are fully consistent with effects) documented and reported in the scientific literature for soybeans. The strategy taken for measurement of compositional parameters was to focus as much as possible on the raw agricultural commodity, the soybean seed. It is reasonable to infer that if the line 40-3-2 soybean seeds are not materially different from control soybean seeds, then products derived from the line 40-3-2 seeds also will not be materially different from the products derived from control seeds. In order to provide test material for these analyses, line 40-3-2 and the parental variety were grown at nine field locations in 1992 and four field locations in 1993 under Good Laboratory Practice guidelines. Although we focused our analyses on the seed, we also manufactured several selected important soy protein products for additional analysis. Toasted meal was chosen because it is the main soybean protein product used in animal feed. Defatted meal (flour) was made because that is the starting material for a large number of soybean protein products used in food. Protein isolate and concentrate were manufactured from the defatted meal due to the human food use of these two fractions. In addition, crude lecithin and refined, bleached, deodorized oil were manufactured. Therefore, while we focused on the seeds, we also obtained data on several of the most important processed commercial soybean products.

Analyses were performed on both nutrients and antinutrients. The nutrients measured in the soybean seeds included proximates (percentages of protein, fat, fiber, ash, carbohydrates, moisture), amino acids, and fatty acids. Antinutrients (or purported antinutrients) measured in either the seed or the toasted meal were trypsin inhibitor, lectins, phytosterols (genistein, daidzein), stachyose, raffinose, and phytate. These analyses are listed in Table 3. The toasted meal, defatted (nontoasted) meal, protein isolate, and protein concentrate processing fractions were also analyzed for proximate composition, while fatty acid analysis was performed on the refined, bleached deodorized oil. Compositional analysis was performed on the crude lecithin fraction. The results of these analyses showed that line 40-3-2 soybean seeds and processed fractions were not materially different from the control seeds or processed fractions. It is also important to point out that processing (to toasted meal) inactivated trypsin inhibitor and lectin in line 40-3-2, as expected. The levels of these purported antinutrients in line 40-3-2 toasted meal were not materially different from the levels in the control toasted meal.

It is worthwhile to note that there were no statistically significant differences in any amino acid, including aromatic amino acids, between the 40-3-2 seeds and the control seeds (from the nine-site 1992 field study). This is important because EPSPS is an enzyme of the aromatic amino biosynthetic pathway, and one could speculate that additional EPSPS activity could cause an increase in aromatic amino acids. This clearly was not the case for the line 40-3-2 seeds expressing small amounts of CP4 EPSPS activity.

The primary use of soybeans is as a protein supplement for animal feeds, which accounts for almost 97% of the use of the processed protein products. Insertion of the CP4 EPSPS gene into soybeans was not expected to have any effect on wholesomeness, which was supported by the compositional quality analyses summarized above. However, to provide data to support the expectation that line 40-3-2 is equivalent to the parental soybean variety as an animal feed, several animal feeding studies were performed. Feeding studies were conducted on rat, chicken, cow, catfish, and quail (Hammond et al. 1995). These studies were designed to compare the wholesomeness (ability to support typical growth and well-being) of line 40-3-2 to the parental line. No statistically significant difference in feed conversion was shown for line 40-3-2 relative to the parental variety. Thus, based on the extensive compositional and feeding studies, we concluded that no unexpected effects resulted from the insertion of the CP4 EPSPS gene in line 40-3-2.

To assess the expected or intended effects in food and feed of the added CP4 EPSPS protein, levels of the protein were determined in whole seed, toasted meal (feed ingredient), defatted meal,

TABLE 3 Summary of Compositional Analyses Performed on Line 40-3-2

	Soybeans	
Component	**Meal**	**Toasted**
Proximate analysis	NMD	NMD
Amino acid composition	NMD	ND
Fatty acid composition	NMD	ND
Trypsin inhibitors	NMD	NMD
Lectins	NMD	NMD
Phytosterols	NMD	NMD
Stachyose, raffinose	ND	NMD
Phytate	ND	NMD

NMD = not materially different from the control; ND = not determined.

protein isolate, and protein concentrate. The mean expression of CP4 EPSPS in line 40-3-2, based on enzyme-linked immunosorbent assay analysis of seeds from nine field sites (1992), was 0.288 μg/mg tissue fresh weight (~0.08% of the total protein) (Padgette et al. 1995a). In the protein processing fractions manufactured from line 40-3-2 soybeans the level of CP4 EPSPS was no greater than 0.1% of the total protein in any of the fractions. CP4 EPSPS enzymatic activity was not detectable in toasted meal, isolate, or concentrate due to processing, but was measurable in the defatted, nontoasted meal. Based on these results, CP4 EPSPS is expected to be present at low levels in the food and feed derived from these GTS.

The donor of CP4 EPSPS, *Agrobacterium* sp. strain CP4, is not a food source, and therefore the question of its allergenicity as a food is not directly applicable. We conclude that the allergenic potential of soybeans expressing CP4 EPSPS and the products derived from these soybeans is not significantly different than other soybean varieties. This conclusion of lack of allergenic potential for CP4 EPSPS was further addressed from two perspectives. First, the endogenous protein allergens and their levels present in commercially available soybeans were compared to those present in soybeans with the CP4 EPSPS gene to assess whether the composition and relative quantities of these allergens are comparable (Burkes and Fuchs 1995). No differences were observed. Second, the biochemical properties of known allergenic proteins were compared to the CP4 EPSPS protein as an approach to assess the allergenic potential for the CP4 EPSPS (Table 4). It is important to remember in this light that EPSPS proteins are a diverse set of related proteins typically present in foods and feeds derived from plants and microbes and are not known to be commonly allergenic. Although there are no predictive assays available to assess the allergenic potential of proteins (U.S. Food and Drug Administration 1992), the biochemical profile of the CP4 EPSPS protein provides a basis for allergenic assessment when compared to known protein allergens. While CP4 EPSPS at 47,600 Da fits the mass criteria of 10,000 to 70,000 Da, CP4 EPSP synthase possesses none of the other characteristics common to protein allergens. CP4 EPSPS is not heat stable, and all detectable functional activity and enzyme-linked immunosorbent assay reactivity is lost after the processing/toasting procedure. More importantly, CP4 EPSPS was shown to be extremely labile to digestion by the proteases present in the mammalian digestive system (Padgette et al. 1995), which supports the prediction that the CP4 EPSPS protein will not survive the peptic and tryptic conditions of the mammalian digestive system. Because most protein allergens are glycosylated, the CP4 EPSPS, as purified from glyphosate-tolerant soybean seed, was examined for glycosylation and shown not to be glycosylated (Harrison et al. 1995). Because there is no glycosylation of the CP4 EPSPS, it does not fit the profile of some common allergenic proteins.

CP4 EPSPS is functionally similar to the endogenous soybean EPSPS and shows no significant homology to any known protein allergen. CP4 EPSPS and endogenous soybean EPSPSs show 51% similarity and 26% identity in amino acid sequence. In contrast, CP4 EPSPS showed no significant

TABLE 4 Characteristics of Known Allergenic Proteins[a]

Characteristic	Allergens	CP4 EPSPS
Molecular wt 10–70 kDa	Yes	Yes
Glycosylated	Yes[b]	No
Stable to digestion	Yes	No
Stable to processing	Yes	No
Similar to known allergens	—[c]	No
Similar to soybean proteins	—	Yes
Prevalent protein in food	Yes	No

[a] As described in Taylor (1992) and Taylor et al. (1987, 1992).
[b] Typically, but not absolutely.
[c] Implicit for allergenic proteins from soybeans.

homology to any of the 121 amino acid sequences reported for the allergens in the three current protein databases (Harrison et al. 1995) (Genpept, Pir protein, and Swissprot databases). There was no greater homology of the native CP4 EPSPS to any of the 121 amino acid sequences for the allergenic proteins than for a scrambled sequence of the same amino acids that comprise the CP4 EPSPS.

Finally, most food allergens are present as major protein components in the specific food. This is true for the allergens in soybean (Shibasaki et al. 1980; Burks et al. 1988; Pederson *et al.* 1989). In contrast to this generality for common allergenic proteins, CP4 EPSPS is present in soybean seed at low levels, approximately 0.03% of fresh weight of the soybean seed and approximately 0.08% of the total protein. The low levels of CP4 EPSPS in soybeans, combined with the thermal and digestive lability of this protein relative to that for known food allergens, suggests a very low probability of CP4 EPSPS surviving to be absorbed via the intestinal mucosa during consumption, and thus potentially triggering production of antibodies, including the immunoglobulin E antibodies responsible for allergenicity.

CP4 EPSPS does not show meaningful amino acid sequence homology when compared to 1935 known protein toxins present in the Pir protein, Swissprot, and Genpept protein databases. The homology search study of CP4 EPSPS to known protein toxins was based on the fact that patterns of amino acid sequence or regions of strong homology shared between two or more proteins may provide insight to the biological activity of the protein. Results from this search indicates that using the best methods available today, there are no biologically significant homologies between the CP4 EPSPS protein sequence and the protein sequence of all known toxins in the available protein databases. This conclusion that CP4 EPSPS is not expected to be toxic is supported by both the wholesomeness studies described above and an acute mouse gavage study (Harrison et al. 1995). The acute mouse gavage study with CP4 EPSPS as the test material was performed in order to directly assess any potential toxicity associated with the protein. Results from this study confirmed that the CP4 EPSPS protein is not toxic. The CP4 EPSPS protein was overproduced and purified from *E. coli*, demonstrated to be equivalent to the soybean seed-produced CP4 EPSPS, and administered by gavage to mice in an acute toxicity test (Harrison et al. 1995). There were no adverse effects in mice administered CP4 EPSPS protein by oral gavage at dosages up to 572 mg/kg. Acute administration was considered appropriate to assess the safety of CP4 EPSPS because proteins that are toxic act via acute mechanisms (Sjoblad et al. 1992; Pariza et al. 1983; Jones et al. 1991).

The absence of unexpected or unintended effects due to the CP4 EPSPS gene in line 40-3-2 was therefore demonstrated by classification of the donor organism as the common soil microorganism *Agrobacterium*, and characterization and complete sequencing of the EPSPS gene transferred to soybean; establishing that the recipient organism, soybean, has a history of safe use; extensive compositional analysis of line 40-3-2, focusing on both nutrients and antinutrient compounds; and animal feeding wholesomeness studies in rat, broiler chicken, dairy cow, catfish, and quail. We have also demonstrated that there are no deleterious effects due to the insertion of the CP4 EPSPS gene and protein in line 40-3-2. This was shown by the similarity of CP4 EPSPS to EPSPS enzymes found in food and microbes; the lack of allergenic potential of CP4 EPSPS; the lack of homology of CP4 EPSPS with known protein toxins; the lack of acute toxicity of CP4 EPSPS as determined by a mouse gavage study; the rapid digestion of CP4 EPSPS in simulated gastric and intestinal fluids; the lack of macroconstituent status and similar level of predicted consumption of CP4 EPSPS to soybean EPSPS; and the lack of expected alterations in carbohydrates, fats, or oils in line 40-3-2, which is supported by extensive analytical data. Consequently, Monsanto confirmed that soybeans modified to be tolerant to glyphosate through addition of the CP4 EPSPS gene and protein are not materially different in composition, safety, or any relevant parameter from soybeans currently on the market. The FDFL agreed with this conclusion, under the Food Policy (U.S. Food and Drug Administration 1992), in November 1994.

9. CONCLUSION

The information contained in this chapter captures the extensive research required to bring soybeans with a Roundup Ready™ gene to commercialization. The development of this product has yielded valuable biochemical information on the target site of glyphosate action, EPSPS, as well as the enzyme which degrades glyphosate to AMPA, GOX. The basic biochemical information described herein provides insight into several important aspects regarding the important herbicide glyphosate, such as development of glyphosate-tolerant EPSPSs and possibilities why the glyphosate target site, EPSPS, is not readily mutated *in planta* to yield glyphosate-resistant weeds. In addition, the work on the GOX enzyme sheds light on the metabolism of glyphosate in nature. Commercialization of GTS, however, depends on much more than generation of agronomically sound plant lines with stable tolerance to glyphosate. Extensive efforts have been required in plant breeding to transfer the herbicide tolerance insert in line 40-3-2 to other elite soybean varieties; herbicide systems development to optimize this new use of glyphosate for the farmer; and in regulatory science to assure the environmental, food and feed safety of the crop, as well as the new herbicide use. Only after all of these tasks are satisfactorily completed and the crop is accepted into standard agricultural pipelines will the real impact of glyphosate-tolerant soybeans in agriculture be able to be assessed.

1. ACKNOWLEDGMENTS

We would like to acknowledge our many colleagues at Monsanto whose efforts have allowed the glyphosate tolerance projects to reach their current stages. Special thanks go to William Stallings, Marcia Weldon, Brad LaVallee, Barbara Wells, Denise Bertrand, Kathryn Kolacz, Mary Taylor, Debbie Nida, Tom Nickson, and Nancy Biest, and as well as many Monsanto scientists, field product development specialists, and academic collaborators. We also thank Dr. Reid Palmer for providing information on soybean outcrossing.

REFERENCES

Alberg, D.G. and Bartlett, P.A. 1988. Potent inhibition of 5-enolpyruvylshikimate-3-phosphate synthase by a reaction intermediate analogue. *J.A.C.S.* 111: 2337–2338.

Alberg, D.G., Lauhon, C.T., Nyfeler, R., Fassler, A., and Bartlett, P.A. 1992. Inhibition of EPSP synthase by analogues of the tetrahedral intermediate and of EPSP. *J.A.C.S.* 114: 3535–3546.

Anderson, K.S., Sikorski, J.A., Benesi, A.J., and Johnson, K.A. 1988a. Isolation and structural elucidation of the tetrahedral intermediate in the EPSP synthase enzymatic pathway. *J. A. C. S.* 110: 6577–6579.

Anderson, K.S., Sikorski, J.A., and Johnson, K.A. 1988b. Evaluation of 5-enolpyruvylshikimate-3-phosphate synthase substrate and inhibitor binding by stopped-flow and equilibrium fluorescence measurements. *Biochemistry* 27: 1604–1610.

Anderson, K.S. and Johnson, K.A. 1990. Kinetic and structural analysis of enzyme intermediates: lessons from EPSP synthase. *Chem. Rev.* 90: 1131–1149.

Avila, L.Z., Loo, S.H., and Frost, J.W. 1987. Chemical and mutagenic analysis of aminomethylphosphonate biodegradation. *J.A.C.S.* 109: 6758–6764.

Baird. 1971. Introduction of a new broad spectrum postemergence herbicide class with utility for herbaceous perennial weed control. *North Cent. Weed Control Conf.* 26: 64–68.

Baker, H.G. 1974. The evolution of weeds. *Annu. Rev. of Ecol. Systematics* 5: 1–24.

Barry, G., Kishore, G., Padgette, S., Taylor, M., Kolacz, K., Weldon, M., Re, D., Eichholtz, D., Fincher, K., and Hallas, L. 1992. In *Biosynthesis and Molecular Regulation of Amino Acids in Plants.* Ed. by Singh, B.K., Flores, H.E., and Shannon, J.C. American Society of Plant Physiologists, Madison, WI, pp. 139–145.

Bartlett, S.G., Grossman, A.R., and Chua, N.H. 1982. In *Methods in Chloroplast Molecular Biology.* Ed. by Edelman, M., Hallick, R.B., and Chua, N.H. Elsevier, Amsterdam, pp. 1081–1091.

Benbrook, C. 1991. Racing against the clock. *Agrichem. Age* 35: 30–33.

Boocock, M.R. and Coggins, J.R. 1983. Kinetics of 5-enolpyruvylshikimate-3-phosphate synthase inhibition by glyphosate. *FEBS Lett.* 154: 127–133.

Burks, A.W., Brooks, J.R., and Sampson, H.A. 1988. Allergenicity of major component proteins of soybean determined by enzyme-linked immunosorbent assay (ELISA) and immunoblotting in children with atopic dermatitis and positive soy challenges. *J. Allergy Clin. Immunol.* 81: 1135–1142.

Burks, A.W. and Fuchs, R.L. 1995. Assessment of the endogenous allergens in glyphosphate-tolerant and commercial soybean varieties. *J. Allergy Clin. Immunol.* In press.

Charles, I.G., Keyte, J.W., Brammar, W.J., Smith, M., and Hawkins, A.R. 1986. The isolation and nucleotide sequence of the complex AROM locus of *Aspergillus nidulans*. *Nucl. Acids Res.* 15: 2201–2213.

Chen, C.-M., Ye, Q.-Z., Zhu, Z., Wanner, B.W., and Walsh, C.T. 1990. Molecular biology of carbon-phosphorus bond cleavage: cloning and sequencing of the *phn* (*psiD*) genes involved in alkylphosphonate uptake and C-P lyase activity in *Escherichia coli*. *J. Biol. Chem.* 265: 4461–4471.

Christou, P., McCabe, D.E., and Swain, W.F. 1988. Stable transformation of soybean callus by DNA-coated gold particles. *Plant Physiol.* 87: 671–674.

Comai, L., Sen, L., and Stalker, D.M. 1983. An altered *aroA* gene product confers resistance to the herbicide glyphosate. *Science* 221: 370–371.

Comai, L., Facciotti, D., Hiatt, W.R., Thompson, G., Rose, R.E., and Stalker, D.M. 1985. Expression in plants of a mutant *aroA* gene from *Salmonella typhimurium* confers tolerance to glyphosate. *Nature* 317: 741–744.

Considine, D.M. and Considine, G.D., Eds. 1982. *Foods and Food Production Encyclopedia*, Van Nostrand Reinhold, New York, pp. 1856–1870.

Coppock, J. 1974. Soy proteins in foods-retrospect and prospect. *J. Am. Oil Chem. Soc.* 51: 59A–62A.

Daughton, C.G., Cook, A.M., and Alexander, M. 1979. Biodegradation of phosphonate toxicants yields methane or ethane on cleavage of the C-P bond. *FEMS Microbiol. Lett.* 5: 91–93.

Della-Cioppa, G., Bauer, S.C., Klein, B.K., Shah, D.M., Fraley, R.T., and Kishore, G. 1986. Translocation of the precursor of 5-enolpyruvylshikimate-3-phosphate synthase into chloroplasts of higher plants *in vitro*. *Proc. Natl. Acad. Sci. U.S.A.* 83: 6873–6877.

Della-Cioppa, G., Bauer, S.C., Taylor, M.T., Rochester, D.E., Klein, B. K., Shah, D.M., Fraley, R.T., and Kishore, G.M. 1987. Targeting a herbicide-resistant enzyme from *Escherichia coli* to chloroplasts of higher plants. *Bio/Technology* 5: 579–584.

Devine, M., Duke, S.O., and Fedtke, C. 1993. *Physiology of Herbicide Action*. Prentice Hall, Englewood Cliffs, NJ, pp. 251–294.

Duncan, K., Lewendon, A., and Coggins, J.R. 1984a. The complete amino acid sequence of *Escherichia coli* 5-enolpyruvylshikimate-3-phosphate synthase. *FEBS Lett.* 170: 59–63.

Duncan, K., Lewendon, A., and Coggins, J.R. 1984b. The purification of 5-enolpyruvylshikimate-3-phosphate synthase from an overproducing strain of *Escherichia coli*. *FEBS* 165: 121–127.

Duncan, K., Edwards, R.M., and Coggins, J.R. 1987. The pentafunctional *arom* enzyme of *Saccharomyces cerevisiae* is a mosaic of monofunctional domains. *Biochem. J.* 246: 375–386.

Dunham, D. 1993. *Food Costs...From Farm to Retail in 1992*. Agric. Info. Bull. No. 669, USDA Economic Research Service, Washington, D.C.

Erdman, J.W. 1988. *Soybean Utilization Alternatives*. Ed. by McCann, L. Center for Alternative Crops and Products, University of Minnesota, St. Paul, pp. 169–177.

Erickson, E.H. 1984. Soybean pollination and honey production — a research progress report. *Am. Bee J.* 124: 775–779.

Fillatti, J.J., Kiser, J., Rose, R., and Comai, L. 1987. Efficient transfer of a glyphosate tolerance gene into tomato using a binary *Agrobacterium tumefaciens* vector. *Bio/Technology* 5: 726–730.

Garbe, T., Jones, C., Charles, I., Dougan, G., and Young, D. 1990. Cloning and characterization of the *aroA* gene from *Mycobacterium tuberculosis*. *J. Bacteriol.* 172: 6774–6782.

Gasser, C. 1989. Genetically engineering plants for crop improvement. *Science* 244: 1293–1299.

Gasser, C.S., Winter, J.A., Hironaka, C.M., and Shah, D.M. 1988. Structure, expression, and evolution of the 5-enolpyruvylshikimate-3-phosphate synthase genes of petunia and tomato. *J. Biol. Chem.* 263: 4280–4289.

Genetics Computer Group. 1991. Program Manual for the GCG Package, Version 7, University of Wisconsin Genetics Computer Group, Madison, WI.

Gianessi, L.P. and Puffer, C. 1991. *Herbicide Use in the United States*, Resources for the Future, Washington, D.C.

Gougler, J.A. and Geiger, D.R. 1981. Uptake and distribution of N-phosphono-methylglycine in sugar beet plants. Plant *Physiol.* 68: 668–672.

Gruys, K.J., Walker, M.C., and Sikorski, J.A. 1992. Substrate synergism and the steady-state kinetic reaction mechanism for EPSP synthase from *E. coli*. *Biochemistry* 31: 5534–5544.

Gruys, K.J., Marzabadi, M.R., Pansegrau, P.D., and Sikorski, J.A. 1993. Steady-state kinetic evaluation of the reverse reaction for *Escherichia coli* 5-enolpyruvylshikimate-3-phosphate synthase. *Arch. Biochem. Biophys.* 304: 345–351.

Hallas, L.E., Hahn, E.M., and Korndorfer, C. 1988. Characterization of microbial traits associated with glyphosate biodegradation in industrial activated sludge. *J. Ind. Microbiol.* 3: 377–385.

Haughn, G., Smith, J., Mazur, B., and Somerville, C.R. 1988. An *Arabidopsis* acetolactate synthase gene in tobacco confers resistance to sulfonylurea herbicides. *Mol. Gen. Genet.* 211: 266–271.

Haughn, G.W. and Somerville, C. 1986. Sulfonylurea-resistant mutants of *Arabidopsis thaliana*. *Mol. Gen. Genet.* 204, 430-434.

Haughn, G.W. and Somerville, C.R. 1987. In *Applications of Biotechnology to Agricultural Chemistry*. Ed. by LeBaron, H., Mumma, R.O., Honeycutt, R.C., and Duesing, J.H. American Chemical Society, Washington, D.C. pp. 98–108.

Haughn, G.W. and Somerville, C.R. 1990. A mutation causing imidazolinone resistance maps to the *csr*1 locus of *Arabidopsis thaliana*. *Plant Physiol.* 92: 1081–1085.

Hawkins, A.R. and Smith, M. 1991. Domain structure and interaction within the pentafunctional arom polypeptide. *Eur. J. Biochem.* 196: 717–724.

Hinchee, M.A.W., Padgette, S.R., Kishore, G.M., Delannay, X., and Fraley, R.T. 1993. Herbicide-tolerant crops. In *Transgenic Plants Vol. 1*. Ed. by Kung, S. and Wu Ray, Academic Press, Orlando, FL, pp. 243–263.

Holt, J.S., Powles, S.B., and Holtum, J.A.M. 1993. Mechanisms and agronomic aspects of herbicide resistance. *Annu. Rev. Plant Physiol. Plant Mol. Biol.* 44: 203–229.

Horan, F.E. 1974. Soy protein products and their production. *J. Am. Oil Chem. Soc.* 51: 67A–73A.

Huynh, Q.K., Kishore, G.M., and Bild, G.S. 1988. 5-enolpyruvyl shikimate-3-phosphate synthase from *Escherichia coli*: identification of Lys-22 as a potential active site residue. *J. Biol. Chem.* 263: 735–739.

Isogai, T., Ono, H., Ishitani, Y., Kojo, H., Ueda, Y., and Kohsaka, M. 1990. Structure and expression of cDNA for D-amino acid oxidase active against cephalosporin C from *Fusarium solani*. *J. Biochem.* 108: 1063–1069.

Jacob, G.S., Garbow, J.R., Hallas, L.E., Kimack, N.M., Kishore, G.M., and Schaefer, J. 1988. Metabolism of glyphosate in *Pseudomonas* sp. strain LBr. *Appl. Environ. Microbiol.* 54: 2953–2958.

Johnston, D.T., Van Wijk, A.J.P., and Kilpatrick, D. 1989. Selection for Tolerance to Glyphosate in Fine-Leaved Festuca Species, 6th International Turfgrass Research Conference, Tokyo, July 31 to August 5, 1989.

Johnston, D.T. and Faulkner, J.S. 1991. In *Herbicide-Resistant Weeds and Crops*. Ed. by Caseley, J.C., Cussans, G.W., and Atkin, R.K., Butterworth-Heinemann, Oxford, pp. 319–330.

Jones, D.D. and Maryanski, J.H. 1991. In *Risk Assessment in Genetic Engineering*. Ed. by Levin, M.A. and Strauss, H.S. McGraw-Hill, New York.

Kishore, G. and Shah, D. 1988. Amino acid biosynthesis inhibitors as herbicides. *Annu. Rev. Biochem.* 57: 627–663.

Kishore, G. et al. 1988. EPSP synthase: from biochemistry to genetic engineering of glyphosate tolerance. In *Biotechnology for Crop Protection*, ACS Symp. Ser. No. 379. Ed. by Hedin, P.A., Menn, J.J., and Hollingsworth, R.M. American Chemical Society, Washington, D.C., pp. 37–48.

Kishore, G.M., Brundage, L., Kolk, K., Padgette, S.R., Rochester, D., Huynh, Q.K., and Della-Cioppa, G. 1986. Isolation, purification, and characterization of a glyphosate-tolerant mutant *E. coli* EPSP synthase. *Fed. Proc.* 45: 1506.

Kishore, G.M. and Jacob, G.S. 1987. Degradation of glyphosate by *Pseudomonas* sp. PG2982 via a sarcosine intermediate. *J. Biol. Chem.* 262: 12164–12168.

Klee, H.J., Muskopf, Y.M., and Gasser, C.S. 1987. Cloning of and *Arabidopsis thaliana* gene encoding 5-enolpyruvylshikimate-3-phosphate synthase: sequence analysis and manipulation to obtain glyphosate-tolerant plants. *Mol. Gen. Genet.* 210: 437–442.

Kwon, S.H., Im, K.H., and Kim, J.R. 1972 Studies on the diversity of seed weight in the Korean soybean land races and wild soybean. *Korean J. Breed.* 4: 70–74.

LeBaron, H.M. 1991. Herbicide resistant weeds continue to spread. *Resistant Pest Manage. Newslett.* 3: 36–37.

Leiner, I.E. 1972. In *Soybeans: Chemistry and Technology*. Ed. by Smith, A.K. and Circle, S.J. Avi Publishing, Westport, CT, pp. 203–277.

Loo, S.H., Peters, N.K., and Frost, J. 1987. Genetic characterization of an *Escherichia coli* mutant deficient in organophosphonate biodegradation. *Biochem. Biophys. Res. Commun.* 148: 148-152.

Makino, K., Kim, S.-K., Shinegawa, H., Amemura, M., and Nakata, A. 1991. Molecular analysis of the cryptic and functional *phn* operons for phosphonate use in *Escherichia coli* K-12. *J. Bacteriol.* 173: 2665–2672.

Malik, J., Barry, G., and Kishore, G. 1989. The herbicide glyphosate. *BioFactors* 2: 17–25.

Maskell, D.J., Morrissey, P., and Dougan, G. 1988. Cloning and nucleotide sequence of the *aroA* gene of *Bordetella pertussis*. *J. Bacteriol.* 170: 2467–2471.

Mazur, B. 1989. The development of herbicide resistant crops. *Annu. Rev. Plant Physiol. Plant Mol. Biol.* 40: 441–470.

McCabe, D.E., Swain, W.F., Martinell, B.J., and Christou, P. 1988. Stable transformation of soybean (*Glycine max*) by particle acceleration. *Bio/Technology.* 6: 923–926.

McMullan, G., Watkins, R., Harper, D.B., and Quinn, J.P. 1991. Carbon-phosphorus bond cleavage activity in cell-free extracts of *Enterobacter aerogenes* ATCC 15038 and *Pseudomonas* sp. 4ASW. *Biochem. Int.* 25: 271–279.

Mounts, T.L., Wolf, W.J., and Martinez, W.H. 1987. Processing and utilization. In *Soybeans: Improvement, Production, and Uses*, 2nd ed. Ed. by Wilcox, J.R. American Society of Agronomy, Crop Science Society of America, and Soil Science Society of America, Madison, WI, pp. 819–866.

Mounts, T.L. 1988. Edible soybean oil products. In *Soybean Utilization Alternatives*. Ed. by McCann, L. Center for Alternative Crops and Products, University of Minnesota, St. Paul, pp. 43–56.

Mousdale, D.M. and Coggins, J.R. 1984. Purification and properties of 5-enolpyruvylshikimate-3-phosphate synthase from seedlings of *Pisum sativum* L. *Planta* 160: 78–83.

Nomura, N.S. and Hilton, H.W. 1977. The adsorption and degradation of glyphosate in five Hawaiian sugarcane soils. *Weed Res.* 17: 113–121.

O'Connell, C., Pattee, P.A., and Foster, T.J. 1993. Sequence and mapping of the *aroA* gene of *Staphylococus aureus* 8325-4. *J. Gen. Microbiol.* 139: 1449–1460.

Padgette, S.R., Huynh, Q.K., Borgmeyer, J., Shah, D.M., Brand, L.A., Re, D.B., Bishop, B.F., Rogers, S.G., Fraley, R.T., and Kishore, G.M. 1987. Bacterial expression and isolation of *Petunia hybrida* 5-enolpyruvylshikimate-3-phosphate synthase. *Arch. Biochem. Biophys.* 258: 564–573.

Padgette, S.R., Huynh, Q.K., Aykent, S., Sammons, R.D., Sikorski, J.A., and Kishore, G.M. 1988a. Identification of the reactive cysteines of *Escherichia coli* 5-enolypyruvylshikimate-3-phosphate synthase and their nonessentiality for enzymatic catalysis. *J. Biol. Chem.* 263: 1798–1802.

Padgette, S.R., Smith, C.E., Huynh, Q.K., and Kishore, G.M. 1988b. Arginine chemical modification of *Petunia hybrida* 5-enolpyruvylshikimate-3-phosphate synthase. *Arch. Biochem. Biophys.* 266: 254–262.

Padgette, S.R., Della-Cioppa, G., Shah, D.M., Fraley, R.T., and Kishore, G.M. 1989. Selective herbicide tolerance through protein engineering. In *Cell Culture and Somatic Cell Genetics of Plants, Vol. 6.* Ed. by Schell, J. and Vasil, I. Academic Press, New York, pp. 441–476.

Padgette, S.R., Re, D.B., Gasser, C.S., Eichholtz, D.A., Frazier, R.B., Hironaka, C.M., Levine, E.B., Shah, D.M., Fraley, R.T., and Kishore, G.M. 1991. Site-directed mutagenesis of a conserved region of the 5-enolpyruvylshikimate-3-phosphate synthase active site. *J. Biol. Chem.* 266: 22364–22369.

Pariza, M.W. and Foster, E.M. 1983. Determining the safety of enzymes used in food processing. *J. Food Protect.* 46: 453–468.

Pederson, H.S. and Djurtoft, R. 1989. Antigenic and allergenic properties of acidic and basic peptide chains from glycinin. *Food Agric. Immunol.* 1: 101–109.

Pipke, R. and Amrhein, N. 1988. Degradation of the phosphonate herbicide glyphosate by *Arthrobacter atrocyaneus* ATCC 13752. *Appl. Environ. Microbiol.* 54: 1293–1296.

Powell, H.A., Kerby, N.W., and Powell, P. 1991. Natural tolerance of cyanobacteria to the herbicide glyphosate. *New Phytol.* 119: 421–426.

Powell, H.A., Kerby, N.W., Rowell, P., Mousdale, D.M., and Coggins, J.R. 1992. Purification and properties of a glyphosate-tolerant 5-enolpyruvylshikimate-3-phosphate synthase from the cyanobacterium *Anabaena variabiliis. Planta* 188: 484–490.

Rackis, J.J. 1974. Biological and physiological factors in soybeans. *J. Am. Oil Chem. Soc.* 51: 161A–173A.

Re, D.B., Padgette, S.R., Delannay, X., Kolacz, K.H., Nida, D.L., Peschke, V.M., Derting, C.W., Rogers, S.G., Edwards, J.W., Barry, G.F., and Biest, N.A. 1993. Petition of Determination of Nonregulated Status: Soybeans with a Roundup Ready™ Gene, submitted to USDA, St. Louis.

Ream, J.E., Steinrucken, H.C., Porter, C.A., and Sikorski, J.A. 1988. Purification and properties of 5-enolpyruvylshikimate-3-phosphate synthase from dark-grown seedlings of *Sorghum bicolor. Plant Physiol.* 87: 232–238.

Rogers, S.G., Brand, L.A., Holder, S.B., Sharps, E.S., and Brackin, M.J. 1983. Amplification of the *aroA* gene from *Escherichia coli* results in tolerance to the herbicide glyphosate. *Appl. Environ. Microbiol.* 46: 37–43.

Rueppel, M.L., Brightwell, B.B., Schaefer, J., and Marvel, J.T. 1977. Metabolism and degradation of glyphosate in soil and water. *J. Agric. Food Chem.* 25: 517–528.

Ruff, T., Eichholtz, D., Re, D., Padgette, S., and Kishore, G. 1991. Effects of amino acid substitutions on glyphosate tolerance and activity of EPSPS. *Plant Physiol. Suppl.* 96: 94.

Schulz, A., Kruper, A., and Amrhein, N. 1985. Differential sensitivity of bacterial 5-enolpyruvylshikimate-3-phosphate synthases to the herbicide glyphosate. *FEMS Microbiol. Lett.* 28: 297–301.

Shah, D., Horsch, R., Klee, H., Kishore, G., Winter, J., Tumer, N., Hironaka, C., Sanders, P., Gasser, C., Aykent, S., Siegel, N., Rogers, S., and Fraley, R. 1986. Engineering herbicide tolerance in transgenic plants. *Science* 233: 478–481.

Shibasaki, M., Suzuki, S., Tajima, S., Nemoto, H., and Kuroume, T. 1980. Allergenicity of major component proteins of soybean. *Int. Arch. Allergy Appl. Immunol.* 61: 441–448.

Shinabarger, D.L. and Braymer, H.D. 1986. Glyphosate catabolism by *Pseudomonas* sp. strain PG2982. *J. Bacteriol.* 168: 702–707.

Sikorski, J.A., Anderson, K.S., Cleary, D.G., Miller, M.J., Pansegrau, P.D., Ream, J.E., Sammons, R.D., and Johnson, K.A. 1990. An enzyme-targeted herbicide design program based on EPSP synthase: chemical mechanisms and glyphosate inhibition studies. In *Chemical Aspects of Enzyme Biotechnology.* Ed. by Baldwin, T.O. Plenum Press, New York, pp. 23–39.

Sjoblad, R.D., McClintock, J.T., and Engler, R. 1992. Toxicological considerations for protein components of biological pesticide products. *Regul. Toxicol. Pharmacol.* 15: 3–9.

Skvortzov, B.V. 1927 The soybean-wild and cultivated in Eastern Asia. *Proc. Manchurian Res. Soc. Pub. Ser. A. Nat. Hist. Sect.* 22: 1–8.

Sost, D. and Amrhein, N. 1990. Substitution of Gly-96 to Ala in the 5-enolpyruvylshikimate-3-phosphate synthase of *Klebsiella pneumoniae* results in a greatly reduced affinity for the herbicide glyphosate. *Arch. Biochem. Biophys.* 282: 433–436.

Soy Stats Reference Guide. 1991. American Soybean Association, St. Louis.

Soy Stats Reference Guide. 1992. American Soybean Association, St. Louis.

Stalker, D.M., Hiatt, W.R., and Comai, L. 1985. A single amino acid substitution in the enzyme 5-enolpyruvylshikimate-3-phosphate synthase confers resistance to the herbicide glyphosate. *J. Biol. Chem.* 260: 4724–4728.

Stallings, W.C., Abdel-Meguid, S.S., Lim, L.W., Shieh, H.-S.D., Leimgruber, N.K., Stegeman, R.A., Anderson, K.S., Sikorski, J.A., Padgette, S.R., and Kishore, G.M. 1991. Structure and topological symmetry of the glyphosate target 5-enolpyruvylshikimate-3-phosphate synthase: a distinctive protein fold. *Proc. Natl. Acad. Sci. U.S.A.* 88: 5046–5050.

Steinrucken, H.C. and Amrhein, N. 1980. The herbicide glyphosate is a potent inhibitor of 5-enolpyruvyl shikimic acid-3-phosphate synthase. *Biochem. Biophys. Res. Commun.* 94: 1207–1212.

Steinrucken, H.C. and Amrhein, N. 1984a. 5-Enolpyruvylshikimate-3-phosphate synthase of *Klebsiella pneumoniae*: inhibition by glyphosate [*N*-(phosphonomethyl)-glycine]. *Eur. J. Biochem.* 143: 351–357.

Steinrucken, H.C. and Amrhein, N. 1984b. 5-Enolpyruvylshikimate-3-phosphate synthase of Kle*bsiella pneumoniae*. I. Purification and properties. *Eur. J. Biochem.* 143: 341–349.

Steinrucken, H.C., Schulz, A., Amrhein, N., Porter, C.A., and Fraley, R.T. 1986. Overproduction of 5-enolpyruvylshikimate-3-phosphate synthase in a glyphosate-tolerant *Petunia hybrida* cell line. *Arch. Biochem. Biophys.* 244: 169–178.

Subramanian, M.V., Loney-Gallant, V., Dias, J.M., and Mireles, L.C. 1991. Acetolactate synthase inhibiting herbicides bind to the regulatory site. *Plant Physiol.* 96: 310–313.

Suzuki, K., Ogishima, M., Sugiyama, M., Inouye, Y., Nakamura, S., and Immura, S. 1992. Molecular cloning and expression of a *Streptomyces* sarcosine oxidase gene in *Streptomyces lividans. Biosci. Biotech. Biochem.* 56: 432–436.

Taylor, S.L., Lemanske, R.F., Jr., Bush, R.K., and Busse, W.W. 1987. Food allergens: structure and immunologic properties. *Ann. Allergy* 59: 93–99.

Taylor, S.L. 1992. Chemistry and detection of food allergens. *Food Technol.* 39: 146–152.

Taylor, S.L., Nordlee, J.A., and Bush, R.K. 1992. Food allergies. In *Food Safety Assessment,* ACS Symp. Ser. 484. Ed. by Finley, J.W., Robinson, S.F., and Armstrong, D.J. American Chemical Society, Washington, D.C.

Timko, M.P., Herdies, L., de Alameida, E., Cashmore, A.R., Leemans, J., and Krebbers, E. 1988. Genetic engineering of nuclear-encoded components of the photosynthetic apparatus in *Arabidopsis.* In *The Impact of Chemistry on Biotechnology — A Mutlidisciplinary Discussion.* Ed. by Phillips, M., Shoemaker, S.P., Middlekauff, R.D., and Ottenbrite, R.M., Eds. ACS Books, Washington, D.C., pp. 279–295.

Torstensson, L. 1985. Behavior of glyphosate and its degradation. In *The Herbicide Glyphosate.* Ed. by Grossbard, E. and Atkinson, D. Butterworths, Boston, pp. 137–150.

United States Food and Drug Administration. 1992. Foods derived from new plant varieties. *Fed. Reg.* 57(104): 22984–23005.

Wackett, L.P., Shames, S.L., Venditti, C.P., and Walsh, C.T. 1987a. Bacterial carbon-phosphorus lyase: products, rates, and regulation of phosphonic and phosphinic acid metabolism. *J. Bacteriol.* 169: 710–717.

Wackett, L.P., Wanner, B.L., Venditti, C.P., and Walsh, C.T. 1987b. Involvement of the phosphate regulon and the *psiD* locus in the carbon-phosphorus lyase activity of *Escherichia coli* K-12. *J. Bacteriol.* 169: 1753–1756.

Waggle, D.H. and Kolar, C.W. 1979. Types of soy protein products. In *Soy Protein and Human Nutrition.* Ed. by Wilke, H.L., Hopkins, D.T., and Waggle, D.H. Academic Press, New York, pp. 19–51.

Wilke, H.L., Hopkins, D.T., and Waggle, D.H. 1979. *Soy Protein and Human Nutrition.* Academic Press, New York.

Wolf, W.J. 1983. Edible soybean protein products. In *CRC Handbook of Processing and Utilization in Agriculture, Vol. II, Part 2.* Ed. by Wolff, I.A. CRC Press, Boca Raton, FL, pp. 23–55.

Zeleznick, L.D., Myers, T.C., and Titchener, E.B. 1963. Growth of *Escherichia coli* on methyl- and ethylphosphonic acids. *Biochim. Biophys. Acta* 78: 546–547.

PLATE 3 A, B and C (Chapter 4) Soybeans with a Roundup Ready™ gene.

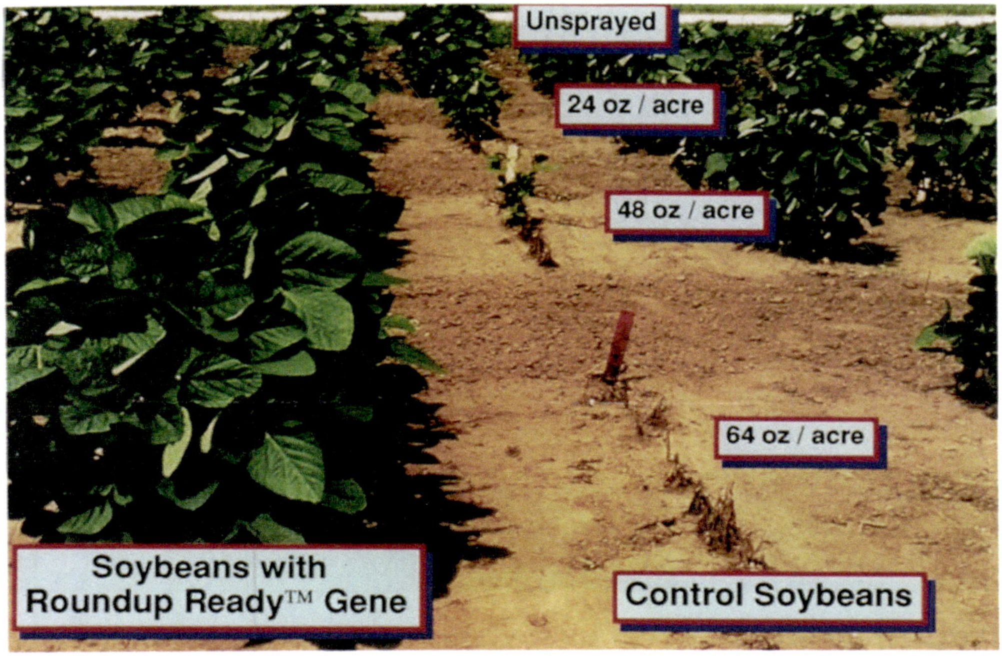

Panel A Line evaluation in Jerseyville, Illinois, in 1991 demonstrating that CP4 EPSPS confers glyphosate tolerance to soybeans. Soybeans with a Roundup Ready™ gene are shown center left, after spraying with the indicated rates of Roundup® herbicide. The parental, non-modified control soybean line is shown center right, also sprayed at the indicated rates.

Panel B Weed control using soybeans with a Roundup Ready™ gene. The plot was sprayed at 0.56 lb/A of Roundup® herbicide, with surfactant added at 0.5% v/v. The parental, non-modified control soybean line is shown at center right. Soybeans with Roundup Ready™ gene are shown at left. The rear of the picture shows the weed pressure present with no Roundup® herbicide.

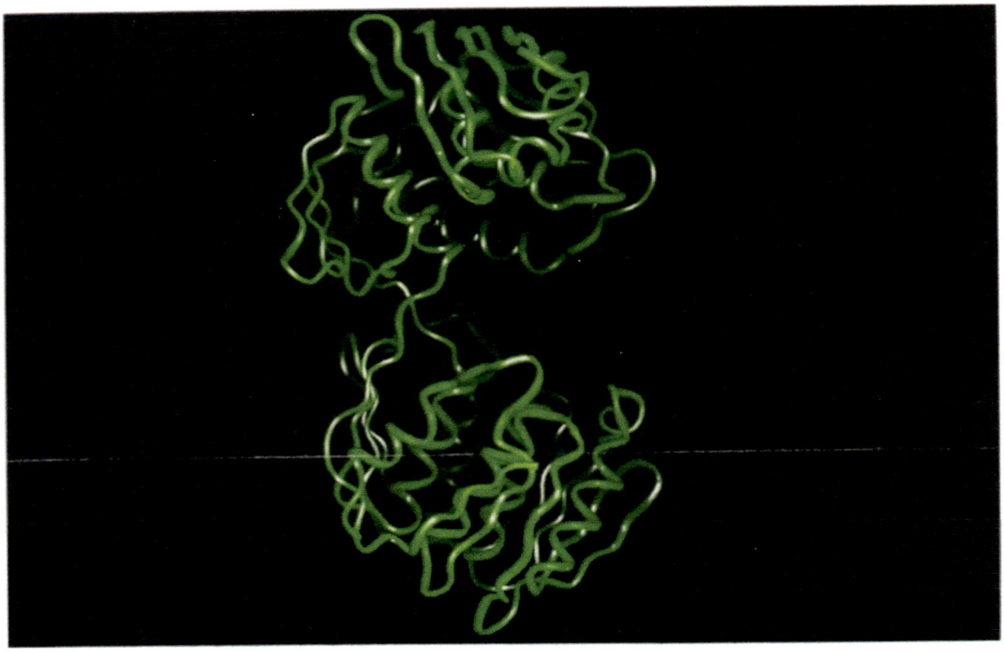

PLATE 1 (Chapter 4) X-ray crystal structure of *E. coli* EPSPS. The model pictured is from data refined from that previously published (Stallings *et al.* 1991.)

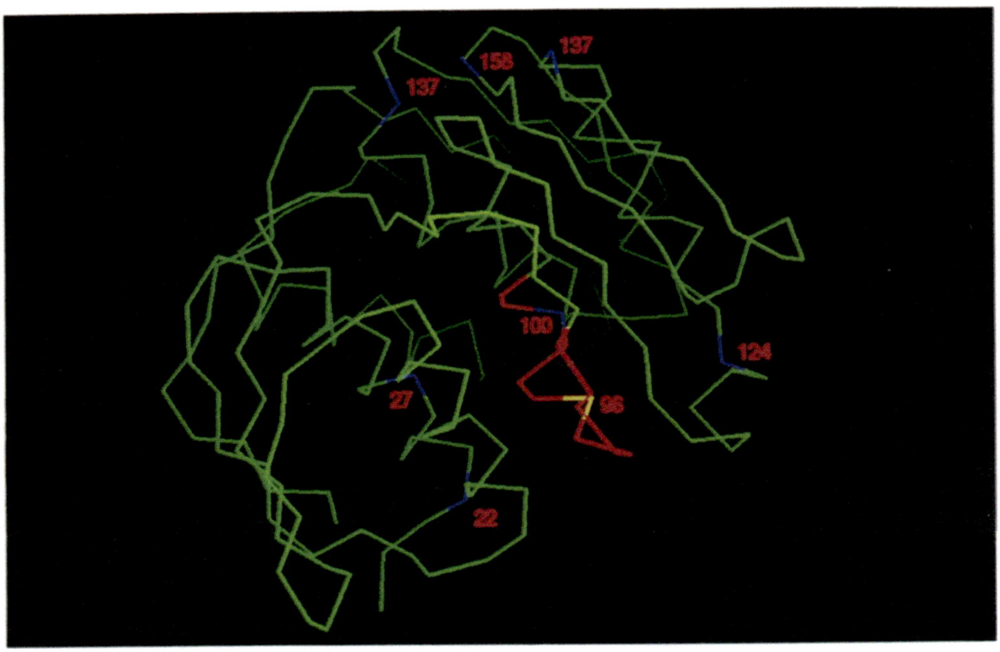

PLATE 2 (Chapter 4) *E. coli* EPSPS X-ray crystal structure backbone plot of the "upper" domain. Upper domain shows placement of active site residues Lys^{22}, Arg^{27}, Arg^{100}, and Arg^{124}. Highlighted is the "conserved region" indicated in Table 1 denoting the Gly^{96} residue which confers glyphosate tolerance when substituted with Ala. Also shown are the "second site" substitution locations of residues 183, 158, and 137, which correspond to residues 192, 167, and 144 of petunia EPSPS.

Panel C *Seed increase showing weed control obtained using Roundup® herbicide treatment over the top of* soybeans with a Roundup Ready™ gene.

PLATE 4 (Chapter 6) Bromoxynil-resistant cotton in a field situation. CE1 refers to the parental cotton variety without the bromoxynil-specific nitrilase gene. CE1-45-1A and CE1-45-5A are two selected lines from an initial transformation event and express the bromoxynil-specific nitrilase. Stv 506 refers to a control variety. All rows were sprayed at the 10–15 leaf stage with a Buctril® formulation containing 1.12 kg a.i./hec bromoxynil. The picture was taken four days after treatment.

PLATE 5 (Chapter 12) Tobacco plants grown mixotrophically in the presence of 3μM fluridone **(A)** or in soil and treated with 3 μM norflurazon **(B)**. Plants at the right are non-transformed controls, plants at the left carry the *crtI* gene-fusion construct.

Chapter 5

PHOSPHINOTHRICIN-RESISTANT CROPS

Indra K. Vasil

CONTENTS

1. INTRODUCTION

Detoxification is an important mechanism by which many plants as well as bacteria develop natural tolerance to a variety of herbicides. During the past few years, novel and successful strategies have been developed for engineering selective herbicide tolerance in plants by isolating, cloning, and characterizing bacterial or plant genes that detoxify specific herbicides, and then transferring them to crops of choice by genetic transformation. This chapter describes how metabolic inactivation has been used as a powerful and effective tool in engineering tolerance to the nonselective, broad-spectrum, postemergent, contact (with some systemic activity) herbicide DL-phosphinothricin (also known as glufosinate). Its herbicidal property is due to L-phosphinothricin (PPT, Figure 1), an analogue of glutamate, as the D-isomer is herbicidally inactive. PPT is the active ingredient of the commercial herbicides, Herbiace and Basta. Bialaphos (sold as Herbiace by Meiji Seika Kaisha, Japan), contains the naturally occurring herbicidal tripeptide antibiotic L-phosphinothricinyl-L-alanyl-L-alanine, an allelochemical produced by *Streptomyces hygroscopicus* (Bayer et al. 1972; Kondo et al. 1973; Figure 1). It has been suggested that the tripeptide facilitates transport across the plasma membrane and that peptidase cleavage *in planta* results in the release of L-phosphinothricin. Glufosinate (sold as basta by Hoechst, Germany), is a truncated synthetic version of bialaphos that lacks the terminal alanine and contains L-phosphinothricin. Both are ideal herbicides as they have high unit activity, are highly effective against a wide variety of plant species, and provide a high degree of human and environmental safety because they are nontoxic and are rapidly biodegraded, resulting in minimal residue persistence in soil.

1-56670-045-0/96/$0.00+$.50
© 1996 by CRC Press Inc.

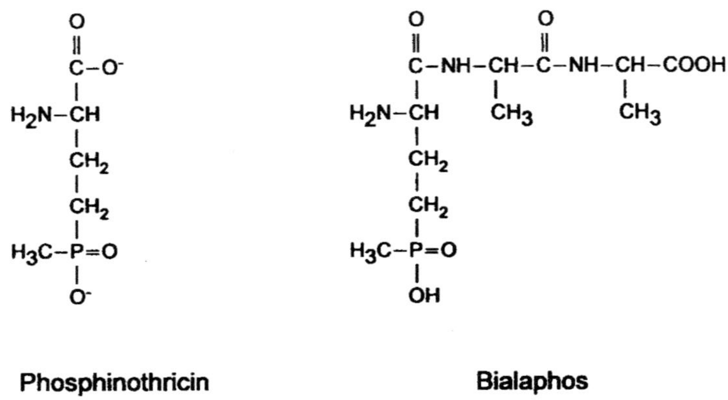

FIGURE 1 L-Phosphinothricin and bialaphos (phosphinothricin-L-alanyl-L-alanine).

2. TARGET SITE OF PPT

PPT is a potent inhibitor of the biosynthetic enzyme glutamine synthetase (GS), which is involved in general nitrogen metabolism in plants, including the assimilation of ammonia accumulated as a result of photorespiration and nitrate reduction (Bayer et al. 1972). GS catalyzes the formation of L-glutamine from L-glutamic acid. The herbicidal activity of PPT is based on its inhibition of GS resulting in the rapid accumulation of intracellular ammonia, cessation of photorespiration and photosynthesis, and chloroplast disruption (Tachibana et al. 1986a, b; Wild et al. 1987, 1989; Devine et al. 1993). Although the accumulation of ammonia (rather than the lack of glutamine) may contribute to the death of plant cells, the primary cause of phytotoxicity is believed to be rapid inhibition of RuBP carboxylase/photorespiration (Devine et al. 1993). Numerous isozymic forms of GS are found in different organs of plants (McNally et al. 1983). Sometimes two different forms may be found in the same organ, such as leaves, where one of the isozymes is present in the cytosol and the other in the chloroplasts. PPT is known to inhibit all known forms of GS that are found in plants.

3. STRATEGIES FOR ENGINEERING RESISTANCE TO PPT

PPT is mobile in xylem as well as in phloem (Shelp et al. 1992). Its uptake is restricted to leaves and roots of plants; it normally has no effect on woody parts. Accordingly, it is used for weed control in vineyards and fruit orchards (Langeludeke et al. 1981), as a directed spray in agronomic crops, as a general preplant herbicide, and for weed control in nonagricultural settings. It is expected that in the future engineered PPT resistance in crops will significantly broaden the use of this important and desirable herbicide.

Efforts to select plants that are naturally resistant to PPT because of mutations resulting in structural changes in GS have not been successful. Therefore, overproduction of GS, the target enzyme, and metabolic inactivation of PPT have been used as possible strategies to engineer resistance to PPT. Some attempts have been made to select PPT-resistant plants through tissue culture. PPT-resistant cell lines of alfalfa were obtained by stepwise selection (Donn et al. 1984). The cell lines were 20- to 100-fold more tolerant to PPT than the nonselected controls. Although no plants could be regenerated from the "mutant" lines, they provided an opportunity to study the mechanism of resistance, which was found to be overproduction of GS due to a 4- to 11-fold amplification of the GS gene (Tischer et al. 1986). The amplified gene is probably localized in the cytosol as it lacks the characteristic chloroplast transit peptide sequences. It is not known if PPT tolerance can be conferred by overproduction of chloroplast-associated GS. The alfalfa GS gene complements a glnA mutation in *Escherichia coli* (DasSarma et al., 1986). This makes it posssible to select PPT-tolerant mutants of the alfalfa GS in *E. coli*, which could be useful in the development of PPT-resistant crops. Overexpression of the alfalfa cytosolic GS gene in transgenic tobacco plants causing a 3-fold increase in the specific activity of GS resulted in a concomitant 20-fold increase in tolerance to PPT *in vitro* (Eckes et al. 1989). Plants sprayed with four times the normal rate of field application were severely damaged. The growth and reproduction of GS overproducing plants was normal, although they contained sevenfold less free NH_3. Nevertheless, it remains to be clearly determined if PPT-resistant plants can be produced in a wide spectrum of crops by the overproduction of cytosolic and/or plastidic GS, and what effect (if any) such overproduction would have on overall plant growth, particularly on cellular metabolism and productivity.

During the last few years much attention has been focused on the metabolic inactivation of PPT. These efforts have resulted in the isolation and characterization of phosphinothricin-*N*-acetyltransferase (PAT), an enzyme that inactivates L-phosphinothricin (as well as bialaphos and glufosinate) by converting it to *N*-acetyl-L-phosphinothricin in the presence of acetyl coenzyme A as a substrate. The acetylated forms of PPT, bialaphos, and glufosinate are stable in plants, but have no herbicidal activity. Two similar genes, *bar* and *pat*, with significant sequence homology and both encoding the acetylating enzyme PAT, have been isolated from *Streptomyces hygroscopicus*, the organism that produces bialaphos (Murakami et al. 1986; Thompson et al. 1987), and from *S. viridochromogenes* (Strauch et al. 1988; Wohlleben et al. 1988), respectively. Bialaphos autotoxicity in the bacteria is prevented by the metabolism of PPT by acetylation into an inactive derivative. Herbicide resistance thus can be achieved by the transfer of either one of these genes into plants. As shown in the following section, almost all of the present success in engineering PPT resistance is based on the transfer and efficient expression of the resistant *bar* gene into crop species.

4. PPT-RESISTANT CROPS

Rapid advances in the genetic transformation of plants have made it possible to introduce the *bar* gene into many important crops during the past few years (Table 1). Transgenic plants have been obtained following indirect gene transfer through *Agrobacterium tumefaciens* or by the direct delivery of DNA into protoplasts by electroporation or polyethylene glycol treatment, or by high velocity bombardment of DNA-coated microprojectiles into intact cells and tissues. One of the first examples of engineering herbicide resistance by the introduction and expression of a detoxifying enzyme was provided by DeBlock et al. (1987). They introduced the *bar* gene, driven by the 35S constitutive promoter of cauliflower mosaic virus (CaMV), into tobacco plants. Plants expressing the resistant gene produced PAT, the acetylating/detoxifying enzyme, and were found to be completely tolerant to four to ten times the level of the herbicide normally applied under field conditions. Varying levels of PAT activity were found in independently transformed plants, but even plants with low levels of PAT activity (0.001% of total leaf protein) were found be herbicide resistant. Herbicide treatment did not change the level of ammonia in the transgenic plants, demonstrating complete and efficient protection of the plant GS from the herbicide. Both the native as well as a synthetic *bar* gene confer resistance to PPT by detoxification of the herbicide. The metabolism of PPT has been shown to be different in transformed and untransformed plants (Droge et al. 1992). For example, it is rapidly acetylated in transgenic tobacco and carrot plants but is successively deaminated and decarboxylated in untransformed plants. The performance of transgenic PPT-resistant tobacco and potato plants in field trials in Europe was similar to nontransformed plants (De Greef et al. 1989). Also, field testing of some transgenic crops for several years has shown that herbicide resistance was stable and that there was no loss of agronomic characteristics (Botterman et al. 1991). In addition, from one of the most exhaustive ecological studies of any transgenic plants

TABLE 1 Phosphinothricin-Resistant Crops

Species	Ref.
Avena sativa (oats)	Somers et al. 1992
Beta vulgaris (beet root)	Botterman and Leemans 1988
Brassica napus (oilseed rape)	DeBlock et al. 1989
B. oleracea (wild cabbage)	DeBlock et al. 1989
Daucus carota (carrot)	Droge et al. 1992
Festuca arundinacea (tall fescue)	Wang et al. 1992
Hordeum vulgare (barley)	Wan and Lemaux 1994
Lycopersicon esculentum (tomato)	DeBlock et al. 1987, 1989
Medicago sativa (alfalfa)	Eckes et al. 1989; D'Halluin et al. 1990
Nicotiana tabacum (tobacco)	DeBlock et al. 1987
Oryza sativa (rice)	Datta et al. 1992; Toki et al. 1992
Populus spp. hybrids (poplar)	DeBlock 1990; Devillard 1992; Chupeau et al. 1994
Secale cereale (rye)	Castillo et al. 1994
Solanum tuberosum (potato)	DeBlock et al. 1987
Sorghum bicolor (sorghum)	Casas et al. 1993
Triticum aestivum (wheat)	Vasil et al. 1992, 1993; Weeks et al. 1993
Zea mays (maize)	Fromm et al. 1990; Gordon-Kamm et al. 1990

to date, Crawley et al. (1993) concluded that herbicide tolerance (PPT resistance) did not increase the invasive potential of transgenic oilseed rape (*Brassica napus*). Indeed, they found that transgenic lines were less invasive and less persistent than their untransformed counterparts, and therefore posed no greater risk to the environment than the conventional crop. Encouraging though these results are, it should be remembered that they are limited to a single gene in a single species. Extreme care is warranted in extrapolating such results to other genes and other crops. In addition, the important question of escape of genes through pollen to weed species should be seriously evaluated. Overall, however, the results of field and environmental safety tests are encouraging and it is expected that several transgenic crops will be commercially introduced before the end of this century.

Production of wheat, the single most important crop for human food, depends heavily on the use of herbicides in all regions of the world. With widespread and long-term herbicide use, and the fact that wheat has a single mechanism for degrading herbicides, many of the most noxious weeds of wheat have developed resistances to the currently available selective herbicides by evolving a similar mechanism (Gressel 1992a). This is particularly true of herbicides that have high residual and high selection pressure. The fact that it has not been possible to select plants for natural structural mutations in the GS gene that confers resistance to PPT-based herbicides suggests that weeds may not readily develop resistance to PPT, which has a low residual. Also, PPT controls most of the problematic herbicide-resistant weeds found in wheat fields. Thus, the recent success in introducing the *bar* gene to produce PPT-resistant wheat (Vasil et al. 1992, 1993; Weeks et al. 1993) is welcome news. The *bar* gene of *S. hygroscopicus* was introduced by high velocity bombardment of DNA-coated gold or tungsten microprojectiles into embryogenic callus or immature embryos of wheat, and regenerating plants on nutrient media containing PPT. Integration of the functional gene into the plant genome was confirmed by PAT activity, molecular probes, and resistance to topical applications of Basta. Male and female transmission of the *bar* gene (and the resistant phenotype), which segregated as a dominant Mendelian trait, was demonstrated.

Wheat is a self-pollinated crop. It does not cross easily with other gramineous species, and its pollen remains viable only for a short time and does not travel long distances. These characteristics minimize any chance for the transfer of the resistance gene to weed species. On the other hand, caution should be exercised in introducing the *bar* gene into those crops which pollinate freely with close relatives, such as oats (Somers et al. 1992) and sorghum (Casas et al. 1993), which cross-pollinate with wild oats and johnsongrass, respectively (Gressel 1992b).

5. OTHER USES OF PPT RESISTANCE

The herbicide-resistant *bar* gene also has proved to be valuable as a selectable marker, especially in the genetic transformation of cereals where other available genes have not been particularly useful. It is not surprising, therefore, that this gene was used to obtain transgenic plants of wheat, rice, maize, barley, etc. (see Table 1). Use of the *bar* gene allows the production of a agronomically useful plant without introducing another nonessential selectable marker gene which otherwise serves no useful purpose. In an interesting study, Uchimiya et al. (1993) found that PPT treatment of transgenic rice expressing the *bar* gene also protected the plants from sheath blight caused by *Rhizoctonia solani*. The mechanism of this added protection is not known. Another novel use of resistance to PPT has been made for hybrid seed production in oilseed rape by linking it to a male sterility gene (Mariani et al. 1992).

Selection of putative transformed lines growing on PPT-containing media is tedious and time consuming, and is complicated because of the presence of many escapes or untransformed lines. Early identification of the transformed lines and weeding out of the escapes is thus desirable. Kramer et al. (1993) described a reliable method which permits early detection of the transformed lines based on the increase in pH of the medium caused by actively growing transformed cells. An increase in the acidity of the medium changes the color of the pH indicator chlorophenol red from red to orange to yellow. It was possible to select maize transformants by this method only after 2 to 4 weeks of selection on PPT, rather than the usual 2 to 3 months.

6. CONCLUSIONS

The use of the *bar* gene of *Streptomyces hygroscopicus,* which encodes the acetylating enzyme PAT, has been of great value in engineering resistance to PPT-based herbicides Basta (glufosinate) and Herbiace (bialaphos) in a number of important crops. Field trials of PPT-resistant crops have confirmed that their growth characteristics are similar to nontransformed plants, and that there are no adverse effects on yield or quality. The broad spectrum and high unit activity of PPT, combined with its nontoxic nature and rapid biodegradation, make it a suitable candidate for engineering herbicide resistance in crop plants.

REFERENCES

Bayer, E., Guge, K.H., Hagele, K., Hogenmajer, H., Jessipow, S., Konig, W.A., and Zahner, H. 1972. Phosphinothricin and phosphinothricin-alanyl-alanin, *Helv. Chim. Acta* 55: 224.

Botterman, J. and Leemans, J. 1988. Engineering herbicide resistance into plants. *Trends Genet.* 4: 219.

Botterman, J., D'Halluin, K., De Block, M., De Greef, W., and Leemans, J. 1991. Engineering of glufosinate resistance and evaluation under field conditions. In *Herbicide Resistance in Weeds and Crops*. Ed. by Caseley, J.C., Cussans, G.W., and Atkin, R.K. Butterworth-Heinemann, Oxford, p. 355,

Casas, A.M., Kononowicz, A.K., Zehr, U.B., Tomes, D.T., Axtell, J.D., Butler, L.G., Bressan, R.A., and Hasegawa, P.M. 1993. Transgenic sorghum plants via microprojectile bombardment. *Proc. Natl. Acad. Sci. U.S.A.* 90: 11212.

Castillo, A.M., Vasil, V. and Vasil, I.K. 1994. Rapid production of fertile transgenic plants of rye (*Secale cereale* L.). *Bio/Technology* 12: 1366.

Chupeau, M., Pautot, V., and Chupeau, Y. 1994. Recovery of transgenic trees after electroporation of poplar protoplasts. *Transgen. Res.* 3: 13.

Crawley, M.J., Hails, R.S., Rees, M., Kohn, D., and Buxton, J. 1993. Ecology of transgenic oilseed rape in natural habitats. *Nature* 363: 620.

DasSarma, S., Tischer, E., and Goodman, H.M. 1986. Plant glutamine synthetase complements a glnA mutation in *Escherichia coli. Science* 232: 1242.

Datta, S.K., Datta, K., Soltanifar, N., Donn, G., and Potrykus, I. 1992. Herbicide-resistant Indica rice plants from IRRI breeding line IR72 after PEG-mediated transformation of protoplasts. *Plant Molec. Biol.* 20: 619.

De Block, M. 1990. Factors influencing the tissue culture and the *Agrobacterium tumefaciens* mediated transformation of hybrid aspen and poplar clones. *Plant Physiol.* 93: 1110.

De Block, M., Botterman, J., Vanderwiele, M., Dockz, J., Thoen, C., Gossele, V., Movva, N.R., Thompson, C., van Montagu, M., and Leemans, J. 1987. Engineering herbicide resistance in plants by expression of a detoxifying enzyme. *EMBO J.* 6: 513.

De Block, M., DeBrouwer, D., and Tenning, P. 1989. Transformation of *Brassica napus* and *Brassica oleracea* using *Agrobacterium tumefaciens* and the expression of the *bar* and *neo* genes in the transgenic plants. *Plant Physiol.* 91: 694.

De Greef, W., Delon, R., De Block, M., Leemans, J., and Botterman, J. 1989. Evaluation of herbicide resistance in transgenic crops under field conditions. *Bio/Technology* 7: 61.

D'Halluin, K., Botterman, J., and De Greef, W. 1990. Engineering of herbicide-resistant alfalfa and evaluation under field conditions. *Crop Sci.* 30: 866.

Devillard, C. 1992. Transformation in vitro du tremble (*Populus temula* x *Populus alba*) par *Agrobacterium rhizogenes* et regeneration de plantes tolerantes du basta. *C.R. Acad. Sci. Paris,* 314 (Ser. III): 291.

Devine, M., Duke, S.O., and Fedke, C. 1993. *Physiology of Herbicide Action.* Prentice-Hall, Englewood Cliffs, NJ.

Donn, G., Tischer, E., Smith, J.A., and Goodman, H.M. 1984. Herbicide-resistant alfalfa cells: an example of gene amplification in plant cells. *J. Molec. Appl. Genet.* 2: 621.

Droge, W., Broer, I., and Puhler, A. 1992. Transgenic plants containing the phosphinothricin-*N*-acetyl transferase gene metabolize the herbicide L-phosphinothricin (glufosinate) differently from untransformed plants. *Planta* 187: 142.

Eckes, P., Vijtewaal, B., and Donn, G. 1989. Synthetic gene confers resistance to the broad spectrum herbicide L-phosphinothricin in plants. *J. Cell. Biochem.* Suppl. 13D.

Fromm, M.E., Morrish, F., Armstrong, C., Williams, R., Thomas, J., and Klein, T.H. 1990. Inheritance of chimeric genes in the progeny of transgenic maize plants. *Bio/Technology* 8: 833.

Gordon-Kamm, W.J., Spencer, T.M., Mangano, M.L., Adams, T.R., Daines, R.J., Start, W.G., O'Brien, J.V., Chambers, S.A., Adams, W.R., Jr., Willetts, N.G., Rice, T.B., Mackey, C.J., Krueger, R.W., Kausch, A.P., and Lemaux, P.G. 1990. Transformation of maize cells and regeneration of fertile transgenic plants. *Plant Cell* 2: 603.

Gressel, J. 1992a. The needs for herbicide-resistant crops. In *Resistance '91: Achievements and Developments in Combating Pesticide Resistance.* Ed. by Denholm, I., Devonshire A.L., and Hollomon, D.W. Elsevier, London, pp. 283.

Gressel, J. 1992b. Genetically-engineered herbicide-resistant crops: a moral imperative for world food production. *Agro-Ind. Hi-Tech.* 3: 3.

Kondo, Y., Samura, T., Ogawa, Y., Tsuruoka, T., Watanabe, A., Totukaya, K., Suzuki, T., Moriyama, C., Yoshima, J., and Inouye, S. 1973. Studies of a new antibiotic, SF1293. I. Isolation and physio-chemical and biological characterization of SF-1295 substances. *Sci. Rep. Seika* 13: 34.

Kramer, C., DiMaio, J., Carswell, G.K., and Shillito, R.D. 1993. Selection of transformed protoplast-derived *Zea mays* colonies with phosphinothricin and a novel assay using the pH indicator chlorophenol red. *Planta* 190: 454.

Langeludeke, P., Krause, D., Rose, E., Wallmullder, F. and Walther, K.H. 1981. Unkrautbekampfung mit Hoe 39866 im Obst- und Weinbau. *Mitt. Biol. Bundesanst. Land. Forstw.* 203: 256.

Mariani, C., Gossele, V., De Beuckeleer, M., De Block, M., Goldberg, R.B., De Greef, W., and Leemans, J. 1992. A chimeric ribonuclease-inhibitor gene restores fertility to male sterile plants. *Nature* 357: 384.

Murakami, T., Anzai, H., Imai, S., Satoh, A., Nagaoka, D. and Thompson, C.J. 1986. The bialaphos biosynthetic genes of *Streptomyces hygroscopicus*: molecular cloning and characterization of the gene cluster. *Molec. Gen. Genet.* 205: 42.

McNally, S.F., Hirel, B., Gadal, P., Mann, A.F., and Stewart, G.R. 1983. Glutamine synthetases of higher plants. *Plant Physiol.* 72: 22.

Shelp, B.J., Swanton, C.J., and Hall, J.C. 1992. Glufosinate (phosphinothricin) mobility in young soybean shoots. *J. Plant Physiol.* 139: 626.

Somers, D.A., Rines, H.W., Gu, W., Kaeppler, H.F., and Bushnell, W.R. 1992. Fertile, transgenic oat plants. *Bio/Technology* 10: 1589,.

Strauch, E., Wohlleben, W., and Puhler, A. 1988. Cloning of the phosphinothricin-*N*-acetyl-transferase gene from *Streptomyces viridochromogenes* Tu 494 and its expression in *Streptomyces lividans* and *Escherichia coli. Gene* 63: 65.

Tachibana, K., Watanabe, T., Sekizawa, Y., and Takematsu, T. 1986a. Inhibition of glutamine synthetase and quantitative changes of free amino acids in shoots of bialaphos-treated Japanese barnyard millet. *J. Pest. Sci.* 11: 27.

Tachibana, K., Watanabe, T., Sekizawa, Y., and Takematsu, T. 1986b. Accumulation of ammonia in plants treated with bialaphos. *J. Pest. Sci.* 11: 33.

Thompson, C.K., Rao, M.N., Tizard, R., Crameri, R., Davies, J.E., Lauwereys, M., and Botterman, J. 1987. Characterization of the herbicide resistance gene *bar* from *Strepromyces hygroscopicus. EMBO J.* 6: 2519.

Tischer, E., DasSarma, S., and Goodman, H.M. 1986. Nucleotide sequence of an alfalfa glutamine synthetase gene. *Molec. Gen. Genet.* 203: 221.

Toki, S., Takamatsu, S., Nojiri, C., Ooba, S., Anzai, H., Iwata, M., Christensen, A.H., Quail, P.H., and Uchimiya, H. 1992. Expression of a maize ubiquitin gene promoter-*bar* chimeric gene in transgenic rice plants. *Plant Physiol.* 100: 1503.

Uchimiya, H., Iwata, M., Nojiri, C., Samarajeewa, P.K., Takamatsu, S., Ooba, S., Anzai, H., Christensen, A.H., Quail, P.H., and Toki, S. 1993. Bialaphos treatment of transgenic rice plants expressing a *bar* gene prevents infection by the sheath blight pathogen (*Rhizoctonia solani*). *Bio/Technology* 11: 835.

Vasil, V., Castillo, A.M., Fromm, M.E., and Vasil, I.K. 1992. Herbicide resistant fertile transgenic wheat plants obtained by microprojectile bombardment of regenerable embryogenic callus. *Bio/Technology* 10: 667.

Vasil, V., Srivastava, V., Castillo, A.M., Fromm, M.E., and Vasil, I.K. 1993. Rapid production of transgenic wheat plants by direct bombardment of cultured immature embryos. *Bio/Technology* 11: 1553.

Wan, Y. and Lemaux, P.G. 1994. Generation of large numbers of independently transformed fertile barley plants. *Plant Physiol.* 104: 37.

Wang, Z., Takamizo, T., Iglesias, V.A., Osusky, M., Nagel, J., Potrykus, I., and Spangenberg, G. 1992. Transgenic plants of tall fescue (*Festuca arundinacea* Schreb.) obtained by direct gene transfer to protoplasts. *Bio/Technology* 10: 691.

Weeks, J.T., Anderson, O.D., and Blechl, A.E. 1993. Rapid production of multiple independent lines of fertile transgenic wheat (*Triticum aestivum*). *Plant Physiol.* 102: 1077.

Wild, A., Sauer, H., and Ruhle, W. 1987. The effect of phosphinothricin (glufosinate) on photosynthesis. I. Inhibition of photosynthesis and accumulation of ammonia. *Z. Naturforsch.* 42C: 263.

Wild, A. and Ziegler, C. 1989. The effect of bialaphos on ammonium-asssimilation and photosynthesis. I. Effect on the enzymes of ammonium-assimilation. *Z. Naturforsch.* 44C: 97.

Wohlleben, W., Arnold, W., Broer, I., Hillemann, D., Strauch, E., and Puhler, A. 1988. Nucleotide sequence of the phosphinothricin-*N*-acetyl-transferase gene from *Streptomyces viridochromogenes* Tu94 and its expression in *Nicotiana tabacum*. *Gene* 70: 25.

Chapter 6

COTTON WEED CONTROL USING THE BXN™ SYSTEM

David M. Stalker
John A. Kiser
Greg Baldwin
Bruce Coulombe
Catherine M. Houck

CONTENTS

1. INTRODUCTION

Herbicides have become an indispensable tool of modern agriculture in that they allow economically superior weed control and are more labor and energy efficient than manual or mechanical cultivation methods. While many agronomically useful compounds are reasonably safe because they are rapidly degraded in the biosphere and are nontoxic to animals and birds, some compounds should be replaced as they pollute the environment. This fact, coupled with information elucidating the mode of action of compounds, the progress achieved in the genetic manipulation of plants and the ability to transfer genetic information from any source organism has opened up an exciting and productive area for new agricultural products. These opportunities are further supported by the observation that many herbicide resistance determinants are dominant single-gene traits, rendering them amenable to gene transfer techniques. The result will be the use of environmentally compatible herbicides that favor a broad weed control spectrum as concern increases over contamination of the environment, toxicity to animals, and persistence of residues in soils and groundwater.

1-56670-045-0/96/$0.00+$.50
© 1996 by CRC Press Inc.

Many herbicides exert their mode of action by inactivating "target proteins" essential for vital plant functions such as the photosynthetic complex or other essential biochemical pathways. As crop plants usually share these processes with weeds, herbicides are often nonselective. Other compounds can be utilized selectively on tolerant crops, mainly as a consequence of differential uptake or metabolism of the herbicide, or by a precise localization and timing of the herbicide application. An alternative is to confer resistance in crops to broad-spectrum compounds with a more favorable environmental profile.

Two approaches have been described generally for the genetic engineering of herbicide resistance in crops: (1) the modification of a plant enzyme or other sensitive biochemical target of herbicide action to render it insensitive to the herbicide or to induce the overproduction of the unmodified target protein permitting normal metabolism to occur, and (2) the introduction of an enzyme or enzyme system that will degrade or detoxify the compound in the plant prior to the herbicide reaching the target site of action. Plants modified in these ways may be obtained either by selection for resistance against a specific compound or by applying gene transfer techniques utilizing genes encoding herbicide resistance determinants. The transfer of heterologous genes to crop plants may be brought about by the use of gene-transfer vectors derived from soil bacteria such as *Agrobacterium tumefaciens* or *A. rhizogenes*. These bacteria are used as hosts to deliver the vector system to plant cells. Genes may also be introduced into plant cells as DNA by various microprojectile devices. In either case the gene is stably integrated into the plant cell genome. Whole plants containing the desired herbicide resistant trait can be derived from either of these two processes.

In this chapter we describe the development of a herbicide resistance determinant that detoxifies the herbicide bromoxynil (marketed under the trade name Buctril®). This gene was introduced into cotton by an *Agrobacterium* vector system resulting in bromoxynil-resistant transgenic cotton plants. We further chronicle the development of these plants through full field tests and the analysis of this potential product as an alternative weed control system for the cotton grower. Finally, we describe the potential of this novel cotton weed control system in providing a nonpersistent, environmentally compatible herbicide and the resulting environmental benefits with respect to cotton weed control practices now in use.

2. CHARACTERIZATION OF THE BROMOXYNIL RESISTANCE GENE AND BROMOXYNIL-RESISTANT TRANSGENIC COTTON

A bacterial degradation/detoxification system has been identified for the herbicide bromoxynil (3,5-dibromo 4-hydroxybenzonitrile). Bromoxynil is a broadleaf weed killer currently registered for use on a variety of tolerant crops, such as wheat, barley, oats, onions, garlic, and certain varieties of corn and alfalfa. Bromoxynil is a contact herbicide that is a potent photosystem II inhibitor and is neither translocated nor concentrated within the plant upon spraying. Early metabolism studies in tolerant wheat (Buckland et al. 1973) and corn revealed that a primary metabolic pathway for bromoxynil is the conversion of the herbicide to 3,5-dibromo 4-hydroxybenzamide (with the following conversion to 3,5-dibromo 4-hydroxybenzoic acid). These metabolites can be further degraded by other enzymes that cleave the bromine atoms and subsequently attack the benzoic acid ring structure. It is known that the two metabolites described above are at least 100-fold less toxic to plant cells than bromoxynil; therefore, conversion of bromoxynil to either of these metabolites should inactivate the compound and result in a bromoxynil-resistant phenotype.

Bromoxynil, when used in field situations, is rapidly converted in soil via pathways similar to those described in naturally tolerant crops (Collins 1973; Ingram and Pullin 1974). Dependent upon the soil type and other environmental conditions, the half-life of bromoxynil in the environment can be as short as 7 days. In soil this conversion appears to bypass the amide derivative of bromoxynil and proceeds directly to 3,5-dibromo 4-hydroxybenzoic acid in a single step reaction (Hsu and Camper 1975).

An enzyme that acts on the cyano moiety of nitrile-containing compounds is termed a nitrilase. Nitrilases appear to be quite ubiquitous in nature and nitrilase enzymes have been described that act on a variety of aliphatic, cyano-substituted, and aromatic-nitrile containing compounds (Harper 1977; Bandyopadhyay et al. 1986; Kobayashi et al. 1990; Chand Balla et al. 1992). There was also an early report on the identification of a nitrilase activity with a fairly nonspecific substrate profile in crude extracts of barley leaf tissue (Thimann and Mahadevan 1964) and the recent identification and cloning of a nitrilase from *Arabidopsis* (Bartling et al. 1992). None of these enzymes, however, exhibited specificity for the herbicide bromoxynil. A bacterium, identified as *Klebsiella ozaenae*, has been isolated from bromoxynil-contaminated soil. This isolate utilized bromoxynil as a sole nitrogen source and expressed a nitrilase that demonstrated a strict specificity for bromoxynil as substrate in relation to other benzonitrile compounds (McBride et al. 1986). This nitrilase was subsequently shown to be plasmid encoded in *Klebsiella ozaenae* (Stalker and McBride 1987), and a gene encoding a 37,000 mol wt polypeptide was cloned and expressed in *Escherichia coli* (Stalker et al. 1988a). The bacterial gene was spliced to plant promoters such as the tobacco RUBP carboxylase small subunit (SSU) and the CaMV 35S promoters and these chimeric genes introduced into transformed and regenerated tobacco plants (*Nicotiana tabacum*). Bromoxynil resistance (Stalker et al. 1988b) was obtained when these tobacco plants were sprayed with up to tenfold field rates of commercial formulations of bromoxynil (Buctril®) without any symptom of injury. Chimeric genes expressing the bromoxynil-specific nitrilase were subsequently introduced into cotton varieties via *Agrobacterium* transformation and these plants also displayed high levels of resistance to Buctril® when tested in the greenhouse and in the field (Color Plate 4*). The gene encoding the bromoxynil-specific nitrilase is inherited in a Mendelian fashion and exhibits a dominant character when the plants are sprayed with bromoxynil. Bromoxynil-resistant (BXN™) cotton has now been field tested for 5 years and rates as high as 11.2 kg a.i./ha (0.56 kg a.i./ha constitute a normal field use rate) have been sprayed with no damage sustained by the plants. These plants can undergo multiple applications or be sprayed at any time during the plant growing cycle.

The question arises as to the basis for metabolism of bromoxynil in the genetically modified cotton plants expressing the bromoxynil-specific nitrilase. To determine the metabolites present in normal and BXN™ cotton, a metabolism study was conducted utilizing uniformly labeled ^{14}C-bromoxynil, spiked into a normal bromoxynil application. Selected leaves of normal and BXN™ cotton plants were treated with 0.84 and 4.2 kg a.i./ha bromoxynil, respectively, the leaves harvested 87 d after the initial treatment, and the remaining radioactive compounds extracted and measured. Metabolite levels were monitored by high-performance liquid chromatography HPLC analysis, compared to known metabolite standards, and the structure of the metabolites identified in the normal and BXN™ plants indicated in Figure 1. Normal cotton leaves (soluble and insoluble material) contained residues exclusively as bromoxynil with a trace (>0.03%) of the metabolite 3,5 dibromo 4-hydroxybenzoic acid, indicating no metabolism of the herbicide. This trace amount probably occurs via spontaneous conversion of bromoxynil over an extended period of time. Leaves from genetically modified BXN™ cotton plants were extracted and the metabolite pattern indicated that all the bromoxynil had been converted to other compounds, the majority of the applied radioactivity (61%) residing in the primary metabolite 3,5-dibromo 4-hydroxybenzoic acid in addition to small amounts (between 3 and 6%) of other compounds. These metabolites, such as 3-bromo 4-hydroxybenzoic acid, 2,6-dibromophenol, and 3,5-dibromo 4-hydroxybenzoic acid are further breakdown products of the primary metabolite 3,5-dibromo 4-hydroxybenzoic acid and indicate that the bromoxynil-specific nitrilase was efficient in the conversion of bromoxynil to 3,5-dibromo 4-hydroxybenzoic acid and was responsible for the level of resistance demonstrated in BXN™ cotton. Also, approximately 35% of the ^{14}C-labeled bromoxynil applied could not be detected, indicating that complete metabolic conversion of some portion of the application occurred. This was inferred as the radioactive bromoxynil was uniformly labeled within the benzene ring structure and loss of ^{14}C-label from this moiety would be indicative of complete conversion of the

---

* Color plates follow page 84.

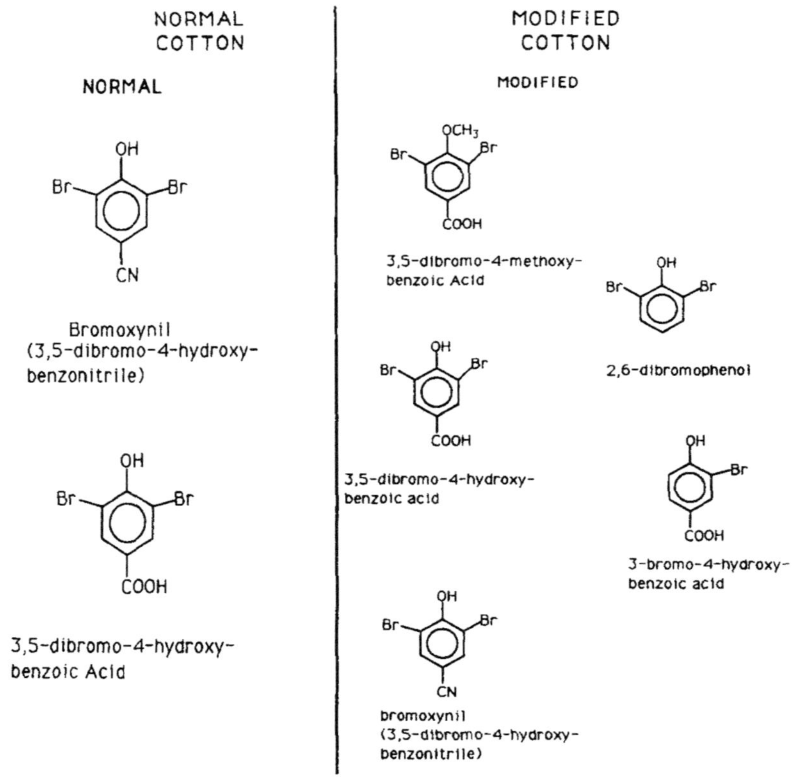

FIGURE 1 Metabolites of control (normal) cotton and modified cotton expressing the bromoxynil-specific nitrilase.

herbicide. Finally, all of the metabolites found in BXN™ cotton were identical to metabolites previously identified in bromoxynil-tolerant crops such as wheat, barley, and corn (Threfall 1969; Schafer and Chilcote 1970; Buckland et al. 1973; Gemma and Wargo 1982; Kovacs 1986), indicating that the environmental introduction of an uncharacterized metabolite is not an issue when BXN™ cotton is utilized in a field situation.

3. DEVELOPMENT OF BXN™ COTTON AS A HERBICIDE-RESISTANT CROP

Resistance to the herbicide bromoxynil is only one agronomic trait and although Buctril® resistance provides significant advantage to growers, it does not substitute for the traits of primary importance to cotton producers: lint yield and fiber quality. Field evaluation of BXN™ cotton lines and selection for agronomically superior strains is an ongoing process necessary to produce premium BXN™ varieties.

Cotton lines containing the BXN™ gene have been extensively field tested for ten growing seasons over the last 5 years in over 102 U.S. sites in 12 states and several overseas locations (Coulombe et al. 1994; Ward et al. 1993; Baldwin et al. 1992; Kiser and Mitchell 1991; Mitten and Kiser 1990).

Figure 2 shows a map of BXN™ cotton field locations for the 1993 summer season. Tests were done in 10 major cotton-producing states at a total of 53 locations. Many tests were done by leading weed scientists and cotton extension agronomists to compare weed management systems and variety performance under different weed management regimes (Sanders and Burch 1994; Murdock 1994;

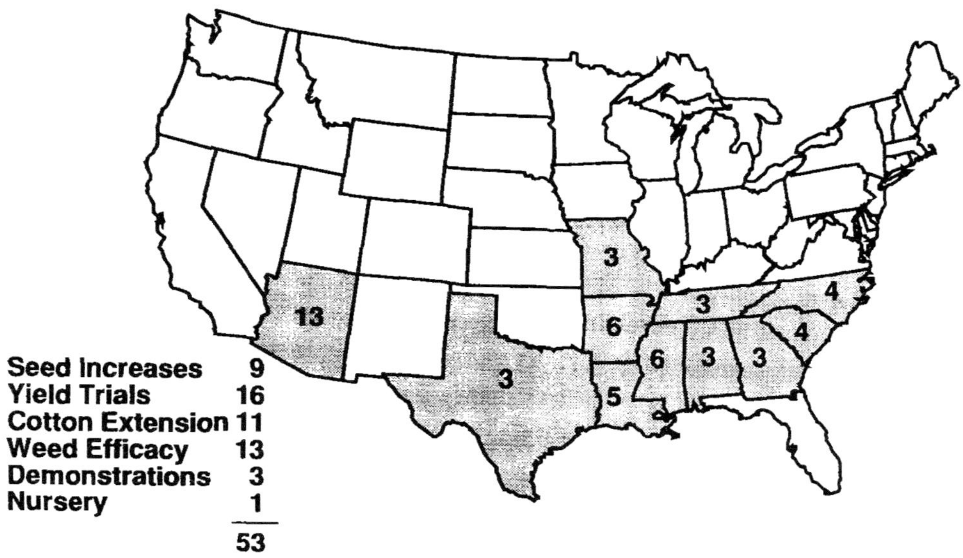

FIGURE 2 U.S. map indicating the locations for and the number of BXN™ cotton field trials for 1993. The number of trials of each type is listed at the left.

Richburg and Wilcut 1994; Keeton et al. 1994; Beaty and Guy 1994; Smith et al. 1993; Wilcut et al. 1992, 1993; Richburg et al. 1993; Grey et al. 1993; McLaughlin 1992).

As in any plant breeding program, individual plant selections were made from BXN™ progeny and strains were tested in replicated trials for lint yield, fiber quality (HVI testing, particularly for fiber length, strength, and fineness), and plant maturity (earliness). Tests were done in multiple years in different locations to allow additional agronomic observations and to allow measurement of yield stability in different environments.

Results of replicated yield trials were presented (Coulombe et al. 1994; Ward et al. 1993) and some of the mid-South yield data are summarized in Table 1. Many BXN™ strains have been selected which yield significantly more lint than the nontransgenic parental variety. Several lines yielded as well as locally adapted commercial varieties, indicating that agronomically superior varieties can be selected from these transgenic strains.

Fiber quality is a key component of a successful cotton variety. BXN™ strains are routinely screened and selected for superior fiber characteristics. Table 1 summarizes the key fiber characteristics found in transgenic strains compared not only to the nontransgenic parent, but to other current commercial varieties grown in the same locations. These trials were carried out with grower cooperators and standard weed control practices specific for each grower location were utilized. No Buctril® was applied in any of the yield trials. Clearly, strains that combine acceptable agronomic characteristics, excellent fiber quality, and resistance to foliar applications of Buctril® herbicide have been selected.

Seed increases for commercial production of BXN™ cotton varieties are currently underway and, pending final action by regulatory agencies, experimental lines are scheduled for small-scale commercial introduction in the 1994 season with commercial release for sale the following year.

Second-generation BXN™ varieties are being produced from crosses using backcross and pedigree breeding methods. It is easy to screen progeny for the presence of the bromoxynil-specific nitrilase gene and, because it is inherited as a simple Mendelian trait, it is not difficult to select homozygous progeny for further development.

TABLE 1 Summary of 1993 BXN™ Yield and Fiber Quality at Five Mid-South Locations

Entry[a]	Entry Type	Lint Yield[b]	Fiber Strength[c]	Fiber Length[d]	Micronaire
ST453	Nontransgenic	1165	26	27.94	4.9
ST LA887	Nontransgenic	1100	30	28.70	4.8
BXN10215-8	Transgenic	1050*	27	28.45	5.1
BXN10215-31	Transgenic	1029*	26	27.18	5.1
DPL5415	Nontransgenic	960	28	27.94	5.0
BXN10217-14	Transgenic	927	25	27.94	5.0
BXN10217-15	Transgenic	904	25	28.45	4.8
BXN10217-1	Transgenic	878	26	28.45	4.9
BXN10211-1	Transgenic	857	26	28.70	4.7
BXN10211-20	Transgenic	847	26	28.96	4.5
BXN10222-1	Transgenic	749	26	28.45	4.5
ST C315	Nontransgenic	735	27	28.96	4.6
Mean		798	26	27.94	4.9
L.S.D.		92	1.2	0.022	0.2

Note: Yields marked with an asterisk are not significantly different from the mean of the three highest yielding control varieties (mean of ST453, ST LA887, and DPL 5415 = 1072 kg/ha).

[a] Entries with BXN prefix are transgenic BXN™ cotton lines; others are commercial varieties grown as controls.
[b] Lint yield expressed as kilograms of lint per hectare.
[c] Strength expressed as grams per tex.
[d] Length of fiber (millimeters).

4. DEVELOPMENT OF THE BXN™ SYSTEM FOR COTTON WEED CONTROL

Herbicide resistant BXN™ cotton is a viable alternative for growers because it provides significant advantages over existing weed management systems. Broadleaf weed control in cotton is currently achieved through a combination of mechanical cultivation between the rows, soil-applied systemic herbicides (preemergent and/or preplant incorporated), and postdirected applications using herbicides that are toxic to cotton if they make significant contact with the cotton plant (Abernathy and McWhorter 1992). There are several problems with these existing systems.

The postdirected applications are extremely dependent on timing because they rely on the ability to apply herbicides when the cotton is tall enough to allow precise placement of the nozzles under the leaves (to avoid injury to the crop), but before the weeds have grown to the same height as the cotton. Weather and field conditions, particularly in the mid-south and southeastern cotton growing regions often do not allow postdirected applications at the desired time, resulting in crop injury and/or inability to effectively control the weeds.

A different issue arises with the preplant incorporated and preemergence herbicide applications. These must be used before the weeds have emerged and are thus applied before it is known whether there will be a weed problem in a particular field. This is prudent in the existing situation because it is difficult to control weeds in a postemergent manner. However, the BXN™/Buctril® system will allow the grower to apply the broadleaf herbicide only if there is a weed problem, and only at the rate necessary to control the weeds that are actually present. This flexibility in application timing and rate, coupled with the ability to respond to the field situation are key components needed to implement integrated pest management (IPM) systems for weed control in cotton (Linker and Coble 1993).

Much of the development of the BXN™/Buctril® weed control system over the last few years has focused on characterizing the herbicide application parameters, in particular, defining the weed species that are susceptible and the minimum herbicide rates that are needed to control different size weeds of the susceptible species (Smith et al. 1993; Richburg et al. 1993; Grey et al. 1993; Wilcut et al. 1992, 1993; McLaughlin 1992). Because the majority of cotton acreage is cultivated between the rows, Buctril® application rates can be refined further by varying the band width and applying the material only to a narrow band over the cotton row, while cultivating between the rows (Murdock 1994; Richburg and Wilcut 1994; Keeton et al. 1994; Beaty and Guy 1994).

Many such studies have been done by weed researchers across the Cotton Belt. We present examples from the second year of a weed management study that was done on the Stoneville Pedigreed Seed Co. farm during the 1992 and 1993 seasons. The objectives of the study were (1) to compare broadleaf weed control and cotton yields in plots managed under several BXN™/Buctril® weed control systems vs. standard weed management systems, and (2) to determine if BXN™ systems can reduce the number of herbicide applications and the quantity of herbicides used in cotton production without reducing lint yield.

Tables 2 and 3 summarize the results of these studies. Each weed management regime was applied to four row plots arranged in a randomized complete block with four replications. Weed

TABLE 2 Yield Comparisons Using BXN™/Buctril® System vs. Standard Weed Management Systems: No Preemergence Herbicide Applications

Weed Management System (Broadleaf)	Rate (kg a.i./H) Broadcast Basis	% Application Time[a]	% Morningglory Control[b]	% Cocklebur Control[b]	Lint Yield[c] (% of Standard)
Treflan®[d] (hand weeded)	0.56	PPI	100.0	100.0	98
Treflan® (no postemergence control)	0.56	PPI	0.0	0.0	2
Treflan®	0.56	PPI			
Buctril® (broadcast)	0.56	Early post	92.5	100.0	101
Treflan®	0.56	PPI			
Buctril® (broadcast)	0.56	Early post	91.3	100.0	
Buctril® (postdirected, 13 in. band)	0.56	Late post	100.0	98.8	106
Treflan®	0.56	PPI			
Buctril® (over-the-top, 13 in. band)	0.56	Early post	93.8	100.0	93
Treflan®	0.56	PPI			
Buctril® (over-the-top, 13 in. band)	0.56	Early post	93.3	100.0	
Buctril® (postdirected, 13 in. band)	0.56	Late post	96.7	100.0	112
Treflan®	0.56	PPI			
Cotoran®[e] + MSMA[f] (postdirected, 16 in. band)	0.47 + 0.94	Early post	81.6	82.5	
Bladex®[g] + MSMA (postdirected 16 in. band)	0.28 + 0.94	Late post	80.0	88.3	90

[a] PPI is preplant incorporated herbicide applied 1 month prior to planting. Early postemergence applications were made when cotton plants were 2.5 in. tall. Late postemergence applications were made when cotton plants were 19 in. tall.

[b] Weed ratings were taken 5 d after the early postemergence treatment and 4 d after the late postemergence treatment.

[c] Yield is reported as % of yield from plots that received the standard herbicide control treatment of Treflan® PPI, Cotoran® PRE, Cotoran® + MSMA early postdirected, Bladex® + MSMA late postdirected.

[d] Treflan® is a registered trademark of DowElanco Chemical Company for trifluralin.

[e] Cotoran® is a registered trademark of CIBA-GEIGY for fluometuron.

[f] MSMA is a monosodium methanearsonate.

[g] Bladex® is a registered trademark of E.I. DuPont de Nemours & Co. for cyanazine.

TABLE 3 Yield Comparisons Using BXN™/Buctril® System vs. Standard Weed Management Systems: Preemergence Herbicide Applications Included

Weed Management System (Broadleaf)	Rate (kg a.i./H) Broadcast Basis	% Application Time[a]	% Morningglory Control[b]	% Cocklebur Control[b]	Lint Yield[c] (% of Standard)
Treflan®	0.56	PPI			
Cotoran® (no postemergence control)	0.47	PRE	0.0	0.0	13
Treflan®	0.56	PPI			
Cotoran®	0.47	PRE			
Buctril® (broadcast)	0.56	Early post	93.8	100.0	106
Treflan®	0.56	PPI			
Cotoran®	0.47	PRE			
Buctril® (broadcast)	0.56	Early post	94.3	100.0	
Buctril® (postdirected, 13 in. band)	0.56	Late post	97.5	97.5	109
Treflan®	0.56	PPI			
Cotoran®	0.47	PRE			
Buctril® (over-the-top, 13 in. band)	0.56	Early post	98.0	100.0	103
Treflan®	0.56	PPI			
Cotoran®	0.47	PRE			
Buctril® (over-the-top, 13 in. band)	0.56	Early post	96.8	100.0	
Buctril® (postdirected, 13 in. band)	0.56	Late post	93.8	95.0	110
Treflan®	0.56	PPI			
Cotoran®	0.47	PRE			
Cotoran® + MSMA (postdirected 16 in. band)	0.47 + 0.94	Early post	86.7	100.0	
Bladex® + MSMA (postdirected 16 in. band)	0.27 + 0.94	Late post	90.0	95.0	100[d]

[a] PPI is preplant incorporated herbicide applied 1 month prior to planting. PRE is preemergence herbicide application applied at planting. Early postemergence applications were made when cotton plants were 2.5 in. tall. Late postemergence applications were made when cotton plants were 19 in. tall.

[b] Weed ratings were taken 5 d after the early postemergence treatment and 4 d after the late postemergence treatment.

[c] Yield is reported as % of yield from plots that received the standard herbicide control treatment of Treflan® PPI, Cotoran® PRE, Cotoran® + MSMA early postdirected, Bladex® + MSMA late postdirected.

[d] Standard weed control treatment, for yield comparison.

control ratings are shown for morningglory and cocklebur, which were seeded into these plots. These two weed species were reported to cause between 30 and 44% of the yield reduction attributable to cotton weeds in the key mid-South cotton-producing states of Mississippi, Louisiana, Arkansas, Tennessee, and Missouri (Chandler and Cooke 1992). Each of the eight Buctril®-containing treatments had better than 90% control of these two pivotal weeds. A standard weed control system consisting of preplant incorporated, preemergence, and two postdirected herbicide sprays (Table 3, last row) also gave over 90% control of cocklebur, but only 86.7 to 90% control of morningglory within 4 to 5 d after postemergence treatments. An important observation is that soil-applied preemergence herbicide treatments were necessary for this level of weed control in the standard system, but were not necessary to achieve over 90% control of these weeds in the Buctril®-containing treatments (Table 2; last row is standard treatment without preemergence herbicide).

Crop injury was not observed in any of the Buctril®-treated plots. In contrast, the plots treated with postdirected applications of Cotoran® plus MSMA and Bladex® plus MSMA had crop injury ratings of 23 to 25% at 5 d post-treatment.

Most importantly, all but one of the Buctril® treatments produced higher lint yield than plots that had standard weed control measures (Tables 2 and 3), although the differences were not statistically significant. This demonstrates that effective weed control leading to consistently high lint yield can be achieved using the BXN™/Buctril® system. The system increases the grower's options for weed control and provides more flexibility in application methods and timing.

5. ENVIRONMENTAL CONSEQUENCES OF THE BXN™ COTTON/BUCTRIL® WEED CONTROL SYSTEM

Several issues have been raised concerning environmental impact resulting from farmers and growers using the BXN™ cotton/Buctril® weed control system. These include the possibility that BXN™ cotton might be more weedy than other cotton varieties; the transfer of the bromoxynil resistance gene to other plant species in the cotton-growing environment; the potential for increased use of Buctril® as BXN™ cotton is introduced; and the impact on total herbicide use for cotton as a crop.

The U.S. Department of Agriculture (USDA) recently conducted a detailed study of the issues under its jurisdiction and determined that there is no reason to regulate BXN™ cotton any differently than traditionally bred cotton varieties. Their ruling included an Environmental Assessment that reached a "Finding of No Significant Impact" in accordance with the National Environmental Policy Act. In making this determination, USDA considered scientific literature, expert opinions, public comments, and data gathered in the 5 years of field trials with cotton containing the BXN™ gene.

Although cotton is not a weed pest, it is useful to address concerns about introduction of genetically modified plants in the environment and the potential to produce a new weed or increase the aggressiveness of existing weeds (Tiedje et al. 1989). It can be demonstrated that BXN™ cotton is no more a weed pest risk than any other cotton variety and that cotton itself is not a weed pest. Three aspects of weediness are of general concern: comparison of transformed crops with exotic species; potential for transformed, domesticated crops to convert to a weedy state; potential for hybridization between domesticated crops; and wild relatives creating or enhancing that crop's weediness potential (Keeler 1989).

The analogy between the introduction of an exotic species into a new environment and the introduction of a genetically modified crop plant is not readily applicable to BXN™ cotton. Introduced exotic plants that are pests bring with them many traits that enhance weediness and in some cases leave behind control organisms (predators) and competitors. Genetically modified plants are altered in only a few characteristics that relate to crop production characteristics, will generally not be released into exotic environments, and will be planted in growing regions typical for that specific crop. Cotton has been grown in areas of Africa, North and South America, Asia, Australia, and southern Europe without becoming a weed pest and the worldwide crop is highly characterized and is a major source of fiber and vegetable oil. Because cotton is an exotic species in the U.S. and has not become a weed pest, the concern of exotics becoming pests on introduction into a new environment is not warranted for BXN™ cotton. Also, there is no increased potential for BXN™ cotton to become a weed pest as the gene encoding the bromoxynil-specific nitrilase confers no selective advantage outside agricultural fields where the herbicide Buctril® will be utilized. Finally, upland cotton, *Gossypium hirsutum*, does not outcross with weedy relatives in the continental U.S., nor does it display significant potential to develop into a weed or contain attributes associated with weed pests (Keeler 1989). Similarly, BXN™ cotton possesses the same agronomic potential as other cotton varieties, which have never been considered weeds, even in agricultural fields. In general, cotton lacks many of the traits characteristic of weed pests and has not been considered a weed pest in the U.S.

Another concern is whether transgenic crops might have increased weediness due to the introduction of herbicide resistance and that the introduced resistance gene resistance may be spread in the environment. Key factors in evaluating the weediness risk of transgenic herbicide resistant plants are the possibility of volunteer plants becoming weeds in a subsequent year or the potential for introgression of herbicide resistance genes into weedy relatives. Corn, wheat, sunflower, and sugarbeets are classical examples of crops which can become volunteer weeds, but are controlled in subsequent crop rotation regimes by cultivation and the use of different herbicide products. Natural resistance for some herbicides already exists in some weed species and this resistance is determined genetically and can be transferred to related species. For a herbicide resistance gene to persist in a weed population, the gene must offer a selective advantage to those plants possessing it. This can occur when the herbicide in question is applied continuously for a number of generations and/or the herbicide persists in the environment to create a continuous selective pressure to maintain the gene (Gressel 1984). One way to control this problem (whether the gene is derived from classical breeding or genetic engineering) is by incorporating appropriate agricultural practices. A second way is to apply herbicides that are rapidly degraded in the environment to ensure that a long-term selective situation is not created. Herbicide resistance is not an environmental concern for BXN™ cotton because BXN™ cotton is not a weed, will be grown only in agricultural fields where Buctril® is utilized, and does not persist in the environment. BXN™ cotton does not have any weedy, sexually compatible relatives that could obtain the gene encoding the bromoxynil-specific nitrilase via outcrossing. Finally, the environmental half-life (Hsu and Camper 1975) of bromoxynil (7 to 10 d depending on the soil type) is such that a field-selective situation is not created. No excessive pressure is exerted to maintain or outcross the bromoxynil-specific nitrilase nor to select for any indigenous weeds resistant to bromoxynil. Bromoxynil, in various formulations, has been extensively used for the past 15 years on crops such as wheat, barley, onion, and garlic. During this period of time there have been no published reports of bromoxynil-resistant weeds appearing in fields sprayed with Buctril® or its derivatives.

Weed control is a major problem in cotton production and depends heavily on the use of herbicides, supplemented by mechanical cultivation methods. It is estimated that cotton crop loss from weeds constitutes $600 million annually despite the expenditure of $200 million/year in weed control costs (Chandler and Cooke 1992). In the weed management trials performed with BXN™ cotton (above) Buctril® was found to effectively control key broadleaf weeds when applied in a band equivalent to 1.7 kg a.i./ha (broadcast amount 0.56 kg a.i./ha). This regime also included Treflan® preplant incorporated and in several studies Cotoran® was used preemergent. These studies suggest that the use of Buctril® in cotton can dramatically reduce overall herbicide use. Eventually, the actual reduction in total amount of herbicides realized with the BXN™ cotton/Buctril® weed control system will depend on a number of components and require several years of actual field use before precise data can be obtained. It is possible, however, to provide estimates based on published reports for potential herbicide reduction using the above system for successful weed control. Two examples are shown in Tables 4 and 5. Table 4 indicates the estimated herbicide reduction when no change in the number of herbicide applications is used. A Treflan®, Cotoran®, Buctril® herbicide regime offers reductions of 41 and 45%, respectively, in the total amount of herbicide a.i. applied per hectare over a standard herbicide regime and an alternative weed management program developed by USDA weed extension scientists in Mississippi (Chandler and Cooke 1992). Table 5 indicates the estimated reduction when the preemergent herbicide application is eliminated altogether. The Treflan®, Buctril® regime offers reductions of 47 and 51%, respectively, over the standard regime and Chandler and Cooke's system. If these projections are realized when enough statistical data become available, they represent a considerable reduction in the total amount of herbicide introduced into the environment over weed control methods now employed.

The use of the BXN™ cotton/Buctril® weed control system could significantly reduce the amount of herbicides applied during a growing season and result in less expense to the cotton grower for controlling weeds. The use of of herbicide resistant crops, such as BXN™ cotton, has the potential

TABLE 4 Example of Herbicide Reduction Using a BXN™ Cotton/Buctril® System: No Change in Number of Herbicide Applications

Treatment	Herbicide	kg a.i./ha/trial (broadcast equivalent)	kg a.i./year in Mississippi[a]
Buctril®	Treflan®	0.56	
	Cotoran® (banded)	0.47	
	Buctril® (banded)	0.56	
	Buctril® (banded)	0.56 = 2.15 total	1,086,000
Standard herbicide regime	Treflan®	0.56	
	Cotoran® (banded)	0.47	
	Cotoran® +	0.47	
	MSMA (banded)	0.94	
	Bladex® +	0.27	
	MSMA (banded)	0.94 = 3.65 total	1,850,000
Chandler and Cooke (1992) Mississippi weed control	Treflan® +	0.56	
	Zorial® (broadcast)	0.54	
	Cotoran® +	0.35	
	Zorial® (banded)	0.13	
	Cotoran® +	0.22	
	MSMA (banded)	0.56	
	Caparol® (banded)	0.15	
	Bladex® (broadcast)	1.40 = 3.91 total	1,975,000

[a] 504,000 ha planted in 1991 in Mississippi (USDA 1992).

TABLE 5 Example of Herbicide Reduction Using a BXN™Cotton/Buctril® System: Elimination of the Preemergence Herbicide Application

Treatment	Herbicide	kg a.i./ha/trial (broadcast equivalent)	kg a.i./year in Mississippi[a]
Buctril®, no preemergent	Treflan®	0.56	
	Buctril® (banded)	0.56	
	Buctril® (banded)	0.56 = 1.68 total	849,000
Standard herbicide regime, no preemergent	Treflan®	0.56	
	Cotoran®	0.47	
	+ MSMA (banded)	0.94	
	Bladex®	0.27	
	+ MSMA (banded)	0.94 = 3.18 total	1,612,000
Chandler and Cooke (1992) (preemergent herbicide deleted)	Treflan® +	0.56	
	Zorial® (broadcast)	0.54	
	Cotoran® +	0.22	
	MSMA (banded)	0.56	
	Caparol® (banded)	0.14	
	Bladex® (broadcast)	1.40 = 3.42 total	1,732,000

[a] 504,000 ha planted in 1991 in Mississippi (USDA 1992).

to increase flexibility and ease of weed control for growers and will promote more cost-effective integrated pest management systems for weeds.

ACKNOWLEDGMENTS

The authors wish to thank the following people for contributing to the success of this project: Kevin McBride for initial gene discovery; JoAnne Fillatti for initiating cotton transformation and the entire cotton transformation team; Sam Yenne at Rhone Poulenc for providing metabolism data; Don Panter, Roger Ward, and Josh Stanton for field trial data; and Keith Redenbaugh and Lori Malyj for regulatory information and discussion.

REFERENCES

Abernathy, J. R. and McWhorter, C. G. 1992. Evaluation of weed control in cotton. In *Weeds of Cotton: Characterization and Control.* Ed. by McWhorter, C.G. and Abernathy, J.R. The Cotton Foundation, Memphis, TN, p. 1.

Baldwin, G., Kiser, J., Mitchell, J., and Germany, A. 1992. Impacts of Buctril® herbicide spray on Stoneville cotton containing the Bromotol™ gene. *Proc. Beltwide Cotton Prod. Res. Conf.* 555.

Bandyopadhyay, A.K., Nagasawa, T., Asano, Y., Fujishiro, K., Tani, Y., and Yamada, H. 1986. Purification and characterization of nitrilases from *Arthrobacter* sp. strain J-1. *Appl. Environ. Microbiol.* 51: 302.

Bartling, D., Seedorf, M., Mithofer, A., and Weiler, E.W. 1992. Cloning and expression on an *Arabadopsis* nitrilase which can convert indole-3-acetonitrile to the plant hormone, indole-3-acetic acid. *Eur. J. Biochem.* 205: 417.

Beaty, I.D. and Guy, C.B. 1994. Weed control in cotton using combinations of Buctril and Staple. *Proc. Beltwide Cotton Prod. Res. Conf.* p. 1695.

Buckland, J.L., Collins, R.F., and Pullin, E.M. 1973. Metabolism of bromoxynil octanoate in growing wheat. *Pestic. Sci.* 4: 149.

Chand Balla, T., Miura, A., Wakamoto, A., Ohba, Y., and Furuhashi, K. 1992. Asymmetric hydrolysis of a-aminonitriles to optically active amino acids by a nitrilase of *Rhodococcus rhodochrous* PA-34. *Appl. Microbiol. Biotechnol.* 37: 184.

Chandler, J.M. and Cooke, F.T., Jr. 1992. Economics of cotton losses caused by weeds. In *Weeds of Cotton: Characterization and Control.* Ed. by McWhorter, C.G., and Abernathy, J.R. The Cotton Foundation, Memphis, TN, p. 85.

Collins, R.F. 1973. Perfusion studies with bromoxynil octanoate in soil. *Pestic. Sci.* 4: 181.

Coulombe, B.A., Panter, D.M., Stanton, J.J., and Ward, R.G. 1994. Yield performance of transgenic cotton strains resistant to Buctril® herbicide. *Proc. Beltwide Cotton Prod. Res. Conf.* p. 651.

Gemma, A.A. and Wargo, J.P. 1982. Metabolism of ^{14}C-Bromoxynil in Field Corn. Reference No. 82/136/BHL/A6,ASD No. 82/029. Rhone-Poulenc Ag Company.

Gressel, J. 1984. *Origins and Development of Adaption.* Ciba Foundation Symposium 102, Pitman Books, London, p. 73.

Grey, T.L., Wilcut, J.W., and Richburg, J.S., III. 1993. Buctril efficacy on annual broadleaf weeds and mixtures with graminicides for grass control in Georgia cotton. *Proc. Beltwide Cotton Prod. Res. Conf.* p. 1530.

Harper, D.B. 1977. Fungal degradation of aromatic nitriles: enzymology of C-N cleavage by *Fusarium solani. Biochem. J.* 167: 685.

Hsu, J.C. and Camper, N.D. 1975. Isolation of ioxynil degraders from soil-enrichment cultures. *Can. J. Microbiol.* 22: 537.

Ingram, G.H. and Pullin, E.M. 1974. Persistance of bromoxynil in three soil types. *Pestic. Sci.* 5: 287.

Keeler, K. 1989. Can genetically engineered crops become weeds? *Biotechnology* 7: 1134.

Keeton, A., Murdock, E.C., Stapleton, G.S., and Toler, J.E. 1994. Response of broadleaf weeds to Staple (DPX-PE350) bromoxynil and glyphosate. *Proc. Beltwide Cotton Production Res. Conf.* p. 1705.

Kiser, J. and Mitchell, J. 1991. Bromoxynil safened cotton: trialing history, yield and quality. *Proc. Beltwide Cotton Prod. Res. Conf.* p. 566.

Kobayashi, M., Yanaka, N., Nagasawa, T., and Yamada, H. 1990. Purification and characterization of a novel nitrilase of *Rhodococcus rhodochrous* K22 that acts on aliphatic nitriles. *J. Bacteriol.* 172: 4807.

Kovacs, M.F. 1986. Regulatory aspects of bound residues. *Res. Rev.* 97: 1.

Linker, H.M. and Coble, H.D. 1993. Row crop weed management strategies in integrated pest management systems. In *Successful Implementation of Integrated Pest Management for Agricultural Crops.* Ed. by Leslie, A.R. and Cuperus, G.W. Lewis Publishers, Boca Raton, FL, chap. 7.

McBride, K.E., Kenny, J.W., and Stalker, D.M. 1986. Metabolism of the herbicide bromoxynil by *Klebsiella pneumoniae* subsp. *ozaenae. Appl. Environ. Microbiol.* 52: 325.

McLaughlin, R. 1992. Review of the 1991 field trial results on bromoxynil-tolerant cotton. *Proc. Beltwide Cotton Production Res. Conf.* p. 1316.

Mitten, D. and Kiser, J. 1990. Genetically-engineered herbicide tolerant cotton. *Proc. Beltwide Cotton Prod. Res. Conf.* p. 97.

Murdock, E.C. 1994. Weed control in bromoxynil-tolerant cotton. *Proc. Beltwide Cotton Prod. Res. Conf.* p. 1700.

Richburg, J.S., III and Wilcut, J.W. 1994. Annual broadleaf weed efficacy with bromoxynil. *Proc. Beltwide Cotton Prod. Res. Conf.* p. 1702.

Richburg, J.S., III, Wilcut, J.W., and Ingram, E.G. 1993. Buctril systems for weed management in transgenic cotton. *Proc. Beltwide Cotton Prod. Res. Conf.* p. 1530.

Sanders, D. and Burch, T. 1994. Taking transgenic cotton to the farm. *Proc. Beltwide Cotton Production Res. Conf.* p. 1697.

Schafer, D.E. and Chilcote, D.O. 1970. Translocation and degradation of bromoxynil in a resistant and a susceptible species. *Weed Sci.* 18: 729.

Smith, M.C., Jordan, D.L., Frans, R.E., and McClelland, M.R. 1993. Influence of application rate and timing on efficacy of bromoxynil. *Proc. Beltwide Cotton Prod. Res. Conf.* p. 1519.

Stalker, D.M. and McBride, K.E. 1987. Cloning and expression in *E. coli* of a *Klebsiella ozaenae* plasmid-borne gene encoding a nitrilase specific for the herbicide bromoxynil. *J. Bacteriol.* 169: 955.

Stalker, D.M., Malyj, L.D., and McBride, K.E. 1988a. Purification and properties of a nitrilase specific for the herbicide bromoxynil and corresponding nucleotide sequence analysis for the *bxn* gene. *J. Biol. Chem.* 263: 6310.

Stalker, D.M., McBride, K.E., and Malyj, L.D. 1988b. Herbicide resistance in transgenic plants expressing a bacterial detoxification gene. *Science* 242: 419.

Thimann, K.V. and Mahadevan, S. 1964. Nitrilase 1. Occurrence, preparation, and general properties of the enzyme. *Arch. Biochem. Biophys.* 105: 133.

Threfall, C. 1969. Quantitative Radiochemical Balance Studies on Growing Wheat in an Enclosed System, Herbicides-Bromoxynil Octanoate (M&B 10,731). Rep. No. PRG497, May and Baker Ltd., Dagenham, U.K., p. 37.

Tiedje, J., Colwell, R., Grossman, Y., Hodson, R., Lenski, R., Mack, R., and Regal, P. 1989. The planned introduction of genetically engineered organisms: ecological considerations and recommendations. *Ecology* 70: 298.

Ward, R., Baldwin, G., Stanton, J., Panter, D., and Kiser, J. 1993. Yield potential of Bromotol™ cotton strains containing Buctril® resistance. *Proc. Beltwide Cotton Prod. Res. Conf.* p. 617.

Wilcut, J.W., Eastin, E.F., and Richburg, J.S., III. 1993. Buctril systems and efficacy for transgenic cotton in Georgia. *Proc. Beltwide Cotton Prod. Res. Conf.* p. 1524.

Wilcut, J., Coble, H., Murdock, E., and Snipes, C. 1992. Weed response to bromoxynil in Georgia, North and South Carolina, and Mississippi cotton. *Proc. Beltwide Cotton Prod. Res. Conf.* p. 1317.

Chapter 7

TRIAZINE-RESISTANT CROPS: THE AGRONOMIC IMPACT AND PHYSIOLOGICAL CONSEQUENCES OF CHLOROPLAST MUTATION

J. Christopher Hall
Michael J. Donnelly-Vanderloo
David J. Hume

CONTENTS

1-56670-045-0/96/$0.00+$.50
© 1996 by CRC Press Inc.

1. INTRODUCTION

The problem of weed resistance to herbicides first became apparent in the late 1960s when populations of *Senecio vulgaris,* showing no susceptibility to simazine, appeared in a nursery that had been treated repeatedly for 10 years (Bandeen et al. 1982; Ryan 1970). Since that time, over 50 weed species worldwide have been identified which exhibit triazine resistance (Fuerst and Norman 1991).

The rapid development of triazine resistance can be attributed to three causal factors (Gressel 1991). First, the triazines have a single mode of action, the blockage of electron transport in photosystem II (PS II). This single mode of action implies that a simple mutation may confer resistance. Second, the triazines typically have a long half-life in the soil, thereby increasing the selection pressure on weed populations. Finally, the triazines have been used continuously under some situations, causing more selection pressure.

The development of herbicide-resistant crop cultivars in Canada occurred as an after-thought, following the initial detection in corn of chlorotriazine-resistant broadleaved weeds such as *Chenopodium album* in Ontario (Bandeen and McLaren 1976) and bird's rape (*Brassica rapa)* in Québec (Maltais and Bouchard 1978). As a result, a breeding program was set up in the Crop Science Department, University of Guelph, to transfer the source of triazine resistance from *B. rapa* into rapeseed (*B. napus*). This trait was found to exhibit uniparental inheritance through the maternal parent (Souza Machado et al. 1978). Repeated backcrossing, using the crop parents as recurrent pollen parent, resulted in the development of the first atrazine-resistant cultivars. The first atrazine-resistant canola cultivar to be released on a commercial basis was OAC Triton (Beversdorf and Hume 1984), followed by Tribute in 1986 and OAC Triumph in 1987.

The transfer of chlorotriazine resistance into *Brassica* crops was considered a breakthrough during the 1970s and 1980s. Use of triazine-resistant crops could provide an effective and economical method of controlling many related cruciferous weed species such as wild mustard (*Sinapis arvensis* L.) and stinkweed (*Thlaspi arvense* L.) that otherwise were very difficult, if not impossible to control. Because these weeds were high in various antinutritional factors such as erucic acid and thioglucosinolates, crop contamination could lead to outright sales rejection, in addition to reducing overall yield. Triazine-resistant rapeseed also was useful in situations in which the planting of atrazine-susceptible crops was limited due to high levels of atrazine residues already present in the soil. In Ontario, resistant cultivars also permitted control of perennial weeds such as quackgrass (*Agropyron repens* (L.) Beauv.) with atrazine.

Research on triazine resistance led to the discoveries that the site of action was located at the chloroplast level (Radosevich 1977; Souza Machado et al. 1977). The trait was cytoplasmically inherited (Souza Machado et al. 1978), and triazine tolerance was transferred to crops such as canola, rutabaga (Swede), and cole crops (Souza Machado et al. 1979, 1980; Souza Machado and Hume 1987). The trait was transferred to cultivated spring rapeseed, *B. napus* (Beversdorf and Hume 1984), and broccoli, *B. oleracea* (Ayotte et al. 1989), and other cultivated *Brassica* oilseed species (Forhan 1987).

The mode and mechanism of action of the triazines, the basis of resistance to the triazines by plants, as well as the agronomic performance of triazine-resistant crops have been discussed by numerous researchers (Andreasson and Vaungard 1988; Ashton and Crafts 1981; Bandeen et al. 1982; Beaversdorf et al. 1988; Gressel 1985, 1991; Gressel and Ben-Sinai 1985; Mets and Thiel 1989; Souza Machado and Hume 1987; Trebst 1991; and van Oorschot 1991).

2. MODE OF ACTION OF THE SYMMETRICAL TRIAZINES

2.1. CHLOROPLAST AND PHOTOSYNTHETIC ELECTRON FLOW

Photosynthetic electron flow occurs in the stacked and unstacked thylakoid membranes within the chloroplast organelles (Fuerst and Norman 1991). The electron transport from water to $NADP^+$ requires the participation of three membrane-spanning multiprotein complexes: (1) the photosystem II (PS II) complex, (2) the cytochrome b_6/f complex, and (3) the photosystem I (PS I) complex (Anderson and Anderson 1988). A fourth component of the thylakoid membrane is the ATP synthase complex which is not directly involved in the transport of electrons in photosynthesis. Photosystem II is mainly confined to the appressed lamellae, while PS I and the ATP synthase complex are found in the stroma lamellae, (Fuerst and Norman 1991). Cytochrome b_6/f is evenly distributed between appressed and stroma lamellae (Anderson and Anderson 1988). The three electron transport complexes mentioned above are interconnected by the lateral mobility of the plastoquinone (PQ) pool and plastocyanin (PC).

2.1.1. Photosystem II Electron Transfer

The PS II complex, which contains chlorophyll (Chl), is capable of light-induced oxygen evolution, thereby reducing PQ (Haehnel 1984). This system is comprised of an oxygen evolving complex (OEC), a light-harvesting chlorophyll (LHC) antenna protein, and a reaction center complex (Figure 1; Fuerst and Norman 1991).

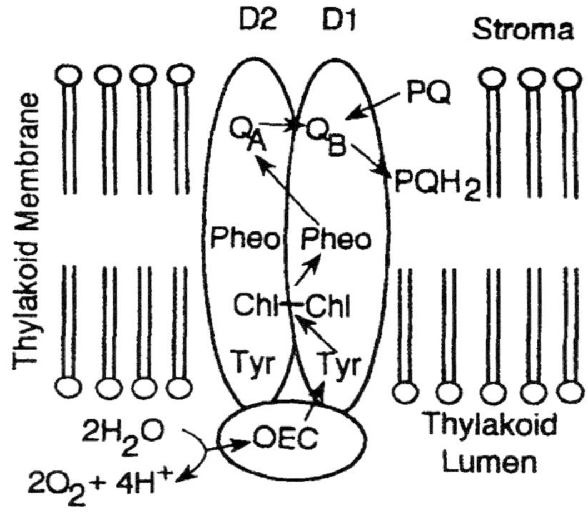

FIGURE 1 PS II reaction center complex. OEC, oxygen-evolving complex; Tyr, tyrosine amino acid residue 160; Chl-Chl, chlorophyll *a* dimer, also known as P680; Pheo, pheophytin; Q_A and Q_B, bound plastoquinone; PQH_2, plastohydroquinone. (Reproduced with permission from Fuerst, E.P. and Norman, M.A. 1991. *Weed Sci.* 39: 461.

Light-induced excitation energy is transferred by the antenna pigments of the LHC to the chlorophyll *a* dimer, known as P680. This induces a charge separation in the P680 that results in the transfer of the excited electron to pheophytin (Pheo). The loss of this electron causes P680 to become positively charged, thereby acquiring an electron to neutralize this charge imbalance (Fuerst and Norman 1991). This electron is provided through the splitting of water by the OEC, producing molecular oxygen and hydrogen ions (Figure 1). H^+ ions contribute to the proton concentration gradient for ATP synthesis. One excited electron is passed from pheophytin to the first stable plastoquinone, Q_A, which is bound in a special niche of the D2 protein (Andreasson and Vaungard 1988). The electron is then transferred from the reduced Q_A to Q_B via magnetic interaction between the Q_A^- and Fe^{2+} (Andreasson and Vaungard 1988). Once Q_A gives up its electron to Q_B, Q_B is released from the D1 protein binding site, allowing another reduced Q_B to bind. This allows a required second electron to be transferred to the Q_B, which is bound in a special niche of the D1 protein. Q_B now accept two protons from the stroma side of the thylakoid membrane and then is released from the D1 binding site as plastohydroquinone (PQH_2). Another PQ molecule enters the D1 niche and, when bound, becomes known as Q_B (Fuerst and Norman 1991).

2.1.2. Cytochrome b_6/f Complex

The cytochrome b_6/f complex is composed of a polypeptide of cytochrome *f*, a polypeptide containing two cytochrome b_6 hemes, and a Fe_2S_2 protein (Haehnel 1984). This complex transports electrons from PS II to PS I. The cytochrome b_6/f complex oxidizes PQ, which occurs when cytochrome f/Fe_2S_2 withdraws two electrons from PQH_2 (Fuerst and Norman 1991; Haehnel 1984). Once PQH_2 is oxidized to PQ, it is recycled to PS II to be reused. Cytochrome *f* donates the electrons to PC, after which PC carries the electrons on the lumen side of the thylakoid membrane to the PS I reaction center (Fuerst and Norman 1991). During electron transport from PS II to PS I, the cytochrome b_6/f complex is utilized to transport two protons per electron across the membrane from the stroma to the lumen (Fuerst and Norman 1991; Andreasson and Vaungard 1988).

2.1.3. Photosystem I Electron Transport

The PS I electron transport system is a multiprotein complex which functions as a light-driven process that photoreduces ferredoxin using PC as an electron donor (Fuerst and Norman 1991, Golbeck 1987). The PS I complex is comprised of a reaction center complex, which contains P700, along with eight protein subunits and LHC antenna proteins (Fuerst and Norman 1991). The LHC transfers light energy to P700, which is probably a chlorophyll *a* dimer. The transfer of energy induces a charge separation in P700, resulting in the excitation of one of its electrons. The excited electron is donated to a series of five known acceptors which include A_0 (a chlorophyll *a* monomer), A1 [possibly phylloquinone (vitamin K_1)], F_x, and F_a/F_b (both are membrane-bound, Fe-S-centered proteins) (Fuerst and Norman 1991). The properties and organization of these five electron acceptors facilitate efficient sequential electron transfer from the primary acceptor, A_0, to the terminal acceptor, F_a/F_b. This results in little or no reverse electron flow (Anderson and Anderson 1988). Ferredoxin acts as a carrier of electrons from the reduced F_a/F_b acceptor to ferredoxin:$NADP^+$ oxidoreductase (FNR). Ferredoxin:$NADP^+$ is found only on the stroma lamellae surface and catalyses the reduction of $NADP^+$ to NADPH. Electrons being donated from ferredoxin to cytochrome b_6 result in cyclic electron transport, which facilitates the conversion of PQ to PQH_2, creating the proton gradient required for ATP synthesis via the transfer of protons from the stroma to the lumen (Fuerst and Norman 1991).

2.2. ABSORPTION, TRANSLOCATION, SELECTIVITY, AND INJURY SYMPTOMS

Triazine herbicides are generally soil applied, being taken up by the roots, swept acropetally in the apoplast, and finally accumulated in the rapidly transpiring leaf margins and tips of broadleaf and grass plants, respectively (Hathway 1989; Moody et al. 1970; Vostral et al. 1970). The rate of root absorption and translocation is related to the transpiration rate of the plant (Vostral et al. 1970). The appearance of triazine in the tops of plants has occurred in as little as 30 min after application (Hay 1976). Foliar applications of triazines also are effective if the entire plant surface is covered and surfactants are used to enhance the foliar uptake (Dexter et al. 1966; Hay 1976). The triazines are known as pseudoapoplastic herbicides because they move in the apoplast but possess the ability to penetrate the symplast in order to enter the chloroplast (Edgington and Peterson 1977).

The selectivity of atrazine in some tolerant species such as corn and sorghum is based on the ability of these species to carry on conjugation of atrazine to glutathione due to high levels of glutathione *s*-transferase (GST) (Ashton and Crafts 1981). Plants that are susceptible to atrazine first exhibit foliar chlorosis, generally in the interveinal regions. Loss of membrane integrity, chloroplast destruction, and leaf senescence are also associated with the phytotoxic action of atrazine. In some isogenic cotton strains the triazines are compartmentalized by lysigenous glands and are prevented from reaching the target site (Hathway 1989). In nonglanded strains the triazines are significantly more toxic. Much of this chapter describes a third group of tolerant species, in which mutations at the atrazine binding site result in herbicide tolerance.

2.3. MECHANISM OF ACTION OF THE TRIAZINES AND FREE RADICAL PRODUCTION

The triazines block electron transport in PS II by displacing Q_B from its binding site on the D1 protein (Figure 2) (Fuerst and Norman 1991). This action prevents the reduction of Q_B to PQH_2. In the absence of the triazines Q_B is bound to the D1 protein by two hydrogen bonds: one between the Q_B carbonyl group and the hydroxyl group of serine (Ser) 264 on the D1 protein and the other between a second carbonyl group on Q_B and histidine (His) 215 of the D1 protein (Fuerst and Norman 1991). Photosystem II electron transport inhibitors, such as atrazine, block the Q_B binding site by binding to the D1 protein in the Q_B binding niche. Atrazine, by means of hydrogen bonding, van der Waal's forces, and hydrophobic interactions, binds to the D1 protein at Ser 264 and phenylalanine (Phe) 265. Phe 255 also is involved in hydrophobic interactions in the binding of both Q_B and atrazine (Fuerst and Norman 1991).

Trebst (1987) divided PS II herbicides into two families: those that bind to the D1 protein near Ser 264, such as the triazines and ureas, and those that bind at His 215, such as the nitrophenols. The nitrophenols do not appear to bind with Ser 264, but instead are oriented toward His 215. Because the bond between His 215 and the nitrophenols is a stronger bond than that of triazines and the ureas with Ser 264, nitrophenols have a longer resident time on the thylakoid membrane.

As a result of electron transport inhibition of PS II by atrazine, there is a lower concentration of protons on the lumen side of the thylakoid membrane. This reduces the quantity of ATP produced by the ATP synthase complex. In addition, PS I is inhibited due to the reduction in electron flow from PS II, which in turn decreases the production of NADPH. It was once thought that this decrease in reducing power and ATP resulted in lower energy yields, thereby decreasing the CO_2 fixation needed for carbohydrate production, and resulting in death by starvation (Dodge 1983). However, the typical symptoms of triazine injury described above are not readily accounted for by the starvation hypothesis (Moreland and Hilton 1976). Rather, the phytotoxic effect of the triazines appears to be related to the increase in the concentration of free radicals due to inhibition of PS II.

The first stage in triazine-induced free radical formation involves the production of singlet oxygen. Singlet oxygen is an excited form of molecular oxygen and is readily created when ground state O_2 receives excitation energy from triplet Chl (Foote 1976). Chlorophyll is a photosensitizer molecule, meaning that it has two types of electronically excited states, singlet and triplet Chl (Foote 1976). Both types are formed when ground state Chl is excited by a photon but the formation of

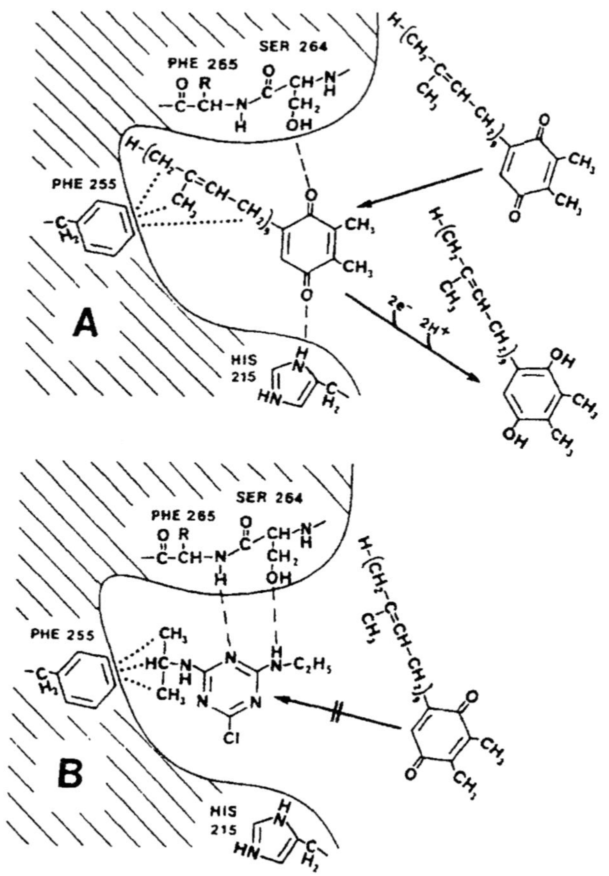

FIGURE 2 Binding site within the D1 protein for plastoquinone and the triazines. Dashed lines represent hydrogen bonds; dotted lines represent hydrophobic interactions. (A) Plastoquinone binds to the D1 protein, accepts two electrons and two protons, and is released as plastohydroquinone. (B) Atrazine binds to the D1 protein and prevents the binding of plastoquinone. (From Fuerst, E.P. and Norman, M.A. 1991. *Weed Sci.* 39: 462. With permission.)

singlet Chl is the most common occurrence. Under normal photosynthetic conditions, triplet Chl and singlet O_2 can be quenched by free radical scavenging molecules such as carotenes, ascorbate, and/or glutathione (Fuerst and Norman 1991). Furthermore, in a healthy photosynthetic system approximately only 4 in 10,000 quantum events lead to the formation of triplet Chl (Foote 1976). When atrazine blocks the normal electron flow in PS II, it also indirectly interferes with the transfer of excitation energy, thereby spontaneously producing abnormally high levels of triplet Chl from singlet Chl in a process known as intersystem crossing (Dodge 1983; Fuerst and Norman 1991).

The two forms of Chl react quite differently in the chloroplast. Whereas singlet Chl transfers its excitation energy to adjoining cells and ultimately to the photosynthetic reaction center, triplet Chl either reacts with O_2 to form singlet oxygen or it reacts directly with other molecules by donating electrons or abstracting hydrogen (Foote 1976). Without the normal quenching of singlet Chl by the reaction center, two deleterious reactions (type I and II) are known to occur: In type I reactions hydrogen is abstracted from unsaturated fatty acids in photosynthetic membranes by triplet Chl, whether or not oxygen is present. In addition, if $^3O_2$ is present, lipid peroxidation occurs. With type II reactions, triplet Chl reacts with molecular oxygen, forming singlet oxygen. Both singlet oxygen and triplet Chl contribute to lipid peroxidation of linolenic and linoleic acid, which are the predominant polyunsaturated fatty acids of the thylakoid membrane. The result of lipid peroxidation is loss of membrane integrity, a loss of chloroplast structure, chlorosis, and finally, cell death (Fuerst and Norman 1991).

2.4. MECHANISMS OF RESISTANCE TO THE TRIAZINES

The D1 polypeptide consists of five hydrophobic transmembrane helical spans and two extrinsic parallel helices (Trebst 1987). The D1 protein has a folded conformation. The portion of the protein which is involved with Q_B herbicide binding exists between the fourth and fifth membrane spanning helices and includes a parallel helix (Trebst 1991). The most significant mechanism of resistance to atrazine in weeds is the alteration of the binding site on the D1 protein (Fuerst and Norman 1991; Hathway 1989; Mets and Thiel 1989). The alteration is a consequence of a mutation occurring on the psbA gene which leads to the substitution of Ser 264 by glycine (Gly) 264 in D1 protein. The alteration results in reduction, by several orders of magnitude, in the binding of atrazine to the D1 protein, due largely to the absence of the hydrogen bond between the amino alkyl side chain of atrazine and Ser 264. However, binding of Q_B still occurs because it can bind via a hydrogen bond to an adjacent amino acid (Fuerst and Norman 1991; Trebst 1987). Mutation of the psbA gene was reported to occur in all naturally occurring, triazine-resistant weeds, including *Amaranthus hybridus*, *Chenopodium album*, *Phalaris paradoxa*, and *Solanum nigrum* (Fuerst and Norman 1991), as well as artificially bred *Brassica napus,* which originates from atrazine-resistant *B. rapa* (Reith and Straus 1987). This mutation results in impaired electron transport in PS II, which may explain some of the yield reduction found in triazine-resistant canola, but other mechanisms also appear to be involved.

There are many other types of mutations that create triazine resistance, but these are mainly produced under artificial conditions. For example, researchers in Japan have isolated a triazine-resistant mutation in a tobacco cell culture in which Ser 264 is replaced by threonine (Thr) 264 (Sigematsu et al. 1989). In a comparative study between plants with the Gly 264 and Thr 264 mutations it was determined that the Thr 264 conferred a broader cross-resistance to other herbicides than did the Gly 264 mutation. This difference was attributed to conformational differences in the secondary structure of the protein. Whereas the Gly caused an α-helical conformation in the protein, the Thr forced the protein sequence to take on a ß-sheet conformation. It has been hypothesized that Thr 264 change in protein structure modified the binding site of other PS II herbicides more drastically than the Gly 264 mutation. To date, the Thr 264 mutation has been found only in cultured plant cells.

Anderson and Gronwald (1991) found an atrazine-resistant biotype of *Abutilon theophrasti* that was resistant to atrazine because of an enhanced capacity to detoxify the herbicide by glutathione conjugation. They found that GST activity was about four times higher in the resistant biotype when atrazine was used as the substrate, thereby explaining the susceptibility differences between the two biotypes. The effectiveness of this GST metabolic mechanism as a mode of resistance is, however, much lower than that conferred by chloroplastic mutation. A resistance factor ranging between 100 and 1000 would describe chloroplastic resistance while factors of 4 to 8 describes metabolic resistance (van Oorschot 1991).

2.4.1. Cross-Resistance

Varying degrees of cross-resistance to the uracils, pyridazinones, and certain ureas exist in triazine-resistant plants. Most triazine-resistant weeds show the same pattern of cross resistance because they have the same mutation. For example, van Oorschot (1988) found that the cross-resistance factor for triazine-resistant weeds was 3 to 7 for metamitron, 2 to 9 for bromacil, 3 to 5 for monolinuron, and 1 for diuron. The binding site has been described as consisting of two domains, the essential-element domain, and the specificity-determining domain (Gressel and Segel 1982). The essential element domain recognizes a common structural element that is essential for interacting with the binding site. This element consists of an electron-deficient sp^2 hybrid carbon adjacent to an oxygen or nitrogen electron-withdrawing group (Gressel and Segel 1982, Mets and Thiel 1989). The specificity-determining domain is the tertiary structure of the binding site that is formed by the various amino acid side chains of the D1 protein (Mets and Thiel 1989). It is here that the selectivity of the different PS II inhibitors is determined by the ability of their hydrophobic side chains to occupy the various niches of the specificity-determining domain (Gressel and Segel 1982; Mets and Thiel 1989).

Gressel (1985) proposed a more simplistic model of the different binding sites described above for the PS II electron inhibitors. In this allosteric model of the binding sites on the D1 protein the urea herbicides are thought to bind to sites 1 and 2, the triazines to sites 2 and 3, the phenolics to site 4, and bicarbonate to site 4 and 5 (Gressel 1985). Some of the different herbicide binding sites overlap, and, therefore, alteration of one binding site can affect the binding of more than one herbicide group, resulting in cross-resistance.

Another type of cross-resistance involves photosynthesis-inhibiting herbicides which have completely different sites of action. Examples exist of triazine-resistant weeds which also exhibit resistance to paraquat. In the case of *Conyza canadensis* a triazine-resistant biotype was shown to have a resistant factor of 450 to paraquat (Polos et al. 1991). In another study Clay and Hadleigh (1991) found triazine–paraquat co-resistance in *Poa annua* biotypes. A likely explanation for this cross-resistance may lie in the fact that both triazines and paraquat have a similar mode of action. Paraquat toxifies the chloroplast directly with free radicals, whereas free radical production from triazine application occurs indirectly. Free-radical scavengers which would detoxify paraquat would also detoxify the side effects produced by triazine binding.

2.4.2. Negative Cross-Resistance

Negative-cross resistance occurs when herbicides, other than the one to which the plant has developed resistance, become more toxic to the resistant biotype than to the susceptible biotype. Gressel (1991) listed several herbicides to which negative-cross resistance has occurred in atrazine-resistant weeds. These herbicides include those that act at or near the atrazine-binding site as well as those that act on other photosynthetic processes, such as bentazon and paraquat. Gressel (1991) cites two references which indicate that there are *B. napus* biotypes that have shown negative-cross resistance to bromoxynil, dinoseb, DNOC, ioxinyl, medinoterb, and pyridate.

3. INCORPORATION OF TRIAZINE RESISTANCE INTO CROPS

3.1. *BRASSICA* SPECIES

A triazine-resistant bird's rape biotype (*Brassica rapa* L.), synonymous with turnip rape (*B. campestris* L.), was found in two corn fields near Sherbrooke, Québec (Maltais and Bouchard, 1978). Genetic analysis of this biotype demonstrated reciprocal differences for triazine resistance in F_1 crosses, indicating uniparental inheritance through the female parent (Souza Machado et al. 1978). Analysis of backcross progeny for triazine-resistance segregation confirmed that this trait was cytoplasmically inherited (Beversdorf et al. 1980a; Souza Machado and Bandeen 1982). A breeding program at the Crop Science Department, University of Guelph, released triazine-resistant *B. rapa* germplasm in 1979. This germplasm, referred to as ATR-5Ca, was derived from the bulked seeds of the fourth backcross using the weed biotype as the donor parent and "Candle", a triple zero (low erucic acid, low thioglucosinolate, low fiber) Canadian *B. rapa* cultivar as the recurrent pollen parent (Beversdorf et al. 1980c).

Triazine-resistant *B. napus* germplasm referred to as ATR-5Tw was also released. Triazine resistance was incorporated into ATR-5Tw by crossing the bird's rape biotype ($2n = 20$) with Tower ($2n = 38$), a double zero (low erucic acid and low thioglucosinolate content, hereafter referred to as canola) Canadian *B. napus* cultivar. Backcrossing the triploid F_1 plants ($2n = 29$) produced three 38-chromosome plants identified by root tip squashes. ATR-5Tw was derived from seeds of the fourth backcross of bird's rape by Tower (Beversdorf et al. 1980b). Although this interspecific transfer was stabilized by the second backcross, additional backcrosses were required to eliminate high levels of erucic acid and glucosinolates to meet canola quality standards. OAC Triton, the first licensed, triazine-resistant canola cultivar, represents a single population from bulked seeds of ATR-5Tw. This cultivar was selected following the fourth backcross for uniform maturity and height as well as vigor (Beversdorf and Hume 1984).

ATR-5Tw was also used to continue the transfer of triazine resistance into other *Brassica* crops of commercial significance, including rutabaga (*B. napus* L.) and Chinese cabbage (*B. rapa* L.) (Souza Machado et al. 1983). Later, triazine-resistant canola cultivars were used to introduce this trait into two cultivated mustard species, *B. juncea* and *B. carinata* (Forhan 1987) and broccoli (*B. oleracea*) (Ayotte et al. 1989). The backcrossing technique used in these programs effectively combined virtually all the nuclear genome from the domesticated *Brassica* crop with the cytoplasmic genome (chloroplast and mitochondrial genes) from triazine-resistant bird's rape (*B. rapa* L.) Protoplast fusion also was used to transfer triazine resistance into cytoplasmic male sterile lines as a means of hybrid seed production (Barsby et al. 1987).

Although chloroplast genomes of triazine-resistant and triazine-susceptible *B. napus* are similar, they are not identical. This occurs because the cytoplasmic genome of susceptible *B. napus* is thought to originate from *B. oleracea*. Conversely, the cytoplasmic genome of resistant *B. napus* originates from *B. rapa* because of the backcrossing to *B. rapa* to incorporate resistance (Erickson et al. 1983). Therefore, in addition to the gene conferring triazine resistance there are small nuclear and cytoplasmic genome differences between triazine-resistant cultivars and their triazine-susceptible recurrent parents. This may explain why some *Brassica* species are less affected by incorporation of triazine resistance than others.

3.2. OTHER SPECIES

Since the incorporation of triazine resistance into cultivated *Brassica* sp., several other mutations conferring resistance in other species have been transferred from wild relatives into closely related crops. Atrazine has been used for weed control in foxtail millet (*Setaria italica* L.) production, although some crop damage occurs. The observation of triazine resistance at the chloroplast level in green foxtail (*Setaria viridis* L. Beauv.) provided an opportunity to increase the tolerance of millet to this herbicide (Gasquez and Compoint 1981). Darmency and Pernes (1985) made the cross using *Setaria viridis* (L.) as the female parent and a Chinese cultivar of *Setaria italica* (L.) (Beauv.) as the pollen donor. Four backcrosses were used to make the introgression into the foxtail millet cultivar. In contrast to classical breeding programs Austin and Helgeson (1987) used somatic fusion to transfer triazine-resistant cytoplasm from *Solanum brevidens* to *S. tuberosum*.

4. AGRONOMIC CHARACTERISTICS

4.1 PLANT PRODUCTIVITY AND MATURITY

Early studies suggested that triazine-resistant canola may be associated with certain negative factors such as reduced yield and delayed maturity, although it was suggested that some of these effects could have resulted from repeated backcrossing without selection for yield (Weiss 1982).

However, extensive field trials have since shown that triazine-resistant cultivars selected for yield are inferior to susceptible cultivars. In Ontario, Triton yields averaged 87% of those for the susceptible cultivar Regent over 19 locations during 1981 to 1983. There was also a delay in maturity and increased lodging compared to Regent. In western Canada yield losses were more severe, with Triton averaging 79% of Regent over 40 trials in 1982 to 1983 (Beversdorf and Hume 1984). To eliminate any potentially confounding differences in the nuclear genome of backcrosses, reciprocal crosses were made, which resulted in the same nuclear genome in the progeny, with or without the triazine-resistant cytoplasm. In field trials during 1984 and 1985 the triazine-resistant progeny exhibited 21 and 25% lower yields than susceptible progeny, flowered 2 days later, and plant height was reduced by 8%. Maturity also was delayed by 1 day in the resistant progeny (Beversdorf et al. 1988). The results indicated that seed yields were reduced in triazine-resistant cultivars, thereby limiting their usefulness. Whole plant studies found dry matter accumulation was reduced (although not always as severely as seed yield) in triazine-resistant *B. napus* cultivars. For example, Juras (1984) found 15% less biomass dry weight and 22% less seed weight in resistant cultivars. The reduction in dry matter accumulation appeared more pronounced during seed filling (Donnelly and Hume 1984).

Yield decreases associated with incorporation of triazine resistance into *B. rapa* canola germplasm were greater than those in susceptible *B. napus*, even though cytoplasmic DNA divergence was less than in *B. napus* because the weedy biotype and cultivar are the same species. Reductions of 60 and 47% of F_1 hybrid seed yield (Forhan 1987) and above-ground dry matter (Hobbs 1987), respectively, have been reported. For these reasons, triazine-resistant *B. rapa* canola cultivars were never licensed in Canada.

Triazine-resistant rutabaga (*B. napus*) showed no significant difference in yield relative to susceptible cultivars, provided that they were planted early in May. However, delayed plantings resulted in significant yield reductions (Souza Machado and Hume 1987). Triazine-resistant *B. juncea* and *B. carinata* also had reduced yields (Forhan 1987). Furthermore, triazine-resistant *Setaria italica* had 22% lower grain yields than its triazine-susceptible counterpart, although early-season growth rates were similar between these two lines at cool temperatures (Darmency and Pernes 1989).

4.2. FITNESS AND COMPETITIVE ABILITY

In competitive fitness studies resistant and susceptible lines are interplanted to ascertain differences during growth. One such study conducted by Gressel and Ben-Sinai (1985) found that total plant dry weight, number and weight of siliques, and seed yield were reduced by over two thirds in the resistant cultivar. A competition experiment using resistant and susceptible lines of foxtail millet also found susceptible individuals were better competitors (Reboud and Till-Bottraud 1991).

Another type of competition study involves comparing the weed thresholds of triazine-resistant and susceptible cultivars. This enables the calculation of weed densities at which it would be economically advantageous to grow lower yielding triazine-resistant varieties in the presence of atrazine. Using wild oats as the model weed, Forcella (1987) found that when atrazine was applied at 1.5 kg/ha, triazine-resistant biomass and yields were stabilized at the same level as occurred when 25 to 30 wild oats per square meter reduced yields of triazine-susceptible plants. High weed pressures often make even moderate yield reductions tolerable because of a lack of alternate weed control measures.

4.3. SEED QUALITY

Triazine-resistant canola has consistently had a lower seed oil concentration than normal, triazine-susceptible cultivars. For example, in Manitoba three recommended triazine-resistant cultivars (OAC Triton, OAC Triumph, and Tribute) averaged 41.2% oil, compared to an average of

43.3% oil in eight triazine-susceptible cultivars (Thomas 1990). Comparisons were made across five growing zones and 5 years. In Ontario between 1987 and 1989 three triazine-resistant cultivars averaged 38.6% oil and 38.2% protein in the meal as compared to 40.7% oil and 37.5% meal protein in 11 triazine-susceptible cultivars (OOPSCC 1990). Most cultivar trial results reviewed indicated a consistent 2% decrease in seed oil content in triazine-resistant cultivars as compared to their normal counterparts.

Seedling vigor of triazine-resistant cultivars also was consistently lower than that of triazine-susceptible cultivars (Thomas 1990). This reduction in seedling vigor also exists in the bird's rape source of triazine resistance. Mapplebeck et al. (1982) compared germination and seedling growth of triazine-susceptible and triazine-resistant bird's rape. They found susceptible seedlings germinated and grew faster, emerged sooner, emerged from greater depths, and had higher water contents. No differences in respiration rates were found between the two types; however, Brine (1989) compared 17 seedlots from four different triazine-susceptible *B. napus* cultivars with 12 seedlots from two different triazine-tolerant cultivars. Triazine-susceptible seedlots had higher germination (89.5 vs. 84.6%) and higher average stands in the field (87.3 vs. 82.8 plants per square meter). Donnelly-Vanderloo and Hume (1993) measured lower relative growth rates in seedlings of triazine-resistant *B. napus* lines grown in the field than in seedlings of triazine-susceptible cultivars (unpublished results). Slower growth would be consistent with lower photosynthetic rates frequently observed in triazine-resistant *B. napus* cultivars (see, e.g., Hobbs 1987).

When reciprocal crosses were made between triazine-resistant and triazine-susceptible cultivars, the presence of the triazine-resistant trait delayed flowering in the progeny by an average of 2 days, but maturity by only 1 day (Beversdorf et al. 1988).

Higher than normal levels of free fatty acids (FFA) occur in the oil of spring canola (*B. napus*), grown in Ontario during years in which there is heat and drought stress during and after flowering (May 1993). Triazine-resistant *B. napus* cultivars are particularly susceptible to high FFA levels. Three triazine-resistant cultivars (Stallion, OAC Triton, and OAC Triumph) averaged 4.5% FFA as compared to 3.3% FFA in three triazine-susceptible cultivars (Delta, Westar, and Global). All cultivars were averaged across six locations in 1988, a year in which FFA levels were very high. In 1991, another stress year, triazine-resistant cultivars averaged 2.5% FFA as compared to 1.3% for triazine-susceptible lines. In nonstress years (e.g., 1990) differences were still significant (0.42% FFA in triazine-resistant cultivars as compared to 0.26% for triazine-susceptible cultivars) (all data on FFA levels are from May 1993). May (1993) interpreted his results to indicate that production of oxygen free radicals, as a result of impaired electron flow in the triazine-resistant cultivars, resulted in more stress effects than in most other cultivars, and that these stress effects resulted in impaired incorporation of FFAs into triglycerides during seed development. His results also indicated that more FFAs were cleaved from triglycerides of triazine-resistant cultivars during seed maturation than from triglycerides of triazine-susceptible cultivars.

4.4. TOLERANCE TO ENVIRONMENTAL STRESS

Studies examining the effect of temperature on plant productivity suggest that yield reductions associated with triazine resistance may change under different environments. Resistant *B. napus* cultivars grown at 25°/20°C (day/night) had a 21 and 14% decrease in dry matter and photosynthetic rate as compared to the normal parent. However, resistant plants grown at 15°/10°C showed no decrease in photosynthetic rate or dry matter accumulation. Field analysis showed no difference in vegetative dry matter, but there was decreased dry matter accumulation later in the season during pod filling (Donnelly and Hume 1984). This suggests that triazine-resistant plants may exhibit greater yield reductions when grown under certain environmental conditions.

Data collected from cultivar licensing trials support this hypothesis. For example, OAC Triton (resistant) seed yields averaged 94% of Regent (susceptible) in six trials in Ontario during 1981 and

1982, when weather conditions were normal. In contrast, seed yields were only 73% of Regent in seven trials in 1983, when conditions in June and July were hot and dry (Beversdorf and Hume 1984). Field trials conducted in southern Ontario from 1986 to 1990 are plotted (Figure 3) to show the percent reduction in yield due to triazine resistance (*y* axis) vs. the mean yield of the susceptible cultivars (*x* axis). Low yields suggest years with more stress. There is a significant negative slope indicating that the percent yield reduction was greater in lower yielding years (Donnelly-Vanderloo and Hume 1993). This phenomena may have a bearing on the extent to which primary and secondary effects of the mutation affect the performance of triazine-resistant crops grown in different environments.

5. PHOTOSYNTHETIC CAPABILITY

5.1. PRIMARY EFFECT ON PHOTOSYSTEM II

The primary effect of the mutation which replaces Ser 264 with Gly is a less efficient electron transfer. The absence of Gly not only prevents the binding of atrazine but also reduces the binding of PQ, which, in turn, slows electron transport from Q_A to Q_B as shown by altered Chl fluorescence kinetics (Ali et al. 1986; Bowes et al. 1980; Vermaas et al. 1984). This results in lower photochemical quenching, quantum yield, oxygen flash yield, and increased nonphotochemical quenching at low irradiance. Sundby et al. (1993b) showed this intrinsic impairment of PS II at low, medium, and high irradiance in triazine-resistant *B. napus*. This reduction in PS II electron transport was initially considered to be the direct cause of the yield reductions associated with the Ser 264 mutation (Jursinic and Pearcy 1988). However, if this were the only consequence of the mutation, it makes it difficult to explain the reductions found under conditions in which photosynthesis was light saturated. Such reductions in the yield of triazine-resistant *B. napus* at high irradiances (Hobbs 1987; van Oorschot and van Leeuwen 1989) suggest that the mutation results in pleiotropism (multiple effects produced by a single gene).

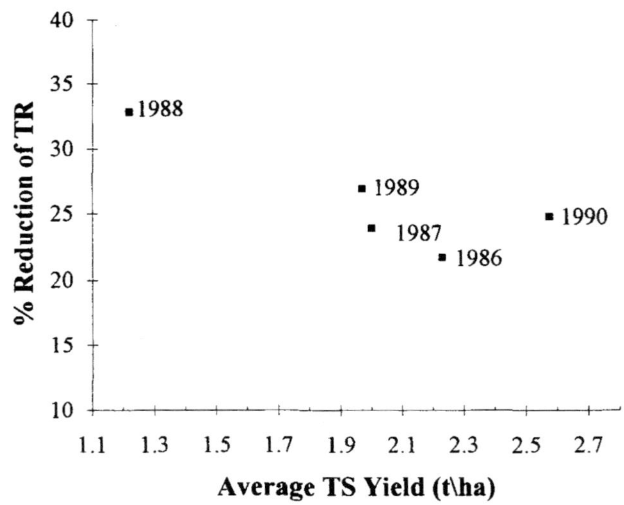

FIGURE 3 Yield data from field trials comparing triazine-resistant (TR) and triazine-susceptible (TS) cultivars at various locations across southern Ontario from 1986 to 1990. The percent reduction in the yield of triazine-resistant cultivars when compared to the susceptible cultivars (*y* axis) is plotted against the mean yield of the susceptible cultivars (averaged across all locations) for each year . Growing conditions were typical for southern Ontario in all years except 1988, which was unusually hot and dry.

5.2. PLEIOTROPIC EFFECTS

5.2.1. Photoinhibition

Photoinhibition of photosynthesis results when excessive excitation of the photosynthetic apparatus occurs (Osmond 1981). Functional characteristics of photoinhibition include a reduction in quantum yield and a decreased rate of light-saturated photosynthesis (Powles 1984). Although not fully understood, one type of photoinhibition involves inhibition of the PS II reaction center (Ogren and Oquist 1984). Recovery typically involves replacement of the damaged D1 protein (Kyle 1987). Another type of photoinhibition involves the formation of zeaxanthin which quenches excited antenna Chl (Demmig and Bjorkman 1987). Both types of photoinhibition quench excited Chl through release of heat.

There is evidence that photoinhibition may be one of the major pleiotropic consequences of the mutation conferring triazine resistance. Hart and Stemler (1990b) showed that under high irradiance there was a greater decrease in photon yield in triazine-resistant *B. napus* than in its susceptible counterpart; this was attributed to photoinhibition. In their study the slow electron transfer that characterizes resistant plants was evident, however, photon yield was similar between both *B. napus* lines. In addition, differences between resistant and susceptible plants in terms of growth and biomass allocation were relatively small under low irradiances (Hart et al. 1992). These studies suggest that an intrinsic reduction in PS II electron transport due to the D1 mutation directly accounts for only a portion of the yield reduction. It has been suggested that a secondary effect, photoinhibition, is also responsible for some of the yield reduction (Hart and Stemler 1990a). Sundby et al. (1993b) stated that both the intrinsic lowering of light-limited photosynthetic efficiency and the increased sensitivity to photoinhibition probably contribute to reduced crop yields in the field, to different extents, depending on growth conditions. Field studies showing a proportionately greater yield reduction under stress conditions (heat and drought) would tend to support this hypothesis because these conditions, especially drought, are known to favor photoinhibitory processes (Powles 1984, Havaux 1992). In conclusion, it is important to consider the contribution of photoinhibition and PS II efficiency under various field conditions because the growth environment determines the relative importance each of these components has on yield.

5.2.2. Protective Mechanisms

Triazine resistance may adversely affect protective mechanisms within the plant. For example, the xanthophyll cycle is one way by which plants can safely dissipate excess energy, employing conversion of violoxanthan to zeaxanthan, which then quenches excited Chl (Demmig et al. 1987). If reduced PQ is a signal for zeaxanthan production (Siefermann and Yamamoto 1975), then it seems plausible that triazine-resistant plants with a larger oxidized PQ pool, would delay zeaxanthan formation, which could lead to more serious photosystem damage. This is supported by Sundby et al. (1993a), who found triazine-resistant cultivars that had reduced photoprotective quenching mechanisms. Alternatively, D1 protein repair is another protective mechanism. It appears resistant plants have a limited ability to respond to increased photodamage by repairing D1 protein (Sundby et al. 1993b). Until these mechanisms can be individually isolated, it will be difficult to determine their relative importance.

5.2.3. Reorganization of the Chloroplast

The point mutation in D1 leads to an extensive reorganization of functional units within the chloroplast. In general, triazine-resistant *B. napus* and *B. rapa* chloroplasts exhibit increased granal stacking and a lower proportion of starch and stromal lamellae. Characters associated with increased granal stacking such as lower Chl *a/b* ratio and increased PS II antenna protein were also found in resistant lines (Vaughn and Duke 1984; Vaughn 1986). Triazine-resistant plastids also have a slight increase in unsaturated lipid content and an increase in the proportion of Δ-3-*trans*-hexadecenoic acid. These lipid profile changes may be related to the increased stacking of triazine-resistant thylakoids (Lemoine et al. 1986). Many of these changes are similar to those found in shade-adapted leaves (Boardman 1977). However, gross anatomical parameters and many levels of functional proteins were similar when susceptible and resistant lines were compared. Therefore, the mutation does not confer a shade-type anatomy (Vaughn 1986). Susceptible types could be modified to resemble resistant plastids by applying sublethal doses of PS II inhibitors. These results suggest that ultrastructural modifications are secondary effects of impaired PS II electron flow (Vaughn and Duke 1984) and probably attempts to compensate for an increase of the oxidation state of the PQ pool (Mattoo et al. 1984). Increasing the size of the PS II antenna via state transitions compensate for an excitation imbalance under light-limited conditions (Horton and Hague 1988). Under high light intensity, state transitions reduce PS II size; however, triazine-resistant plants maintain a large antenna around PS II under all irradiances which may be detrimental under high light intensity.

5.2.4. Heat Sensitivity

The organization and composition of chloroplast membranes is altered to optimize electron transport in a wide range of environments (Melis et al. 1985). Because lateral and transverse diffusion of electron carriers is vital for electron transport, the fluidity of the chloroplast thylakoid membranes plays an important role in controlling the photosynthetic light reactions (Yanamoto et al. 1981). High temperatures can cause uncoupling of electron transport, which may result from increased thylakoid permeability to protons, resulting in decreased ATP production and CO_2 fixation (Stidham et al. 1982). Low temperatures decrease thylakoid fluidity and may limit electron transport by restricting diffusion of PQ between PS II and cytochrome b_6/f (Oquist 1982) or by reducing the rate of interaction of PQ with its site of oxidation on the cytochrome b_6/f complex (Mitchell and Barber 1986).

Photosynthetic acclimation to low temperatures is associated with significant increases in unsaturated chloroplast lipid and an increased lipid/protein ratio (Yoshida 1984, Aro et al. 1987). Low temperature adaptation has been shown to increase O_2 evolution and whole-chain electron transport, measured under low temperatures relative to unadapted plants (Mitchell and Barber 1986). It seems likely that a regulatory mechanism exists to maintain an optimal fluidity of thylakoid membranes at various temperatures.

In triazine-resistant thylakoids it is plausible that the decreased level of lipid saturation may have pleiotropic effects at high and low temperatures. Increased heat sensitivity of PS II in a triazine-resistant *B. napus* cultivar was associated with altered fluorescence kinetics and reduced electron transfer activity (Ducruet and Lemoine 1985). Similar effects were noted with *B. rapa* chloroplasts, except that the decreased activity began at lower temperatures.

Even without invoking pleiotropic effects such as the aforementioned changes in thylakoid lipid composition, it is possible for resistant germplasm to show heat sensitivity. At low temperatures when there was little or no electron transport limitation, carbon assimilation rates of triazine-resistant and -susceptible *B. napus* were similar. As temperatures increased, carbon assimilation became limited by electon transport in both resistant and susceptible *B. napus*. However, there was a greater limitation on carbon assimilation as effected by electron transport in the triazine-resistant biotype (Dekker and Sharkey 1992). This may account for the near-normal photosynthetic capability of resistant and susceptible *B. napus* at temperatures of 10° to 15°C and a decline in triazine-resistant performance at 25°C (Donnelly and Hume 1984).

5.2.5. Diurnal and Ontogenetic Patterns

Another pleiotropism may manifest itself as altered diurnal and ontogenetic patterns of photosynthesis. Triazine-resistant canola (*B. napus*) seedling leaves fixed more carbon relative to a susceptible cultivar early and late in the photosynthetic diurnal pattern at 25°C. During the rest of the cycle, the susceptible cultivar fixed more carbon than the resistant cultivar (Dekker and Burmester 1990). During vegetative growth, carbon assimilation of resistant plants was lower than that of susceptible plants at most times of the day. Resistant plant leaves were always cooler than susceptible leaves under similar growing conditions, and this was correlated with periods of increased water conductance (Dekker and Burmester 1992). The authors speculated that these traits may confer an ecological advantage to the triazine-resistant biotype in certain unfavorable niches such as cool, low-light environments.

6. CURRENT STATUS OF TRIAZINE-RESISTANT CROPS

6.1. ATTEMPTS TO OVERCOME YIELD REDUCTION

The triazine-resistant cultivars OAC Triton and OAC Triumph were released in 1984 and 1987, respectively, and the breeding program for improved yield and quality in *B. napus* continued at the University of Guelph until 1990. All of the material in that breeding program was screened using leaf fluorescence measurements (Hume and Braunberger 1993), searching for lines which were triazine resistant but which did not have impaired photosynthesis. Preliminary trials indicated that photosynthetic efficiency, as measured by leaf fluorescence, was distinctly lower in most triazine-resistant lines. However, a number of lines were found with higher photosynthetic efficiency and these were advanced to F_7 and F_8 generations, tested for yield, and significant (p <0.05) positive correlations were found between photosynthetic efficiency and seed yield. In fact, some lines with high photosynthetic efficiency were even entered in provincial yield trials. It was observed, however, that these more efficient lines had begun to lose their tolerance to triazine herbicides. Braunberger (1987) used a flow cytometer to show that lines with improved photosynthetic efficiency contained mixtures of triazine-resistant and triazine-susceptible chloroplasts. As generations advanced, the more efficient triazine-susceptible chloroplasts increased in frequency, resulting in increased photosynthetic efficiency, but they lost much of their tolerance to triazine herbicides. The mechanism resulting in the expression of triazine-susceptible chloroplasts in the progeny when triazine-resistant females were crossed with susceptible males is not understood. It is unlikely that plastid transmission occurs through pollen transfer from the male.

In a final attempt to identify triazine-resistant lines with higher photosynthetic efficiency, Mihajlovic (1990) selected lines from the breeding program which appeared to have stable triazine resistance and improved photosynthetic efficiency. These lines were subjected to microspore culture and resulting seedlings were treated with colchicine to create doubled haploids, which were homozygous and true breeding. These doubled haploid lines were field tested for yield and subjected to carbon exchange rate measurements using a portable CO_2 analyzer. The selected lines did not perform any better than the cultivars OAC Triton and OAC Triumph.

6.2. REPLACEMENT OF TRIAZINE-RESISTANT CANOLA

Some commercial production of the triazine-resistant *B. napus* cultivar Stallion continued in Ontario in 1993, but this technology was largely replaced for a number of reasons. First, the herbicide ethametsulfuron-methyl (Muster®) has been registered in Canada. It is capable of effectively controlling wild mustard and other cruciferous weeds in cultivated canola (*B. napus* and *B. rapa*). This means that higher-yielding cultivars can be grown in wild mustard-infested fields and Muster® can be used for postemergence control of wild mustard. A second reason for replacement of the triazine-resistant canola is that triazine herbicides, particularly atrazine, have come under close environmental scrutiny because of long-lived residues and detection of atrazine in groundwater (Goodrich et al. 1991). A third reason for phasing out triazine-resistant canola cultivars has been their inherently lower quality than triazine-susceptible cultivars. Triazine-resistant cultivars have seed which is consistently lower in oil content and higher in FFAs, as discussed previously. Triazine-resistant cultivars also are lower yielding and have poorer seedling vigor than triazine-susceptible cultivars. Apparently, the only reasons some still are grown is that atrazine is an inexpensive herbicide and the yield penalty for triazine-resistant cultivars was not substantial in 1992 and 1993, during which cool, moist growing conditions prevailed.

7. CONCLUSION

Triazine-resistant canola was one of the first commercially successful herbicide-resistant crops to be developed. However, in spite of a comprehensive plant breeding program throughout the 1970s and 1980s, no triazine-resistant cultivars of *B. napus* or *B. rapa* with significantly improved photosynthetic efficiency were ever developed. Furthermore, under most growing conditions their yield, oil quality and content, and seedling vigor were reduced when compared to their triazine-susceptible counterparts. Currently throughout Canada there is little triazine-resistant canola grown. As an example, all the registered triazine-resistant cultivars of canola in the province of Ontario were removed recently from the list of recommended crops because of unacceptably low levels of oil and high FFA content.

Further research is needed on the extent to which photoinhibition is responsible for yield reductions in triazine-resistant crops and how this phenomenon interacts with environmental conditions during growth. This may have broader implications as to the extent to which crop performance is determined by photosynthetic efficiency at low light vs. tolerance to environmental stress (Tollenaar 1991).

Although classical plant breeding has failed to circumvent the deleterious effects associated with the triazine-resistance trait, direct genetic manipulation may be more successful. For example, D1 mutations other than Ser 264 to Gly were found in prokaryotes and *Chlamydomonas reinhardii* that have moderate resistance to several PS II inhibitors, yet still maintain apparently normal PS II kinetics (Brown et al. 1988, Wildner et al. 1989). Another possible solution may include a second mutation in the Q_B binding pocket, which by itself does not confer resistance, but modifies the Ser 264 to Gly mutation so that it does not affect Q_B binding kinetics. This may explain why triazine resistance in *Phalaris paradoxa* is not linked to inferior photosynthetic and growth performance. This species, unlike all plants and algae examined so far, has a lysine at position 238 instead of arginine (Arg). Therefore, it is possible that substitution of Arg at 238, which is close to the binding site, ameliorates the effect that the Ser 264 to Gly mutation has on PQ binding kinetics (Schonfeld et al. 1987).

Altering the D1 binding site to prevent reduced PQ binding should alleviate primary and pleiotropic effects due to triazine resistance. However, should certain pleiotropic effects prove to be responsible for a large portion of the inferior performance associated with triazine resistance there would be other options to improve agronomic traits. For example, if excessive production of free radicals was overloading native protective mechanisms it should be possible to link a highly active gene promoter sequence to produce free radical scavengers. An analogous situation arises in triazine-resistant velvetleaf, which has elevated levels of glutathione (Anderson and Gronwald 1991). Until the exact nature of pleiotropism and its relevance under varying environments is known it is difficult to predict the success of this strategy.

Development of triazine-resistant crops provided an insight into the technical, political, and ethical problems involved in the use and registration of herbicide-resistant crops. Although development of triazine-resistant crops can be considered technically successful, these crops have been short-lived successes because of rapid technological changes such as the development of new herbicides, the concern about pesticides in the environment, and lower crop yield and quality due to the physiological changes brought about by the resistance trait.

ACKNOWLEDGMENTS

We would like to thank J. Parnell, J. Thompson, and R. Tshanz for providing technical assistance in the preparation of this manuscript and for their assistance in surveying the literature on this topic.

REFERENCES

Ali, A., Fuerst, E.P., Arntzen, C.J., and Souza Machado, V. 1986. Chloroplastic triazine-resistance in rutabaga. *Plant Physiol.* 80: 511.

Anderson, J.M. and Anderson, B. 1988. The dynamic photosynthetic membrane and regulation of solar energy conversion. *Trends Biochem. Sci.* 13: 351.

Anderson, M.P. and Gronwald, J.W. 1991. Atrazine-resistance in a velvetleaf *(Abutilon theophrasti)* biotype due to enhanced glutathione s-transferase activity. *Plant Physiol.* 96: 104.

Andreasson, L.E. and Vaungard, T. 1988. Electron transport in photosystems I and II. *Annu. Rev. Plant Phys. Plant Mol. Biol.* 39: 379.

Aro. E., Somersalow, S., and Karunen, P. 1987. Membrane lipids of *Ceratodon pururcus* protonemata grown at high and low temperatures. *Physiol. Plant.* 69: 65.

Ashton, F.M. and Crafts, A.S. 1981. Triazines. In *Mode of Action of Herbicides,* 2nd ed. John Wiley & Sons, New York, chap. 19.

Austin, S. and Helgeson, J.P. 1987. Interspecific somatic fusions between *Solanum brevidens* and *S. tuberosum.* In *Plant Molecular Biology.* Ed. by Von Wettstein, D. and Chua, N.H. Plenum Press, New York, p. 209.

Ayotte, R., Harney, P.N., and Souza Machado, V. 1989. The transfer of triazine resistance from *Brassica napus* to *B. oleracea* L. IV. Second and third backcrosses to *B. oleracea* and recovery of an 18-chromosome, triazine-resistant backcross. *Euphytica* 40: 15.

Bandeen, J.D. and McLaren, R.D. 1976. Resistance of *Chenopodium album* to triazine herbicides. *Can. J. Plant Sci.* 56: 411.

Bandeen, J.D., Stephenson, G.R., and Cowett, E.R. 1982. Discovery and distribution of herbicide-resistant weeds in North America. In *Herbicide Resistance in Plants.* Ed. by LeBaron, H.M. and Gressel, J. John Wiley & Sons, New York, chap. 2.

Barsby, T.L., Kemble, R.J., and Yarrow, S.A. 1987. *Brassica* hybrids and their utility in plant breeding. *Plant Mol. Biol.* 140: 223.

Beversdorf, W.D., Weiss-Lerman, J., Erickson, L.R., and Souza Machado, V. 1980a. Transfer of cytoplasmically-inherited triazine-resistance from bird's rape to cultivated oilseed rape *Brassica campestris* and *B. napus. Can. J. Genet. Cytol.* 22: 167.

Beversdorf, W.D. and Hume, D.J. 1984. OAC Triton spring rapeseed, *Can. J. Plant Sci.* 64: 1007.

Beversdorf, W.D., Hume, D.J., and Donnelly-Vanderloo, M.J. 1988. Agronomic performance of triazine-resistant and susceptible reciprocal spring canola hybrids. *Crop Sci.* 28: 932.

Beversdorf, W.D., Weiss-Lerman, J., and Erickson, L.R. 1980b. Registration of triazine-resistant *Brassica campestris* germplasm. *Crop Sci.* 20: 289.

Beversdorf, W.D., Weiss-Lerman, J., and Erickson, L.R. 1980c. Registration of triazine resistant *Brassica napus* germplasm. *Crop Sci.* 20: 289.

Boardman, N.K. 1977. Comparative photosynthesis in sun and shade plants. *Annu. Rev. Plant Physiol.* 28: 355.

Bowes, J., Crofts, A.R., and Arntzen, C.J. 1980. Redox reactions on the reducing side of photosystem II in chloroplasts with altered herbicide-binding properties. *Arch. Biochem. Biophys.* 200: 303.

Braunberger, P.G. 1987. Determination of Chloroplast Inheritance in Crosses Involving Triazine — Resistant and Triazine-Susceptible Canola (*Brassica napus* L.) with the Use of a Fluorescence-Activated Flow Cytometer. Undergraduate research report, Crop Science Dept., University of Guelph, Guelph, Canada.

Brine, R.P. 1989. Evaluation of Seed Vigor in Rapeseed and the Effect of Seed Vigor on Yield and Yield Components, M.Sc. thesis, University of Guelph, Guelph, Canada.

Brown, A.E., Truelove, B., Highfill, C.T., and Smith, S.G. 1988. Physiological competence of atrazine-resistant strains of the photosynthetic bacterium *Rhodobacter sphaeroides* (Rhodospirillaceae). *Weed Sci.* 36: 703.

Clay, D.V. and Hadleigh, R. 1991. Cross-resistance of *Poa annua* biotypes to paraquat and triazines. In *Herbicide Resistance in Weeds and Crops.* Long Ashton Int. Symp. Ed. by Caseley, J.C., Cussans, G.W., and Atkin, R.K. Butterworth-Heinemann, Boston, p. 433.

Darmency, H. and Pernes, J. 1989. Agronomic performance of a triazine-resistant foxtail millet (*Setaria italica* (L.) Beauv.). *Weed Res.* 29: 147.

Darmency, H. and Pernes, J. 1985. Use of wild *Setaria italica* (L.) Beauv.) to improve triazine resistance in cultivated *S. italica* (L.) by hybridization. *Weed Res.* 25: 175.
Dekker, J. and Burmester, R.G. 1990. Differential pleiotropy in a *psb*A gene mutant of *Brassica napus* implies altered temporal photosynthesis and thermal tolerance. *Z. Naturforsch.* 45c: 474.
Dekker, J.H. and Burmester, R.G. 1992. Pleiotropy in triazine-resistant *Brassica napus. Plant Physiol.* 100: 2052.
Dekker, J.H. and Sharkey, T.D. 1992. Regulation of photosynthesis in triazine-resistant and -susceptible *Brassica napus. Plant Physiol.* 98: 1069.
Demmig, B. and Bjorkman, O. 1987. Comparison of the effect of excessive light on chlorophyll fluorescence (77K) and photon yield of O_2 evolution in leaves of higher plants. *Planta* 171: 171.
Demmig, B., Winter, K., Kruger, A., and Czygan, F.-C. 1987. Photoinhibition and zeaxanthin formation in intact leaves. *Plant Physiol.* 84: 218.
Dexter, A.G., Burnside, O.C., and Lavy, T.L. 1966. Factors influencing the phytotoxicity of foliar applications of atrazine. *Weeds* 14: 222.
Dodge, A.D. 1983. Toxic oxygen species and herbicide action. In *Pesticide Chemistry: Human Welfare and the Environment, Proc. 5th Int. Cong. Pestic. Chem.* Ed. by Miyamoto, J. and Kearney, P.C. Pergamon Press, Oxford, p. 55.
Donnelly, M.J. and Hume, D.J. 1984. Photosynthetic rates and growth analysis of triazine-tolerant and normal rapeseed. *Can. J. Plant Sci.* 64: 432.
Donnelly-VanderLoo, M.J. and Hume, D.J. 1993. Unpublished data.
Ducruet, J.-M. and Lemoine, Y. 1985. Increased heat sensitivity of the photosynthetic apparatus in triazine-resistant biotypes from different plant species. *Plant Cell Physiol.* 26: 419.
Edgington, L.V. and Peterson, C.A. 1977. Systemic fungicides: theory, uptake and translocation. In *Antifungal Compounds*, Vol. 2. Siegel, M.R. and Sisler, H.D., Eds. Marcel Dekker, New York, chap. 2.
Erickson, L.R., Straus, N.A., and Beversdorf, W.D. 1983. Restriction patterns reveal origins of chloroplast genomes in *Brassica* amphiploids. *Theor. Appl. Genet.* 65: 201.
Foote, C.S. 1976. Photosensitized oxidation and singlet oxygen. In *Free Radicals in Biology*, Vol. 2. Academic Press, New York, p. 85.
Forcella, F. 1987. Herbicide-resistant crops: yield penalties and weed thresholds for oilseed rape (*Brassica napus* L.). *Weed Res.* 27: 31.
Forhan, M.A. 1987. The Effect of the Triazine Resistant Cytoplasm in Selected Backcrosses and F_1 Hybrids of *Brassica campestris, Brassica juncea,* and *Brassica napus,* M.Sc. thesis, University of Saskatchewan, Saskatoon, Canada.
Fuerst, E.P. and Norman, M.A. 1991. Interactions of herbicides with photosynthetic electron transport. *Weed Sci.* 39: 458.
Gasquez, J. and Compoint, J.P. 1981. Observation de chloroplastes resistants aux triazines chez une panicoides. *Setaria viridis* L. *Agronomie* 1: 923.
Golbeck, J.H. 1987. Structure function and organization of the photosystem I reaction center complex. *Biochim. Biophys. Acta* 895: 167.
Goodrich, J.A., Lykins, B.W., and Clark, R.M. 1991. Drinking water from agriculturally contaminated groundwater. *J. Environ. Qual.* 20: 707.
Gressel, J. 1991. Why get resistance? It can be prevented or delayed. In *Herbicide Resistance in Weeds and Crops*. Long Ashton Int. Symp. Ed. by Caseley, J.C., Cussans, G.W., and Atkin, R.K. Butterworth-Heinemann, Boston, p. 1.
Gressel, J. 1985. Herbicide tolerance and resistance: alteration of site of activity. In *Weed Physiology*. Vol. 2. Ed. by Duke, S.O. CRC Press, Boca Raton, FL, p. 159.
Gressel, J. and Segel, L.A. 1982. Interrelating factors controlling the rate of appearance of resistance: the outlook for the future. In *Herbicide Resistance In Plants*. Ed. by LeBaron, H.M. and Gressel, J. John Wiley & Sons, New York, p. 325.
Gressel J. and Ben-Sinai, G. 1985. Low intraspecific competitive fitness in a triazine-resistant, nearly nuclear-isogenic line of *Brassica napus. Plant Sci.* 38: 29.
Haehnel, W. 1984. Photosynthetic electron transport in higher plants. *Annu. Rev. Plant Physiol.* 35: 659.
Hart, J.J., Radosevich, S.R., and Stemler, A. 1992. Influence of light intensity on growth of triazine-resistant rapeseed (*Brassica napus*). *Weed Res.* 32: 349.
Hart, J.J. and Stemler, A. 1990a. Similar photosynthetic performance in low-light grown isonuclear triazine-resistant and -susceptible *Brassica napus* L. *Plant Physiol.* 94: 1295.
Hart, J.J. and Stemler, A. 1990b. High light-induced reduction and low light-enhanced recovery of photon yield in triazine-resistant *Brassica napus* L., *Plant Physiol.* 94: 1301.
Hathway, D.H. 1989. *Molecular Mechanisms of Herbicide Selectivity*. Oxford University Press, New York.
Havaux, M. 1992. Stress tolerance of photosystem II in vivo. *Plant Physiol.* 100: 424.
Hay, J.R. 1976. Herbicide transport in plants. In *Herbicides: Physiology, Biochemistry, Ecology*. Vol.1. Ed. by Audus, L.J., Academic Press, London, chap. 12.
Hobbs, S.L.A. 1987. Comparison of photosynthesis in normal and triazine-resistant *Brassica. Can. J. Plant Sci.* 67: 457.
Horton, P. and Hague, A. 1988. Studies of the induction of chlorophyll fluorescence in isolated barley chloroplasts. IV. Resolution of non-photochemical quenching. *Physiol. Plant.*, 74: 566.

Hume, D.J. and Braunberger, P.G. 1993. Unpublished data.

Juras, L. 1984. Selective Weed Contriazine-Resistantiazine-Resistantol in Triazine-Resistant Rapeseed. M.Sc. thesis, University of Manitoba, Winnipeg, Canada.

Jursinic, P.A. and Pearcy, R.W. 1988. Determination of the rate limiting step for photosynthesis in a nearly isonuclear rapeseed (*Brassica napus* L.) biotype resistant to atrazine. *Plant Physiol.* 88: 11.

Kyle, D.J. 1987. The biochemical basis for photoinhibition of photosystem II. In *Topics in Photosynthesis. Photoinhibition,* Vol. 9. Ed. by Kyle, D.J., Osmond, C.B., and Arntzen, C.J. Elsevier, Amsterdam, p. 197.

Lemoine, Y., Dubacq, J.P., Zabnlon, G., and Ducret, J.M. 1986. Organization of the photosynthetic apparatus from triazine-resistant and -susceptible biotypes of several plant species. *Can. J. Bot.* 64: 2999.

Maltais, B. and Bouchard, C.J. 1978. Une moutarde des oiseaux (*Brassica rapa* L.) resistante a l'atrazine. *Phytoprotection* 59: 117.

Mapplebeck, L.R., Souza Machado, V., and Grodzinski, B. 1982. Seed germination and seedling growth characteristics of atriazine-susceptible and resistant biotypes of *Brassica campestris. Can. J. Plant Sci.* 62: 733.

Mattoo, A.K., St. John, J.B., and Wergin, W.P. 1984. Adaptive reorganization of protein and lipid components in chloroplast membranes as associated with herbicide binding. *J. Cell Biochem.* 24: 163.

May, W.E. 1993. Free Fatty Acids in Ontario — Grown Spring Canola, M.Sc. thesis, University of Guelph, Guelph, Ontario, Canada.

Melis, A., Manodori, A., Glick, R.E., Ghirardi, M.L., McCauley, S.W., and Neale, P.J. 1985. The mechanism of photosynthetic membrane adaptation to environmental stress conditions: a hypothesis on the role of electron transport capacity and of ATP/NADPH pool in the regulation of thylakoid membrane organization and function. *Physiol. Veg.* 23: 757.

Mets, L. and Thiel, A. 1989. Biochemistry and genetic control of the photosystem II herbicide target site. In *Target Sites of Herbicide Action.* Ed. by Böger, P. CRC Press, Boca Raton, FL, chap. 1.

Mihajlovic, M. 1990. Effects of Selection for Low Leaf Fluorescence on Photosynthesis and Yield of Triazine-Resistant Rapeseed (*Brassica napus* L.), M.Sc. thesis, University of Guelph, Guelph, Ontario, Canada.

Mitchell, R.A.C. and Barber, J. 1986. Adaption of photosynthetic electron-transport rate to growth temperature in pea. *Planta* 169: 429.

Moody, K., Kust, C.A., and Buchholtz, K.P. 1970. Uptake of herbicides by soybean roots in culture solutions. *Weed Sci.* 18: 642.

Moreland, D.E. and Hilton, J.L. 1976. Actions on photosynthetic systems. In *Herbicides: Physiology, Biochemistry, Ecology,* Vol. 1, 2nd ed. Ed. by Audus, L.J., Academic Press, London, chap. 16.

Ogren, E. and Oquist, G. 1984. Photoinhibition of photosynthesis in *Lemna gibba* as induced by the interaction between light and temperature. III. Chlorophyll fluorescence at 77K. *Physiol. Plant.* 62: 193.

OOPSCC. 1990. Eastern Canadian Co-operative Spring Canola Trial Results. Unpublished report of the Ontario Oil and Protein Seed Crop Committee, Guelph, Ontario, Canada.

Oquist, G. 1982. Effects of low temperature on photosynthesis. *Plant Cell Environ.* 6: 281.

Osmond, C.B. 1981. Photorespiration and photoinhibition. Some implications for the energetics of photosynthesis. *Biochim. Biophys. Acta* 639: 77.

Polos, E., Szigeti, Z., Varadi, G., and Lehoczki, E. 1991. Diquat resistance in paraquat/atrazine co-resistant *Conyza canadensis.* In *Herbicide Resistance in Weeds and Crops.* Long Ashton Int. Symp. Ed. by Caseley, J.C., Cussans, G.W., and Atkin, R.K. Butterworth-Heinemann, Boston, p. 468.

Powles, S.B. 1984. Photoinhibition of photosynthesis induced by visible light. *Annu. Rev. Plant Physiol.* 35: 15.

Radosevich, S.R. 1977. Mechanism of atrazine resistance in lambs-quarters and pigweed. *Weed Sci.* 25: 316.

Reboud, X. and Till-Bottraud, I. 1991. The cost of herbicide resistance measured by a competition experiment. *Theor. Appl. Genet.* 82: 690.

Reith, M. and Staus, N.A. 1987. Nucleotide sequence of the chloroplast gene responsible for triazine resistance in canola. *Theor. Appl. Genet.* 73: 357.

Ryan, G.F. 1970. Resistance of common groundsel to simazine and atrazine. *Weed Sci.* 18: 614.

Schonfeld, M., Yaacoby, T., Ben-Yehuda, A., Rubin, B., and Hirschberg, J. 1987. Triazine resistance in *Phlaris paradoxa*; physiological and molecular analyses. *Z. Naturforsch.* 42c: 779.

Siefermann, D. and Yamamoto, H.Y. 1975. Light-induced de-epoxidation of violaxanthin in lettuce chloroplasts. *Biochim. Biophys. Acta.* 387: 149.

Sigematsu,Y., Sato, F., and Yamada, Y. 1989. The mechanism of herbicide resistance in tobacco cells with a new mutation in the Q_b protein. *Plant Physiol.* 89: 986.

Souza Machado, V., Bandeen, J.D., and Adlington, T. 1979. Transfer of Triazine-Resistance to Rutabaga. Research Report of the Expert Committee on Weeds — East Section, Agriculture Canada, p. 410.

Souza Machado, V., Bandeen, J.D., Stephenson, G.R., and Jensen, K.I.N. 1977. Differential atrazine interference with the Hill reaction of isolated chloroplasts from *Chenopodium album* L. biotypes. *Weed Res.* 17: 407.

Souza Machado, V., Bandeen, J.D., Stephenson, G.R., and Lavigne, P. 1978. Uniparental inheritance of chloroplast atrazine tolerance in *Brassica campestris. Can. J. Plant Sci.* 58: 977.

Souza Machado, V., Beversdorf, W., and Switzer, C.M. 1980. Potential for breeding triazine resistance into *Brassica* spp. *Proc. Br. Crop Prot. Conf. - Weeds* 3: 855.

Souza Machado, V. and Hume, D.J. 1987. Breeding herbicide-tolerant cultivars — a Canadian experience. *Proc. Br. Crop Prot. Conf. - Weeds.* 2: 473.

Souza Machado, V. and Bandeen, J.D. 1982. Genetic analysis of chloroplast atriazine-resistantiazine-resistantazine resistance in *Brassica campestriis* — cytoplasmic inheritance. *Weed Sci.* 30: 281.

Souza Machado, V., Ali, A., and Shupe, J. 1983. Breeding chloro-triazine herbicide resistance into rutabaga (*Brassica napus* L.) and Chinese cabbage (*Brassica campestris* L.) Eucarpia. *Crucifers Newsl.* 8: 21.

Stidham, M.A., Uribe, E.G., and Williams, G.J. 1982. Temperature dependence of photosynthesis in *Agropyron smithii* Rydb. *Plant Physiol.* 69: 929.

Sundby, C.. McCaffery, S., Chow, W.S., and Anderson, J.M. 1993a. Photosystem II function, photoinhibition, and turnover of the D1 protein at different irradiances in normal and atrazine resistant plants with an altered QB binding site. In *Research in Photosynthesis,* Vol. 4. Ed. by Murata, N. Kluwer Academic, Dordrecht, Netherlands, p. 443.

Sundby, C., Chow, W.S., and Anderson, J.M. 1993b. Effects on photosystem II function, photoinhibition, and performance of the spontaneous mutation of serine-264 in the photosystem II reaction centre D1 protein in triazine-resistant *Brassica napus* L. *Plant Physiol.* 103: 105.

Thomas, P. 1990. *Update to Canola Growers Manual.* Canola Council of Canada, Winnipeg, Manitoba, Canada, p. 200.

Tollenaar, M. 1991. The physiological basis of the genetic improvement of maize hybrids in Ontario from 1959 to 1988. *Crop Sci.* 31: 119.

Trebst, A. 1987. The three-dimensional structure of the herbicide binding niche on the reaction center polypeptides of photosystem II. *Z. Naturforsch.* 42c: 742.

Trebst, A. 1991. The molecular basis of resistance of photosystem II herbicides. In *Herbicide Resistance in Weeds and Crops,* Long Ashton Int. Symp. Ed. by Caseley, J.C., Cussans, G.W., and Atkin, R.K. Butterworth-Heinemann, Boston, p. 145.

van Oorschot, J.L.P. and van Leeuwen, P.H. 1989. Photosynthetic capacity of intact leaves of resistant and susceptible cultivars of *Brassica napus* L. to atrazine. *Weed Res.* 29: 29.

van Oorschot, J.L.P. 1988. Inhibition of photosynthesis in intact plants of biotypes resistant or susceptible to atrazine and cross-resistance to other herbicides. *Weed Res.* 28: 223.

van Oorschot, J.L.P. 1991. Chloroplastic resistance of weeds to triazines in Europe. In *Herbicide Resistance in Weeds and Crops,* Long Ashton Int. Symp. Ed. by Caseley, J.C., Cussans. G.W., and Atkin, R.K. Butterworth-Heinemann, Boston, p. 87.

Vaughn, K.C. 1986. Characterization of triazine-resistant and -susceptible isolines of canola (*Brassica napus* L.). *Plant Physiol.* 82: 859.

Vaughn, K.C. and Duke, S.O. 1984. Ultrastructural alterations to chloroplasts in triazine-resistant weed biotypes. *Physiol. Plant.* 62: 510.

Vermaas, W.F.J., Renegerand, G., and Dohnt, G. 1984. The reduction of the oxygen-evolving system in chloroplasts by thylakoid components. *Biochim. Biophys. Acta* 347: 194.

Vostral, H. J., Buchholtz, J.K.P., and Kust, C.A. 1970. Effect of root temperature on absorption and translocation of atrazine in soybeans. *Weed Sci.* 18: 115.

Weiss, J. 1982. Transfer of Cytoplasmically Inherited Triazine Resistance into Oilseed Rape (*Brassica campestris* and *B. napus*) and Preliminary Comparisons of *In Vitro* and *In Vivo* Performance in Near Isogenic Lines, M.Sc. thesis, University of Guelph, Guelph, Canada.

Wildner, G.F., Heistercamp, U., Bodner, U., Johanningmeier, U., and Haehnel, W. 1989. An amino acid substitution in the Q_B-protein causes herbicide resistance without impairing electron transport from Q_A to Q_B. *Z. Naturforsch.* 44c: 431.

Yanamoto, Y., Ford, R.C., and Barber, J. 1981. Relationship between thylakoid membrane fluidity and the functioning of pea chloroplasts. *Plant Physiol.* 67: 1069.

Yoshida, S. 1984. Chemical and biophysical changes in the plasma membrane during cold acclimation of mulberry bark cells (*Marus bombycis* Koidz. cv. Goroji). *Plant Physiol.* 76: 257.

Chapter 8

SULFONYLUREA HERBICIDE-RESISTANT CROPS

Leonard L. Saari
C. Jeffry Mauvais

CONTENTS

1. INTRODUCTION

To be successful, commercial herbicides must control weeds, display crop safety, and have low toxicity toward other organisms. The sulfonylurea herbicides fulfill these requirements admirably, and, in addition, are applied at very low use rates (a few grams per hectare) and are available for use in several crops (Saari et al. 1994; Brown and Cotterman 1994). For the most part, the

1-56670-045-0/96/$0.00+$.50
© 1996 by CRC Press Inc.

identification of crop-selective sulfonylurea herbicides has been accomplished by screening thousands of chemical analogs for the desired properties.

Another approach to obtaining selectivity is to develop herbicide-resistant crops. That is, resistance to sulfonylurea herbicides can be selected by screening for mutant plants exposed to the herbicide or by introducing a resistance gene into the crop by genetic engineering. Development of sulfonylurea-resistant crops has not been pursued to the same extent as resistance to other herbicide chemistries, in part due to the high degree of crop selectivity found via traditional chemical screening. However, concerns about crop selectivity in particular market segments and residual soil activity for rotational crops have resulted in some sulfonylurea-resistant crop development.

The topic of sulfonylurea herbicide-resistant crops is extensively covered through 1989 (Mazur and Falco 1989; Falco et al. 1989), especially the molecular methods that are used to obtain transformed, resistant crops. Several reviews are available on resistance to acetolactate synthase (ALS) inhibitors (Saari et al. 1994; Thill et al. 1991; Reed et al. 1989; Shaner 1991; Hartnett et al. 1990, 1991; Newhouse et al. 1990, 1991). This chapter summarizes the previous work on sulfonylurea herbicide resistance in crops and reviews the progress since 1989 in developing new resistant crops through selection and transformation. Chapter 9 covers the development of imidazolinone-resistant crops.

2. BIOCHEMISTRY AND GENETICS

2.1. MODE OF ACTION

The molecular target of the sulfonylurea herbicides is the enzyme ALS, also known as acetohydroxyacid synthase (AHAS) (LaRossa and Schloss 1984; Chaleff and Mauvais 1984; Ray 1984, 1989). Acetolactate synthase catalyzes the first common steps in the biosynthesis of the branched-chain amino acids isoleucine, valine, and leucine in plants and microbes. This pathway is not present in animals.

Several lines of biochemical and genetic evidence demonstrate that inhibition of ALS is the mechanism of action of the sulfonylurea herbicides. First, it was shown that sulfonylureas inhibit ALS activity isolated from bacteria (LaRossa and Schloss 1984), yeast (*Saccharomyces cerevisiae*) (Falco and Dumas 1985), and plants (Chaleff and Mauvais 1984; Ray 1984). Second, genetic methods were used to isolate microbial and plant mutants that are resistant to the sulfonylurea herbicides, and in most cases the ALS from these organisms is also insensitive to inhibition by sulfonylureas (Chaleff and Ray 1984; Chaleff and Mauvais 1984; Falco and Dumas 1985). In resistant plants the sulfonylurea-insensitive ALS activity cosegregates in genetic crosses with herbicide resistance observed at the organism level (Chaleff and Mauvais 1984; Haughn and Somerville 1986). Finally, the transformation of sensitive plant species with genes encoding resistant forms of ALS further implicates this enzyme as the target site of sulfonylureas (Lee et al. 1988, Haughn et al. 1988). The imidazolinone (Shaner et al. 1984), triazolopyrimidine (Gerwick et al. 1990), and pyrimidinyl thiobenzoate (Takahashi et al. 1991) herbicides also inhibit ALS.

2.2. ACETOLACTATE SYNTHASE GENE AND PROTEIN STRUCTURE

The quaternary structure of ALS is apparently different between prokaryotes and eukaryotes. Enterobacteria have multiple ALS isozymes with each isozyme composed of small and large subunits combined in a native $\alpha_2\beta_2$ arrangement (Schloss 1990). Acetolactate synthase isozyme II from *Salmonella typhimurium*, for instance, has an overall mol wt of 138,000 and comprises two subunits, each of mol wt 9700 and 59,300. In yeast and plants only monomers and oligomers of a single subunit of ALS are found (Bekkaoui et al. 1993a; Singh and Schmitt 1989; Durner and Böger 1990; Singh et al. 1991a, b; Muhitch et al. 1987; Falco and Dumas 1985). The reported mol wt of monomeric plant ALS varies from 58,000 (Durner and Böger 1988) to 72,877 (Mazur et al. 1987). Comparison of the deduced amino acid sequences of yeast and plant ALS to the *Escherichia coli* ALS large subunit (Falco and Dumas 1985; Mazur et al. 1987; Mazur and Falco 1989) indicates that the bacterial and eukaryotic proteins are homologous.

No direct evidence exists for an ALS small subunit in yeast or plants, but no conclusive documentation for its absence exists either. There is, however, indirect evidence suggesting the presence of a small subunit of ALS in eukaryotes. First, small subunits are required in combination with large subunits for valine feedback sensitivity of ALS isolated from enteric bacteria (Weinstock et al. 1992; Eoyang and Silverman 1986). Leucine and valine also feedback regulate ALS in plants (Miflin 1969, Miflin and Cave 1972). However, it was shown that ALS isolated from *E. coli* expressing the *Arabidopsis thaliana* (L.) Heynh. ALS gene is not inhibited by valine and leucine, whereas ALS isolated from *A. thaliana* plants is inhibited by these amino acids (Singh et al. 1992). This suggests, among several possibilities, that a regulatory element may be present in plant but not *E. coli* ALS extracts. Second, in tobacco (*Nicotiana tabacum* L.) transformed with *A. thaliana* ALS there is a discrepancy between the 25-fold elevated messenger RNA (mRNA) levels as a result of ALS genes expressed under control of the cauliflower mosaic virus 35S (CaMV35S) promoter and the twofold increase in ALS specific activities (Odell et al. 1990). Limited availability of a second subunit may explain this result. These observations, while not proving the existence of a small subunit of ALS, suggest that some type of regulatory element may be missing.

Some plant species express more than one form of ALS. Maize (*Zea mays* L.), soybean (*Glycine max* (L.) Merrill), canola (*Brassica napus* L.), and tobacco contain two or more ALS genes per haploid genome (Mazur et al. 1987; Rutledge et al. 1991; Falco et al. 1989; Wiersma et al. 1989), whereas sugarbeet (*Beta vulgaris* L.) and the cruciferous weed, *A. thaliana*, contain single ALS genes (Mazur et al. 1987; Keeler et al. 1993).

There is approximately 85% conservation of amino acids in the deduced ALS protein and 75% conservation of nucleotides in the ALS genes from *A. thaliana* and tobacco (Mazur et al. 1987). Three large conserved domains exist among the deduced amino acid sequences of yeast and bacterial ALS enzymes (Falco et al. 1985). Similar conserved domains are found in the plant ALS-deduced sequences (Mazur and Falco 1989). Several eukaryotic ALS genes have now been shown to lack introns (Mazur et al. 1987; Bekkaoui et al. 1993b; Wiersma et al. 1989), which facilitates the expression of eukaryotic ALS in bacteria.

In plants ALS is encoded in the nucleus, but the final gene product resides in the chloroplast (Jones et al. 1985; Mifflin 1974; Chaleff and Ray 1984). Acetolactate synthase is transported into chloroplasts via a transit peptide which is cleaved from the mature protein (Mazur et al. 1987; Mazur and Falco 1989).

3. METHODS FOR INTRODUCING RESISTANCE

Sulfonylurea herbicide resistance has been introduced into crops using two methods. One method involves the selection of mutant cells or seeds on sulfonylurea-impregnated media, and the other method involves the introduction of genes encoding insensitive forms of ALS through transformation.

3.1. MUTANT SELECTION

The first successful production of a sulfonylurea-resistant plant was accomplished by selecting tobacco cells on chlorsulfuron-supplemented medium followed by regeneration of mutant plants (Chaleff and Ray 1984). The resistance was due to a sulfonylurea-insensitive form of ALS (Chaleff and Mauvais 1984). Additional resistant tobacco lines were selected on media containing several different sulfonylurea herbicides, and these too were found to be resistant as a result of sulfonylurea-insensitive ALS (Chaleff and Bascomb 1987; Harms et al. 1992; Creason and Chaleff 1988) (Table 1). Where examined, resistance was inherited as a semidominant trait (Chaleff and Bascomb 1987; Chaleff and Ray 1984).

Selection also resulted in sulfonylurea-resistant lines of sugarbeet (Saunders et al. 1992; Hart et al. 1992, 1993), canola (Kenyon et al. 1987; Swanson et al. 1988; Saxena et al. 1990; Tonnemaker et al. 1992), soybean (Sebastian et al. 1989; Sebastian and Chaleff 1987), flax (*Linum usitatissimum* L.; Jordan and McHughen 1987), lettuce (*Lactuca sativa* L.; Mallory-Smith et al. 1993), and birdsfoot trefoil (*Lotus corniculatus* L.; Pofelis et al. 1992) (Table 1). All of these resistant crops were obtained using a sulfonylurea herbicide for selection, with chlorsulfuron being the most common selection agent.

In most cases the observed resistance is due to the selection of crop lines having an ALS that is less sensitive to inhibition by sulfonylurea herbicides (Table 1). Where determined, the resistance is caused by a semidominant or dominant mutation. The only exception is a single-gene recessive mutation resulting in sulfonylurea resistance in soybean that is not due to an altered sensitivity of ALS to inhibition by sulfonylureas (Sebastian and Chaleff 1987).

Callus or suspension cultures are the starting material for the selection of most resistant lines (Table 1). However, mutagenized seeds were used for selection of resistant soybean lines (Sebastian and Chaleff 1987; Sebastian et al. 1989), and a sulfonylurea-resistant weed, prickly lettuce (*Lactuca serriola* L.) (Mallory-Smith et al. 1990), was used in the development of sulfonylurea-resistant commercial lettuce. Prickly lettuce, a weed found in cereal-growing regions of Idaho, became resistant to sulfonylurea herbicides as a result of their agronomic use. Because both prickly and commercial lettuce species have the same number of chromosomes, it was possible to transfer the resistance gene from the resistant prickly lettuce biotype to the commercial lettuce line through a genetic cross (Mallory-Smith et al. 1993).

Mutagens such as ethyl methane sulfonate, *N*-nitroso-*N*-methylurea, ethylnitrosourea, and sodium azide were used to increase the probability of selecting resistant mutants in canola (Saxena et al. 1990; Tonnemaker et al. 1992; Swanson et al. 1988), soybean (Sebastian and Chaleff 1987; Sebastian et al. 1989), tobacco (Chaleff and Bascomb 1987; Chaleff and Ray 1984), and flax (Jordan and McHughen 1987). However, resistant lines of canola (Kenyon et al. 1987), sugarbeet (Saunders et al. 1992), lettuce (Mallory-Smith et al. 1993), birdsfoot trefoil (Pofelis et al. 1992), and tobacco (Harms et al. 1992; Creason and Chaleff 1988) were also found without the use of mutagens.

3.2. TRANSFORMATION

Sulfonylurea-resistant plants also have been obtained through transformation with genes encoding sulfonylurea-resistant forms of ALS. Tobacco has been transformed more frequently than other crops (Table 2), due to the relative ease, at least initially, of transformation and regeneration. Crops transformed to exhibit sulfonylurea-herbicide resistance now include canola (Wiersma et al. 1989, Miki et al. 1990), chicory (*Cichorium intybus* L.; Vermeulen et al. 1992), cotton (*Gossypium hirsutum* L.; Mauvais 1994), flax (McSheffrey et al. 1992; McHughen and Holm 1991), maize (Fromm et al. 1990), rice (*Oryza sativa* L.; Li et al. 1992), sugarbeet (D'Halluin et al. 1992), and tomato (*Lycopersicon esculentum* Mill.; Hartnett et al. 1990) (Table 3).

The plant materials most often used for tobacco transformation are herbicide-sensitive leaf discs or protoplasts (Table 2), which are co-cultivated with *Agrobacterium tumefaciens* (Horsch et al. 1985; Rogers et al. 1986; Bevan 1984) carrying plasmids containing a sulfonylurea-insensitive ALS gene, and sometimes antibiotic markers. Transformed cells are selected on either sulfonylurea- or antibiotic-impregnated media, and then plants are regenerated from these cell lines. Other dicots have been transformed (Table 3) using similar methods. For instance, sugarbeet transformants were obtained by incubating friable embryogenic callus with *A. tumefaciens* (D'Halluin et al. 1992).

TABLE 1 Herbicide-Resistant Crops Obtained via Selection with Sulfonylurea Herbicides

Crop Species	Starting Material	Insensitive ALS?	Selection Agent	Fold-Increase in Resistance (Whole Plant Level)	Mutagen[a]	Inheritance	Ref.
Beta vulgaris	Callus	Yes	Chlorsulfuron	144–377 for homozygous	None	Semidominant	Saunders et al. 1992; Hart et al. 1992, 1993
Brassica napus	Microspores[b]	Unknown	Chlorsulfuron		None		Kenyon et al. 1987
	Microspores[b]	Yes	Chlorsulfuron	12	ENU	Semidominant	Swanson et al. 1988
	Embryogenic suspension cultures	Yes	Chlorsulfuron	10–1000 (calli)	EMS		Saxena et al. 1990
B. napus[c]	Seeds	Unknown	Chlorsulfuron/ metsulfuron methyl	25	EMS		Tonnemaker et al. 1992
Glycine max	Seeds	No	Chlorsulfuron		ENU	Recessive	Sebastian and Chaleff 1987
	Seeds	Yes	Chlorsulfuron	100	EMS or NMU	Semidominant	Sebastian et al. 1989
Lactuca sativa	*Lactuca serriola*[d]	Yes[e]	Thifensulfuron methyl/ tribenuron methyl		None		Mallory-Smith et al. 1993
Linum usitatissimum	Callus	Unknown	Chlorsulfuron		NaN_3	Inconclusive	Jordan and McHughen 1987
Lotus corniculatus	Callus	Probably	Thifensulfuron methyl	4	None		Pofelis et al. 1992
Nicotiana tabacum[f]	Suspension culture	Yes	Cinosulfuron, then primisulfuron		None		Harms et al. 1992
	Callus	Yes	Chlorsulfuron		ENU	Semidominant	Chaleff and Bascomb 1987
	Callus	Yes	Sulfometuron methyl		None	Dominant	Creason and Chaleff 1988
	Callus	Yes	Chlorsulfuron	>100	ENU	Semidominant	Chaleff and Ray 1984

[a] EMS, ethyl methane sulfonate; NMU, *N*-nitroso-*N*-methylurea; ENU, ethylnitrosourea.

[b] Selection took place at the haploid level; chromosome doubling using colchicine was performed by Swanson et al. (1988) to restore fertility and normal morphology.

[c] Resistant to soil-residual levels of sulfonylureas but not to postemergent rates.

[d] A naturally occurring resistant biotype of *L. serriola* was selected through agronomic use of sulfonylurea herbicides. The resistant trait was backcrossed into commercial lettuce.

[e] Mallory-Smith et al. (1990).

[f] Two rounds of selection with sulfonylurea herbicides resulted in two resistance mechanisms: a first selection with a lethal concentration of cinosulfuron presumably selected a mutant expressing a *SurB* gene encoding an ALS with a Pro 196 to Ser change; a second selection with primisulfuron apparently resulted in amplification of the resistant gene.

TABLE 2 Sulfonylurea Herbicide-Resistant Tobacco Obtained via Transformation

Method of Transformation	Gene Introduced	Amino Acid Change in ALS	Promoter Used	Resistant to:	Ref.
Agrobacterium tumefaciens	Canola *ahas3r*, *Arabidopsis csr1-1*, or both[a]	Trp 557 to Leu (*ahas3r*) and Pro 197 to Ser (*csr1-1*)	CaMV35S[b]	Chlorsulfuron DPX-R9674[c]	Brandle et al. 1994
	Arabidopsis csr1-1	Pro 197 to Ser	CaMV35S	Chlorsulfuron Amidosulfuron	Brandle and Miki 1993 Brandle et al. 1992
	Arabidopsis csr1-1	Pro 197 to Ser	CaMV35S	Chlorsulfuron	Charest et al. 1990
	Arabidopsis csr1-1	Pro 197 to Ser	ALS	Chlorsulfuron, ethametsulfuron methyl, thifensulfuron methyl, tribenuron methyl, not triasulfuron	Gabard et al. 1989
	Arabidopsis csr1-1	Pro 197 to Ser	ALS	Chlorsulfuron	Haughn et al. 1988
	Arabidopsis csr1-1	Pro 197 to Ser	ALS or CaMV35S	Chlorsulfuron	Odell et al. 1990
	Arabidopsis csr1-1	Pro 197 to Ser	ALS or p70	Chlorsulfuron	Tourneur et al. 1993
	Arabidopsis hybrid	Pro 197 to Ser and Ser 653 to Asn	CaMV35S	Chlorsulfuron and imazapyr	Hattori et al. 1992
	Tobacco *SurB-Hra*	Pro 196 to Ala and Trp 573 to Leu	CaMV35S	Chlorsulfuron (and imidazolinones[d])	Lee et al. 1988
Direct	Tobacco *SurB*	Pro 196 to Ser	ALS	Chlorsulfuron, triasulfuron, primisulfuron, imazaquin	Harms et al. 1992

[a] The *ahas3r* and *csr1-1* genes were introduced separately into tobacco lines via transformation. Crossing two homozygous lines carrying the individual *csr1-1* and *ahas3r* genes resulted in a resistant hemizygous line that carried both genes.

[b] Cauliflower mosaic virus 35S promoter.

[c] Growth of seedlings homozygous for the *csr1-1* gene was slightly inhibited by chlorsulfuron and highly inhibited by DPX-R9674. Growth of seedlings homozygous for the *ahas3r* gene was not inhibited by either herbicide. When treated with DPX-R9674, the line hemizygous for both genes had an intermediate growth response between lines with either gene alone.

[d] Mazur and Falco 1989.

TABLE 3 Sulfonylurea Herbicide-Resistant Crops (Other than Tobacco) Obtained via Transformation

Crop Species	Method of Transformation	Gene Introduced	Amino Acid Change in ALS	Promoter Used	Resistant to	Field Tested?	Fold-Increase in Resistance	Ref.
Beta vulgaris	*Agrobacterium tumefaciens*	Tobacco *SurB/SurA* ALS or sugarbeet ALS	Pro 191 to Ala and Trp 568 to Leu Unknown[a]	ALS or CaMV35S[b]	Chlorsulfuron Trifensulfuron methyl Tribenuron methyl	No		D'Halluin et al. 1992
Brassica napus	*Agrobacterium tumefaciens*	Canola ALS	Pro 173 to Ser	ALS	Chlorsulfuron	No		Wiersma et al. 1989
	Agrobacterium tumefaciens	*Arabidopsis csr1-1*	Pro 197 to Ser	CaMV35S	Chlorsulfuron	No		Miki et al. 1990
Cichorium intybus	*Agrobacterium tumefaciens*	*Arabidopsis csr1-1*	Pro 197 to Ser	ALS	Chlorsulfuron	No		Vermeulen et al. 1992
Gossypium hirsutum	*Agrobacterium tumefaciens*	Tobacco *SurB-Hra*	Pro 196 to Ala and Trp 573 to Leu	ALS	Various sulfonylureas	Yes		Mauvais 1994
Linum usitatissimum	*Agrobacterium tumefaciens*	*Arabidopsis csr1-1*	Pro 197 to Ser	ALS	Chlorsulfuron Metsulfuron methyl	Yes	25–260 to chlorsulfuron (root growth)	McSheffrey et al. 1992; McHughen and Rowland 1991; McHughen and Holm 1991; McHughen 1989
Lycopersicon esculentum	*Agrobacterium tumefaciens*	Tobacco *SurB-Hra*	Pro 196 to Ala and Trp 573 to Leu	ALS	Various sulfonylureas	Yes		Hartnett et al. 1990
Oryza sativa	PEG-mediated[c]	*Arabidopsis csr1-1*	Pro 197 to Ser	CaMV35S	Chlorsulfuron	No	200 (Calli)	Li et al. 1992
Zea mays	DNA-coated microprojectiles	Maize ALS	Pro to Ser	CaMV35S	Chlorsulfuron	No		Fromm et al. 1990

[a] Mutant tobacco ALS gene consists of the ALS *SurB* gene, with part of the coding region substituted by a homologous fragment of the ALS *SurA* gene which contains two mutations resulting in a Pro to Ala and Trp to Leu change in the ALS protein. Sugarbeet ALS mutation not identified.

[b] Mutants having a mutant tobacco ALS gene under control of its native promoter were selected with kanamycin because chlorsulfuron resistance could not be used as a selectable marker in this construct. Mutants with the sugarbeet ALS gene under CaMV35S (cauliflower mosaic virus 35S promoter) promoter control were selected with chlorsulfuron.

[c] Polyethylene glycol-mediated.

The transformation and regeneration of monocots requires different procedures. Rice and maize were transformed by bombardment of protoplasts and embryogenic callus, respectively, with DNA-coated microprojectiles followed by selection and plant regeneration (Li et al. 1992, Fromm et al. 1990).

4. RESISTANCE AS A FUNCTION OF ACETOLACTATE SYNTHASE EXPRESSION AND ACETOLACTATE SYNTHASE MUTATIONS

The degree of sulfonylurea herbicide resistance generated by transformation or mutant selection depends on the specific mutation(s) in the ALS gene, the level of resistant gene expression, and the herbicide used for screening. That is, different amino acid changes in ALS result in varying resistance phenotypes. Also, the degree to which a sulfonylurea-sensitive or -resistant ALS gene is expressed affects the resistance observed in whole plants. Finally, the degree of resistance in a given crop line may vary with regard to different sulfonylurea herbicides.

4.1. GENES AND MUTATIONS IN TOBACCO, YEAST, AND *ARABIDOPSIS THALIANA*

Sulfonylurea herbicide resistance that has been introduced into crops through transformation is the result of earlier work on bacteria, yeast, tobacco, and *A. thaliana*. Studying yeast mutants, for example, provides an estimate of the number of mutations in the ALS gene that result in a sulfonylurea herbicide resistance phenotype. Forty-one spontaneous mutants of yeast were independently isolated and analyzed by DNA sequencing to determine amino acid substitutions in ALS resulting in resistance (Falco et al. 1989). Twenty-four different amino acid changes at 10 sites throughout the ALS protein were identified via the mutations. Amino acid substitutions in ALS resulting in herbicide resistance occur in three conserved domains in yeast, bacteria, and plants (Mazur and Falco 1989). Site-directed mutagenesis was used to obtain additional mutations at these sites in the yeast *ILV2* gene to identify even more substitutions resulting in the herbicide resistance phenotype (Falco et al. 1989; Mazur and Falco 1989). These experiments demonstrate that substitution of nearly any amino acid for Ala 117, Pro 192, or Trp586 results in a herbicide-resistant ALS. Clearly, sulfonylurea-herbicide resistance is a function of several possible mutations in the ALS gene.

Mutations to sulfonylurea resistance define two unlinked genetic loci for ALS in tobacco, *SurB* and *SurA* (Chaleff and Ray 1984; Chaleff and Bascomb 1987). The resistant tobacco line, HRA, contains two mutations in the *SurB* gene (Pro 196 to Ala and Trp 573 to Leu) and was obtained by two successive rounds of selection with sulfonylurea herbicides (Lee et al. 1988; Creason and Chaleff 1988). This line is 1000-fold more resistant to chlorsulfuron than the sensitive biotype (Falco et al. 1989). The C3 tobacco line contains one mutation in the *SurA* gene, Pro 196 to Gln (Lee et al. 1988). Transformation of wild-type tobacco with either the *SurA-C3* or *SurB-Hra* gene results in sulfonylurea-resistant tobacco (Lee et al. 1988; Falco et al. 1989). The double mutation in the HRA line yields an approximately five-fold greater resistance in transformed tissue relative to the original single mutant lines (Creason and Chaleff 1988) and cross-resistance to both sulfonylurea and imidazolinone herbicides (Lee et al. 1988).

One of the ALS genes frequently used in transformation to produce sulfonylurea herbicide resistant crop lines is the *csr1-1* gene from *A. thaliana*. The *csr1-1* gene differs from the wild-type gene by only a single base pair change, resulting in the substitution of a Ser for Pro 197 in the ALS protein (Haughn et al. 1988, Mazur et al. 1987). Changes in the analogous Pro to Ser in yeast ALS (Yadav et al. 1986), to Gln in C3 tobacco ALS (Lee et al. 1988), to Ala in *S4-Hra* tobacco ALS (this gene has a second mutation: Trp 573 to Leu) (Lee et al. 1988), and to Ser in canola ALS (Wiersma et al. 1989) have resulted in herbicide-resistant phenotypes.

4.2. USE OF RESISTANT ACETOLACTATE SYNTHASE GENES IN CROPS

Mutant ALS genes used for plant transformations have been derived often from *A. thaliana* or tobacco, although ALS genes from canola, maize, and sugarbeet also have been utilized (Tables 2 and 3). The production of sulfonylurea-resistant tobacco, canola, chicory, flax, and rice via transformation has employed the mutant *A. thaliana csr1-1* ALS gene (Haughn and Somerville 1986; Haughn et al. 1988) (Tables 2 and 3). In addition, a mutant sugarbeet ALS gene with an unidentified mutation (D'Halluin et al. 1992) and a mutant canola ALS gene (Wiersma et al. 1989) were used to transform those respective plant species. The canola ALS gene was mutated *in vitro* (which resulted in a Pro 173 to Ser change in ALS) and reintroduced into plants using *A. tumefaciens* vectors. Maize was transformed with a mutant maize ALS gene, again with a nucleotide change that results in a Pro to Ser change (equivalent to Pro 197 to Ser in *A. thaliana*) in the ALS protein (Fromm et al. 1990). Sulfonylurea-resistant tobacco cell lines were obtained using mutant chimeric tobacco ALS genes coding for either Trp 568 to Leu or Ala 199 to Asp changes (Hartnett et al. 1990).

As noted above, the frequently used *csr1-1* gene (Tables 2 and 3) results in a Pro 197 to Ser change in the *A. thaliana* ALS enzyme. Pro 197 (and equivalent changes in other plant ALS enzymes) in the ALS protein have been characterized generally as selectively resistant to sulfonylurea but not imidazolinone herbicides (Mazur and Falco 1989; Haughn and Somerville 1986; Lee et al. 1988; Mourad and King 1992). Closer inspection of these studies combined with more precise descriptions of the results changes the interpretations slightly. It is more accurate to describe proline changes in ALS as resulting in plants resistant to most sulfonylureas with a slight cross-resistance to some imidazolinones (Saari et al. 1994).

In addition to the single point mutations, double mutations in the ALS gene have been used to create sulfonylurea-resistant crops. The *SurB-Hra* gene, which encodes a mutant ALS protein with two amino acid changes relative to wild-type, Pro 196 to Ala and Trp 573 to Leu, was introduced into tobacco, tomato, and cotton (Lee et al. 1988; Hartnett et al. 1990; Mauvais 1994). Either the Pro 196 or Trp 573 changes alone confers resistance in tobacco plants (Hartnett et al. 1990), but the two mutations together afford high levels of resistance to sulfonylurea and imidazolinone herbicides (Mazur and Falco 1989; Creason and Chaleff 1988).

A different double set of ALS mutations also was introduced into tobacco. A chimeric *A. thaliana* ALS gene containing mutations that encode ALS with Pro 197 to Ser and Ser 653 to Asn changes was introduced into tobacco, generating plants that are resistant to both sulfonylurea and imidazolinone herbicides (Hattori et al. 1992). Transformation with genes carrying either mutation alone demonstrated that the Ser 653 to Asn change confers resistance to imidazolinone but not sulfonylurea herbicides (Haughn and Somerville 1990; Sathasivan et al. 1991), and that the Pro 197 to Ser change gives high resistance to sulfonylurea and low resistance to some imidazolinone herbicides (Saari et al. 1994).

Finally, two individual resistant ALS genes were introduced into tobacco separately and in combination to examine their relative effects in conferring resistance (Brandle et al. 1994). These two genes were the *csr1-1* gene, and a newly reported gene, the *ahas3r* gene from canola, which contains a single base pair change resulting in a Trp 557 to Leu change in the ALS protein. The individual genes were introduced into separate plants, and a doubly hemizygous line with both resistant ALS genes was produced by crossing the homozygous, single resistant gene lines. Both single gene lines are resistant to chlorsulfuron, but the line carrying the Trp mutation possesses greater resistance to DPX-R9674 (a mixture of thifensulfuron methyl and tribenuron methyl) than the line with the Pro mutation. The line with both mutant genes is less resistant to DPX-R9674 than the line expressing ALS with the single Trp 557 to Leu change. This difference may be due to variations in gene expression because the line carrying both genes is hemizygous.

The above-mentioned studies show that introduction of ALS genes with various single mutations or combinations of mutations results in qualitatively and quantitatively diverse patterns of resistance to ALS-inhibitor herbicides.

4.3. EXPRESSION AS A FUNCTION OF THE PROMOTER

The expression of resistant ALS genes in transformed crops is dependent, in part, on the promoter used to regulate gene expression. In most cases the CaMV35S promoter, which generally results in high levels of constitutive gene expression, is used. In other cases resistance was achieved using native ALS promoters (Tables 2 and 3). For example, expression of the mutant *A. thaliana* gene, *csr1-1*, under control of its own promoter, confers sulfonylurea resistance in transgenic chicory and flax plants (Vermeulen et al. 1992; McSheffrey et al. 1992). However, this gene does not confer sufficient resistance in rice (Li et al. 1992). There are also cases of resistant tobacco ALS genes under control of their native promoters that were used to establish sulfonylurea resistance in crops such as sugarbeet and cotton (D'Halluin et al. 1992; Mauvais, 1994).

The relationship between ALS gene expression and whole-plant resistance was examined in detail with transgenic tobacco (Odell et al. 1990; Charest et al. 1990). Overexpression in tobacco of the *A. thaliana* wild-type gene via the CaMV35S promoter generates no more than a three-fold increase in resistance to chlorsulfuron relative to nontransgenic tobacco (Odell et al. 1990; Charest et al. 1990). In contrast, expression of the mutant *A. thaliana csr1-1* gene under the control of its native promoter (Odell et al. 1990; Haughn et al. 1988; Gabard et al. 1989), or under control of the CaMV35S promoter (Odell et al. 1990; Charest et al. 1990), confers high levels of resistance to tobacco plants. The resistance gene itself and the resistance gene under control of the CaMV35S promoter increased resistance to chlorsulfuron by 300- and 3000-fold, respectively (Odell et al. 1990). In these experiments the CaMV35S promoter elevated the level of ALS mRNA up to 25-fold. This increase in gene expression was correlated with a 10-fold higher level of resistance.

Paradoxically, the specific activity of ALS isolated from transgenic tobacco expressing the *csr1-1* gene under control of the CaMV35S promoter is only twofold higher than that from plants transformed with the resistance gene with its own promoter (Odell et al. 1990). This slight increase in specific activity does not correlate with the increase in ALS mRNA described above. Possible explanations for the discrepancy between increases in ALS specific activity and ALS mRNA levels include blocks in mRNA translation, impediments to ALS transport into the chloroplast, lessened protein stability, and the requirement for a second "small" subunit, similar to the ALS structure found in enteric bacteria (Odell et al. 1990).

In another experiment the same resistance gene (*A. thaliana csr-1*) was expressed in transgenic tobacco using either the native or a p70 promoter (CaMV35S promoter with a double enhancer) (Tourneur et al. 1993). Results were similar to those reported by Odell et al. (1990) in that there were 300- and 1500-fold increases in resistance to chlorsulfuron of plants transformed with the *csr-1* gene under control of the native and enhanced CaMV35S promoter, respectively. In this case the ALS activity was 12-fold and the whole plant resistance 1500-fold greater when the p70 promoter was used. Northern blot analysis of RNA isolated from wild-type and transformed plants confirmed that the p70 promoter increased *csr-1* mRNA levels, but the increase was not quantitated and therefore cannot be compared to the Odell et al. (1990) study. In another study using resistant tobacco with native genes a closer correlation between ALS specific activity and gene expression was seen (Harms et al. 1992).

4.4. ROLE OF THE CHLOROPLAST TRANSIT PEPTIDE

It is unclear what impact the transport of resistant ALS protein into the chloroplast has on whole plant resistance. Most, if not all, of the ALS is located in the chloroplast in untransformed plant species (Jones et al. 1985; Mifflin 1974). Transport into the chloroplast is facilitated by the chloroplast transit peptide on the pre-ALS protein (Mazur and Falco 1989; Mazur et al. 1987). The sequences which encode the transit peptide are not highly conserved between *A. thaliana* and tobacco ALS genes.

Rice, a monocot, transformed with a dicotyledonous *A. thaliana csr1-1* gene under control of the CaMV35S promoter, is sulfonylurea resistant (Li et al. 1992). This suggests that the mechanism of ALS transport into the chloroplast is a conserved function among diverse plant species, that ALS can function normally outside of the chloroplast, or both. Several other examples of crops transformed with chimeric ALS genes suggest similar possibilities (Tables 2 and 3).

A somewhat related area is the expression of plant ALS in bacteria. Efficient expression of canola ALS in and functional complementation of *S. typhimurium* devoid of ALS activity grown on minimal media requires that at least part of the chloroplast transit peptide sequence be removed (Wiersma et al. 1990). However, plant ALS gene constructs with no truncations, or with truncations which eliminate the transit sequence very close to the purported processing site result in no complementation. In two other studies *E. coli* strains deficient in ALS are complemented by the expression of *A. thaliana* ALS genes which include the transit peptide sequence (Smith et al. 1989; Singh et al. 1992). In both cases the *E. coli* somehow processes the *A. thaliana* ALS protein to a nearly correct molecular weight because the ALS proteins isolated from *A. thaliana* and transformed *E. coli* migrate equally on sodium dodecyl sulfate gels. It is unknown whether *E. coli* cleaves the *A. thaliana* protein precisely at the processing site normally recognized by plants. The necessity for precise processing of pre-ALS protein has not been resolved by these studies.

5. HERBICIDES AND CROSS-RESISTANCE

For the majority of sulfonylurea-resistant crops, only the selection agent is reported in evaluating resistance. However, some degree of cross-resistance to additional sulfonylurea herbicides was noted for sulfonylurea-resistant lines of tobacco (Gabard et al. 1989; Brandle et al. 1992; Brandle and Miki 1993; Harms et al. 1992; Mazur and Falco 1989), sugarbeet (Hart et al. 1992, 1993; D'Halluin et al. 1992), soybean (Sebastian and Chaleff 1989), lettuce (Mallory-Smith et al. 1993), and flax (McSheffrey et al. 1992; McHughen and Holm 1991). For example, a sugarbeet line selected with chlorsulfuron is cross-resistant to primisulfuron, thifensulfuron methyl, and chlorimuron ethyl (Hart et al. 1992, 1993). A soybean line, also selected with chlorsulfuron, is cross-resistant to thifensulfuron methyl, tribenuron methyl, and chlorimuron ethyl (Sebastian et al. 1989). Tobacco, first selected with cinosulfuron and then primisulfuron, is also resistant to sulfometuron methyl, triasulfuron, chlorsulfuron, and imazaquin (Harms et al. 1992). Some resistant crops were obtained using combinations of sulfonylureas such as lettuce selected with thifensulfuron methyl and tribenuron methyl (Mallory-Smith et al. 1993), and canola selected with chlorsulfuron and metsulfuron methyl (Tonnemaker et al. 1992).

Cross-resistance to other sulfonylureas, as reported above, is not unexpected. Acetolactate synthase mutations which confer resistance to one sulfonylurea herbicide (as a result of target-site sensitivity changes) generally accord some resistance to most other sulfonylurea herbicides (Saari et al. 1994). Three exceptions, where cross-resistance to certain other sulfonylurea herbicides was not observed, include sulfonylurea-resistant tobacco (Gabard et al. 1989; Brandle et al. 1992) and sugarbeet (Hart et al. 1992). First, tobacco resistant to amidosulfuron and chlorsulfuron is minimally resistant to DPX-R9674 (Brandle et al. 1992). Because the ALS isolated from this resistant tobacco is insensitive to inhibition by DPX-R9674, metabolism of DPX-R9674 to a phytotoxic product is a possible explanation for the whole plant results. Second, a resistant sugarbeet line was reported as resistant to several sulfonylureas but not to metsulfuron methyl (Hart et al. 1992). As resistance to metsulfuron methyl was determined at only one rate of this herbicide, some degree of cross-resistance probably would have been observed if the study had included lower rates. Third, a transgenic tobacco line transformed with the *A. thaliana csr1-1* gene was resistant to chlorsulfuron and several other sulfonylureas, but not triasulfuron (Gabard et al. 1989). This result is unusual because some cross-resistance to sulfonylureas is generally seen for proline changes. For example, tobacco transformed with a tobacco ALS gene containing a corresponding proline change displays a relatively high degree of cross-resistance to triasulfuron (Mauvais and Saari 1989).

In general, the resistance factors (amount of herbicide required for the control of the herbicide-resistant line divided by the amount required for control of the herbicide-sensitive line) obtained for resistant crops were quite large (25-fold or higher). In two separate studies selected sugarbeet lines are 300- to 1000-fold (Saunders et al. 1992) and 144- to 377-fold (Hart et al. 1992, 1993) more resistant to chlorsulfuron than sensitive lines. Birdsfoot trefoil (Pofelis et al. 1992), canola (Tonnemaker et al. 1992), soybean (Sebastian et al. 1989), and tobacco (Chaleff and Ray 1984) lines are more resistant to thifensulfuron methyl (4-fold), chlorsulfuron/metsulfuron methyl (25-fold), chlorsulfuron (100-fold), and chlorsulfuron (>100-fold), respectively. However, quantitative comparisons of herbicide resistance for mutant and nonmutant plants must be viewed cautiously because of the dependence of plant response on growth conditions and developmental stage at the time of herbicide treatment.

6. FIELD TESTING FOR AGRONOMIC PERFORMANCE

Successful development of resistant crops depends on their ability to perform, on a yield and quality basis, equally well in the field relative to elite cultivars. The published reports of sulfonylurea-resistant crops generally cite greenhouse or laboratory studies rather than field studies. However, there are a few field studies of sulfonylurea-resistant crops. For example, in the absence of herbicide, two transgenic tobacco lines harboring the *csr1-1* gene are lower yielding than untransformed controls in field tests (Brandle and Miki 1993). In contrast, transgenic flax expressing the same *csr1-1* gene perform equally well as controls in field tests in the absence of herbicide (McHughen and Holm 1991; McSheffrey et al. 1992). Cotton transformed with the tobacco *SurB-Hra* gene exhibited yields equivalent to the non-transgenic parental line in numerous field trials from 1990 through 1994 (Mauvais 1994).

Several studies also examined field performance under herbicide treatment. In birdsfoot trefoil selected for sulfonylurea resistance herbicide application affects overall forage yield, but less in resistant than nonselected lines (Pofelis et al. 1992). The mutant gene from the HRA line of tobacco was used to transform both tobacco and tomato, and these transformed lines were uninjured relative to sensitive lines, by application of weed-controlling rates of several sulfonylurea herbicides (Hartnett et al. 1990). From these studies it is apparent that the yield of sulfonylurea herbicide resistant crops can equal the yield of sensitive lines under field conditions.

7. SUMMARY

The high degree of crop selectivity found in the sulfonylurea herbicide class has reduced the need for developing sulfonylurea-resistant crops. However, scientific interest along with concerns about selectivity in market niches and residual activity in crop rotations led to the development of at least 12 sulfonylurea-resistant crop species, some of which may be developed into commercial cultivars. Both selection of resistant mutants and introduction of resistant ALS genes via transformation are successful avenues for producing sulfonylurea-resistant crops. Nearly all resistant crops that are reported express herbicide-insensitive forms of ALS. To date, no crops have been developed with an increased capacity to metabolize sulfonylurea herbicides to nonphytotoxic compounds. Selection for sulfonylurea-resistant mutants was accomplished both with and without the aid of mutagens, and included both tissue culture and seeds as starting material. Transformation of sensitive crop lines with genes encoding sulfonylurea-resistant ALS was performed mainly by *Agrobacterium*-mediated transformation, but other methods such as bombardment with DNA-coated microprojectiles also were used. The degree of sulfonylurea herbicide resistance generated by transformation or mutant selection depends on the specific mutation(s) in the ALS gene, the level of resistant gene expression, and the herbicide used. The genes most frequently employed in transformation encode ALS proteins with single amino acid changes at a proline equivalent to Pro 197 in *A. thaliana*. Genes that encode ALS with two amino acid changes were also used, and transgenic plants harboring these genes usually have higher levels of sulfonylurea resistance than lines transformed with genes encoding ALS with single amino acid changes. In general, ALS genes expressed by the cauliflower mosaic virus 35S promoter confer higher levels of sulfonylurea resistance than ALS genes regulated by their own promoters. A given resistant line can express varying levels of resistance to different sulfonylurea and nonsulfonylurea herbicides. Field testing has shown that the yield of resistant lines can equal that of nontransformed lines.

ACKNOWLEDGMENTS

We thank Drs. John Pierce and S. Carl Falco for reviewing the manuscript.

REFERENCES

Bekkaoui, F., Schorr, P., and Crosby, W.L. 1993a. Acetolactate synthase from *Brassica napus*: immunological characterization and quaternary structure of the native enzyme. *Physiol. Plant.* 88: 475.

Bekkaoui, F., Nadin-Davis, S.A., and Crosby, W.L. 1993b. Isolation and structure of an acetolactate synthase gene from *Schizosaccharomyces pombe* and complementation of the *ilv2* mutation in *Saccharomyces cerevisiae*. *Curr. Genet.* 24: 544.

Bevan, M. 1984. Binary *Agrobacterium* vectors for plant transformation. *Nucleic Acids Res.* 12: 8711.

Brandle, J.E. and Miki, B.L. 1993. Agronomic performance of sulfonylurea-resistant transgenic flue-cured tobacco grown under field conditions. *Crop Sci.* 33: 847.

Brandle, J.E., Morrison, M.J., Hattori, J., and Miki, B.L. 1994. A comparison of two genes for sulfonylurea herbicide resistance in transgenic tobacco seedlings. *Crop Sci.* 34: 226.

Brandle, J.E., Labbe, H., Zilkey, B.F., and Miki, B.L. 1992. Resistance to the sulfonylurea herbicides chlorsulfuron, amidosulfuron, and DPX-R9674 in transgenic flue-cured tobacco. *Crop Sci.* 32: 1049.

Brown, H.M. and Cotterman, J.C. 1994. Recent advances in sulfonylurea herbicides. In *Herbicides Inhibiting Branched-Chain Amino Acid Biosynthesis*. Ed by Stetter, J. Springer-Verlag, New York, pp. 47–81.

Chaleff, R.S. and Mauvais, C.J. 1984. Acetolactate synthase is the site of action of two sulfonylurea herbicides in higher plants. *Science* 224: 1443.

Chaleff, R.S., and Bascomb, N.F. 1987. Genetic and biochemical evidence for multiple forms of acetolactate synthase in *Nicotiana tabacum*. *Mol. Gen. Genet.* 210: 33.

Chaleff, R.S. and Ray, T.B. 1984. Herbicide-resistant mutants from tobacco cell cultures. *Science* 223: 1148.

Charest, P.J., Hattori, J., DeMoor, J., Iyer, V.N., and Miki, B.L. 1990. *In vitro* study of transgenic tobacco expressing *Arabidopsis* wild type and mutant acetohydroxyacid synthase genes. *Plant Cell Rep.* 8: 643.

Creason, G.L. and Chaleff, R.S. 1988. A second mutation enhances resistance of a tobacco mutant to sulfonylurea herbicides. *Theor. Appl. Genet.* 76: 177.

D'Halluin, K., Bossut, M., Bonne, E., Mazur, B., Leemans, J., and Botterman, J. 1992. Transformation of sugarbeet (*Beta vulgaris* L.) and evaluation of herbicide resistance in transgenic plants. *Bio/Technology* 10: 309.

Delannay, X. et al. 1995. Yield evaluation of a glyphosate-tolerant soybean line after treatment with glyphosate. *Crop Science*, in press.

Durner, J. and Böger, P. 1990. Oligomeric forms of plant acetolactate synthase depend on flavin adenine dinucleotide. *Plant Physiol.* 93: 1027.

Durner, J. and Böger, P. 1988. Acetolactate synthase from barley (*Hordeum vulgare* L.): purification and partial characterization. *Z. Naturforsch.* 43c: 850.

Eoyang, L. and Silverman, P.M. 1986. Role of small subunit (IlvN polypeptide) of acetohydroxyacid synthase I from *Escherichia coli* K-12 in sensitivity of the enzyme to valine inhibition. *J. Bacteriol.* 166: 901.

Falco, S.C. and Dumas, K.S. 1985. Genetic analysis of mutants of *Saccharomyces cerevisiae* resistant to the herbicide sulfometuron methyl. *Genetics* 109: 21.

Falco, S.C., McDevitt, R.E., Chui, C.-F., Hartnett, M.E., Knowlton, S., Mauvais, C.J., Smith, J.K., and Mazur, B.J. 1989. Engineering herbicide-resistant acetolactate synthase. *Dev. Ind. Microbiol.* 30: 187.

Fromm, M.E., Morrish, F., Armstrong, C., Williams, R., Thomas, J., and Klein, T.M. 1990. Inheritance and expression of chimeric genes in the progeny of transgenic maize plants. *Bio/Technology* 8: 833.

Gabard, J.M., Charest, P.J., Iyer, V.N., and Miki, B.L. 1989. Cross-resistance to short residual sulfonylurea herbicides in transgenic tobacco plants. *Plant Physiol.* 91: 574.

Gerwick, B.C., Subramanian, M.V., Loney-Gallant, V.I., and Chandler, D.P. 1990. Mechanism of action of the 1,2,4-triazolo[1,5-*a*]pyrimidines. *Pestic. Sci.* 29: 357.

Hammond, B.G., Vicini, J.L., Hartnell, G.F., Naylor, M.W., Knight, C.D., Robinson, E., Fuchs, R.L., and Padgette, S.R. 1995. The feeding value of soybeans fed to rats, poultry, catfish, and dairy cattle are not altered by incorporation of glyphosate tolerance. *J. Nutrition*, in press.

Harms, C.T., Armour, S.L., DiMaio, J.J., Middlesteadt, L.A., Murray, D., Negrotto, D.V., Thompson-Taylor, H., Weymann, K., Montoya, A.L., Shillito, R.D., and Jen, G.C. 1992. Herbicide resistance due to amplification of a mutant acetohydroxyacid synthase gene. *Mol. Gen. Genet.* 233: 427.

Harrison, L.A., Bailey, M.R., Naylor, M.W., Ream, J., Hammond, B.G., Nida, D.L., Burnette, B., Nickson, T.E., Taylor, M.L., Fuchs, R.L., and Padgette, S.R. 1995. The expressed protein in glyphosate-tolerant soybeans, 5-enolypyruvylshikimate-3-phosphate synthase from *Agrobacterium* sp. strain CP4, is rapidly digested and is not toxic to mice upon acute gavage administration. *J. Nutrition*, in press.

Hart, S.E., Saunders, J.W., and Penner, D. 1993. Semidominant nature of monogenic sulfonylurea herbicide resistance in sugarbeet (*Beta vulgaris*). *Weed Sci.* 41: 317.

Hart, S.E., Saunders, J.W., and Penner, D. 1992. Chlorsulfuron-resistant sugarbeet: cross-resistance and physiological basis of resistance. *Weed Sci.* 40: 378.

Hartnett, M.E., Chui, C., Mauvais, C.J., McDevitt, R.E., Knowlton, S., Smith, J.K., Falco, S.C., and Mazur, B.J. 1990. Herbicide-resistant plants carrying mutated acetolactate synthase genes. In *Managing Resistance to Agrochemicals. From Fundamental Research to Practical Strategies.* Ed. by Green, M.B., LeBaron, H.M., and Moberg, W.K. American Chemical Society, Washington, D.C., chap. 31.

Hartnett, M.E., Chui, C.-F., Falco, S.C., Knowlton, S., Mauvais, C.J., and Mazur, B.J. 1991. Molecular analysis of sulfonylurea herbicide-resistant ALS genes. In *Herbicide Resistance in Weeds and Crops.* Ed. by Caseley, J.C., Cussans, G.W., and Atkin, R.K. Butterworth-Heinemann, Oxford, pp. 343–353.

Hattori, J., Rutledge, R., Labbe, H., Brown, D., Sunohara, G., and Miki, B. 1992. Multiple resistance to sulfonylureas and imidazolinones conferred by an acetohydroxyacid synthase gene with separate mutations for selective resistance. *Mol. Gen. Genet.* 232: 167.

Haughn, G.W., Smith, J., Mazur, B., and Somerville, C. 1988. Transformation with a mutant *Arabidopsis* acetolactate synthase gene renders tobacco resistant to sulfonylurea herbicides. *Mol. Gen. Genet.* 211: 266.

Haughn, G.W. and Somerville, C.R. 1990. A mutation causing imidazolinone resistance maps to the *csr1* locus of *Arabidopsis thaliana. Plant Physiol.* 92: 1081.

Haughn, G.W. and Somerville, C. 1986. Sulfonylurea-resistant mutants of *Arabidopsis thaliana. Mol. Gen. Genet.* 204: 430.

Horsch, R.B., Fry, J.E., Hoffman, N.L., Eichholtz, D., Rogers, S.G., and Fraley, R.T. 1985. A simple and general method for transferring genes into plants. *Science* 227: 1229.

Jones, A.V., Young, R.M., and Leto, K.J. 1985. Subcellular localization and properties of acetolactate synthase, target site of the sulfonylurea herbicides. *Plant Physiol.* 77: S-293.

Jordan, M.C. and McHughen, A. 1987. Selection for chlorsulfuron resistance in flax (*Linum usitatissimum*) cell cultures. *J. Plant Physiol.* 131: 333.

Keeler, S.J., Sanders, P., Smith, J.K., and Mazur, B.J. 1993. Regulation of tobacco acetolactate synthase gene expression. *Plant Physiol.* 102: 1009.

Kenyon, P.D., Marshall, G., and Morrison, I.N. 1987. Selection for sulfonylurea herbicide tolerance in oilseed rape (Br*assica napus*) using microspore culture. In *Brighton Crop Protection Conference - Weeds.* British Crop Protection Council, Surrey, pp. 871–877.

LaRossa, R.A. and Schloss, J.V. 1984. The sulfonylurea herbicide sulfometuron methyl is an extremely potent and selective inhibitor of acetolactate synthase in *Salmonella typhimurium. J. Biol. Chem.* 259: 8753.

Lee, K.Y., Townsend, J., Tepperman, J., Black, M., Chui, C.-F., Mazur, B., Dunsmuir, P., and Bedbrook, J. 1988. The molecular basis of sulfonylurea herbicide resistance in tobacco. *EMBO J.* 7: 1241.

Li, Z., Hayashimoto, A., and Murai, N. 1992. A sulfonylurea herbicide resistance gene from *Arabidopsis thaliana* as a new selectable marker for production of fertile transgenic rice plants. *Plant Physiol.* 100: 662.

Mallory-Smith, C., Thill, D.C., and Dial, M.J. 1993. ID-BR1: sulfonylurea herbicide-resistant lettuce germplasm. *HortScience* 28: 63.

Mallory-Smith, C.A., Thill, D.C., and Dial, M.J. 1990. Identification of sulfonylurea herbicide-resistant prickly lettuce (*Lactuca serriola*). *Weed Technol.* 4: 163.

Mauvais, C.J. 1994. Unpublished data.

Mauvais, C.J. and Saari, L.L. 1989. Unpublished data.

Mazur, B.J. and Falco, S.C. 1989. The development of herbicide resistant crops. *Annu. Rev. Plant Physiol. Plant Mol. Biol.* 40: 441.

Mazur, B.J., Chui, C.-F., and Smith, J.K. 1987. Isolation and characterization of plant genes coding for acetolactate synthase, the target enzyme for two classes of herbicides. *Plant Physiol.* 85: 1110.

McHughen, A. 1989. *Agrobacterium* mediated transfer of chlorsulfuron resistance to commercial flax cultivars. *Plant Cell Rep.* 8: 445.

McHughen, A. and Holm, F. 1991. Herbicide resistant transgenic flax field test: agronomic performance in normal and sulfonylurea-containing soils. *Euphytica* 55: 49.

McHughen, A. and Rowland, G.G. 1991. The effect of T-DNA on the agronomic performance of transgenic flax plants. *Euphytica* 55: 269.

McSheffrey, S.A., McHughen, A., and Devine, M.D. 1992. Characterization of transgenic sulfonylurea-resistant flax (*Linum usitatissimum*). *Theor. Appl. Genet.* 84: 480.

Miflin, B.J. 1974. The location of nitrite reductase and other enzymes related to amino acid biosynthesis in the plastids of roots and leaves. *Plant Physiol.* 54: 550.

Miflin, B.J. 1969. Acetolactate synthase from barley seedlings. *Phytochemistry* 8: 2271.

Miflin, B.J. and Cave, P.R. 1972. The control of leucine, isoleucine, and valine biosynthesis in a range of higher plants, *J. Exp. Bot.* 23: 511.

Miki, B.L., Labbe, H., Hattori, J., Ouellet, T., Gabbard, J., Sunohara, G., Charest, P.J., and Iyer, V.N. 1990. Transformation of Brassica *napus* canola cultivars with *Arabidopsis thaliana* acetohydroxyacid synthase genes and analysis of herbicide resistance. *Theor. Appl. Genet.* 80: 449.

Mourad, G. and King, J. 1992. Effect of four classes of herbicides on growth and acetolactate-synthase activity in several variants of *Arabidopsis thaliana*. *Planta* 188: 491.

Muhitch, M.J., Shaner, D.L., and Stidham, M.A. 1987. Imidazolinones and acetohydroxyacid synthase from higher plants. Properties of the enzyme from maize suspension culture cells and evidence for the binding of imazapyr to acetohydroxyacid synthase *in vivo*. *Plant Physiol.* 83: 451.

Newhouse, K.E., Wang, T., and Anderson, P.C. 1991. Imidazolinone-resistant crops. In *The Imidazolinone Herbicides*. Ed. by Shaner, D.L. and O'Connor, S.L. CRC Press, Boca Raton, FL, pp. 139–150.

Newhouse, K.E., Shaner, D.L., Wang, T., and Fincher, R. 1990. Genetic modification of crop responses to imidazolinone herbicides. In *Managing Resistance to Agrochemicals. From Fundamental Research to Practical Strategies*. Ed. by Green, M.B., LeBaron, H.M., and Moberg, W.K. American Chemical Society, Washington, D.C., chap. 32.

Odell, J.T., Caimi, P.G., Yadav, N.S., and Mauvais, C.J. 1990. Comparison of increased expression of wild-type and herbicide-resistant acetolactate synthase genes in transgenic plants, and indication of posttranscriptional limitation on enzyme activity. *Plant Physiol.* 94: 1647.

Padgette, S.R., Kolacz, K.H., Delannay, X., Re, D.B., LaVallee, B.J., Tinius, C.N., Rhodes, W.K., Otero, Y.I., Barry, G.F., Eichholtz, D.A., Peschke, V.M., Taylor, N.B., and Kishore, G.M. 1995a. Development, identification, and characterization of a glyphosate-tolerant soybean line. *Crop Science*, in press.

Padgette, S.R., Taylor, N.B., Nida, D.L., Bailey, M.R., MacDonald, J., Holden, L.R., and Fuchs, R.L. 1995b. The composition of glyphosate-tolerant soybeans is equivalent to conventional soybeans. *J. Nutrition*, in press.

Pofelis, S., Le, H., and Grant, W.F. 1992. The development of sulfonylurea herbicide-resistant birdsfoot trefoil (*Lotus corniculatus*) plants from *in vitro* selection. *Theor. Appl. Genet.* 83: 480.

Ray, T.B. 1989. Herbicides as inhibitors of amino acid biosynthesis. In *Target Sites of Herbicide Action*. Ed. by Böger, P. and Sandmann, G. CRC Press, Boca Raton, FL, chap. 6.

Ray, T.B. 1984. Site of action of chlorsulfuron. Inhibition of valine and isoleucine biosynthesis in plants. *Plant Physiol.* 75: 827.

Reed, W.T., Saladini, J.L., Cotterman, J.C., Primiani, M.M., and Saari, L.L. 1989. Resistance in weeds to sulfonylurea herbicides. In *Brighton Crop Protection Conference - Weeds*. British Crop Protection Council, Surrey, pp. 295–300.

Rogers, S.G., Horsch, R.B., and Fraley, R.T. 1986. Gene transfer in plants: production of transformed plants using Ti plasmid vectors. *Methods Enzymol.* 118: 627.

Rutledge, R.G., Oullet, T., Hattori, J., and Miki, B.L. 1991. Molecular characterization and genetic origin of the *Brassica napus* acetohydroxyacid synthase multigene family. *Mol. Gen. Genet.* 229: 31.

Saari, L.L., Cotterman, J.C., and Thill, D.C. 1994. Resistance to acetolactate synthase inhibiting herbicides. In *Herbicide Resistance in Plants: Biology and Biochemistry*. Ed. by Powles, S. and Holtum, J. CRC Press, Boca Raton, FL, pp. 83–140.

Sathasivan, K., Haughn, G.W., and Murai, N. 1991. Molecular basis of imidazolinone herbicide resistance in *Arabidopsis thaliana* var Columbia. *Plant Physiol.* 97: 1044.

Saunders, J.W., Acquaah, G., Renner, K.A., and Doley, W.P. 1992. Monogenic dominant sulfonylurea resistance in sugarbeet from somatic cell selection. *Crop Sci.* 32: 1357.

Saxena, P.K., Williams, D., and King, J. 1990. The selection of chlorsulfuron-resistant cell lines of independent origin from an embryogenic cell suspension culture of *Brassica napus* L. *Plant Sci.* 69: 231.

Schloss, J.V. 1990. Acetolactate synthase, mechanism of action and its herbicide binding site, *Pestic. Sci.* 29: 283.

Sebastian, S.A. and Chaleff, R.S. 1987. Soybean mutants with increased tolerance for sulfonylurea herbicides. *Crop Sci.* 27: 948.

Sebastian, S.A., Fader, G.M., Ulrich, J.F., Forney, D.R., and Chaleff, R.S. 1989. Semidominant soybean mutation for resistance to sulfonylurea herbicides. *Crop Sci.* 29: 1403.

Shaner, D.L. 1991. Mechanisms of resistance to acetolactate synthase/acetohydroxyacid synthase inhibitors. In *Herbicide Resistance in Weeds and Crops*. Ed. by Caseley, J.C., Cussans, G.W., and Atkin, R.K. Butterworth-Heinemann, Oxford, pp. 187–198.

Shaner, D.L., Anderson, P.C., and Stidham, M.A. 1984. Imidazolinones. Potent inhibitors of acetohydroxyacid synthase. *Plant Physiol.* 76: 545.

Singh, B.K. and Schmitt, G.K. 1989. Flavin adenine dinucleotide causes oligomerization of acetohydroxyacid synthase from Black Mexican sweet corn cells. *FEBS Lett.* 258: 113.

Singh, B., Szamosi, I., Hand, J.M., and Misra, R. 1992. *Arabidopsis* acetohydroxyacid synthase expressed in *Escherichia coli* is insensitive to the feedback inhibitors. *Plant Physiol.* 99: 812.

Singh, B., Schmitt, G., Lillis, M., Hand, J.M., and Misra, R. 1991a. Overexpression of acetohydroxyacid synthase from *Arabidopsis* as an inducible fusion protein in *Escherichia coli*. *Plant Physiol.* 97: 657.

Singh, B.K., Lumanglas, A., and Wang, B.S. 1991b. Production of a monocot-specific monoclonal antibody against acetohydroxyacid synthase and its use in the purification and characterization of the enzyme. *Proc. Natl. Acad. Sci. U.S.A.* 88: 4572.

Smith, J.K., Schloss, J.V., and Mazur, B.J. 1989. Functional expression of plant acetolactate synthase genes in *Escherichia coli*. *Proc. Natl. Acad. Sci. U.S.A.* 86: 4179.

Swanson, E.B., Coumans, M.P., Brown, G.L., Patel, J.D., and Beversdorf, W.D. 1988. The characterization of herbicide tolerant plants in *Brassica napus* L. after *in vitro* selection of microspores and protoplasts. *Plant Cell Rep.* 7: 83.

Takahashi, S., Shigematsu, S., and Morita, A. 1991. KIH-2031, a new herbicide for cotton. In *Brighton Crop Protection Conference - Weeds*. British Crop Protection Council, Surrey, pp. 57–62.

Thill, D.C., Mallory-Smith, C.A., Saari, L.L., Cotterman, J.C., Primiani, M.M., and Saladini, J.L. 1991. Sulfonylurea herbicide resistant weeds: discovery, distribution, biology, mechanism, and management. In *Herbicide Resistance in Weeds and Crops*. Ed. by Caseley, J.C., Cussans, G.W., and Atkin, R.K. Butterworth-Heinemann, Oxford, pp. 115–128.

Tonnemaker, K.A., Auld, D.L., Thill, D.C., Mallory-Smith, C.A., and Erickson, D.A. 1992. Development of sulfonylurea-resistant rapeseed using chemical mutagenesis. *Crop Sci.* 32: 1387.

Tourneur, C., Jouanin, L., and Vaucheret, H. 1993. Over expression of acetolactate synthase confers resistance to valine in transgenic tobacco. *Plant Sci.* 88: 159.

Vermeulen, A., Vaucheret, H., Pautot, V., and Chupeau, Y. 1992. *Agrobacterium* mediated transfer of a mutant *Arabidopsis* acetolactate synthase gene confers resistance to chlorsulfuron in chicory (*Cichorium intybus* L.). *Plant Cell Rep.* 11: 243.

Weinstock, O., Sella, C., Chipman, D.M., and Barak, Z. 1992. Properties of subcloned subunits of bacterial acetohydroxy acid synthases. *J. Bacteriol.* 174: 5560.

Wiersma, P.A., Schmiemann, M.G., Condie, J.A., Crosby, W.L., and Maloney, M.M. 1989. Isolation, expression and phylogenetic inheritance of an acetolactate synthase gene from *Brassica napus*. *Mol. Gen. Genet.* 219: 413.

Wiersma, P.A., Hachey, J.E., Crosby, W.L., and Moloney, M.M. 1990. Specific truncations of an acetolactate synthase gene from *Brassica napus* efficiently complement *ilvB*/*ilvG* mutants of *Salmonella typhimurium*. *Mol. Gen. Genet.* 224: 155.

Yadav, N., McDevitt, R.E., Benard, S., and Falco, S.C. 1986. Single amino acid substitutions in the enzyme acetolactate synthase confer resistance to the herbicide sulfometuron methyl. *Proc. Natl. Acad. Sci. U.S.A.* 83: 4418.

Chapter 9

IMIDAZOLINONE-RESISTANT CROPS: SELECTION, CHARACTERIZATION, AND MANAGEMENT

Dale L. Shaner
Newell F. Bascomb
Wendy Smith

CONTENTS

1-56670-045-0/96/$0.00+$.50
© 1996 by CRC Press Inc.

1. INTRODUCTION

The imidazolinones are a class of herbicides that control a broad spectrum of weeds at low rates and are used throughout the world in legumes, cereals, forests, and plantation crops. These herbicides are widely used, not only because of their efficacy, but also because of their low mammalian toxicity and low environmental impact.

Early in the development of imidazolinones, researchers began to select for imidazolinone resistance in crops with no natural tolerance to the herbicides in this class (Newhouse et al. 1991) because imidazolinone-resistant varieties greatly increase the options for weed control in these crops, which include corn, wheat, and canola. In this chapter we review the history, selection, and development of imidazolinone-resistant crops and their role in efficient and cost-effective weed management.

The development of imidazolinone-resistant crops began in 1982 as a collaboration between American Cyanamid and Molecular Genetics, Inc. (MGI). This instance is probably the first in which selection for a herbicide-resistant crop began so early in the development of a class of herbicides. Field tests in 1980 and 1981 demonstrated the efficacy of imazaquin and imazethapyr on the complex of weeds that infest soybeans and maize, including such difficult-to-control weeds as shattercane (*Sorghum bicolor*) and johnsongrass (*Sorghum halepense*). Although these herbicides could be used in soybeans, they were not selective in maize. Finding an imidazolinone that would control these noxious grasses in maize seemed remote, so the decision was made to attempt to select for resistant maize lines.

Unlike instances of selection for resistance to other classes of herbicides, in which the initial research was often done on species that were easily regenerated (e.g., tobacco) or on tissue cultures from which plants could not be obtained, the first species to be used in this effort to select for imidazolinone resistance was maize. The goal was to achieve resistance in an economically important crop by selecting for resistance in tissue culture and then regenerating plants.

At the inception of this research, the mode of action of the imidazolinones was unknown, although it was already known that the imidazolinones act primarily at the meristems of plants. However, it was unclear if whole-plant resistance could be selected via tissue culture techniques. Initial research by Anderson and Hibberd (1985) showed that the dose-response curve of maize in culture was the same as for seedlings and suggested that perhaps resistance selected at the tissue culture level would be manifested in intact plants. (In fact, the discovery of the mode of action of the imidazolinones came from the collaborative research between the scientists at American Cyanamid and MGI while developing this methodology for selecting for resistance [Shaner and Anderson 1985].)

One of the problems facing Anderson and Georgeson (1989) while they were developing the selection system for imidazolinone resistance was maintaining the regeneration capacity of maize tissue culture during the selection procedure. Their work showed that concentrations of imazaquin as low as 96 n*M* completely prevented the regeneration of plants from corn callus. They also demonstrated that concentrations of herbicide that inhibited cell culture growth >50% also caused the loss of somatic embryo formation. The selection protocol that was developed involved exposing friable, embryogenic maize cell cultures (initiated from immature embryos of A188 × B73 hybrids) to sublethal levels of imazaquin that inhibited growth by 30%. Vigorously growing callus that retained embryogenic morphology was subcultured three times in maintenance media containing 96 n*M* imazaquin followed by subculturing twice in media with 320 n*M* (Anderson and Georgeson 1989). Using this protocol, Anderson and Georgeson isolated eight cell lines which showed various levels of resistance to imidazolinones.

As soon as these imidazolinone-resistant cell lines were identified, the cultures were transferred to plant regeneration media, and plants were successfully obtained from five of the lines (Anderson and Georgeson 1989, unpublished data). Subsequent research showed that the mechanism of resistance in these five lines resulted from an alteration in the gene for acetohydroxyacid synthase, the site of action of the imidazolinones and several other herbicides including the sulfonylureas, triazolopyrimidine sulfonamides, and pyrimidyl-oxy-benzoates (Stidham 1991). This resistance was expressed at the whole-plant level and was inherited as a semidominant trait.

Thus, the first attempt to select for imidazolinone resistance proved successful. This early work not only proved that imidazolinone-resistant crops could be selected but also led to the discovery of the site of action of this potent class of herbicides and to the development of other imidazolinone-resistant crops.

2. SELECTION PROCEDURES FOR OBTAINING IMIDAZOLINONE-RESISTANT CROPS

2.1. IMIDAZOLINONE-RESISTANT CORN FROM TISSUE CULTURE SELECTION

Although the imidazolinone herbicides control many of the weeds that are a problem in corn, the use of these herbicides on corn is not recommended because of damage to the crop. Corn is often planted in rotation with soybeans, which can be safely treated with imidazolinones, but an additional margin of safety is desirable in order to prevent any damage to corn from the application of the herbicide to soybeans the previous year. To allow for extended use of this herbicide and to provide the additional safety, the program described above was initiated between Cyanamid and MGI.

The procedure employed was based on tissue culture selection of maize cell lines that could regenerate whole plants (Newhouse et al. 1991). The tissue culture lines were embryogenic callus lines derived from a cross between A188 (University of Minnesota, Crop Improvement Association) and B73 (Iowa State University). The callus was selected on agar media plates containing a concentration of imazaquin that inhibited growth by <40%. After the cells were subcultured several times, a uniform cell line was obtained that was able to grow on media containing 300 times more imazaquin than the initial culture. Although several lines were eventually obtained, three lines were initially characterized, and these were designated XA17, XI12, and QJ22.

The AHAS activity of XA17, XI12, and QJ22 was analyzed for resistance to imidazolinone and sulfonylurea herbicides. The extractable AHAS activity from the XA17 line was at least 100 times more resistant to imidazolinone and sulfonylurea herbicides than the wild-type (B73) enzyme (Figure 1). These results confirmed that AHAS was the target site of the herbicide and that resistance was due to an altered form of the enzyme. The herbicide resistance of the enzyme did not alter the enzyme's feedback sensitivity to valine and leucine (Figure 2). Genetic analysis indicated that the mutations in the XI12 and XA17 lines were at a different locus than the mutation in QJ22.

Biochemical analysis of XI12 and QJ22 showed that the amount of resistant AHAS activity from the XI12 line was approximately 80% of the total enzyme activity, whereas the resistant activity from the QJ22 line was only 20% of the total extractable AHAS activity (Figure 3). These results suggest different levels of expression from the different loci.

2.2. IMIDAZOLINONE-RESISTANT CORN FROM POLLEN MUTAGENESIS

In addition to using cell-culture selection, a pollen-mutagenesis procedure was also used to obtain imidazolinone-resistant corn. This work was accomplished by ICI Seeds and resulted in the first commercially available herbicide-resistant crop seed (Bright et al. 1990).

Pollen mutagenesis was accomplished by treating 3 mg of freshly isolated pollen grains with 1% ethyl methanesulfonate (EMS) in light paraffin oil for 40 min. The treated grains were then painted onto the silks of detassled, inbred female parents. The plants were grown to maturity and the M_1 seeds were harvested.

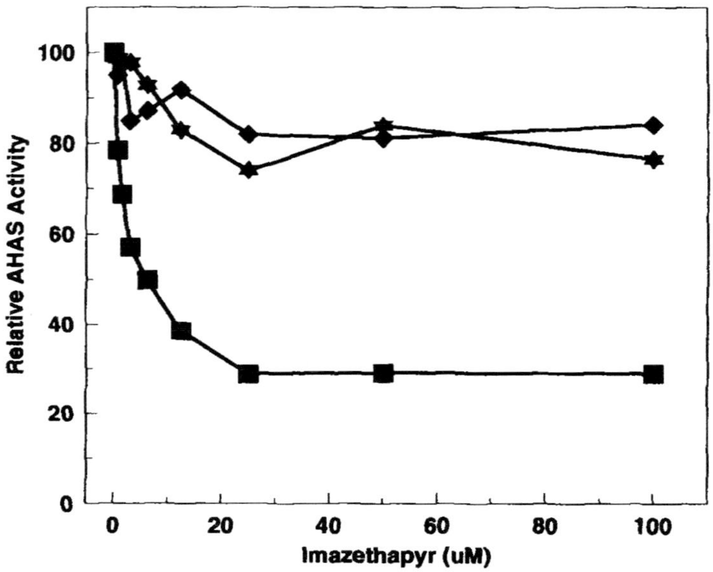

FIGURE 1 Inhibition of acetohydroxyacid synthase from wild-type corn, B73 (■) and mutant lines XA17 (◆) and XI12 (✱) by imazethapyr. AHAS was extracted from corn seedlings using standard assay conditions.

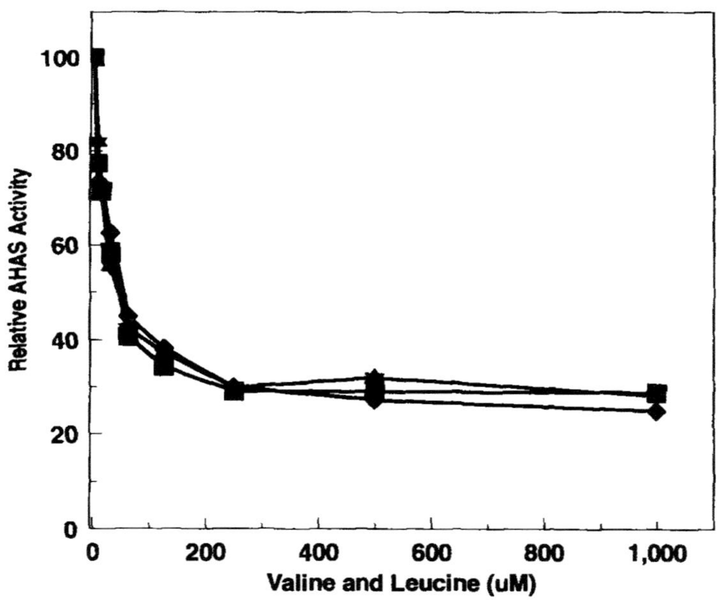

FIGURE 2 Inhibition of acetohydroxyacid synthase from wild-type corn, B73 (■) and mutant lines XA17 (◆) and XI12 (✱) by valine plus leucine. AHAS was extracted from corn seedlings using standard assay conditions.

The seeds (100,000 total) were screened by planting 100 seeds per tray and spraying with the equivalent of 250 g/ha imazethapyr. This application rate allowed germination of the seeds, but sensitive plants were severely affected. Ten plants survived this treatment and nine were found to have true resistance to the herbicide. Five of these plants showed a phenotype that was similar to

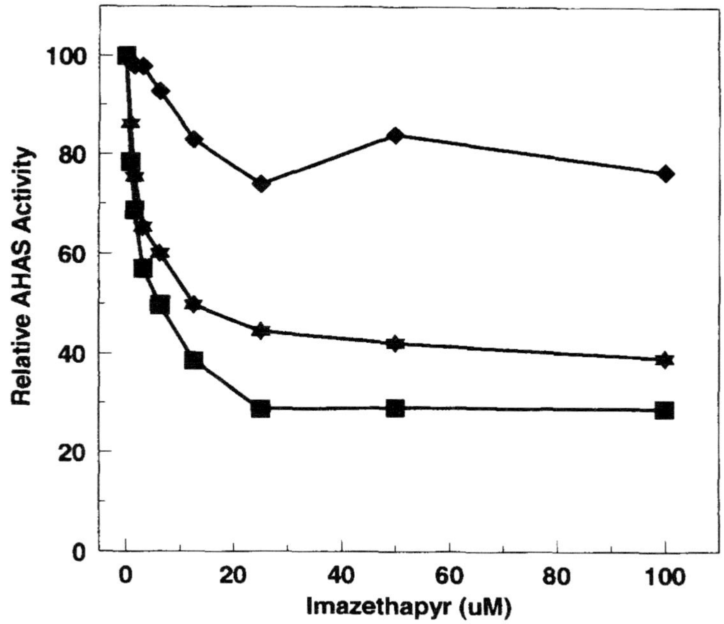

FIGURE 3 Inhibition of acetohydroxyacid synthase from wild-type corn, B73 (■) and mutant lines XI12 (◆) and QJ22 (✱) by imazethapyr. AHAS was extracted from corn seedlings using standard assay conditions.

that of the control plants (untreated), and four showed some damage from the herbicide application. Genetic analysis of these plants showed that resistance resulted from a single dominant gene.

Biochemical analysis of extracted AHAS activity against imidazolinone, triazolopyrimidine, and sulfonylurea herbicides showed a 2- to 660-fold increased resistance to imazethapyr and essentially no increased resistance to the sulfonylurea or triazolopyrimidine classes of herbicides.

2.3. IMIDAZOLINONE-RESISTANT WHEAT FROM SEED MUTAGENESIS

Wheat is inherently more tolerant to the imidazolinone herbicides than are many other commercially important crops, but is nevertheless stunted and damaged by high concentrations. To increase the tolerance of wheat to imidazolinone herbicides, a project was initiated to obtain imidazolinone-resistant wheat from a mutagenized seed population (Newhouse et al. 1992).

Mutagenesis and selection was a four-step procedure which included (1) seed mutagenesis, (2) a seed soak in herbicide, (3) preemergence herbicide application, and (4) screening and selection of herbicide-resistant mutants. The mutants from this screen were then characterized biochemically and genetically. The mutagenized population used for this procedure was obtained from 5000 seeds of a French winter-wheat variety (*Triticum aestivum* L. em Thell cv. Fidel). Seeds were imbibed overnight and mutagenized with 1 m*M* sodium azide for 2 h. The plants from this generation were designated M_1. The M_2 seed was harvested from these plants at maturity.

The first step of the screen was a seed soak as described by Sebastian and Chaleff (1987). The seeds were disinfected and each lot of 25 seeds was treated with 25 ml of 1 *M* imazethapyr for 3 d. In the second step the soaked seeds were drained, planted at a density of 1000 seeds per flat (15 × 20 cm), and then watered and sprayed immediately with imazethapyr at a concentration of 300 g/ha. The seedlings that emerged 4 weeks after planting were transplanted to untreated soil and grown to maturity. From 120,000 M_2 seeds derived from the initial 5000 M_1, this procedure resulted in four lines that were confirmed to be imidazolinone resistant. The lines were designated FS1, FS2, FS3, and FS4 (for Fidel Selection).

Genetic characterization showed that the M_3 generations of FS1, FS2, and FS4 were homozygous for herbicide tolerance because the plants did not segregate for tolerance, whereas FS3 required an additional generation of selfing to identify nonsegregating progeny. The F_1 populations of hybrids produced by crossing each tolerant line with wild-type Fidel were found to be less resistant to the herbicide. The F_2 progeny from a cross with wild-type Fidel all segregated in a 3/1 tolerant-to-susceptible ratio, which indicated that the tolerance was due to a single semidominant gene.

Biochemical analysis of FS1, FS2, FS3, and FS4 was performed to determine whether resistance was attributable to resistant AHAS. Initial analysis of the extractable AHAS activity showed only a relatively low level of resistant activity. To clarify the AHAS-based resistance, the resistant line FS4 and the wild-type were sprayed with herbicide 10 d after planting, and leaf material was harvested 3 d later for AHAS extraction. The spray treatment inhibited the endogenous sensitive AHAS and provided a clear demonstration of resistant AHAS in the selected wheat lines (Newhouse et al. 1992).

2.4. IMIDAZOLINONE-RESISTANT CANOLA FROM MICROSPORE MUTAGENESIS

The development of imidazolinone-resistant canola was done as a collaborative research program between American Cyanamid and Allelix Crop Technologies (Swanson et al. 1989). Imidazolinone-resistant canola was developed for several reasons. First, it would allow farmers to use a herbicide that has very low use rates and low mammalian toxicity. In addition, because canola is grown in rotation with other crops, resistant canola would permit the use of imidazolinone herbicides on other crops used in the rotation without fear of damaging the canola.

Microspore mutagenesis was chosen for the selection of imidazolinone-resistant canola for several reasons. Microspores are a large population of single-cell haploids that show a low level of somaclonal variation; therefore, they can be used for efficient selection of recessive traits by mutagenesis procedures that produce a final product that breeds true for the selected trait.

Microspore mutagenesis was performed as described in an earlier paper (Swanson et al. 1988) with 20 *M* ethyl nitrosourea (ENU) as the mutagen and selection on 40 g/l of imazethapyr. Surviving embryos were transferred to medium containing 50 g/l of imazethapyr and regenerated plantlets were transferred to soil. Ten putative mutant embryos were selected and produced plants, which were then selfed. Half produced progeny with some degree of resistance to imazethapyr. Seed from these five plants were treated preplant incorporated with increasing concentrations of imazethapyr, and two lines (P1 and P2) were selected for further analysis based on their high level of resistance. The parental Topas canola line was killed by 20 g/ha of imazethapyr and were severely damaged at much lower rates. The line P1 tolerated 50 g/ha, and line P2 tolerated up to 500 g/ha.

Biochemical analysis of the AHAS extracted from these lines showed only marginal imidazolinone-resistant AHAS activity in the P1 line, but the P2 line produced AHAS that was over 500 times more tolerant than AHAS from the wild-type Topas. It is probable that the two AHAS loci contribute different proportions of the total AHAS activity, as observed in maize. Crossing the resistant lines to wild-type Topas resulted in progeny that had intermediate levels of resistance. Crossing P1 with P2 resulted in progeny that were more resistant than either heterozygote and showed an additive effect of the tolerance between these lines.

2.5. IMIDAZOLINONE-RESISTANT PLANTS FROM TRANSFORMATION

Many crop species that have been recalcitrant to transformation are beginning to yield to the continuing efforts of plant tissue culture. The ability to transform important crops allows the introduction of modified herbicide target genes as a mechanism to introduce herbicide resistance. For example, an *Arabidopsis* AHAS gene was isolated from a genomic library prepared from a mutant *Arabidopsis* that was selectively resistant to imidazolinone herbicides and did not confer resistance to the sulfonylurea or triazolopyrimidine classes of herbicides (Sathasivan et al. 1990). The genomic DNA fragment containing the entire AHAS structural gene, more than 2 kb of 5′ untranslated DNA and approximately 1.5 kb of 3′ untranslated DNA, was isolated and cloned into a plasmid for maintenance in *E. coli*. The mutation was identified by sequencing and was found to be a substitution in the codon for amino acid 653, a serine (AGT) that was changed to asparagine (AAT).

Transformed tobacco was selected on kanamycin to obtain kanamycin-resistant calluses. In addition, material was analyzed by callus induction from regenerated plantlets in the presence of various concentrations of imazapyr. Southern blot analysis of DNA from the herbicide-resistant plants confirmed the presence of the *Arabidopsis* AHAS gene in this material.

Enzyme assays confirmed production by transgenic plants of a herbicide-insensitive form of AHAS that is 1000-fold more resistant to imazapyr than is the wild-type enzyme. Because AHAS activity in nontransformed plants is inhibited by imidazolinone herbicides, the existence of this herbicide-tolerant activity also confirms expression of the introduced gene.

3. CHARACTERIZATION OF IMIDAZOLINONE-RESISTANT CROPS

3.1. IMIDAZOLINONE-RESISTANT CORN

The resistance expressed in corn plants was characterized both genetically and phenotypically in greenhouse and field studies. Genetic analysis was performed on nonsegregating progeny of resistant plants regenerated from XA17 by crossing with the imidazolinone-susceptible inbred, B73. The inheritance of imazaquin tolerance by the progeny demonstrated the genetic stability of the trait. These plants were further analyzed via selfing and backcrossing to B73. Selfed progeny exhibited the expected 3/1 ratio of resistant plants to susceptible plants, and the backcrossed progeny segregated in a 1/1 ratio as expected for a single, dominant, nuclear mutation. Subsequent detailed genetic studies demonstrated that resistance was inherited as a single, semidominant allele. The four alleles examined (XA17, XI12, QJ22, and XS40) indicate that at least two loci for the AHAS enzyme are present in corn. The tolerance genes in XA17 and XI12 appear to be alleles at one locus while the tolerance genes from QJ22 and XS40 are at a second locus. The two loci were mapped to chromosomes using genetic marker stocks in which reciprocal translocations were linked to the waxy endosperm locus. XA17 and XI12 define a locus that from the genetic-mapping results appears to be on the long arm of chromosome 5. The locus defined by QJ22 and XS40 seems to be on chromosome 4 and may be located close to the centromere.

Both the level and spectrum of tolerance were evaluated for each of five lines from which plants were regenerated. The first plants produced were regenerated from the callus selection designated XA17 and were resistant not only to the imidazolinone used in the selection protocol but were also cross-resistant to other imidazolinones (including imazapyr and imazethapyr) as well as to an entirely different class of chemistry, the sulfonylureas. The level of tolerance was determined based on the rates required for 50% growth inhibition with postemergence herbicide applications and was more than 1000 times greater than that of the wild-type parental lines. Plants were also regenerated from XI12, QJ22, and XS40 as well as one other line, ZA54. While plants regenerated from these cell lines were cross-resistant to other imidazolinones, none were cross-tolerant to the sulfonylureas (or other AHAS-inhibiting herbicides tested). In addition, the relative degree of tolerance of the resistant cell lines to the different imidazolinones varied. For example, XI12 was 1000 times more tolerant to imazethapyr but was only 60 times more tolerant to imazaquin, and QJ22 was only 300 times more tolerant to imazethapyr and 15 times more tolerant to imazaquin.

The imidazolinone resistance conferred by the tolerance gene from XA17 was introduced by Pioneer Hi-Bred International into their elite inbred lines via a backcrossing program. Imidazolinone-resistant hybrids have been tested extensively in field trials that evaluated both efficacy and crop tolerance over several years. Tolerance of the imidazolinone-resistant hybrids is routinely tested by postemergence applications of imazethapyr at one, two, and four times the commercial use rate (70 g/ha) of this product. Yields for hybrids containing the imidazolinone-resistance trait were measured in both the absence and presence of the herbicide, and there were no negative effects on yield (Figure 4). To date, no deleterious effects have been attributed or linked to the imidazolinone-resistance trait.

Other companies have since developed imidazolinone-resistant/tolerant corn (IMI-Corn™ * hybrids) via methods described elsewhere in this chapter. Genetic analysis of these plants demonstrated resistance due to one dominant gene. These companies also introduced the trait into elite inbred lines for the production of imidazolinone-resistant/-tolerant hybrids. In 1994 seed for an estimated 2.2 million acres of IMI-Corn will be commercially available representing a total of 25 different hybrids from nine different seed companies.

3.2. IMIDAZOLINONE-RESISTANT CANOLA

Plants regenerated from canola microspore cultures were haploid and nonfertile, but normal plant morphology and fertility were restored following colchicine treatment. As for corn, the

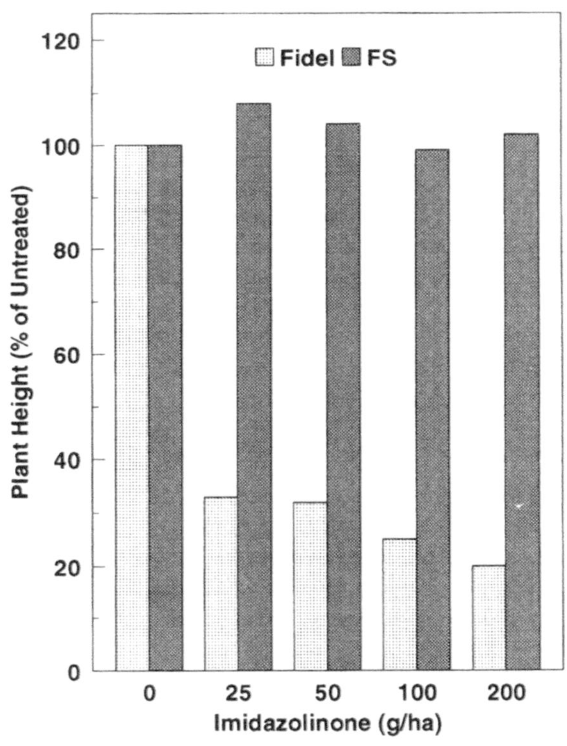

FIGURE 4 Effect on yield of imidazolinone resistance gene in corn. Yields are indicated as kg/ha and represent the summation of four field trials conducted in 1992. 3377SU is a Pioneer hybrid which is not resistant to the imidazolinones, and 3377IR is a comparable hybrid which contains the gene conferring imidazolinone resistance (homozygous). Imazethapyr was applied at one (70 g/ha), two, and four times the commercial use rate. All tests were conducted with the soil insecticide, terbufos (Counter® 15G; registered trademark of American Cyanamid Company, Wayne, NJ), in-furrow.

*IMI-Corn is a trademark of American Cyanamid Company.

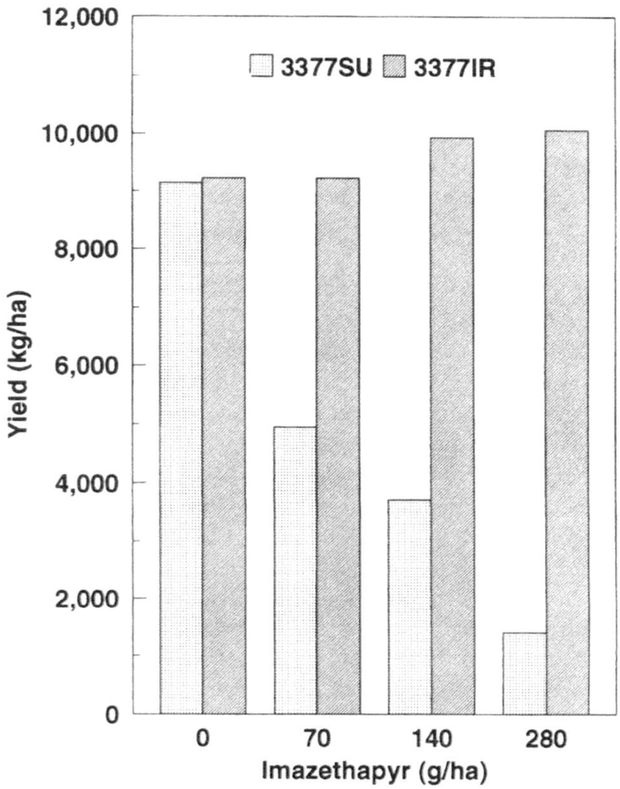

FIGURE 5 Crop response of imidazolinone-resistant and imidazolinone-susceptible canola varieties 14 d after treatment. Results are presented as the percent of crop injury following postemergence application of imazethapyr at 100, 200, or 400 g/ha (almost six times the use rate) and represent a greenhouse trial conducted in 1992. P1 HE and P2 HE are double heterozygotes, and P2 HO is homozygous.

herbicide resistance was characterized genetically and phenotypically at the whole-plant level. Genetic crosses of two of the selected imidazolinone-resistant plant lines, P1 and P2, demonstrated that the imidazolinone resistance was inherited as a simple, Mendelian, semidominant trait. Interestingly, crosses between P1 and P2 resulted in progeny heterozygous for both loci and with greater imidazolinone resistance than plants heterozygous for either resistance trait alone. This result indicates that in this cross the herbicide resistance is additive. If the relative resistances to the imidazolinones are compared, the genotypes would be ranked: P1 homozygous <P1 × P2 heterozygous <P2 homozygous. The preliminary breeding program introduced this resistance trait into traditional canola lines as well as into hybrid canola lines in development. Commercialization of the first imidazolinone-resistant canola varieties in conjunction with the appropriate chemical registration is anticipated by 1996. Resistance at the whole-plant level was initially evaluated in the greenhouse by spraying the plants with various rates of imazethapyr. P1 was resistant to 50 πg/ha imazethapyr, which is two to three times the rate required to kill susceptible parental type plants (20 g/ha), while P2 was resistant to rates up to 500 g/ha. (Figure 5 shows the increase in tolerance of the initial canola selections to 400 g/ha imazethapyr postemergence.) Similar to the case with corn, one line, P1, exhibited resistance to imidazolinones other than the compound used in the selection process, while the other, P2, was cross-resistant to both imidazolinones and sulfonylureas. Field tests have confirmed both the relative degree of tolerance for each selection as well as the cross-resistant nature of the trait. Preliminary field data indicate that there is no

reduction in yield associated with the imidazolinone resistance nor were any other deleterious effects observed with either of these lines.

3.3. IMIDAZOLINONE-RESISTANT WHEAT

Seed (M_3 generation) from each of the four plants initially selected was screened for imidazolinone resistance and demonstrated that the herbicide-tolerant trait was transmitted to the progeny. FS1, FS2, and FS4 were homozygous for herbicide tolerance, whereas FS3 required an additional generation of selfing to identify nonsegregating progeny. Genetic analysis of these lines indicated that resistance was inherited as a single dominant or semidominant trait. An F_1 population, produced by crossing each tolerant line with wild-type Fidel, was found to be less resistant to the herbicide, while the F_2 progeny segregated in a 3:1 resistant-to-susceptible ratio. Preliminary genetic analysis indicates that the four initial selections are either allelic at a single locus or very tightly linked and that there are no maternal effects on inheritance. The data to date support the likelihood that all four selections were derived from the same mutational event.

The imidazolinone-resistant wheat was screened for the level of resistance to imidazolinones as well as to sulfonylureas in both greenhouse and field tests. This trait also confers tolerance to the imidazolinones imazapyr and imazaquin; however, in contrast to the XA17 gene in corn, only marginal tolerance to the sulfonylurea herbicide sulfometuron methyl was observed. Preliminary greenhouse experiments tested tolerance of the putative imidazolinone-resistant wheat to four different imidazolinones, imazethapyr, imazaquin, imazapyr, and an experimental compound, and one sulfonylurea (sulfmeturon methyl) applied either preemergence or postemergence. The imidazolinone-tolerant wheat selections were approximately 8 times more tolerant to imazethapyr and 16 times more tolerant to imazaquin, imazapyr, and the experimental compound, but only 2 times more tolerant to sulfmeturon methyl than the susceptible parent. Further testing with higher rates of imazethapyr demonstrated 40- to 100-fold increased tolerance in the selections. Cross-resistance to other sulfonylureas and other AHAS inhibitors was examined, but only twofold increased tolerance was observed, as seen with sulfmeturon methyl.

Field testing to evaluate crop tolerance of the imidazolinone-resistant wheat selections began in 1989 and is ongoing; efficacy trials were initiated in 1993. These tests confirmed the resistance of the wheat to imazethapyr (Figure 6) as well as to several other imidazolinones at up to five times commercial use rates and reconfirmed the minimal tolerance (two times) to the sulfonylureas afforded by the trait. Preliminary data indicate that yield is unaffected by the imidazolinone resistance trait as was seen with the imidazolinone-resistant corn and canola. Programs have been initiated to introduce this trait into commercial wheat varieties in the U.S. as well as in other countries. The first wheat lines containing the imidazolinone resistance trait are expected to become available in the late 1990s.

4. MANAGEMENT OF IMIDAZOLINONE-RESISTANT CROPS

The availability of imidazolinone-resistant crops offers many benefits and advantages to the grower by allowing the development of a very flexible weed management program. However, as with any innovation, there are risks associated with imidazolinone-resistant crops that must be considered along with the benefits. With proper management, these risks can be dealt with effectively. These crops will be used only as part of an integrated weed management program. Such a program will not depend on the imidazolinones alone for weed control, but rather these herbicides will be part of a multifaceted weed-control program. An integrated program will include using tank mixtures, rotating imidazolinones with other herbicides having different modes of action, combining mechanical, cultural and biological weed-control methods with imidazolinones, and using the minimal amounts of herbicide needed to control weeds economically.

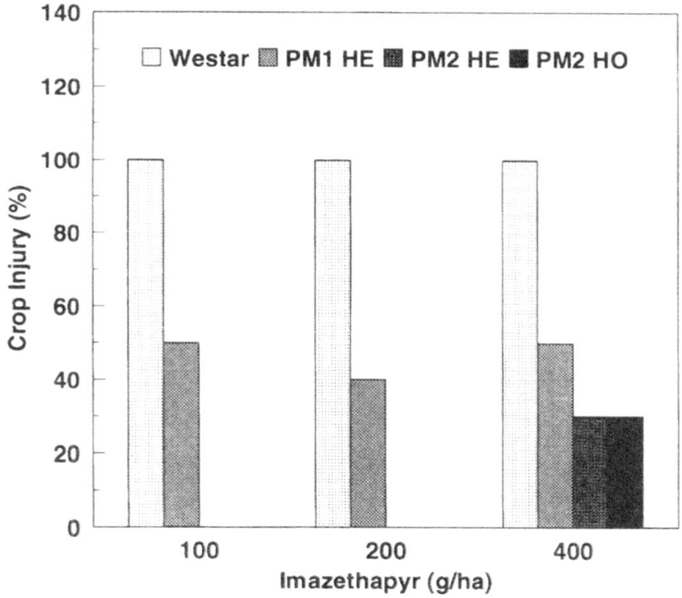

FIGURE 6 Crop response of imidazolinone-resistant and wild-type (imidazolinone susceptible) wheat varieties 6 weeks after treatment. Results are presented as plant height as a percentage of the untreated controls (100%) and represent a field trial conducted in 1992. Fidel is the "wild-type" cultivar which was mutagenized and screened to identify the four imidazolinone-resistant selections; FS in this graph is the selection FS2. Note the relatively high level of tolerance of the Fidel to imazethapyr; the plants are stunted but not usually killed.

4.1. RISKS ASSOCIATED WITH IMIDAZOLINONE-RESISTANT CROPS

There have been several concerns raised over the development of herbicide-resistant crops. These include increased use of herbicides, increased herbicide residue in food and groundwater, spread of resistance from the crop to associated weeds, and increased selection of herbicide-resistant weed populations (Duke et al. 1991). All of these areas of concern have been addressed during the development of imidazolinone-resistant crops and pose no serious problems.

4.1.1. *Increased Herbicide Use*

The availability of imidazolinone-resistant crops will not result in increased herbicide use in those crops. Most of the areas in which an imidazolinone-resistant crop will be used already receive herbicides as part of the weed-control program. For example, >96% of the soybean and maize grown in the U.S. receives at least one herbicide application yearly (Ellis 1992). Total herbicide use in the U.S. has decreased in recent years (Figure 7). Because of the low use rates of the imidazolinones, this trend will continue, and the actual quantity of herbicide applied to the land will be less than what is currently being used. Thus, the low use rates of the imidazolinones, coupled with their low toxicity to invertebrates and vertebrates, will result in even lower environmental impact of herbicides than in the current situation.

Another of the concerns about herbicide-resistant crops is that the farmer will indiscriminately apply high rates of the herbicide. However, this practice is extremely unlikely with imidazolinone-resistant crops because it is not cost effective nor is it allowed by the label for the herbicide on the crop.

4.1.2. Increased Herbicide Residues in Food

The use of imidazolinones on resistant crops will not result in increased herbicide residue in food or fiber. In order for imidazolinones to be registered for use on resistant crops, they must

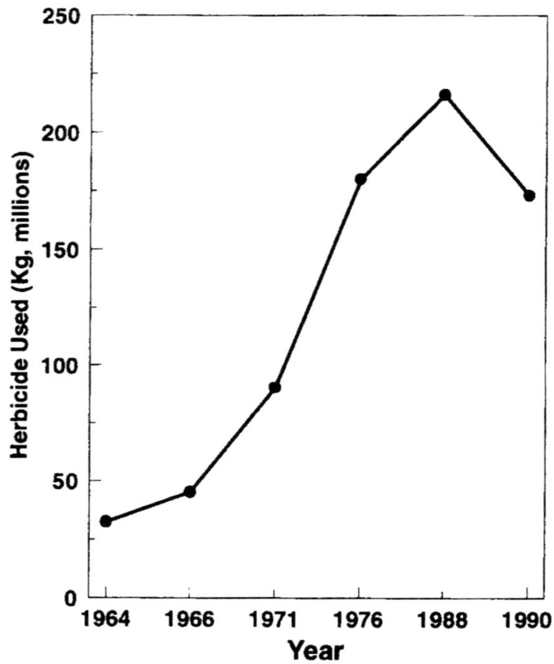

FIGURE 7 The total amount of herbicide used in the U.S. from 1962 to 1990. (Data adapted from Ellis 1992.)

undergo all of the rigorous and extensive residue and metabolism studies required for any herbicide. These multiyear studies determine the route of metabolism in the crop and the level of residue in the commodity. Although the mechanism of resistance in imidazolinone-resistant crops results from the altered sensitivity of the target site for the herbicide, these crops also can metabolize these herbicides (unpublished data). In the susceptible counterparts of these crops the rate of metabolism is not rapid enough to make the crop tolerant to the herbicide. In the resistant crops that are not adversely affected by the herbicide the rate of metabolism is fast enough to eventually detoxify the herbicide and sequester the metabolites in nonharvested parts of the plant. Residue levels in the commodity are often undetectable and represent little risk to the consumer.

4.1.3. Increased Groundwater Contamination

Extensive field tests showed that the imidazolinones stay within the top layers of the soil (Mangels 1991). The lack of downward movement is due to rapid degradation of the herbicide and its binding characteristics to the soil. The use of these herbicides in resistant crops will not change the behavior of the herbicide in the soil and should not pose any threat to the groundwater.

4.1.4. Spread of Resistance to Associated Weeds

Other concerns involving herbicide-resistant crops are that a resistant crop will become a weed or that the trait will be transferred to associated weeds. It is highly unlikely that imidazolinone-resistant crops will become serious weed problems because most crops are noncompetitive under natural conditions without the intervention of humans. Imidazolinone-resistant crops will not be more competitive than susceptible cultivars and thus will not become any more of a serious weed than their susceptible counterparts.

The spread of the resistant trait to associated weeds is also highly unlikely. There are no known cases in which genes from maize have crossed into any associated weeds. This event would be highly improbable because none of the weeds are closely related to maize. There may be a slightly higher chance of crosses between canola and weedy crucifers, but studies have shown that such crosses are extremely rare events and are not likely to happen under agricultural conditions because the susceptible weeds within the crop are being controlled by the herbicide (Cherfas 1991). This low probability taken in conjunction with the integrated weed management program that will be used with imidazolinone-resistant crops means that movement of the resistant trait to the weed populations is extremely unlikely and thus will not be a serious problem.

One measure that will be necessary with the introduction of imidazolinone-resistant crops will be the use of herbicides with different modes of action from that of the imidazolinones to control volunteer crops in succeeding crops. For example, volunteer corn can be a serious problem in soybeans that are planted after corn, but it can be effectively controlled with imazaquin. However, if the previous crop was imidazolinone-resistant corn, then imazaquin will be ineffective. Under these circumstances, another herbicide, such as an ACCase inhibitor, will be needed. However, the need for a herbicide having an alternative mode of action to control the imidazolinone-resistant volunteer crop is not detrimental because it ensures that herbicides with different modes of action will be included in the crop rotation.

4.1.5. Increased Development of Herbicide-Resistant Weed Populations

The introduction of imidazolinone-resistant crops requires that the farmer use them in an integrated weed management program. If the use of these crops results in the continuous use of imidazolinones alone for weed control, then there will be rapid selection for imidazolinone-resistant weed populations.

The selection of imidazolinone-resistant weed populations would defeat the utility of these crops. Thus, these crops will be used only as part of an integrated weed management program. Such a program will not depend on the imidazolinone alone for weed control but will include the use of tank mixtures, rotation with herbicides having other modes of action, and combinations of mechanical, cultural, and biological weed-control methods.

The development program for imidazolinone-resistant crops includes extensive training of salespersons, dealers, and users in the principles of herbicide-resistant weed management to ensure that these new tools are not abused. Furthermore, herbicide premixes that include herbicides having different modes of action from the imidazolinones but that have overlapping weed spectra are being developed for use on resistant crops. These premixes will further reduce the selection pressure on the weed populations.

4.2. BENEFITS OF IMIDAZOLINONE-RESISTANT CROPS

4.2.1. Cost-Effective Weed Control

Imidazolinone-resistant crops will benefit the farmer and consumer by providing cost-effective weed control, particularly in areas in which the present control practices are not wholly satisfactory. For example, one of the serious weed problems in canola is weedy crucifers. The use of imidazolinones in canola will allow effective control of these closely related weed species without affecting the crop. Triazine-resistant canola was introduced for the same purpose, but this crop suffered from the decreased yield associated with the resistant trait (Beversdorf et al. 1988). Imidazolinone resistance is not associated with decreased yield and thus gives the farmer the opportunity to control weeds without sacrificing yield.

4.2.2. Flexible Weed Management Options

Imidazolinone-resistant crops will give the farmer more flexibility in weed management. The farmer has the option of growing a number of crops that will provide an economic return (depending on the current market) and of developing a weed management program that can use a number of different herbicides. Because of the broad-spectrum activity and flexible application techniques of the imidazolinones, the weed management program that utilizes a resistant crop can be based on the weeds that need to be controlled with less concern about the relative selectivity of the herbicide.

Under conditions of extreme drought, the rate of dissipation of the imidazolinones decreases; thus, the possibility increases that there will be enough residue left in the soil to affect a highly sensitive follow crop. With the availability of imidazolinone-resistant crops, the farmer will have the option of planting a normally sensitive follow crop without worry of carryover from the previous year's soil residue.

4.2.3. Management of Herbicide-Resistant Weeds

The availability of imidazolinone-resistant crops will provide an effective tool in preventing and managing herbicide-resistant weeds. In recent years populations of *Avena fatua, Lolium multiflorum, Sorghum halepense,* and *Setaria viridis* have developed resistance to ACCase-inhibiting herbicides (Holt et al. 1993). In all of these cases these herbicides were the only means used to control grassy weeds. If the ACCase inhibitors had been used in combination or rotation with herbicides with different modes of action, this situation might have been avoided. With the introduction of imidazolinone-resistant crops, the farmer now has the option of using another herbicide mode of action in combinations with other selective herbicides to prevent the selection of resistant weed populations.

Furthermore, populations of *Lolium multiflorum* and *Alopecurus myosuroides* have developed that are resistant to many different classes of herbicides because of the enhanced ability of these populations to metabolize these herbicides via mixed-function oxidases (Holt et al. 1993). Because the mechanism of selectivity of the imidazolinone-resistant crops is not dependent upon metabolism (unlike the selectivity of naturally tolerant crops), different analogs that are not detoxified via mixed-function oxidases could be used. The imidazolinones could be able to control these multiply resistant populations and can be an effective weed management tool.

REFERENCES

Anderson, P.C. and Georgeson, M. 1989. Herbicide tolerant mutants of corn. *Genome* 31: 994.

Anderson, P.C. and Hibberd, K.A. 1985. Evidence for the interaction of an imidazolinone herbicide with leucine, valine, and isoleucine metabolism. *Weed Sci.* 33: 479.

Beversdorf, W.D., Hume, D.J., and Donnelly-Vanderloo, M.J. 1988. Agronomic performance of triazine-resistant and susceptible reciprocal spring canola hybrids. *Crop Sci.* 28: 932.

Bright, S.W.J., Chang, M.T., Evans, I.J., and MacDonald, M.J. 1990. International Patent Appl. WO 90/14000, *Herbicide Resistant Maize.*

Cherfas, J. 1991. Transgenic crops get a test in the wild. *Science* 251: 878.

Duke, S.O., Christy, A.L, Hess, F.D., and Holt, J.S. 1991. Herbicide-resistant crops. Comments from CAST.

Ellis, J.F. 1992. Herbicide development and marketing of weed control in the United States of America. *Proc. 1st Int. Weed Control Cong. 1: 74.*

Holt, J.S., Powles, S.B., and Holtum, J.A.M. 1993. Mechanisms and agronomic aspects of herbicide resistance. *Annu. Rev. Plant Physiol. Mol. Biol.* 44: 203.

Mangels, G. 1991. Behavior of the imidazolinone herbicides in soil—a review of the literature. In *The Imidazolinone Herbicides.* Ed. by Shaner, D.L. and O'Connor, S.L. CRC Press, Boca Raton, FL, chap. 16.

Newhouse, K., Singh, B., Shaner, D., and Stidham, M. 1991. Mutations in corn (*Zea mays* L.) conferring resistance to imidazolinone herbicides. *Theor. Appl. Genet.* 83: 65.

Newhouse, K., Smith, W.A., Starret, M.A., Schaefer, T.J., and Singh, B.K. 1992. Tolerance to imidazolinone herbicides in wheat. *Plant Physiol.* 100: 882.

Newhouse, K.E., Wang, T., and Anderson, P.C. 1991. Imidazolinone resistant crops. In *The Imidazolinone Herbicides.* Ed. by Shaner, D.L. and O'Connor, S.L. CRC Press, Boca Raton, FL, chap. 10.

Sathasivan, K., Haughn, G.W., and Murai, N. 1990. Nucleotide sequence of a mutant acetolactate synthase gene from an imidazolinone-resistant *Arabidopsis thaliana* var. Columbia. *Nucl. Acids Res.* 18: 2188.

Sebastian, S.A. and Chaleff, R.S. 1987. Soybean mutants with increased resistance to sulfonylurea herbicides. *Crop Sci.* 27: 948.

Shaner, D.L. and Anderson, P.C. 1987. Mechanism of action of the imidazolinones and cell culture selection of tolerant maize. In *Biotechnology in Plant Science.* Ed. by Zaitlin, M., Day, P.R., and Hollaender, A. Academic Press, Orlando, FL, p. 287.

Stidham, M.A. 1991. Herbicides that inhibit acetohydroxyacid synthase. *Weed Sci.* 39: 428.

Swanson, E.B., Coumans, M.P., Brown, G.L., Patel, J.D., and Beversdorf, W.D. 1988. The characterization of herbicide tolerant plants in *Brassica napus* L. after in vitro selection of microspores and protoplasts. *Plant Cell Rep.* 7: 83.

Swanson, E.B., Herrgesell, M.J., Arnoldo, M., Sippell, D.W., and Wong, R.S.C. 1989. Microspore mutagenesis and selection: canola plants with field tolerance to the imidazolinones. *Theor. Appl. Genet.* 78: 525.

Chapter 10

GENETIC ENGINEERING OF CROPS FOR TOLERANCE TO 2,4-D

Danny Llewellyn
David Last

CONTENTS

1. PHENOXYACETIC ACID HERBICIDES

2,4-Dichlorophenoxyacetic acid (2,4-D) is a member of a class of herbicides with activity against many dicotyledonous weeds. Although it is an older-generation herbicide, introduced in the 1940s, it is still in general use in many countries. The chemical structure of 2,4-D and some of its more common relatives such as MCPA (2-methyl-4-chlorophenoxyacetic acid) are shown in Figure 1. The group includes butyric acid derivates such as 2,4-DB that are metabolized to active 2,4-D by ß-oxidation, giving them a degree of selectivity different from that of 2,4-D. Because many legumes lack the ß-oxidation activity, 2,4-DB can be used for weed control on legume-containing pastures or crops. The most notable member of the group is 2,4,5-T, which, as a mixture with 2,4-D, formed the notorious "Agent Orange" herbicide used for massive forest defoliation during the Vietnam War. The phenoxy herbicides such as 2,4-D are active as plant growth regulators or hormones with auxin activity. They are absorbed through the leaves and are translocated initially for short distances by

1-56670-045-0/96/$0.00+$.50
© 1996 by CRC Press Inc.

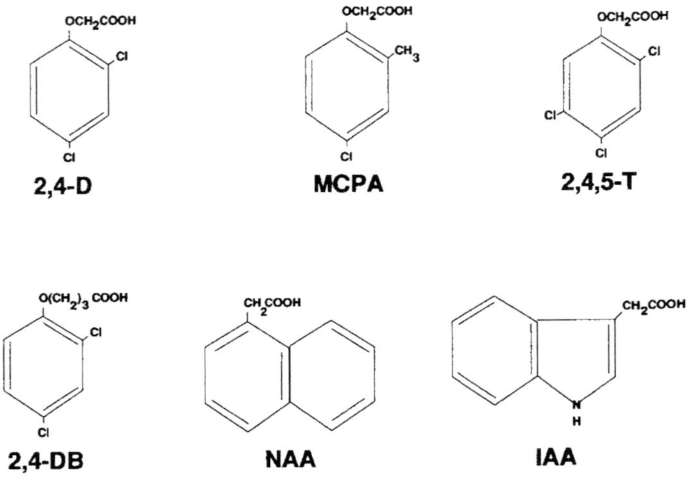

FIGURE 1 Chemical structures of phenoxyacetic acid herbicides and some natural plant auxins. 2,4-D, 2,4-dichlorophenoxyacetic acid; MCPA, 2-methyl,4-chlorophenoxyacetic acid; 2,4,5,-T, 2,4,5,-trichlorophenoxyacetic acid; 2,4-DB, 4-(2,4-dichlorophenoxy)butyric acid; IAA, indole-3-acetic acid; NAA, 1-naphthaleneacetic acid.

intercellular movement to the phloem and then over long distances in the phloem to the shoot meristems. Phloem-translocated herbicides such as the growth regulators move through the sieve tube system similar to sugars, from a source (region of sugar synthesis) to a sink (region of sugar utilization) (Singh and Muller 1979), although the exact mechanisms are still poorly understood. Stimulation of sink activity has been observed with 2,4-D (Kirkwood 1979), and this presumably potentiates its herbicidal activity. At the meristem 2,4-D and other phenoxy herbicides interfere with cell development, presumably through saturation effects arising from mimicry of auxin action on cell elongation (Dodge 1989). Characteristic damage to plants by low levels of 2,4-D is severe twisting and spindly growth of the vegetative and floral tips and organs (Figure 2), followed by the death of the meristems and eventually the plant.

Most commercial formulations contain salts, amine salts, or esters of the phenoxyacetic herbicides rather than the free acids. The esters are believed to be active only after hydrolysis to the free acid in the plant (Loos 1969). En route to the site of action the phenoxy herbicides are also exposed to the potential for modification to less active metabolites or in some cases to the active herbicide, e.g., 2,4-DB. Detoxification of the phenoxy herbicides may play an important part in the selectivity of these chemicals for different plant species. 2,4-D, for example, can undergo a variety of chemical modifications in insensitive plants (reviewed by Pillmoor and Gaunt 1981 and Loos 1969), including side chain degradation, side chain elongation, ring hydroxylation, and conjugation with proteins, amino acids, and glucose (Ashton and Crafts 1981).

Monocotyledonous plants are considerably less sensitive to phenoxyacetic herbicides and hence these herbicides are of particular use for broadleaf weed control in monocot crops such as cereals and sugarcane. 2,4-D is an important auxin in the tissue culture of cereals and is an active growth regulator for these species. It is not surprising, therefore, that cereals are sensitive to 2,4-D at particular developmental stages and can be seriously affected if this herbicide is used inappropriately (Evans 1968). However, the major scope for the production of crops resistant to 2,4-D is among the dicotyledonous crops that are sensitive to low concentrations of phenoxyacetic acid herbicides, particularly in multicropping systems where they may be exposed inadvertently to 2,4-D as spray drift or from residues in the soil or irrigation water. Among the dicot crops, the main focus of genetic engineering research worldwide has been into 2,4-D-tolerant cotton, a crop of considerable economic value that is extremely sensitive to 2,4-D.

FIGURE 2 2,4-D spray drift damage on cotton. Note the twisted distorted leaves and bracts and dead growing points.

2. CURRENT 2,4-D USAGE PATTERNS IN COTTON CROPPING

Weed control remains an important component of cotton production, primarily because of the noncompetitive nature of cotton seedlings (Keeley et al. 1986). Weeds can reduce cotton yields by direct competition for light, nutrients, and water, but can also affect quality parameters through contamination of the harvested lint. Weed spectrum will be determined by geographical location and weed management practices, but the problems are likely to be similar in all highly mechanized production systems such as those found in the U.S. and Australia. Charles (1991) surveyed the weed species and herbicide usage patterns on Australian cotton farms and found that most of the serious weeds were dicotyledonous. Generally, these were weeds of pastures and are either native or naturalized species. *Xanthium occidentale* (Noogoora burr), *X. spinosum* (Bathurst burr), and *Datura* spp. (thornapples) are large plants that are effective competitors with cotton and have been a long-term problem in cotton production. *Cyperus* spp. are perennial grasses adapted to moist soils and can be a problem in fields and drainage channels. Like the dicot weeds *Haloragis glauca* and *Polymeria longifolia,* these grasses have rhizomatous roots and are resistant to cultivation as a form of control. All of these species, except *Cyperus,* are sensitive to 2,4-D, but cannot be controlled effectively with this herbicide, except in fallow fields, because of the serious risk of damaging the cotton crop. 2,4-D is used on about one third of the cotton farms from ground rigs at any time after defoliation through to sowing the next season, but its popularity is reduced by fears of damage that could result from residues in the soil (Charles 1991). Current weed control programs rely heavily on preemergence cultivation and use of nonselective herbicides such as glyphosate before planting, residual herbicides such as trifluralin and atrazine, and hand removal of weeds from postemergence cotton. 2,4-D is more commonly used on cereal crops (oats, barley, sorghum, and maize) in the same valleys in which cotton is grown or on fallow fields in wheat/fallow/cotton rotations to control broadleaf summer weeds. The majority of spraying of 2,4-D on cereals is done from the air, so spray drift, even on relatively calm days, away from the cereals into the cotton growing areas is almost inevitable. There are also instances in which 2,4-D residues are not removed from the tanks and have caused serious damage to cotton in subsequent spraying of an unrelated herbicide, miticide, or insecticide. In recent years there have been some isolated incidences of 2,4-D damage appearing in cotton as a result of its use on fallows. Such damage occurs when weather conditions do not favor growth of 2,4-D-degrading soil microbes, such as when it is unusually cold or dry (Hammond 1992). 2,4-D tolerant cotton would be useful both in the prevention of spray drift damage and perhaps for the control of some of the dicotyledonous weeds that currently trouble the industry by using this relatively cheap compound as an effective postemergence herbicide.

3. GENETIC ENGINEERING OPTIONS FOR 2,4-D RESISTANCE

Genetically engineered herbicide resistances are no longer new. Effective resistances have been developed to a variety of potent herbicides such as glyphosate and the sulfonylureas (Comai et al. 1985, Shah et al. 1986, Haughn et al. 1988). In these cases the molecular target of the herbicide was well characterized and resistances were developed through the manipulation or replacement of the target gene. Comai et al. selected a strain of *Salmonella typhimurium* that was resistant to glyphosate because of a mutation in the gene for 5-enolpyruvylshikimate 3-phosphate synthase (EPSP synthase), the target enzyme for glyphosate. When expressed in plants, the mutant enzyme could substitute for the endogenous plant enzyme that had been inactivated by the herbicide. Shah et al. used a different approach to achieve the same result. They selected a petunia cell line that was resistant to glyphosate through the overproduction of the EPSP synthase enzyme. This petunia gene was cloned and then reintroduced into petunia with a new, much stronger promoter. The new chimeric gene resulted in the over-production of the enzyme, and hence in resistance at a whole plant level. Haughn et al. achieved a similar result with an acetolactate synthase gene from *Arabidopsis,* which confers sulfonylurea herbicide resistance on transgenic tobacco plants. While all of these resistance genes are certainly effective, they may have some deleterious effects, especially in the absence of the herbicide.

Principles similar to those described above cannot be applied to developing a resistance gene against 2,4-D because its primary target is still unknown. 2,4-D is an analog of the plant hormone group known as auxins and at low levels is often used as a substitute for natural auxins in plant tissue culture systems. Some of the symptoms of 2,4-D damage to cotton such as twisting of shoots are suggestive of a hormone-induced mechanism such as perturbed rates of cell division. However, it is likely that there are other, more drastic and undetermined phytotoxic effects.

A second approach used to generate herbicide resistant plants is the engineering of detoxification pathways in susceptible species. A number of herbicides are toxic to some plant species but not others. The difference in sensitivity is often not associated with a difference in susceptibility of the molecular target, but to an ability of the resistant plant to detoxify or break down the herbicide before it has a chance to inactivate the susceptible target gene product. Two examples now exist in which a microbial gene encoding an enzyme that detoxifies a herbicide was modified for expression in plants and introduced into a model species by transformation (De Block et al. 1987, Stalker et al. 1988). The novel detoxification enzymes were expressed and conferred resistance to relatively high levels of phosphinothricin (Basta® or glufosinate) and 3,5-dibromo-4-hydroxybenzonitrile (bromoxynil), respectively. These artificial systems mimic detoxification systems found in plants and are likely to place the least genetic, physiological, or yield burden on the plant, both in the presence or absence of the herbicide. Furthermore, the mode of action of the herbicide need not be known, provided that the appropriate detoxification genes can be found. This second strategy was the most promising for 2,4-D because a variety of microorganisms are known that naturally degrade the herbicide in the environment.

4. A MICROBIAL SOURCE OF 2,4-D DEGRADATION GENES

2,4-D has a relatively short residual life in soils, decreasing by about one half every 14 d or so (Noble and Hargreaves 1992). Other phenoxyacetic acid herbicides such as MCPA and 2,4,5-T break down much more slowly in the soil and can cause serious residual problems (Loos 1969). The breakdown of phenoxyacetic acid herbicides in the soil is performed by a variety of microorganisms including bacteria, yeasts, and fungi from several taxonomic groups. Loos (1969) lists many of these and the particular herbicides that can be used as substrates: a variety of *Pseudomonas* spp., *Achromobacter*, *Flavobacterium*, *Corynebacterium*, and *Arthrobacter* spp. and some Actinomycetes, such as *Nocardia* spp. and *Streptomyces viridochromogenes*, which will all degrade 2,4-D as a carbon source after appropriate enrichment of this substrate. The most well-characterized organisms are strains of *Alcaligenes eutrophus*, which are relatively simple Gram-negative rod bacteria found in most aerated soils. Several different strains of *A. eutrophus* were isolated from soils in fields commonly sprayed with 2,4-D and all are able to grow on synthetic media with 2,4-D as the sole source of carbon (Don and Pemberton 1981) (Figure 3). The common denominator between all the strains is a large (75 kb) plasmid that was shown to encode many of the enzymes necessary for the breakdown of 2,4-D. The overall pathway of 2,4-D degradation (Don et al. 1985) in *A. eutrophus* is shown in Figure 4.

It may be considered impractical to engineer all six *tfd* genes to function in plants, but data on the phytotoxicity of the first intermediate in the pathway, 2,4-dichlorophenol (DCP) suggested that only the first gene in the pathway may be necessary to confer a useful degree of tolerance on transgenic crop plants. Perkins et al. (1987) showed that DCP was toxic to cell suspension cultures of *Nicotiana tabacum* with a 50% inhibition at 10 mg/l and inhibited leaf and root development of tobacco seedlings at the same concentration. 2,4-D at 1 mg/l blocked all plant development. Streber and Willmitzer (1989) also tested the toxicity of DCP to tobacco callus and showed that DCP at 10 mg/l caused no obvious toxic effects, while 2,4-D at 2 mg/l severely reduced growth rates and turned the callus brown. Seed germination studies carried out by Llewellyn et al. (1990) in the presence of 2,4-D or DCP indicated that, in tobacco at least, DCP was some 50- to 100-fold less phytotoxic than 2,4-D. These results were encouraging with respect to expressing just the first step in the pathway for protection against 2,4-D damage.

Streber et al. (1987) were able to clone the *tfdA* gene encoding the first enzymatic step of 2,4-D degradation by Tn5 mutagenesis of the degradative plasmid of *A. eutrophus*. From >1000 kanamycin-resistant clones that had Tn5 insertions, two were identified that lacked the ability to cleave the acetate side chain of 2,4-D. A large *Hin*dIII fragment containing the wild-type plasmid DNA in the region of the tagged *tfdA* gene both complemented the mutations and allowed another strain of *A. eutrophus* (JMP222) to grow on 2,4-D analogs such as phenoxyacetic acid. JMP222 is able to grow on phenol as a carbon source by using chromosomal genes for a meta-cleavage pathway, but cannot normally grow on phenoxyacetate. The *tfdA* gene product is able to accept different substituted and unsubstituted phenoxyacetates as a substrate and can cleave phenoxyacetate to phenol. This assay was used to localize the gene by successive deletions of the large fragment down to 2 kb that could be sequenced. The *tfdA* gene was originally thought to encode a 2,4-D monooxygenase but subsequent biochemical analysis has indicated that the gene product is a ferrous iron dependent dioxygenase that uses α-ketoglutarate as a cosubstrate (Fukomori and Hausinger 1993). The α-ketoglutarate is converted to succinate and carbon dioxide concomitant with the conversion of 2,4-D to DCP and glyoxylate.

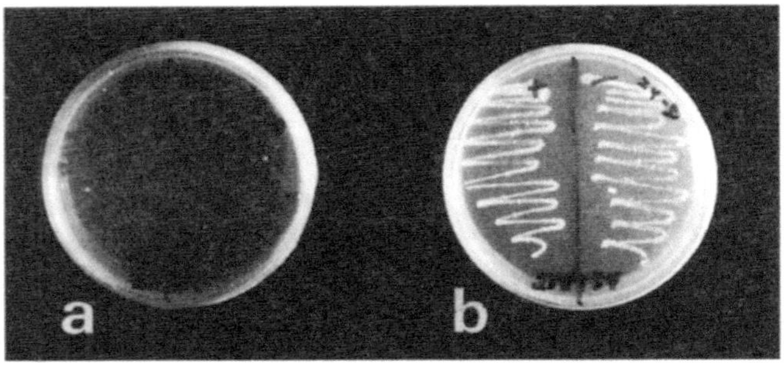

FIGURE 3 *Alcaligenes eutrophus* utilization of 2,4-D as a carbon source. Bacterial plates contain minimal medium supplemented with 125 µg/ml 2,4-D sodium salt as sole source of carbon. (a) JMP228R, a cured strain of *A. eutrophus* lacking the 2,4-D degradative plasmid is unable to grow on 2,4-D. (b) JMP134, a strain of *A. eutrophus* which has the degradative plasmid pJP4, grows on 2,4-D in the absence of any other carbon source.

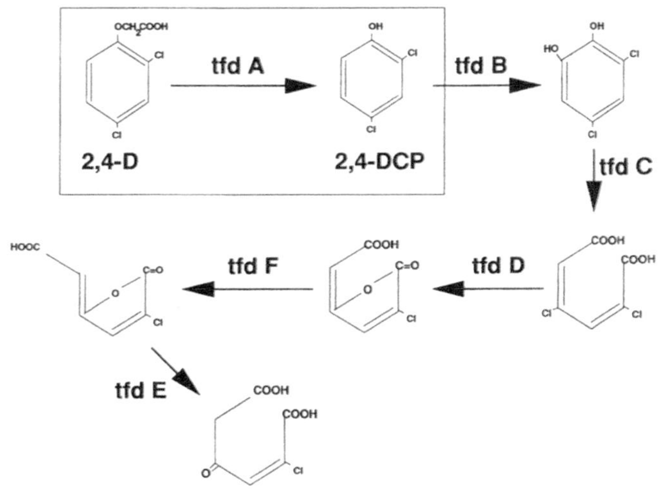

FIGURE 4 Biochemical pathway of 2,4-D degradation encoded by the degradative plasmid pJP4 of *Alcaligenes eutrophus*. 2,4-D degradative genes are designated *tfdA-F*, and the biochemical reaction catalyzed by their gene product is noted at each step. The 2,4-D dioxygenase reaction catalyzing the first breakdown step is boxed.

5. GENETIC ENGINEERING OF THE *ALCALIGENES TFDA* GENE FOR EXPRESSION IN PLANTS

Three research groups have engineered the *tfdA* gene for expression in plants (Streber and Willmitzer 1989, Lyon et al. 1989, Bayley et al. 1992). Sequencing of the *tfdA* gene revealed that it had an unusual translational start codon. This bacterial coding region began with a GTG codon rather than the usual ATG, a situation found in some other bacterial genes such as *tfdC* and *-D* (Perkins et al. 1990) and the *bar* gene from *Streptomyces hygrosopicus* (Thompson et al. 1987). The novel codon is not correctly recognized in plants (Lyon et al. 1989) and had to be modified to an ATG before reasonable expression levels could be attained. The modification of the translation start appeared to have no effect on 2,4-D dioxygenase enzyme activity in *Escherichia coli* (Lyon et al. 1989). Streber and Willmitzer and Lyon et al. also modified the region upstream of the ATG to more closely resemble the consensus sequence compiled from a large number of plant genes (Joshi 1987). Bayley et al. (1992) used a *tfdA* gene derived from the *A. eutrophus* 2,4-D degradative plasmid pJP5 (Don and Pemberton 1981) rather than the pJP4 plasmid used by the earlier researchers. pJP5 has some restriction site polymorphisms relative to pJP4, but is very similar in size. No sequence data were presented for the pJP5 *tfdA* gene, but it is highly likely that it is nearly identical to the gene from pJP4, or at least had a common origin.

In order to express the 2,4-D dioxygenase gene in plants it was necessary to provide a strong plant promoter and transcription termination and polyadenylation signals. The latter were provided by a fragment of the nopaline synthase gene originally from the T-DNA of *Agrobacterium tumefaciens* (Herrera-Estrella et al. 1983) or a similar region from the T-DNA octopine synthase gene (Dhaese et al. 1983). All three groups used the promoter sequences from the 35S promoter of the cauliflower mosaic virus (CaMV), a strong and essentially constitutive promoter in many dicot plants (Odell et al. 1985). Two additional modified 35S promoters were used by Lyon et al. (1989) in an attempt to improve on the expression of the viral promoter. In the first, two additional copies of the so-called upstream enhancer region of the 35S promoter (the region between -90 and -340) were ligated onto the 35S promoter at -340 relative to the transcription start site. All three enhancer regions were in the same orientation. Kay et al. (1987) suggested that multiple copies of this enhancer increase the level of expression of the promoter more than arithmetically. In the second construction an additional 45 bp was added to the small leader region of the 35S promoter fragment (essentially only 2 bp). The addition consisted of the leader region of another viral gene, the coat protein gene of the alfalfa mosaic virus (AMV). This AMV leader has been suggested to improve the translational competitiveness of viral genes at least *in vitro* and perhaps also heterologous genes to which it is attached (Jobling and Gerhke 1987). Streber and Willmitzer also constructed a chimeric *tfdA* gene with the promoter from a light-inducible gene ST-LS1 (Stockhaus et al. 1987) that is expressed in all photosynthetically active tissues of a plant.

All chimeric *tfdA* genes were cloned either directly into binary plant transformation vectors such as pGA470, developed by An et al. (1985), or pBin19, developed by Bevan (1984), which was conjugally transferred into disarmed strains of *Agrobacterium tumefaciens*, or they were cloned into an intermediate vector that was transferred to *Agrobacterium* and recombined into a disarmed Ti plasmid. The binary plasmids are self-replicating and carry left and right T-DNA borders that direct integration into the plant genome, and a selectable kanamycin resistance gene that allows the recovery of transformed plant cells. The transfer functions are provided by modified Ti plasmids deleted for their own T-DNA. Lyon et al. used *Agrobacterium* strains LBA4404 (Hoekema et al. 1983) for transformation of tobacco and AGL1 (Lazo et al. 1991) for transformation of cotton. Bayley et al. used the octopine *Agrobacterium* strain GV3111 (containing a disarmed pTiB6S3) for both tobacco and cotton transformation. The system used by Streber and Willmitzer also used a kanamycin resistance gene as a selectable marker for tobacco transformation, but the chimeric *tfdA* gene, selectable marker, and T-DNA borders were present on the disarmed Ti plasmid GV3850Kan (Jones et al. 1985) rather than on a separate replicon in the *Agrobacterium* strain.

6. EXPRESSION OF THE 2,4-D DIOXYGENASE IN TRANSGENIC TOBACCO

Although none of the researchers were interested in 2,4-D-tolerant tobacco *per se*, tobacco provides a useful model system with which to test the function and specificity of introduced transgenes. Transgenic tobacco plants were produced by *Agrobacterium* infection of leaf pieces using relatively standard protocols with selection on toxic levels of kanamycin sulfate. Streber and Willmitzer (1989) screened 25 plants for production of nopaline (produced from an intact nopaline synthase gene present on the T-DNA) and for expression of the *tfdA* transcript in Northern blot analyses. Of the 25 plants, 20 were producing detectable levels of the expected 1.3-kb chimeric *tfdA* transcript. Lyon et al. (1989) produced >30 transgenic tobacco from each of their three *tfdA* constructs and carried out preliminary screening with a radioactive enzyme assay using ^{14}C-labeled 2,4-D. While this assay worked well with bacterial extracts, it was only semiquantitative with plant extracts, but allowed the selection of the plants expressing the highest levels of the *tfdA* gene product. Since then the designation of the *tfdA* gene product has changed from a monooxygenase to a dioxygenase, new cofactor requirements have been discovered (Fukumori and Hausinger 1993), and it is likely that the original assay could be improved with the addition of ferrous ions and α-ketoglutarate. Bayley et al. (1992) also used a biochemical assay to measure expression of the *tfdA* gene product in transgenic tobacco, but they used whole cells (protoplasts) where endogenous cofactors would be present. They demonstrated an increase in $^{14}CO_2$ production from degraded ^{14}C-labeled 2,4-D in transgenic plants, ranging from 11 to 85 times higher than that from untransformed control tobacco tissues.

2,4-D has a drastic effect on plant growth and development, and when incorporated into tissue culture media can drastically perturb the effects of other hormones in the medium. When added to regeneration medium shoot development can be inhibited, and this provided a convenient assay procedure for ranking the 2,4-D tolerance of different transgenic plants. Streber and Willmitzer placed explanted leaf disks from their transgenic tobacco onto a medium containing 1 mg/l benzyl amino purine supplemented with varying amounts of 2,4-D from 0.1 to 4.0 mg/l. Nontransgenic plants would produce shoots on medium without 2,4-D, but shoot regeneration was inhibited by 0.2 mg/l 2,4-D. Depending on the individual transgenic plant, shoot regeneration was not inhibited by as much as 2 to 4 mg/l 2,4-D for plants transformed with a 35S-*tfdA* gene construct, indicating at least a 20-fold increase in tolerance to this herbicide. Plants transformed with the *tfdA* gene driven by the light-inducible promoter showed no more tolerance to 2,4-D than nontransformed tissues in this shoot regeneration assay, probably because the light-inducible promoter was not functional during the early stages of shoot morphogenesis, when growth is essentially nonphotosynthetic. In a similar assay Lyon et al. (1989) were able to produce plants with a tolerance to as high as 12.8 mg/l 2,4-D, representing a 32-fold increase in tolerance over a nontransformed control. The increased resistance in different transgenic plants varied from 10- to 30-fold over the control plants. In most cases the resistance in the leaf disk assay correlated well with the intensity of the DCP band in the radioactive enzyme assay. Genomic Southern analysis on some of the plants indicated that most contained single-copy, or at most two-copy, insertions of the 35S-AMV-*tfdA*-Nos chimeric gene described in this study. Bayley et al. (1992) used the shoot inhibition assay to select for the highest expressing tobacco plants and were able to recover seven primary transformants that could produce shoots on 6 mg/l 2,4-D and two others that could produce shoots on 4 mg/l 2,4-D.

In order to look at the segregation of the 2,4-D resistance gene in progeny Streber and Willmitzer tested the germination of 2,4-D-tolerant lines on medium containing 0 to 0.1 mg/l 2,4-D, but observed no significant difference in the sensitivity between transformed and nontransformed plants. Lyon et al. (1989), however, germinated F1 progeny seeds on a suitable germination medium containing from 0 to 3.2 mg/l 2,4-D and saw significant differences between controls and plants that showed high tolerance in the shoot inhibition assay. Seeds of control plants failed to germinate, or, if cotyledons emerged, failed to elongate on as little as 0.2 mg/l 2,4-D. Transgenic plants expressing 2,4-D dioxygenase activity, on the other hand, germinated and elongated on as much as 3.2 mg/l 2,4-D. In some progeny sets >90% of the seeds germinated and continued to grow, suggesting multiple gene insertions, while in others there seemed to be the expected three-to-one segregation of a single-copy insertion of a dominant character. At higher levels of 2,4-D there were some seeds obviously more resistant than the rest, and it seems likely that these were the homozygotes.

Selected transgenic tobacco lines or progeny from plants showing increased tolerance in the shoot inhibition assay were clonally or sexually propagated and tested in the glasshouse for increased tolerance to the herbicidal action of 2,4-D sprayed to simulate a field application. Streber and Willmitzer (1989) sprayed clonally propagated plants after 2 weeks' growth in the soil with 2,4-D at doses equivalent to 1, 3, and 10 kg/ha (10 kg/ha was 4 ml of a 1% aqueous solution of 2,4-D sodium salt). Plant damage was assessed after 2 to 3 weeks. A dose of 1 kg/ha totally inhibited growth of nontransgenic control plants. A plant transformed with the 35S-*tfdA* construct was resistant to 10 kg/ha, while a plant transformed with the light-inducible promoter-driven *tfdA* gene was resistant to 1 kg/ha 2,4-D even though it showed no increased tolerance in the shoot inhibition assay. Lyon et al (1989) sprayed F1 progeny seed germinated in soil in small pots (four seedlings per pot) and grown in a containment glasshouse until they were 6 weeks old. Groups of 16 plants each were sprayed with a commercial preparation of 2,4-D isopropyl ester in 50% acetone at concentrations from 0 to 3000 ppm (Figure 5). Control plants sprayed with as little as 30 ppm began to show symptoms after only 3 d when they began to wilt and show abnormal twisting of the stems. Many of the transgenic plants only began to show these same symptoms at 3000 ppm. Some of the transgenic plants were affected more than others, presumably due to the segregation of the 2,4-D resistance gene among the F1 progeny. After about 3 weeks control plants sprayed with >100 ppm were dead or dying, while many of the clearly resistant transgenic plants had grown out of the effects of the herbicide even when sprayed at 3000 ppm. While it is difficult to estimate the equivalent field application rate corresponding to our experimental sprayings we estimate that resistance to 3000 ppm in our system corresponds to at least 3 to 6 kg/ha, which is at least 6- to 12-fold above normal field application rates for 2,4-D on cereals (around 0.5 kg/ha). Bayley et al. (1992) sprayed clones of their primary transformants with 10 ml of either 0.2, 0.5, and 1% aqueous solutions of 2,4-D with estimated dosages of 4, 10, and 20 kg/ha, although this may be on the high side. Many of their plants survived even the 20 kg/ha dose when assessed after 3 weeks.

A. eutrophus JMP134 is able to degrade other substituted phenoxyacetate compounds than 2,4-D, so Streber and Willmitzer (1989) tested their transgenic plants for cross-resistance to some other phenoxyacetate herbicides using the shoot inhibition tissue culture assay. They saw a tolerance similar to that seen with 2,4-D when 4-chlorophenoxyacetic acid was included in the medium and

FIGURE 5 2,4-D dioxygenase gene expressed in transgenic tobacco confers tolerance to spraying with 2,4-D. Plants were sprayed with 1000 ppm 2,4-D isopropyl ester in 50% acetone until dripping wet. Damage was scored after 3 weeks. Transgenic 2,4-D-tolerant tobacco line is on the left and control sensitive tobacco on the right.

only a slight increase in tolerance with MCPA, but no increased tolerance to 2,4,5-T or any of the phenoxypropionic acid herbicides [2-(4-chlorophenoxy)propionic acid. 2-(2,4- dichlorophenoxy)-propionic acid, or 2-(2,4,5-trichlorophenoxy)propionic acid]. Similar results were reported by Lyon et al. (1989) using a seed germination assay. These data are consistent with the known specificity of the *tfd* degradative plasmid of *A. eutrophus* for these differently substituted phenoxyacetates. The enzyme specificity also means that the 2,4-D dioxygenase will not degrade natural plant auxins, such as indole acetic acid (Figure 1), and hence is not toxic to the plants.

7. TRANSGENIC COTTON SHOWING TOLERANCE TO 2,4-D

While there are several crops that could benefit from being made tolerant to the herbicide 2,4-D, little interest has been shown in any but cotton. Perkins et al. (1987) expressed interest in using a *tfdA* gene to confer 2,4-D tolerance on the legume cowpea (*Vigna sinensis*), but although they cloned a duplicated *tfdA* gene (*tfdAII*) from the *Alcaligenes* plasmid pJP4 (Perkins and Lurquin, 1988), they have not continued the project (Lurquin, personal communication), mainly because of concerns over the safety of 2,4-DCP. With this in mind, they cloned and sequenced the genes encoding later steps of the 2,4-D breakdown pathway, *tfdB* (encoding 2,4-DCP hydroxylase) and the chlorocatechol oxidative operon *tfdCDEF* (Perkins et al. 1990) with the idea of using these genes in transgenic plants.

Cotton has been the only commercial crop that has been engineered for 2,4-D tolerance, although it has been difficult, as has the regeneration of cotton plants from tissue culture, let alone transformation. Regenerability is highly genotype dependent and it was not until recently that reliable regeneration via somatic embryogenesis could be achieved (Gawel et al. 1986, Trolinder and Goodin 1987), and then predominantly with "Coker" varieties of cotton. Transformation was reported for Coker cottons using *Agrobacterium*-mediated gene transfer and regeneration from either seedling cotyledon (Firoozabady et al. 1987) or hypocotyl (Umbeck et al. 1987) explants. Transformed callus tissue was selected on media containing kanamycin at relatively low levels (25 to 35 μg/ml) and somatic embryos induced from the callus. These embryos were germinated on appropriate media and grown to maturity in soil. The frequency of transformation even with Coker cotton is still very low but managable. Cousins et al. (1991) developed a tissue culture system and transformation procedures for two commercial Australian cotton cultivars, offering the possibility for direct transformation of elite lines.

Bayley et al. (1992) transformed Coker 312 cotton with the *tfdA* gene from pJP5 driven by the 35S promoter, and regenerated 300 transgenic plantlets from 30 independent cell lines. They prescreened them for increased tolerance to 2,4-D by selection for rooting on medium containing 0.05 mg/l 2,4-D or by treating meristems with 0.01 mg/l 2,4-D. These levels inhibited new root and shoot development on nontransformed control plants. Three transformed lines that survived these treatments were transferred to soil for further analysis, and vegetatively propagated clones also tested for increased tolerance to commercial formulations of 2,4-D. Many of the primary transformants were unaffected by a dosage (1.5 kg/ha) of three times the normal field application rate, whereas control plants were severely affected by a level 300 times less than the recommended dose. Selfed progeny of some lines were assayed for 2,4-D degradation using the protoplast whole cell assay for 2,4-D dioxygenase and were shown to have between 17- and 38-fold the activity present in nontransformed cotton protoplasts. Germination assays showed that these plants could form roots on medium containing 0.1 mg/l 2,4-D, a level that inhibited germination of nontransformed plants. This assay also was used to show three-to-one segregation of the resistance phenotype in one of the transformed lines. Seedlings from the same line when germinated in the absence of 2,4-D, were sprayed with 2,4-D at 1.5 kg/ha, and 83 of the 107 plants sprayed survived this treatment. Transgenic seeds are being distributed to public and private breeders in the U.S. for incorporation of the trait into other commercial germplasm (Trolinder, personal communication).

Lyon et al. (1993) transformed Coker 315 cotton with one of the *tfdA* gene constructs used to transform tobacco and regenerated 201 plantlets from 11 different tissue culture lines. Only four of these produced self-fertile plants when transferred to the glasshouse. Southern analysis and progeny testing indicated that only one of the four lines had a single T-DNA insert, while the others had two copies of the T-DNA. There was no direct correlation between copy number and the level of resistance shown by spraying plants with 2,4-D. 2,4-D dioxygenase expressing progeny from all four lines was undamaged by 300 ppm 2,4-D isopropyl ester in 50% acetone (Figure 6). The line containing a single copy of the chimeric gene was shown to be the most tolerant and was resistant to over 600 ppm 2,4-D. T2 and T3 plants shown to be either hemizygous or homozygous indicated that at discriminating doses the resistance phenotype is semidominant (Lyon et al. 1993). When sprayed with 600 ppm 2,4-D the hemizygous plants containing a single gene copy are slightly damaged, but eventually grow out of the damage and have a normal yield and appearance. Homozygous plants are completely unaffected by this level of 2,4-D, but show some minor damage at 900 ppm and more severe damage at 1200 ppm, although they survive this higher dose. The best transgenic plants were estimated to be about 100-fold more tolerant to 2,4-D than control nontransformed cotton and were tolerant to about twice the normal field application rate. CSIRO is currently backcrossing the 2,4-D tolerance developed by Lyon et al. (1993) into important commercial Australian cultivars to provide them with spray drift protection from 2,4-D.

None of the 2,4-D-tolerant plants have progressed to field trials. In Australia the research group at CSIRO is continuing to develop new gene constructs that may have better 2,4-D tolerance levels in cotton. Because the site of action of the herbicide is at the shoot meristem they are investigating

FIGURE 6 2,4-D-tolerant cotton cultivars engineered to express the *A. eutrophus* 2,4-D dioxygenase gene. Primary Coker transformants and controls were sprayed with 300 ppm 2,4-D isopropyl ester in 50% acetone. The plant on the left contains a single T-DNA insertion of the *tfdA* gene and was unaffected by 2,4-D at the concentration tested. The plant on the right is a nontransformed control cotton that was rapidly killed by the herbicide.

using gene promoters other than the 35S promoter, which has relatively poor expression in the meristem. Promoters from genes, such as histones and ubiquitin that are expressed at high levels in actively dividing cells, were linked to the *tfdA* gene and are currently being assessed in transgenic tobacco. Expression of the second gene in the 2,4-D degradation pathway, 2,4-D hydroxylase (*tfdB*) (Perkins et al. 1990), is also being considered to detoxify the DCP produced by the 2,4-D dioxygenase. Increased tolerance to 2,4-D will provide a greater protection margin and may allow 2,4-D to be used directly on cotton to control problem broadleaf weeds.

8. 2,4-D-TOLERANT COTTON PLANTS — AN ETHICAL DILEMMA?

The ability to produce 2,4-D-tolerant crops such as cotton raises the ethical dilemma of whether these plants should be used in agriculture. Opponents of herbicide-tolerant crops warn that herbicide-tolerant plants will promote indiscriminant usage of herbicides with concomitant human health and environmental impacts (Goldburg et al. 1990; Phelps 1990), but are these fears justified with 2,4-D?

As 2,4-D has been used in agriculture for almost 50 years it sometimes has been labeled as an "older chemical," and like several other older chemicals it has acquired a tainted image as a suspected environmental and human health hazard. The name 2,4-D has become associated in the public mind with less benign chlorinated hydrocarbons such as DDT (dichlorodiphenyl-trichloroethane). Unlike DDT, which accumulates in body fat, 2,4-D is rapidly excreted in an unchanged form from the human body. The biological half-life of 2,4-D in humans is approximately 18 to 20 h (reviewed by Ibrahim et al. 1991). Similarly, 2,4-D is unlikely to accumulate in aquatic organisms, as exemplified by the rapid renal elimination of 2,4-D administered to channel catfish (Plakas et al. 1992). This contrasts markedly with DDT, which is concentrated manyfold in the tissues of aquatic organisms, especially shellfish. 2,4-D is also rapidly broken down in the soil by bacteria and fungi, typically becoming undetectable about 5 to 6 weeks after application (Nobel and Hargreaves 1992). Soils with a history of 2,4-D treatment were shown to have an increased population of 2,4-D-degrading microorganisms (Holben et al. 1992). The unpopularity of 2,4-D in some circles stems in part from its inclusion along with 2,4,5-T in Agent Orange. However, the human health hazard posed by Agent Orange was caused by a contaminant, 2,3,7,8-tetrachlorodibenzo-*p*-dioxin, an extremely toxic compound which causes a severe form of acne and is suspected of causing birth defects and specific cancers. Modern 2,4-D formulations do not contain this dangerous contaminant nor do they contain significant quantities of any other known carcinogens (Ibrahim et al. 1991).

2,4-D has a low acute oral or dermal toxicity. The Western Australian Department of Agriculture lists 2,4-D as "slightly toxic" (poison schedule S5), placing it in the same category as glyphosate, sodium chlorate, and kerosene. The LD_{50} values for 2,4-D (reviewed in Ibrahim et al. 1991) range from 100 mg/kg in dogs (which are known to excrete organic acids more slowly than rodents or humans) to 1000 mg/kg in guinea pigs. Chronic exposure to 2,4-D in rats did not produce any observable detrimental effects at levels of up to 30 mg/kg/d. Studies reviewed by Mullison (1988) concluded that 2,4-D is not carcinogenic in mice at up to 45 mg/kg/d, nor is it teratogenic in rats even at a dosage that caused toxic symptoms in pregnant rats (75 mg/kg/d). There has been a suggestion, based on epidemiological evidence, that exposure to 2,4-D may be linked with the occurrence of the human cancer non-Hodgkin's lymphoma (Woods et al. 1987; Zahm et al. 1990), and this has been the subject of much debate. A panel of 13 scientists reviewing the literature were unable to reach a single conclusion as to whether 2,4-D is a human carcinogen (Ibrahim et al. 1991). Of the 13, five thought that it was a possible but unlikely human carcinogen, while one considered that it was a possible carcinogen. Two panelists thought that it was unlikely that 2,4-D can cause cancer in humans. Following a report of a possible link between 2,4-D and canine malignant lymphoma (Hayes et al. 1991), a scientific review panel examined the available evidence and concluded that the use of 2,4-D by dog owners did not show an association with the occurrence of canine malignant lymphoma (Carlo et al. 1992). In other words, there is no conclusive evidence to implicate 2,4-D as a carcinogen.

The *Agrochemicals Handbook* of the Royal Society of Chemistry (Hartley and Kidd 1983) lists 2,4-D as nontoxic to fish (average lethal concentration of sodium salt, 1160 mg/l/24 h) and nontoxic to bees. 2,4-D is only toxic to waterfowl at very high concentrations (LD_{50} = 2000 mg/kg; Weiss 1986), which would never be approached by the use of it in agriculture.

2,4-D has many of the desirable qualities sought in a herbicide, most notably its low toxicity to animal life and high biodegradability. Furthermore, it is very inexpensive and effective. Its use over many years has been relatively untroubled by problems of resistance development in weeds. There are only six sites in the world where 2,4-D-resistant weeds are thought to have evolved as a result of spraying, compared to over 1000 sites for herbicides such as the triazines and acetolactate synthase inhibitors (Holt et al. 1993). 2,4-D does, however, have a drawback in that it does not adsorb readily to some soils and can leach into groundwater. Precautions should be taken to ensure that pollution of ground water does not occur.

Like any agricultural chemical, 2,4-D should not be used carelessly, but the large body of toxicological data gathered over the past 40 or so years suggests that it can be used according to label directions without endangering humans or the environment.

9. CONCLUSIONS

The herbicide detoxification systems developed for the protection of cotton against 2,4-D have proven to be extremely effective in at least two plant species, and we expect that it also will be useful in other crops that are sensitive to 2,4-D. The plants containing the engineered resistance gene appear to grow and flower normally, producing just as many viable seeds as nontransformed plants, and there appears to be no yield penalty in the absence of any applied herbicide, at least in glasshouse trials. When sprayed at relatively high levels of 2,4-D there is a slight retardation of growth, but the plants recover quickly. Because most commonly used herbicides are broken down by soil microorganisms it would seem feasible to apply the same principles to those herbicides and to quickly develop a whole range of herbicide resistance genes that could, once transformation procedures become routine, revolutionize farming practices in high-input crops such as cotton.

Herbicide-tolerant crop projects, including the 2,4-D-resistant cotton projects described here, have been criticized by environmental groups, such as the Australian Conservation Foundation (Phelps 1990), on the grounds that they could lead to an increase in the use of synthetic chemicals in agriculture. This fear should be balanced with the environmental benefits of being able to extend the spectrum of use of relatively benign herbicides such as 2,4-D and thereby reduce our reliance on more environmentally damaging herbicides. Additionally, control of weeds within the crop with 2,4-D could potentially lead to a no-till cropping system which would have great benefits in reducing soil erosion, providing a mulch, and returning organic nitrogen to the soil.

Although widely held public concern about the use of 2,4-D-resistant cotton are not supported by scientific data, it is nevertheless essential that careful studies in the field are carried out in order to evaluate any possible environmental implications. This will enable cropping systems to be designed so that the new technology can be applied responsibly in a way that benefits the farmer, the consumer, and the environment.

REFERENCES

An, G., Watson, B., Stachel, S., Gordon, M., and Nester, E. 1985. New cloning vehicles for transformation of higher plants. *EMBO J.* 4: 277.

Ashton, F.M. and Crafts, A.S. 1981. *Mode of Action of Herbicides,* 2nd ed. John Wiley & Sons, New York.

Bayley, C., Trolinder, C., Morgan, M., Quisenberry, J.E., and Ow, D.W. 1992. Engineering 2,4-D resistance into cotton. *Theor. Appl. Genet.* 83: 645.

Bevan, M. 1984. Binary *Agrobacterium* vectors for plant transformation. *Nucl. Acids Res.* 12: 8711.

Carlo, G.L., Cole, P., Miller, A.B., Munro, I.C., Solomon, K.R., and Squire, R.A. 1992. Review of a study reporting an association between 2,4-dichlorophenoxyacetic acid and canine malignant lymphoma: report of an expert panel. *Regul. Toxicol. Pharmacol.* 16: 245.

Charles, G.W. 1991. A grower survey of weeds and herbicide use in the New South Wales cotton industry. *Aust. J. Exp. Agric.* 31: 387.

Comai, L., Facciotti, D., Hiatt, W., Thompson, G., Rose, R., and Stalker, D. 1985. Expression in plants of a mutant *aroA* gene from *Salmonella typhimurium* confers tolerance to glyphosate. *Nature* 317: 741.

Cousins, Y.L., Lyon, B.R., and Llewellyn, D.J. 1991. Transformation of an Australian cotton cultivar: prospects for cotton improvement through genetic engineering. *Aust. J. Plant Physiol.* 18: 481.

DeBlock, M., Botterman, J., Vandewiele, M., Dockyx, J., Thoen, A., Gossele, V., Rao Movva, N., Thompson, C., Van Montagu, M., and Leemans, J. 1987. Engineering herbicide resistance in plants by expression of a detoxifying enzyme. *EMBO J.* 6: 2513.

Dhaese, P., De Greve, H., Gielen, J., Seurinck, J., Van Montagu, M., and Schell, J. 1983. Identification of sequences involved in the polyadenylation of higher plant nuclear transcripts using *Agrobacterium* T-DNA genes as models. *EMBO J.* 2: 419.

Dodge, A.D. 1989. Principles. In *Weed Control Handbook.* Ed. by Hance, R.J. and Holly, K. Blackwell Scientific, Oxford, p. 201.

Don, R. and Pemberton, J. 1981. Properties of six pesticide degradation plasmids isolated from *Alcaligenes eutrophus* and *Alcaligenes paradoxus. J. Bacteriol.* 145: 681.

Don, R., Weightman, A., Knackmuss, H., and Timmis, K. 1985. Transposon mutagenesis and cloning analysis of the pathways for degradation of 2,4-dichlorophenoxyacetic acid and 3-chlorobenzoate in *Alcaligenes eutrophus* JMP134(pJP4). *J. Bacteriol.* 161: 85.

Evans, S.A. 1968. Application of herbicides. In *Weed Control Handbook,* Vol. 1, 5th ed. Ed. by Fryer, J D. and Evans, S.A. Blackwell Scientific, Oxford, p. 149.

Firoozabady, E., DeBoer, D., Merlo, D., Halk, E., Amerson, L., Rashka, K., and Murray, E. 1987. Transformation of cotton (*Gossypium hirsutum* L.) by *Agrobacterium tumefaciens* and regeneration of transgenic plants. *Plant Mol. Biol.* 10: 105.

Fukumori, F. and Hausinger, R.P. 1993. *Alcaligenes eutrophus* JMP134 "2,4-dichlorophenoxyacetate monooxygenase" is an α-ketoglutarate-dependent dioxygenase. *J. Bacteriol.* 175: 2083.

Gawel, N., Rao, A., and Robacker, C. 1986. Somatic embryogenesis from leaf and petiole callus cultures of *Gossypium hirsutum. Plant Cell Rep.* 5: 457.

Goldburg, R., Rissler, J., Shand, H., and Hasselbrook, C. 1990. Biotechnology's Bitter Harvest: Herbicide Tolerant Crops and the Threat to Sustainable Agriculture. Biotechnology Working Group, Washington, D.C., p. 53.

Hammond, D. 1992. Managing 2,4-D in cotton fallows. *Aust. Cottongrower 13*: 67.

Hartley, D. and Kidd, H. 1983. *Agrochemicals Handbook.* Royal Society of Chemistry, London.

Haughn, G., Smith, J., Mazur, B., and Sommerville, C. 1988. Transformation with a mutant *Arabidopsis* acetolactate synthase gene renders tobacco resistant to sulfonylurea herbicides. *Mol. Gen. Genet.* 211: 266.

Hayes, H.M., Tarone, R.E., Cantor, K.P., Jessen, C.R., McCurnin, D.M., and Richardson, R.C. 1991. Case-control study of canine malignant lymphoma: positive association with dog owner's use of 2,4-dichlorophenoxyacetic acid herbicides. *J. Nat. Cancer Inst.* 83: 1226.

Herrera-Estrella, L., De Block, M., Messens, E., Hernalsteens, J., Van Montagu, M., and Schell, J. 1983. Chimeric genes as dominant selectable markers in plant cells. *EMBO J.* 2: 987.

Hoekema, A., Hirsch, P., Hooykaas, P., and Schilperoort, R. 1983. A binary plant vector strategy based on separation of *vir*- and T-region of *Agrobacterium tumefaciens* Ti-plasmid. *Nature* 303: 179.

Holben, W.E., Schroeter, B.M., Calbrese, V.G.M., Olsen, R.H., Kukor, J.K., Diederbeck, V.O., Smith, A.E., and Tiedje, J.M. 1992. Gene probe analysis of soil microbial populations selected by amendment with 2,4-dichlorophenoxyacetic acid. *Appl. Environ. Microbiol.* 58: 3941.

Holt, J. S. 1993. Mechanisms and agronomic aspects of herbicide resistance. *Annu. Rev. Plant Physiol. Plant Mol. Biol.* 44: 203.

Ibrahim, M.A., Bond, G.G., Burke, T.A., Cole, P., Dost, F.N., Enterline, M., Gough, R.S., Greenberg, W., Halperin, E., McConnell, M., Munro, I.C., Swenberg, J.A., Zahm, S.H., and Graham, J.D. 1991. Weight of evidence on the human carcinogenicity of 2,4-D. *Environ. Health Perspect. 96*: 213.

Jobling, S. and Gerhke, L. 1987. Enhanced translation of chimeric mRNAs containing plant viral untranslated leader sequences. *Nature* 325: 622.

Jones, J., Dunsmuir, P., and Bedbrook, J. 1985. High level expression of introduced chimaeric genes in regenerated transformed plants. *EMBO J.* 4: 2411.

Joshi, C. 1987. An inspection of the domain between putative TATA box and translation start sites in 79 plant genes. *Nucl. Acids Res.* 15: 6643.

Kay, R., Chan, A., Daly, M., and McPherson, J. 1987. Duplication of CaMV 35S promoter sequences creates a strong enhancer for plant genes. *Science* 236: 1299.

Keeley, P.E., Thullen, R.J., and Carter, C.H. 1986. Influence of planting date on growth of ivyleaf morningglory (*Ipomoea hederacea*) in cotton (*Gossypium hirsutum*). *Weed Sci.* 34: 906.

Kirkwood, R.C. 1979. The uptake and translocation of foliar applied herbicides using an explant system. In *Advances in Pesticide Science*. Ed. by Geissbuhler, H. Pergamon Press, Oxford, p. 410.

Lazo, G.R., Stein, P.A., and Ludwig, R.A. 1991. A DNA transformation competent *Arabidopsis* genomic library in *Agrobacterium*. *Bio/Technology* 9: 963.

Llewellyn, D., Lyon, B.R., Cousins, Y.L., Huppatz, J., Dennis, E.S., and Peacock, W.J. 1990. Genetic engineering of plants for resistance to the herbicide 2,4-D. In *Genetic Engineering of Crop Plants*. Ed. by Lycett, G.W. and Grierson, D. Butterworths, London, p. 67.

Loos, M.A. 1969. Phenoxyalkanoic acids. In *Degradation of Herbicides*. Ed. by Kearney, P.C. and Kaufman, D.D. Marcel Dekker, New York, p. 1.

Lyon, B.R., Cousins, Y.C., Llewellyn, D.J., and Dennis, E.S. 1993. Cotton plants transformed with a bacterial degradation gene are protected from accidental spray drift damage by the herbicide 2,4-dichlorophenoxyacetic acid. *Transgenic Res.* 2: 162.

Lyon, B.R., Llewellyn, D.J., Huppatz, J., Dennis, E.S., and Peacock, W.J. 1989. Expression of a bacterial gene in transgenic tobacco confers resistance to the herbicide 2,4-dichlorophenoxyacetic acid. *Plant Mol. Biol.* 13: 533.

Mullison, W.R. 1988. A 2,4-D toxicological update. *Proc. West Soc. Weed Sci.* 41: 20.

Noble, B. and Hargreaves, P. 1992. Are herbicide residues accumulating in our soils? *Austr. Grain* 2: 10.

Odell, J., Nagy, F., and Chua, N.H. 1985. Identification of DNA sequences required for activity of a plant promoter: the CaMV promoter. *Nature* 313: 810.

Perkins, E.J. and Lurquin, P.F. 1988. Duplication of a 2,4-dichlorophenoxyacetic acid monooxygenase gene in *Alcaligenes eutrophus* JMP134(pJP4). *J. Bacteriol.* 170: 5669.

Perkins, E.J., Stiff, C.M., and Lurquin, P.F. 1987. Use of *Alcaligenes eutrophus* as a source of genes for 2,4-D resistance in plants. *Weed Sci.* 35: 12.

Perkins, E.J., Gordon, M.P., Caceres, O., and Lurquin, P.F. 1990. Organization and sequence analysis of the 2,4-dichlorophenol hydroxylase and dichlorocatechol oxidative operons of plasmid pJP4. *J. Bacteriol.* 172: 2351.

Phelps, R. 1990. A leap into the unknown. *Consuming Interests* August, p. 5.

Pillmoor, J.B. and Gaunt, J.K. 1981. The behaviour and mode of action of phenoxyacetic acid herbicides. In *Progress in Pesticide Biochemistry*, Vol. 1. Ed. by Hutson, D.H. and Roberts, T.R. Wiley-Interscience, New York, p. 147.

Plakas, S.M., Khoo, L., and Barron, M.G. 1992. 2,4-Dichlorophenoxyacetic acid disposition after oral administration in channel catfish. *J. Agric. Food Chem.* 40: 1236.

Shah, D., Horsch, R., Klee, H., Kishore, G., Winter, J., Tumer, N., Hironaka, C., Sanders, P., Gasser, C., Aykent, S., Siegel, N., Rogers, S., and Fraley, R. 1986. Engineering herbicide tolerance in transgenic plants. *Science* 233: 478.

Singh, S.P. and Muller, F. 1979. Translocation of 2,4-D, asulam and amitrole in water hyacinth. *Weed Res.* 19: 171.

Stalker, D., McBride, K., and Malyj, L. 1988. Expression in plants of a bromoxynil-specific bacterial nitrilase that confers herbicide resistance. In *Genetic Improvements of Agriculturally Important Crops: Progress and Issues*. Ed. by Fraley, R.T., Frey, N.M., and Schell, J. Cold Spring Harbor Laboratory, Cold Spring Harbor, NY, p. 37.

Stockhaus, J., Eckes, P., Rocha-Sosa, M., Schell, J., and Willmitzer, L. 1987. Analysis of *cis* active sequence involved in leaf specific expression of a potato gene in transgenic plants. *Proc. Natl. Acad. Sci. U.S.A.* 84: 7943.

Streber, W., Timmis, K., and Zenk, M. 1987. Analysis, cloning and high-level expression of 2,4-dichlorophenoxyacetate monooxygenase gene *tfdA* of *Alcaligenes eutrophus* JMP134. *J. Bacteriol.* 169: 2950.

Streber, W.R. and Willmitzer, L. 1989. Transgenic tobacco expressing a bacterial detoxifying gene are resistant to 2,4-D. *Bio/Technology* 7: 811.

Thompson, C.J., Movva, N.R., Tizard, R., Craneri, R., Davies, J.E., Lauwereys, M., and Botterman, J. 1987. Characterization of the herbicide resistance gene *bar* from *Streptomyces hygroscopicus*. *EMBO J.* 6: 2519.

Trolinder, N. and Goodin, J. 1987. Somatic embryogenesis and plant regeneration in cotton (*Gossypium hirsutum* L.). *Plant Cell Rep.* 6: 231.

Umbeck, P., Johnson, G., Barton, K., and Swain, W. 1987. Genetically transformed cotton (*Gossypium hirsutum* L.) plants. *Bio/Technology* 5: 263.

Weiss, G. (Ed.) 1986. *Hazardous Chemicals Data Book*, 2nd Ed. Noyes Data Corporation, Park Ridge, NJ, p. 363.

Woods, J.A., Polissar L., Severson, R.K., Heuser, L.S., and Kulander, B.G. 1987. Soft tissue sarcoma and non-Hodgkin's lymphoma in relation to phenoxyherbicide and chlorinated phenol exposure in western Washington. *JNCI* 78: 899.

Zahm, S.H., Weissenburger, D.D., Babbit, P.A., Saal, R.C., Vaught, J.B., Cantor, K.P., and Blair, A. 1990. A case-control study of non-Hodgkin's lymphoma and the herbicide 2,4-dichlorophenoxyacetic acid (2,4-D) in eastern Nebraska. *Epidemology* 1: 349.

Chapter 11

ARYLOXYPHENOXYPROPIONATE- AND CYCLOHEXANEDIONE-RESISTANT CROPS

David A. Somers

CONTENTS

1. INTRODUCTION

Aryloxyphenoxypropionate and cyclohexanedione herbicides comprise two families of herbicides with several derivatives of each chemistry that all act through inhibition of acetyl-coenzyme A carboxylase (ACCase) in susceptible species. Herbicides from the two families are chemically distinct but are generally considered together because they inhibit this target site enzyme. For the purposes of this review, herbicides derived from both chemistries are referred to as ACCase-inhibiting herbicides. The generic names of specific herbicides within each family end in a diagnostic syllable such that all aryloxyphenoxypropionate herbicides end in "fop" (e.g., diclofop) and cyclohexanedione herbicides end in "dim" (e.g., sethoxydim), enabling determination of the

1-56670-045-0/96/$0.00+$.50
© 1996 by CRC Press Inc.

chemical origin of the specific herbicide. Most monocotyledonous plants are susceptible to ACCase-inhibiting herbicides, which are generally used postemergence in dicotyledonous crops to control gramineous weeds (Beste et al. 1983). The majority of the ACCase-inhibiting graminicides are characterized by postemergence application, low concentrations required for lethality of target species and low toxicity to nontarget organisms. These positive characteristics in part form the rationale for developing cereal or grass varieties that are resistant to ACCase-inhibiting herbicides. In particular, herbicide resistance in a susceptible monocot such as maize may provide increased flexibility in weed control by allowing postemergence control of weeds that exceed acceptable population levels following cultivation. The objective of this chapter is to review the present status of selection and development of crop plants resistant to ACCase-inhibiting herbicides.

2. MECHANISM(S) OF ACTION AND HERBICIDE TOLERANCE/RESISTANCE

2.1. HERBICIDE TARGET SITE

Acetyl coenzyme-A carboxylase (EC 6.4.1.2) catalyzes the ATP-dependent carboxylation of acetyl-CoA to form malonyl-CoA in the first committed step of fatty acid synthesis (Harwood 1988). Malonyl-CoA also is the two-carbon moiety for fatty acid elongation and synthesis of secondary metabolites. Aryloxyphenoxypropionate and cyclohexanedione herbicides are lethal to susceptible monocotyledonous species because they inhibit ACCase (Burton et al. 1987, 1989, 1991; Focke and Lichtenthaler 1987; Rendina and Felts 1988; Secor and Cseke 1988; Stoltenberg et al. 1989). Herbicide inhibition of ACCase inhibits incorporation of acetate into free fatty acids in isolated maize chloroplasts (Hoppe and Zacher 1985; Burton et al. 1987), in root tips (Hosaka and Takagi 1987), and in cell cultures (Parker et al. 1990b). Dicot ACCases and ACCase from naturally tolerant monocots, such as red fescue, are not inhibited by these herbicides (Burton et al. 1987; Secor and Cseke 1988; Stoltenberg et al. 1989).

ACCase is a multifunctional enzyme that catalyzes carboxylation of a enzyme-bound biotin prosthetic group and a transcarboxylation that transfers the carboxyl group from biotin to acetyl-CoA to form malonyl-CoA (Harwood 1989). The ACCase that functions in fatty acid biosynthesis is localized in the plastid, whereas a cytosolic ACCase is postulated for elongation of extra-long chain fatty acids and synthesis of secondary metabolites (Harwood 1989). As in animal systems from which genes encoding ACCase have been characterized, the ACCase holoenzyme exhibits molecular weights in excess of 400,000 in most plants and generally appears to be a homodimer of subunits of >200 kDa (Slabas and Hellyer 1985; Egin-Buhler and Ebel 1983; Charles and Cherry 1986; Hellyer et al. 1986; Slabas et al. 1986; Egli et al. 1993). Nuclear genes are thought to encode these high molecular weight ACCase subunits based on the genetics of herbicide tolerance conferred by altered ACCase (Parker et al. 1990a; Betts et al. 1992) and on work in progress by several groups that have cloned and sequenced nuclear cDNA clones that putatively encode plastid-localized ACCase.

Smaller ACCase subunits detected by the presence of covalently bound biotin were reported in various plants (Nikolau and Hawke 1984; Nikolau et al. 1984; Charles and Cherry 1986; Hellyer et al. 1986; Hawke and Leech 1990). The relationship of these small biotinylated subunits to ACCase is undetermined; however, plants contain other biotinylated enzymes (Wurtele and Nikolau 1990). Recently it was shown that *accD*, a chloroplast gene found in tobacco and *Arabidopsis*, encodes a subunit of ACCase that is similar in structure to a subunit of prokaryotic ACCases (Sasaki et al. 1993). In prokaryotes ACCases are comprised of three different subunits that exhibit the partial activities of the holoenzyme (Harwood 1988). Antibodies produced against the putative plant *accD* gene product inhibited ACCase activity in both species and detected a 87-kDa polypeptide in

SDS-PAGE immunoblots. This gene is truncated in the rice and wheat plastid genomes, suggesting that this ACCase isoform may not be present in monocots. It is not known what role the *accD*-encoded ACCase plays in plant metabolism or if the *accD* ACCase is present in all dicots, thereby conferring herbicide tolerance.

ACCase in maize is comprised of two isoforms. Egli et al. (1993) purified a herbicide-sensitive and a herbicide-resistant form of ACCase from wild-type maize leaves. The major isoform of ACCase comprised 85% of the recovered activity, was herbicide sensitive, and had subunits that were 227 kDa. This isoform was designated ACCase I and is considered to be the herbicide target site. Further analyses indicated that the 227-kDa ACCase I was plastid-localized (Egli et al. 1993), and likely is the isoform involved in the committed steps of fatty acid biosynthesis for both membrane biogenesis and seed lipid deposition (Somers et al. 1993). An aryloxyphenoxypropionate-resistant ACCase activity was detected in maize leaf extracts that comprised about 20% of the total ACCase activity (Howard and Ridley 1990). This minor ACCase activity was also shown to be insensitive to inhibition by haloxyfop and sethoxydim, was poorly recognized by an antiserum produced against ACCase I, and exhibited 219 kDa subunits (ACCase II) (Egli et al. 1993). While it was not possible to determine the cellular localization of the 219 kDa ACCase II, Egli et al. (1993) hypothesized that it is a cytosolic form postulated for elongation of extra-long chain fatty acids and secondary metabolite synthesis (Harwood 1988). The likely presence of these two ACCase isoforms in partially purified ACCase preparations must be taken into account when considering kinetic investigations of herbicide inhibition.

Kinetic analysis of clethodim (cyclohexanedione) inhibition of ACCase activity in a partially purified fraction from barley chloroplasts, which likely was enriched for the plastid-localized ACCase I, indicated that the herbicide exhibited properties of a linear, reversible, noncompetitive inhibitor (Rendina and Felts 1988). Similar kinetic mechanisms were observed for diclofop and haloxyfop inhibition of ACCase isolated from wheat leaves (Rendina et al. 1988). In Black Mexican Sweet maize suspension culture cells kinetic studies of herbicide inhibition were conducted on partially purified ACCase (Burton et al. 1991). This preparation may have contained the 219-kDa ACCase II because procedures had not been developed to resolve the herbicide-resistant isoform. While it is not known what effect the mixture of the two ACCase isoforms has on the kinetic data, the data are in agreement with the results for barley and wheat (Rendina and Felts 1988; Rendina et al. 1988). Herbicide inhibition best fit a linear, noncompetitive model for ATP, HCO_3^- and acetyl-coenzyme A (Burton et al. 1991). Inhibition constants for haloxyfop were lower than those for sethoxydim, which was consistent with the observations of lower I_{50} for haloxyfop inhibition of ACCase in plant extracts and the increased efficacy of haloxyfop in field applications. The herbicide inhibition appeared to be exerted at the transcarboxylation reaction (Burton et al. 1991). Furthermore, Burton et al. (1991) reported that sethoxydim and haloxyfop were mutually exclusive inhibitors of the enzyme, implying that both herbicides bind to the same site on the enzyme or that the two herbicides bind at different sites that through allosteric interactions regulate exclusive binding.

2.2. MECHANISMS OF TOLERANCE/RESISTANCE

Resistance to ACCase inhibiting herbicides may be conferred by at least three different documented mechanisms: altered target site enzyme, metabolism, and an undetermined mechanism associated with altered membrane electrochemical properties. Selectivity of aryloxyphenoxypropionate and cyclohexanedione herbicides for lethality in monocots is due to differential inhibition of ACCase in susceptible monocots (Burton et al. 1987, 1989, 1991; Focke and Lichtenthaler 1987; Rendina and Felts 1988; Secor and Cseke 1988). As mentioned, dicot ACCase is not inhibited by the herbicides. This differential inhibition of monocot vs. dicot ACCases allows the herbicides to be used in dicot crops for grass weed control. ACCase from naturally tolerant red fescue (Stoltenberg et al. 1989) and diclofop-tolerant biotypes of *Lolium multiflorum* (Betts et al. 1992) were shown to be insensitive to herbicide inhibition.

Wheat is naturally resistant to diclofop due to its ability to rapidly metabolize this herbicide (Shimbukuro et al. 1979), but is susceptible to most other aryloxyphenoxypropionate and cyclohexanedione herbicides.

The mechanism of resistance to aryloxyphenoxypropionate and cyclohexanedione herbicides involving altered membrane electrochemical properties was first reported in *Lolium rigidum* biotypes that are resistant to diclofop methyl (Heap and Knight 1989; Matthews et al. 1990). Similar observations are reported for wild oat biotypes (Devine 1992, Holtum 1992). In both cases the herbicide-tolerant biotypes exhibit normal levels of herbicide-sensitive ACCase. Little or no differences in herbicide uptake, translocation, or metabolism are observed in comparisons between tolerant and sensitive biotypes of either species (Holtum et al. 1991; Devine 1992). Herbicide resistance appears to correlate with recovery of the plasma membrane electrogenic potential after herbicide treatment (Shimbukuro and Hoffer 1991; Hausler et al. 1991; Devine 1992).

3. STRATEGIES FOR HERBICIDE-RESISTANT CROP DEVELOPMENT

3.1. SEXUAL HYBRIDIZATION

Preexisting intra- and interspecific variation in resistance to aryloxyphenoxypropionate and cyclohexanedione herbicides may in some cases serve as useful germplasm for herbicide-resistant crop (HRC) development. This appears to be true in the *Lolium*, *Festuca*, and *Avena* sp. Red fescue (*Festuca rubra* L.) is naturally tolerant to ACCase-inhibiting herbicides. Moreover, red fescue could serve as a source of herbicide resistance for its sexually compatible relatives such as tall fescue (*Festuca arundicea*) and the *Lolium* sp. Diclofop-resistant *L. multiflorum* also may be useful for interspecific and intergeneric transfer of herbicide resistance because it is closely related to and sexually compatible with a number of *Festuca* and *Lolium* sp. (Crowder 1953; Love 1947). Resistance in both red fescue and *L. multiflorum* is attributed to alterations of ACCase (Stoltenberg et al. 1989; Betts et al. 1992); however, red fescue exhibits resistance to both aryloxyphenoxypropionate and cyclohexanedione herbicides, whereas the *Lolium* resistance allele characterized by Betts et al. (1992) specifically conferred resistance to diclofop, suggesting that variation among alleles for herbicide resistance spectrum could be useful in specific HRC/weed combinations.

The diclofop resistance of the *L. multiflorum* biotype analyzed by Betts et al. (1992) was inherited as a single partially dominant gene, although other reports, presumably involving other *L. multiflorum* biotypes, indicate evidence for two genes conferring herbicide resistance (Barker et al. 1990). A backcrossing program to transfer the *L. multiflorum* diclofop-resistance trait to *L. perenne* (perennial ryegrass) was initiated (Betts, Ehlke, and Wyse, University of Minnesota, unpublished results). Diclofop resistance was successfully transferred through the fifth generation of backcrossing and resistant, high-quality turf perennial ryegrass types were recovered. It is important to emphasize that the stability of inheritance and expression of the herbicide resistance trait has not been extensively evaluated in advanced generations of these wide crosses, but the results indicate that wide hybridization among the *Festuca* and *Lolium* spp. is a promising approach.

The *Avena* genus exhibits variation in resistance to aryloxyphenoxypropionate and cyclohexanedione herbicides. Kibite and Harker (1992) tested herbicide resistance in 8000 accession lines from six *Avena* spp. from the USDA World Oat Collection. Ninety-one lines, of which 60 were *A. sativa* accessions, exhibited resistance to diclofop-methyl and eight lines showed moderate resistance to fenoxyprop-ethyl. The mode of inheritance of herbicide tolerance was uncharacterized in the population screen (Kibite and Harker 1992). Barr and Tasker (1992) reviewed the inheritance of herbicide resistance in several oat accessions. Interestingly, resistance in oat may be conferred by either partially dominant genes or recessive genes. Partially dominant resistance of *A. sterilis* accession SAS1 is conferred by an altered ACCase, whereas recessive resistance likely involves the mechanism described by Devine (1992) and Holtum (1992) associated with altered membrane electrochemical properties. Herbicide resistance was incorporated into cultivated oat using the partially dominant resistance of the SAS1 in a backcross program (Barr and Tasker 1992). These authors discussed implications and concerns regarding introgression of the herbicide-resistant genes to wild oats, which are readily crossfertile with cultivated oat.

3.2. TISSUE CULTURE SELECTION

Plant cell and tissue culture technologies involve reduction of plant growth to the cell level, thereby maximizing the number of selectable units that can be screened for tolerance to biochemical inhibitors such as herbicides. For example, maize callus comprises up to 10^7 cells per gram fresh weight (Smith et al. 1984). Plant tissue cultures are usually grown in defined media under controlled environmental conditions, enabling optimization of selection conditions for recovery of the desired variant. Tissue culture *per se* causes genetic variation and thus high frequencies of mutations are observed in progeny of plants regenerated from tissue culture (reviewed for maize by Phillips et al. 1988). Several advances in tissue culture of monocotyledenous plants have resulted in the establishment of regenerable tissue culture systems for cereal and grass species that are amenable to selection for development of HRCs.

In theory, if herbicide mode of action and tolerance mechanism(s) are unknown, it may be more efficient to select for tolerance in haploid tissue cultures so that the probability of recovering both dominant and recessive variation for tolerance is equal. Direct or positive tissue culture selection is used to select for cells able to grow in the presence of the herbicide. Tissue culture selection of aryloxyphenoxypropionate-resistant and/or cyclohexanedione-resistant grass or cereal species would most likely utilize diploid tissue cultures because of difficulties encountered in establishing haploid cultures for most of these species (reviewed for maize by Somers and Anderson 1994). Furthermore, diploid callus is often used for direct selection of herbicide resistance known to be conferred by dominant genes as an approach to avoid selection of potential deleterious recessive mutations conferring herbicide resistance. However, the use of diploid callus does not preclude selection of homozygous recessive mutations induced by tissue culture, as indicated by the recovery of homozygous recessive mutations in tomato plants regenerated from tissue culture (Evans and Sharp 1983). Interestingly, there has been little discussion in the literature regarding the relative merits of herbicide resistance due to altered ACCase or altered membrane potential mechanisms for the development of aryloxyphenoxypropionate and cyclohexanedione HRCs. Furthermore, it is not known if resistance due to the altered membrane mechanism was recovered from tissue culture selection.

Direct selection in tissue cultures involves two different approaches to select callus capable of growth on lethal concentrations of a herbicide as described for maize (Hibberd 1984, Somers and Hibberd 1993). Enrichment selection involves initiating selection at sublethal herbicide concentrations, followed by continual subcultures onto medium containing increasing herbicide concentrations. Selection ceases when callus is capable of growth in the presence of herbicide concentrations greater than that which is lethal to unselected callus. Enrichment selection may maximize selection efficiency because selection pressure is gradually applied to the cell population, presumably allowing (at sublethal herbicide concentrations) cells to crossfeed metabolites to minimize negative effects of low population density caused by too-stringent selection and to avoid transfer of negative factors from senescing susceptible cells to the small population of tolerant variant cells. Lethal dose selection involves plating callus on a lethal concentration of herbicide (Somers and Hibberd 1993). Regeneration of stably transgenic maize plants from callus selected for tolerance conferred by transformed genes encoding resistance to phosphinothricin-based herbicides provide a basis for comparison of the efficiency of enrichment vs. lethal dose selection. Fertile, transgenic maize plants resistant to phosphinothricin were regenerated from callus isolated using a modified enrichment selection protocol (Fromm et al. 1990) and a lethal dose selection (Gordon-Kamm et al. 1990). Both selection procedures were successful for recovery of herbicide-resistant maize; however, the selection and plant regeneration efficiencies of each approach are difficult to compare. Therefore, selection strategy must be determined empirically.

Tissue culture selection for the development of aryloxyphenoxypropionate and cyclohexanedione HRCs has been pursued in maize, wheat, and Kentucky bluegrass. Considerable selection and characterization research was conducted in maize at the University of Minnesota (Parker et al. 1990a, b; Gronwald et al. 1989; Marshall et al. 1992; Somers and Anderson 1993) and is reviewed in more detail in the next section of this chapter. Tissue culture selection of cyclohexanedione-resistant winter wheat was pursued in the laboratory of Dr. Phillip Westra (Colorado State University, personal communication). Herbicide-resistant winter wheat would enable effective control of winter annual weeds. An attempt to select cyclohexanedione-tolerant Kentucky bluegrass was apparently unsuccessful due to problems with developing an appropriate tissue culture system for this species (Carlson et al. 1992).

3.3. MUTATION BREEDING

Induced mutations are a likely source of variation for selection of monocots resistant to aryloxyphenoxypropionate and cyclohexanedione herbicides. Successful selections of sulfonylurea-resistant soybean (Sebastian and Chaleff 1987; Sebastian et al. 1989) and imidazolinone-resistant maize (Bright 1992) demonstrate the feasibility of the mutation breeding approach. Furthermore, these studies indicate that mutations conferring altered herbicide target site enzyme and other resistant mechanisms are recovered using this approach. Therefore, mutation breeding for resistance to aryloxyphenoxypropionate or cyclohexanedione herbicides should be considered as a promising strategy to develop HRCs. Preliminary success has been achieved in the selection of aryloxyphenoxypropionate- and cyclohexanedione-resistant small grains using mutation breeding (C. Konzak, Washington State University, personal communication).

4. CASE STUDY OF HERBICIDE-RESISTANT CROP DEVELOPMENT — MAIZE

4.1. TISSUE CULTURE SELECTION

Preliminary investigations to determine whether herbicide tolerance could be selected in maize tissue cultures were conducted using rapidly growing, nonregenerable Black Mexican Sweet (BMS) maize suspension cultures. Cell lines selected that grew in medium containing 10, 50, and 100 μM sethoxydim were designated B10S, B50S, and B100S, respectively (Parker et al. 1990b) (Table 1). Growth rates of the selected herbicide-tolerant tissue cultures were not significantly different from wild-type tissue cultures in the absence of herbicide. The B50S tissue culture was the most herbicide tolerant, exhibiting approximately a 60-fold increase in sethoxydim tolerance and a 90-fold increase in haloxyfop tolerance compared to the unselected BMS tissue cultures. These results strongly suggest that variation for herbicide tolerance was selectable from maize tissue culture cells (Parker et al. 1990b).

ACCase activity of the selected cell lines was characterized to determine if tolerance was due to alterations of the herbicide target site enzyme. When the tissue cultures were grown in the absence of herbicide, ACCase activity of B10S, B50S, and B100S were 1.7-, 2.6-, and 1.9-fold greater than the activity of ACCase from the BMS tissue culture cells (Table 1). Concentrations of sethoxydim and haloxyfop required to inhibit ACCase activity by 50% were not significantly different among

TABLE 1 Herbicide Tolerance and Acetyl-Coenzyme A Carboxylase Activity and Enzyme Protein Levels of Herbicide-Tolerant Maize Tissue Cultures

	-Fold Increase Compared to Wild-Type			
Cell Line	**I_{50} Haloxyfop**	**I_{50} Sethoxydim**	**ACCase Activity**	**ACCase Protein**
B10S	8.3	7.5	1.7	0.7
B50S	90.0	58.3	2.6	4.7
B100S	30.0	43.3	1.9	2.9

ACCase activity of unselected wild-type maize suspension cultures was 6.1 ± 0.8 nmol HCO_3^- incorporated, mg/protein/min. Herbicide at 0.6 μ*M* sethoxydim and 0.06 μ*M* haloxyfop inhibited growth of wild-type tissue cultures by 50%. Concentrations of herbicide required to inhibit wild-type ACCase by 50% were 5.2 μ*M* sethoxydim and 0.75 μ*M* haloxyfop, and were similar to levels required to inhibit ACCase from the herbicide-tolerant tissue cultures. ACCase enzyme protein levels were determined by scanning reflectance quantification of the 200 kDa ACCase subunit on SDS-PAGE Western blots probed with alkaline phosphatase-conjugated strepavdin. Data from Parker et al. 1990b.

the selected and unselected cell lines, indicating that tolerance was conferred by elevated levels of a wild-type ACCase activity (Table 1). Levels of the 220-kDa biotinylated ACCase subunit (Egli et al. 1993) paralleled elevated activity in extracts of the herbicide-resistant tissue cultures (Table 1; Parker et al. 1990b). These data suggest that the ACCase I herbicide-sensitive isoform was elevated in the tolerant tissue cultures; however, because techniques for resolving ACCase I and II had not been developed at the time of this study, it is impossible to rule out the involvement of ACCase II. An association between elevated ACCase activity and increased chromosome numbers was observed in the tolerant tissue cultures (Marshall et al. 1988), suggesting that polyploidy resulted in elevated enzyme activity. These studies indicated that tissue culture selection for sethoxydim tolerance in maize resulted in the recovery of variation in herbicide tolerance and that selection pressure was exerted on the target site enzyme, ACCase I.

Fertile maize plants were regenerated from six different tissue cultures that exhibited substantial tolerance to sethoxydim and/or haloxyfop (Parker 1990; Parker et al. 1990a; Marshall et al. 1992; Somers et al. 1990). Selection was conducted with friable, embryogenic callus established from immature embryos of F1 and F2 kernels of crosses between A188 and B73. Concentrations of the herbicides that inhibited growth of unselected, wild-type maize callus by 50% were 2.6 μ*M* sethoxydim and 0.04 μ*M* haloxyfop. Enrichment selection was conducted in which callus was initially plated on medium containing sublethal herbicide concentrations. At each subculture interval of 2 to 3 weeks, surviving callus was transferred to medium containing an increased herbicide concentration (Parker 1990b).

Herbicide-tolerant callus lines resulting from tissue culture selection on sethoxydim were designated S1, S2, S3, and S4. Haloxyfop-tolerant callus lines were designated H1 and H2. S1, S2, and H1 were independently isolated from the same A188 × B73 F1 tissue culture, whereas S3 and H2 were isolated from a A188 × B73 F2 tissue culture (Parker 1990). S4 was isolated in a later selection experiment (Marshall, Yun, and Somers, unpublished). Herbicide concentrations inhibiting callus growth by 50% were 2.6, >100, >100, and 2.9 μ*M* sethoxydim and 0.04, 0.17, 0.82, and 0.41 μ*M* haloxyfop for the unselected wild-type, S1, S2, and H1 lines, respectively (Parker 1990). Thus, the two S callus lines exhibited tolerance to both herbicides, whereas H1 callus exhibited tolerance primarily to haloxyfop. The ACCase activity level of each callus line was similar to the activity level in the unselected control callus. However, ACCase of the selected callus lines exhibited reduced inhibition by the two herbicides and the reduction in herbicide inhibition of

ACCase activity in the variant callus lines paralleled the callus herbicide tolerance phenotypes (Parker 1990; Parker et al. 1990a). Reduced herbicide inhibition of callus ACCase activity was subsequently observed for the other selections (Parker, Marshall, Yun, and Somers, unpublished results). These results suggested that in these tissue cultures herbicide tolerance was conferred by alteration(s) in ACCase that reduced herbicide inhibition of enzyme activity as opposed to an increase in an unaltered ACCase activity as shown for the BMS selections (Parker et al. 1990a, b). The different callus tolerance phenotypes, e.g., tolerance to both herbicides exemplified by the S lines and tolerance to only haloxyfop as observed for H1, presumably were due to different mutations in ACCase I. Other resistant tissue cultures were characterized that exhibited elevation of unaltered ACCase activity (Gronwald et al. 1989). Unfortunately, plants regenerated from tissue cultures exhibiting elevated ACCase activity were sterile, making it impossible to further characterize that mechanism of resistance.

Herbicide-tolerant plants regenerated from S1, S2, S3, S4, H1, and H2 were self-pollinated and test crossed to wild-type for genetic analyses and to develop homozygous herbicide-tolerant plant families from each selection. Progeny resulting from self-pollination of all the mutants segregated into three phenotypic classes following herbicide treatment. Plants were either killed, injured in a manner similar to R_0 plants, or completely uninjured. Self-pollinated progeny segregated into these classes approximating a ratio of 1:2:1 for all mutants (Parker 1990; Parker et al. 1990a; Marshall 1990; Marshall et al. 1992). Test cross progeny segregated 1:1 for killed and injured plants. The herbicide tolerance phenotypes of all six selections were inherited as partially dominant nuclear encoded traits and the regenerated plants and therefore the selected callus lines were heterozygous for the mutations conferring herbicide tolerance (Parker 1990; Parker et al. 1990a; Marshall 1990).

Homozygous plant families were selected from progeny of self-pollinated plants regenerated from each tissue culture (Marshall et al. 1991). All S lines exhibited high levels of tolerance to both herbicides as compared to wild-type maize as shown for S2 and S3 in Table 2. Homozygous, herbicide-tolerant derivatives of H1 were tolerant only to haloxyfop, whereas H2 was less tolerant to haloxyfop and exhibited intermediate sethoxydim tolerance. Thus, the mutants exhibited three different herbicide-resistance phenotypes in terms of differential tolerance to haloxyfop and sethoxydim. Other representatives of the cyclohexanedione and aryloxyphenoxypropionate herbicides have been tested; however, no further differentiation of the resistance phenotypes was observed based on these tests (Dotray 1993).

ACCase activities in extracts from homozygous-tolerant seedling leaves of each selection were assayed. In the absence of herbicide, seedling leaf ACCase activity levels of the S lines were similar to wild-type ACCase activities, whereas the H line ACCase activities were slightly lower than the wild-type values (Marshall et al. 1992). A more detailed study of the herbicide tolerance phenotype in relation to different aryloxyphenoxypropionate and cyclohexanedione herbicides and their effects on ACCase was conducted by Dotray (1993). For the mutants analyzed, the reduction in herbicide inhibition of ACCase closely parallelled the whole plant herbicide tolerance phenotype (Table 2; Dotray 1993). One exception was H1, which exhibited resistance primarily to aryloxyphenoxypropionate herbicides. ACCase activity of H1 was less inhibited by both haloxyfop and sethoxydim as compared to ACCase from wild-type seedlings. Although the reason for these results is unclear, one explanation is that H1 ACCase I may be an unstable enzyme, as suggested by the reduced activity level detected in H1 seeding extracts (Marshall et al. 1992), and it is possible that procedures used to partially purify ACCase (Dotray 1993) resulted in its total inactivation, leaving behind the ACCase II isoform. This isoform is resistant to both aryloxyphenoxypropionate and cyclohexanediones (Howard and Ridley 1990; Egli et al. 1993).

Herbicide tolerance and reduced inhibition of ACCase cosegregated, confirming that the mutations in ACCase conferred herbicide tolerance (Marshall et al. 1992). An allelism test also was conducted to determine the relationships between mutations conferring the different herbicide tolerance phenotypes (Marshall et al. 1992). The simplest interpretation of this study was that the

TABLE 2 Seedling Herbicide Lethality and Herbicide Inhibition of ACCase Activity Extracted from Homozygous, Tolerant Seedlings Expressed as -Fold Increases Over Values Determined for Wild-Type Seedlings

	-Fold Increase Relative to Wild-Type			
	Herbicide Lethality[a]		ACCase I_{50}[b]	
Line	Sethoxydim	Haloxyfop	Sethoxydim	Haloxyfop
S2	>127	16	350	23
S3	>127	16	350	23
H1	1	8	1.6	16
H2	16	16	36	40

[a] Determined as the lowest herbicide application rate that was lethal to all seedlings in four replications of 5 plants/pot for each genotype.
[b] Concentration of herbicide that reduced ACCase activity by 50%.

Data from Marshall et al. 1992.

S1, S2, S3, and H1 mutations are allelic and that H2 is either allelic or closely linked (<4.4 map units, based on the supposition that 9 plants killed out of 412 plants tested were due to recombination and not because of the low herbicide tolerance of the H2). The results of herbicide tolerance and ACCase studies, cosegregation and parallel relationships of herbicide tolerance and reduction in herbicide inhibition of ACCase, and the observations that the mutations are most likely all allelic support the conclusion that at least three distinct herbicide tolerance phenotypes were isolated that are encoded by alleles of the ACCase I structural gene (Marshall et al. 1992).

4.3. FIELD EVALUATION AND VARIETY DEVELOPMENT

Homozygous tolerant families developed for each mutant were treated with herbicide in the field (Somers et al. 1990, Marshall 1990, Dotray et al. 1992). Homozygous tolerant S2 plants were evaluated at two stages of development at two locations. Sethoxydim injury was not observed on plants treated with 0.88 kg/ha (Figure 1), a rate exceeding that recommended for control of annual grass weeds. The herbicide treatments had no effect on grain yield (Table 3), flowering date, plant height, and grain moisture content (Dotray et al. 1992). The herbicide tolerance alleles were expressed in diverse genetic backgrounds to evaluate genotype effects on trait penetrance. Backcross and F1 populations with inbreds A188, A619, A641, A665, A682, B73, and W15R stably exhibited herbicide tolerance conferred by the S1, S2, S3, and H1 alleles (Marshall 1990). Herbicide treatments did not adversely affect grain yield or quality nor was the trait associated with deleterious effects. These results indicate that efforts to develop herbicide-tolerant maize varieties for further evaluations of the herbicide-tolerant trait per se and its utility in weed control programs are warranted.

Herbicide-tolerant inbred derivatives of S2 were developed by backcrossing the mutant plants regenerated for tissue culture by a number of public inbred lines (Marshall and Somers, unpublished). Progeny of each backcross were treated with 0.22 kg/ha sethoxydim to identify heterozygous, herbicide-resistant individuals. Segregation ratios of backcross progenies in all crosses were 1:1 for partially injured to susceptible types, which was expected for a cross of the heterozygous donor parent to wild-type inbred line. Backcross programs for some inbreds were advanced to the BC4 generation at which the herbicide-resistant lines were expected to be >93% recurrent inbred parent. BC4 lines were selfed to recover homozygous resistant plants. BC4 selfed progeny segregated 1:2:1 for resistant:partially injured:susceptible types, as expected. These results further indi-

FIGURE 1 Homozygous sethoxydim-tolerant corn (left) and B73 corn (right) 21 d after treatment with 0.88 kg/ha sethoxydim applied at the three- (top) and seven- (bottom) leaf stages (Dotray et al. 1993).

cate that the herbicide resistance trait conferred by altered ACCase is stably inherited. Materials were transferred to corn breeding companies in 1990. Backcrossing and inbred development are in progress. Commercialization of sethoxydim-resistant maize varieties is anticipated in the near future.

5. SUMMARY

Aryloxyphenoxypropionate and cyclohexanedione herbicides inhibit ACCase in herbicide-sensitive monocots. Acetyl coenzyme-A carboxylase catalyzes the ATP-dependent carboxylation of acetyl-coenzyme A to form malonyl-coenzyme A. The current understanding of herbicide sensitivity in sensitive species is that the ACCase isoform localized in plastids is inhibited by the herbicides, thereby blocking production of malonyl-coenzyme A, which is required by fatty acid synthesis. In dicots and also in some naturally herbicide-resistant monocots ACCase is not inhibited by these herbicides. Herbicide resistance due to insensitivity of ACCase to herbicide inhibition is the most common mechanism of resistance. Herbicide metabolism confers resistance in wheat to diclofop-methyl, an aryloxyphenoxypropionate. A more recently observed resistance involving altered membrane potential is not yet completely understood.

In most cases the aryloxyphenoxypropionate and cyclohexanedione herbicides effectively control weeds at low concentrations, have low toxicity to nontarget species, have reduced persistence in the soil and are applied postemergence. Development of cereal and grass varieties with resistance to some or all of these herbicides may provide producers with increased flexibility in weed control. Specifically, the herbicides could be used postemergence for grassy weed control or

TABLE 3 Field Test of Sethoxydim-Tolerant Corn Treated at Two Minnesota Locations[a]

Sethoxydim rate[b]	Yield[d]	
	St. Paul	Waseca
kg ha^{-1}	kg ha^{-1}	
0	3393	3010
0.22	3728	3162
0.44	3786	2857
0.88	3864	2953
LSD (0.05)	NS	NS

[a] Data averaged over growth stages and years.
[b] All treatments received 1.25% by vol crop oil concentrate.
[c] Values represent total ear dry weight.

Data from Dotray et al. 1993.

in combination with herbicides that control broadleaf weeds to provide broad spectrum weed control without damage to the herbicide-resistant crop. For example, weed control in maize is primarily accomplished using persistent, preemergence herbicides. Maize lines resistant to aryloxyphenoxypropionate and/or cyclohexanedione herbicides would enable the development of postemergence weed management strategies.

Sources of resistance to aryloxyphenoxypropionate and cyclohexanedione herbicides include:

1. transfer by sexual hybridization with naturally occurring herbicide-resistant related species
2. sexual hybridization with herbicide-resistant biotypes selected following repeated field applications of herbicide
3. selection of mutations conferring herbicide resistance in tissue culture followed by plant regeneration
4. mutation breeding

The author is aware of efforts to develop resistance to these herbicides in maize, wheat, oat, barley, Italian ryegrass, and Kentucky bluegrass. In most cases in which resistance has been characterized it is due to reduced sensitivity of ACCase to herbicide inhibition. This form of resistance is inherited in a partially or completely dominant manner. Maize mutants, selected in tissue culture, have been extensively characterized and are in advanced stages of HRC development. Commercial hybrids are expected to be available in the next few years. These herbicide-resistant varieties appear to offer certain advantages for weed control in maize. However, it is evident that mutations conferring plant resistance are readily selected both *in vitro* and in the field, as demonstrated by the isolation of herbicide-resistant weed biotypes. Therefore, resistance to aryloxyphenoxypropionate and cyclohexanedione herbicide must be managed as a component of integrated weed control strategies to preserve the efficacy of these herbicide families.

ACKNOWLEDGMENTS

I wish to express my gratitude to Dr. N.J. Ehlke, Department of Agronomy and Plant Genetics, University of Minnesota, and Dr. J.W. Gronwald, USDA-Agricultural Research Service, Plant Science Research Unit, St. Paul, MN, for reviewing this chapter.

REFERENCES

Barker, R.E., Mueller-Warrant, G., and Banowetz, G.M. Inheritance of resistance to diclofop-methyl in Italian ryegrass. *Agron. Abstr.* 79.

Barr, A.R. and Tasker, S.D. 1992. Breeding for herbicide resistance in oats: opportunities and risks. In *Wild Oats in World Agriculture. Proc. 4th Int. Oat Conf.* Ed. by Barr, A.R. and Medd, R.W. Adelaide, Australia, pp. 51–56.

Beste, C.E., Humburg, N.E., Kempen, H.M., Radke, R.O., Riggleman, J.D., Stritzke, J.F., and Miller, G.R. 1983. *Herbicide Handbook of the Weed Science Society of America*, 5th ed. Weed Science Society of America, Champaign, IL.

Betts, K.J., Ehlke, N.J., Wyse, D.W., Gronwald, J.W., and Somers, D.A. 1992. The mechanism of inheritance of diclofop-resistance in Italian ryegrass (*Lolium multiflorum*). *Weed Sci.* 40: 184.

Bright, S.W.J. 1992. Herbicide-resistant crops. In *Biosynthesis and Molecular Regulation of Amino Acids in Plants.* Ed. by Singh, B.K., Flores, H.E., and Shannon, J.C. American Society of Plant Physiologists, p. 184.

Burton, J.D., Gronwald, J.W., Keith, R.A., Somers, D.A., Gengenbach, B.G., and Wyse, D.L. 1991. Kinetics of inhibition of acetyl-coenzyme A carboxylase by sethoxydim and haloxyfop. *Pestic. Biochem. Physiol.* 39: 100.

Burton, J.D., Gronwald, J.W., Somers, D.A., Connelly, J.A., Gengenbach, B.G., and Wyse, D.L. 1987. Inhibition of plant acetyl-coenzyme A carboxylase by the herbicides sethoxydim and haloxyfop. *Biochem. Biophys. Res. Commun.* 148: 1039.

Burton, J.D., Gronwald, J.W., Somers, D.A., Gengenbach, B.G., and Wyse, D.L. 1989. Inhibition of corn acetyl-CoA carboxylase by cyclohexanedione and aryloxyphenoxypropionate herbicides. *Pestic. Biochem. Physiol.* 34: 76.

Carlson, J.A., Wyse, D.L., Ehlke, N.J., and Somers, D.A. 1990. Tissue culture selection system in *Poa pratensis. Proc. North Central Weed Sci. Soc.* 45: 11.

Charles, D.J. and Cherry, J.H. 1986. Purification and characterization of acetyl-CoA carboxylase from developing soybean seeds. *Phytochemistry* 25: 1067.

Crowder, L.V. 1953. Interspecific and intergeneric hybrids of *Festuca* and *Lolium*. *J. Hered.* 44: 195.

Devine, M.D. 1992. The biochemical basis of herbicide resistance in wild oat. In *Wild Oats in World Agriculture, Proc. 4th Int. Oat Conf.*, Ed. by Barr, A.R. and Medd, R.W. Adelaide, Australia, pp. 41–44.

Dotray, P.A., Marshall, L.C., Parker, W.B., Wyse, D.L., Somers, D.A., and Gengenbach, B.G. 1992. Herbicide tolerance and weed control in sethoxydim-tolerant corn (*Zea mays*). *Weed Sci.* 41: 213

Dotray, P.A. 1993. Cross Tolerance to Aryloxyphenoxypropionate and Cyclohexanedione Herbicides in Tissue Culture Derived Corn (*Zea mays*) Lines. Ph.D. thesis, University of Minnesota, St. Paul, p. 79.

Egin-Buhler, B. and Ebel, J. 1983. Improved purification and further characterization of acetyl-CoA carboxylase from cultured cells of parsley (*Petroselinum hortense*). *Eur. J. Biochem.* 133: 335.

Egli, M.A., Gengenbach, B.G., Gronwald, J.W., Somers, D.A., and Wyse, D.L. 1993. Characterization of maize acetyl-CoA carboxylase. *Plant Physiol.* 101: 499.

Evans, D.A. and Sharp, W.R. 1983. Single gene mutations in tomato plants regenerated from tissue cultures. *Science* 221: 949.

Focke, M. and Lichtenthaler, H. 1987. Inhibition of acetyl-coenzyme A carboxylase of barley chloroplasts by cycloxydim and sethoxydim. *Z. Naturforsch.* 42c: 1361.

Fromm, M.E., Morrish, F., Armstrong, C., Williams, R., Thomas, J., and Klein, T.M. 1990. Inheritance and expression of chimeric genes in the progeny of transgenic maize plants. *Bio/Technology* 8: 833.

Gordon-Kamm, W.J., Spencer, T.M., Mangano, M.L., Adams, T.R., Daines, R.J., Start, W.G., O'Brien, J.V., Chambers, S.A., Adams, W.R., Willetts, N.G., Rice, T.B., Mackey, C.J., Kreuger, R.W., Kausch, A.P., and Lemaux, P.G. 1990. Transformation of maize cells and regeneration of fertile transgenic plants. *Plant Cell* 2: 603.

Gronwald, J.W., Parker, W.B., Somers, D.A., Wyse, D.L., and Gengenbach, B.G. 1989. Selection for tolerance to graminicide herbicides in maize tissue culture. Proc. Br. Crop Protec. Conf. *Weeds* 3: 1217.

Harwood, J.L. 1988. Fatty acid metabolism. *Annu. Rev. Plant Physiol. Plant Mol. Biol.* 39: 101.

Harwood, J.L. 1989. The properties and importance of acetyl-coenzyme A carboxylase in plants. In Proc. Brighton Crop Protection Conference, *Weeds 1989*, Vol. 1. British Crop Protection Council, Surrey, p. 155.

Hausler, R.E., Holtum, J.A.M., and Powles, S.B. 1991. Cross-resistance to herbicides in annual ryegrass (*Lolium rigidum*). IV. Correlation between membrane effects and resistance to graminicide. *Plant Physiol.* 97: 1034.

Hawke, J.C. and Leech, R.M. 1990. Acetyl coenzyme A carboxylase in species of *Triticum* of different ploidy. *Planta* 181: 543.

Heap, I. and Knight, R. 1989. The occurrence of herbicide cross-resistance in a population of annual ryegrass, *Lolium rigidum*, resistant to diclofop-methyl. *Aust. J. Agric. Res.* 37: 149.

Hellyer, A., Bambridge, H.E., and Slabas, A.R. 1986. Plant acetyl-CoA carboxylase. *Biochem. Soc. Trans.* 14: 565.

Hibberd, K.A. 1984. Induction, selection and characterization of mutants in maize cell cultures, in *Cell Cultures and Somatic Cell Genetics of Plants*, Vol. 1, *Laboratory Procedures and Their Applications*. Ed. by Vasil, I.K. Academic Press, New York, p. 571.

Holtum, J.A.M. 1992. Herbicide resistant wild oats in Australia. In *Wild Oats in World Agriculture, Proc. 4th Int. Oat Conf.* Ed. by Barr, A.R. and Medd, R.W. Adelaide, Australia, pp. 45–48.

Holtum, J.A.M., Matthews, J.M., Hausler, R.E., Liljegren, D.R., and Powles, S.B. 1991. Cross-resistance to herbicides in annual ryegrass (*Lolium rigidium*). III. On the mechanism of resistance to diclofop-methyl. *Plant Physiol.* 97: 1026.

Hoppe, H.H. and Zacher, H. 1985. Inhibition of fatty acid biosynthesis in isolated bean and maize chloroplasts by herbicidal phenoxyl-phenoxypropionic acid derivatives and structurally related compounds. *Pest. Biochem. Physiol.* 24: 298.

Hosaka, H. and Takagi, M.K. 1987. Biochemical effects of sethoxydim in excised root tips of corn (*Zea mays*). *Weed Sci.* 35: 612.

Howard, J.L. and Ridley, S.M. 1990. Acetyl-CoA carboxylase: a rapid novel assay procedure used in conjunction with the preparation of enzyme from maize leaves. *FEBS Lett.* 261: 261.

Kibite, S. and Harker, K.N. 1992. New sources of herbicide resistance in *Avena* spp. In *Wild Oats in World Agriculture, Proc. 4th Int. Oat Conf.* Ed. by Barr, A.R. and Medd, R.W. Adelaide, Australia, pp. 48–50.

Love, R.M. 1947. Interspecific and intergeneric hybridization in forage crop improvement. *Am. Soc. Agron. J.* 39: 41.

Marshall, L.C. 1990. Characterization of Maize Genotypes Tolerant to Herbicides that Inhibit Acetyl-CoA Carboxylase. Ph.D. thesis, University of Minnesota, St. Paul, p. 91.

Marshall, L., Keith, F.R., Parker, W.B., Somers, D.A., Gronwald, J.W., Wyse, D.L., and Gengenbach, B.G. 1988. Chromosome number and fatty acid levels in sethoxydim-tolerant corn cell lines. *Agron. Abstr.* 170.

Marshall, L.C., Somers, D.A., Dotray, P.D., Gengenbach, B.G., Wyse, D.L., and Gronwald, J.W. 1992. Allelic mutations in acetyl-coenzyme A carboxylase confer herbicide tolerance in maize. *Theor. Appl. Genet.* 83: 435.

Matthews, J.M., Holtum, J.A.M., Liljegren, D.R., Furness, B., and Powles, S.B. 1990. Cross resistance to herbicides in annual ryegrass (*Lolium rigidum*). I. Properties of the herbicide target enzymes acetyl-coenzyme A carboxylase and acetolactate synthase. *Plant Physiol.* 94: 1180.

Nikolau, B.J. and Hawke, J.C. 1984. Purification and characterization of maize leaf acetyl-coenzyme A carboxylase. *Arch. Biochem. Biophys.* 228: 86.

Nikolau, B.J., Wurtele, E.S., and Stumpf, P.K. 1984. Tissue distribution of acetyl-coenzyme A carboxylase in leaves. *Plant Physiol.* 75: 895.

Parker, W.B., Marshall, L.C., Burton, J.D., Somers, D.A., Wyse, D.L., Gronwald, J.W., and Gengenbach, B.G. 1990a. Dominant mutations causing alterations in acetyl-coenzyme A carboxylase confer tolerance to cyclohexanedione and aryloxyphenoxypropionate herbicides in maize. *Proc. Natl. Acad. Sci. U.S.A.* 87: 7175.

Parker, W.B., Somers, D.A., Wyse, D.L., Keith, R.A., Burton, J.D., Gronwald, J.W., and Gengenbach, B.G. 1990b. Selection and characterization of sethoxydim-tolerant maize tissue cultures. *Plant Physiol.* 92: 1220.

Parker, W.B. 1990. Selection and Characterization of Sethoxydim- and Haloxyfop-Tolerant Corn (*Zea mays*). Ph.D. thesis, University of Minnesota, St. Paul, p. 88.

Phillips, R.L., Somers, D.A., and Hibberd, K.A. 1988. Tissue culture and *in vitro* genetic manipulations. In *Corn and Corn Improvement*. Ed. by Sprague, G.F. p. 345.

Rendina, A.R. and Felts, J.M. 1988. Cyclohexanedione herbicides are selective and potent inhibitors of acetyl-CoA carboxylase from grasses. *Plant Physiol.* 86: 983.

Rendina, A.R., Felts, J.M., Beaudoin, J.D., Craig-Kennard, A.C., Look, L.L., Paraskos, S.L., and Hagenah, J.A. 1988. Kinetic characterization, stereoselectivity and species selectivity of the inhibition of plant acetyl-CoA carboxylase by the aryloxyphenoxypropionic acid grass herbicides. *Arch. Biochem. Biophys.* 265: 219.

Sasaki, Y., Hakamada, K., Suama, Y., Nagano, Y., Furusawa, I., and Matsuno, R. 1993. Chloroplast-encoded protein as a subunit of acetyl-CoA carboxylase in pea plant. *J. Biol. Chem.* 268: 25118.

Sebastian, S.A., Facer, G.M., Ulrich, J.F., Forney, D.R., and Chaleff, R.S. 1989. Semidominant soybean mutation for resistance to sulfonylurea herbicides. *Crop Sci.* 29: 1403.

Sebastian, S.A. and Chaleff, R.S. 1987. Soybean mutants with increased tolerance for sulfonylurea herbicides. *Crop Sci.* 27: 948.

Secor, J. and Cseke, C. 1988. Inhibition of acetyl-CoA carboxylase activity by haloxyfop and tralkoxydim. *Plant Physiol.* 86: 10.

Shimbukaro, R.H. and Hoffer, B.L. 1991. Metabolism of diclofop-methyl in susceptible and resistant biotypes of *Lolium rigidum, Pestic. Biochem. Physiol.* 39: 251.

Shimbukaro, R.H., Walsh, W.C., and Hoerauf, R.A. 1979. Metabolism and selectivity of diclofop-methyl in wild oat and wheat. *J. Agric. Food Chem.* 27: 615.

Slabas, A.R. and Hellyer, A. 1985. Rapid purification of a high molecular weight subunit polypeptide form of rape seed acetyl CoA carboxylase. *Plant Sci.* 39: 177.

Slabas, A.R., Hellyer, A., and Bambridge, H.E. 1986. The basic polypeptide subunit of rape leaf acetyl-CoA carboxylase is a 220 kDa protein. *Biochem. Soc. Trans. London* 14: 716.

Smith, J.A., Green, C.E., and Gengenbach, B.G. 1984. Feeder layer support of low density populations of *Zea mays* L. suspension cells. *Plant Sci. Lett.* 36: 67.

Somers, D.A. and Anderson, P.C. 1994. Tissue culture selection of herbicide tolerance in maize. In *Biotechnology in Agriculture and Forestry, Maize.* Ed. by Bajaj, Y.P.S. Springer-Verlag, Berlin, Vol. 25, p. 293.

Somers, D.A., Marshall, L.C., Dotray, P., Parker, W.B., Yun, S.J., Wyse, D.L., Gronwald, J.W., and Gengenbach, B.G. 1990. Characterization of sethoxydim-tolerant corn selected in tissue culture. In *Proc. Corn Utilization Conf. III.* St. Louis, MO, p. 9.

Somers, D.A., Keith, R.A., Egli, M.A., Marshall, L.C., Gengenbach, B.G., Gronwald, J.W., and Wyse, D.L. 1993. Expression of the *Acc1* gene-encoded acetyl-coenzyme A carboxylase in developing maize kernels. *Plant Physiol.* 101: 1097.

Somers, D.A. and Hibberd, K.A. 1993. In vitro selection. In *The Maize Handbook.* Ed. by Freeling, M. and Walbot, V. Springer-Verlag, Berlin, p. 685.

Stoltenberg, D.E., Gronwald, J.W., Wyse, D.L., Burton, J.D., Somers, D.A., and Gengenbach, B.G. 1989. The effect of sethoxydim and haloxyfop on acetyl-coenzyme A carboxylase activity in tolerant and susceptible *Festuca* species. *Weed Sci.* 37: 512.

Wurtele, E.S. and Nikolau, B.J. 1990. Plants contain multiple biotin enzymes: discovery of 3-methylcrotonyl-CoA carboxylase, propionyl-CoA carboxylase and pyruvate carboxylase in the plant kingdom. *Arch. Biochem. Biophys.* 278: 179.

Chapter 12

STEPS TOWARDS GENETIC ENGINEERING OF CROPS RESISTANT AGAINST BLEACHING HERBICIDES

Gerhard Sandmann
Norihiko Misawa
Peter Böger

CONTENTS

1. INTRODUCTION

Bleaching herbicides belong to a chemically diverse group of inhibitors. In recent years research has focused mainly on their mode of action and their interaction with the target enzyme phytoene desaturase (Sandmann and Böger 1989). These phytoene desaturase inhibitors are used as preemergence herbicides mainly in cotton. An exception is diflu-fenican, which shows good selectivity as a pre- and postemergence herbicide in weed control of wheat and barley (Cramp et al. 1987). Molecular genetics of this target enzyme, however, started only four years ago (Chamovitz et al. 1991). Considering this short period remarkable progress has been made to engineer resistance against bleaching herbicides. Although a resistant transgenic crop cultivar is not available as yet, transgenic cyanobacteria and tobacco plants with a resistant target site for bleaching herbicides have already been generated. This article will first introduce the bleaching herbicides together with their target enzyme and then review different approaches leading to genetically altered resistant organisms.

1-56670-045-0/96/$0.00+$.50
© 1996 by CRC Press Inc.

2. COMPOUNDS AND TARGETS

The phytotoxic effect of bleaching herbicides is related to the function of the carotenoids in plants. These pigments are associated with membrane-bound chlorophyll–protein complexes of the photosynthetic apparatus. Besides of their function as accessory pigments in light harvesting, they are essential to protect against photosensitized reactions. When the concentration of colored carotenoids is decreased, e.g., in the presence of bleaching herbicides, their protective function cannot be maintained. A direct consequence is the photobleaching (or oxidation) of chlorophyll and a destruction of the photosynthetic membranes. In addition, a decreased carotenoid content limits the assembly of the photosynthetic apparatus, leading to decreased photosynthetic activity of the plant.

For all compounds listed with their chemical structures in Figure 1 a direct interaction with phytoene desaturase of plants or algae could be demonstrated by enzymological studies (Sandmann and Böger 1989; Mitchell and Bartlett 1990; Babczinski et al. 1990, 1992). *In vitro* investigations gave the same results as studies with intact plants: Increasing concentrations of bleaching herbicides decrease the formation of colored carotenoids and the colorless precursor phytoene accumulates. The herbicides norflurazon and fluridone are reversibly-binding inhibitors of phytoene desaturase as demonstrated by enzyme kinetics measurements (Sandmann et al. 1989; Kowalczyk-Schröder and Sandmann 1992). Quantitative structure activity investigations were carried out to optimize their inhibitory activity and also to find common structural elements in their molecules. These studies were reviewed recently (Sandmann and Böger 1992).

A special group of herbicidal compounds, the benzoyl cyclohexanediones, also interfere with phytoene desaturation when applied to growing plants (Mayonado et al. 1989). However, they differ from all the other bleaching herbicides because they neither directly interact with phytoene deaturase nor do they bleach algal or higher plant cell suspension cultures (Sandmann et al. 1990). Although their interaction with *p*-hydroxyphenylpyruvate dioxygenase, an enzyme in the pathway leading to quinones, was demonstrated recently (Schulz et al. 1993), the effect of benzoyl cyclohexanediones on the process of phytoene desaturation is still obscure because the presence of a quinone in phytoene desaturase has not been shown.

3. PHYTOENE DESATURASES: GENE AND ENZYME DIVERSITY

Genes of phytoene desaturase have been cloned and sequenced from several bacteria, the fungus *Neurospora crassa*, from cyanobacteria, a green alga, as well as from higher plant species such as soybean (see Sandmann 1994 for review). Differences in the catalytic action of various phytoene desaturases have been revealed by complementation of this enzyme from different organisms in *Escherichia coli,* which basically is noncarotenogenic. For this purpose, *E. coli* was cotransformed with two different plasmids. One carried all the genes necessary for the synthesis of the substrate phytoene and a second plasmid carrying a phytoene desaturase gene. The resulting *E. coli* transformants were pigmented and high-performance liquid chromatography analysis of accumulated carotenes was performed (Linden et al. 1991). It was shown that the reaction product of the *Synechococcus* phytoene desaturase (the product of the *pds* gene) is ζ-carotene. In contrast, the *Rhodobacter* enzyme introduced three conjugated double bonds into phytoene, yielding mainly neurosporene. In case of the *Erwinia* enzyme lycopene with four newly introduced double bonds was the major product. Very similar results were obtained by cell-free reactions in which phytoene was converted by membranes isolated from *E. coli* transformants carrying phytoene desaturase genes from these different organisms (Sandmann et al. 1993). Complementation of the phytoene desaturase gene from tomato resulted in accumulation of ζ-carotene in *E. coli* (Pecker et al. 1992), as observed in the case of the cyanobacterium. Figure 2 illustrates the desaturation sequence of the carotenoid biosynthetic pathway either in *Erwinia* or in higher plants.

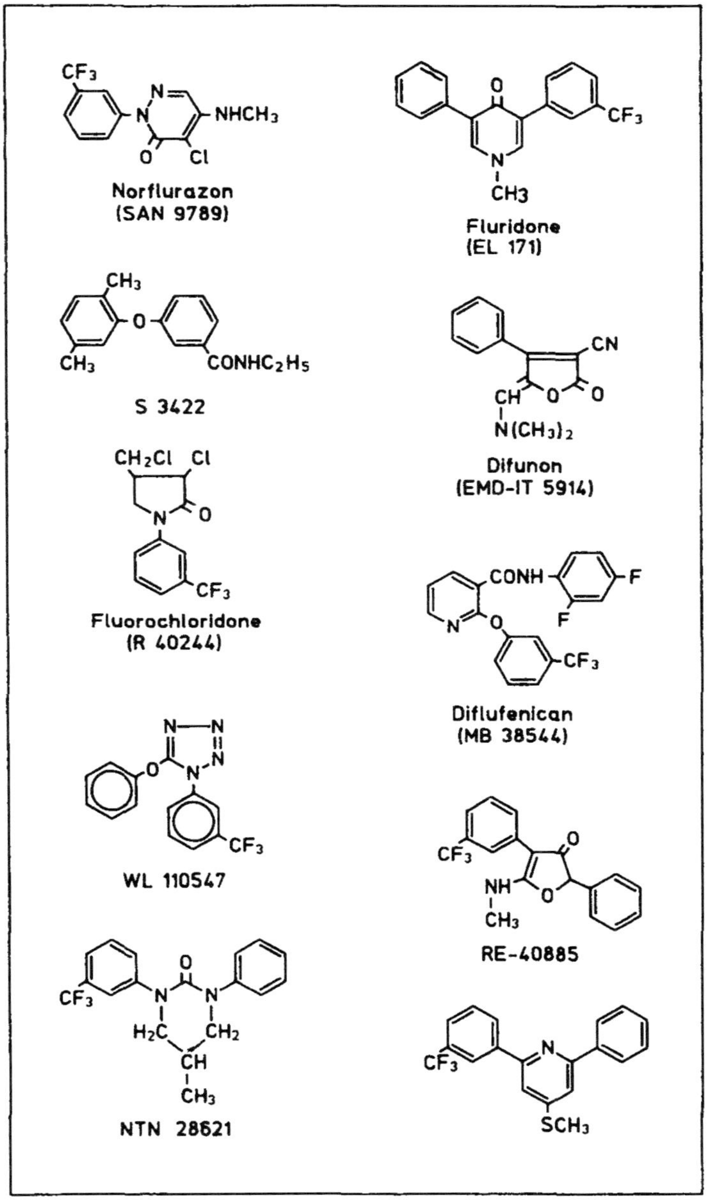

FIGURE 1 Chemical structures of different inhibitors which directly interact with the algal/plant-type phytoene desaturase PDS. No number of trivial name is available for the compound at bottom right. (see Babczinski et al. 1992.)

Comparison of the amino acid sequence deduced from the phytoene desaturase genes revealed a remarkably high homology among the polypeptides from a green alga, a cyanobacterium, or a higher plant (Pecker et al. 1992). All these plant-type phytoene desaturases (PDS, the products of the *psd* genes) share an amino acid identity of >65%. In contrast, homology to bacterial and fungal phytoene desaturases (the products of *crtI* genes) is negligible. Nevertheless, the fungal and bacterial phytoene desaturases (CRTI) are highly conserved, forming a second related group. These results demonstrate that during the course of evolution two completely different and unrelated phytoene desaturase proteins were acquired.

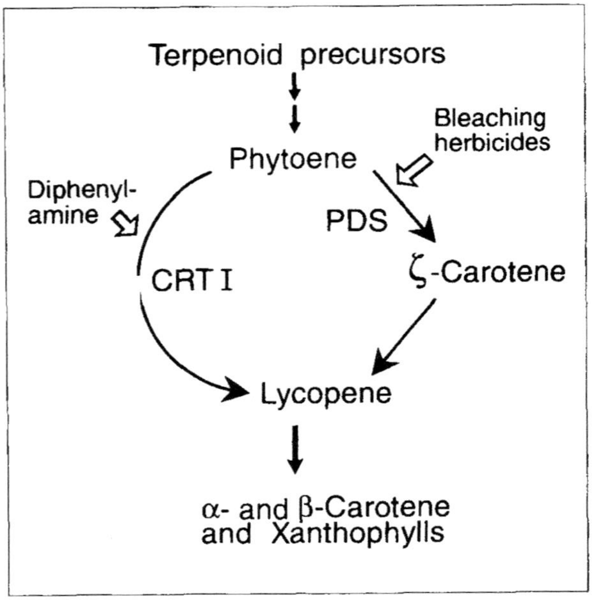

FIGURE 2 Alternative routes and enzymes of the desaturation sequence in the carotenoid biosynthetic pathway. CRTI, the phytoene desaturase found in bacteria and fungi, is inhibited by diphenylamine. The algal/higher plant-type PDS is the target of the bleaching herbicides of Figure 1.

The plant-type enzyme was isolated from *Capsicum* chromoplasts (Hugueney et al. 1992) and from the cyanobacterium *Synechococcus* (Fraser et al. 1993). The latter phytoene desaturase was purified after overexpression of its gene in *E. coli*. This 53-kDa membrane-bound protein could be reactivated after lipid replenishment. The cofactors for this cyanobacterial/algal/plant-type phytoene desaturase were either nicotinamide adenine dinucleotide or its reduced form, NADP. The phytoene desaturase from *Erwinia uredovora* was also overexpressed in *E. coli*, purified to homogeneity, and characterized after reactivation (Fraser et al. 1992). Inhibitor studies were carried out with both nonhomologous types of phytoene desaturases (Sandmann and Fraser 1993). I_{50} values were determined for fluridone, flurtamone, norflurazon, and diphenylamine. The first three herbicides inhibited only PDS, the algal/plant-type phytoene desaturase. The bacterial enzyme CRTI was unaffected. In contrast, CRTI was inhibited by diphenylamine, but PDS was not. This result demonstrates that bacterial phytoene desaturases are insensitive to the conventional bleaching herbicides. The *Erwinia* enzyme with a polypeptide composition completely different from the algal and plant phytoene desaturases is naturally resistant, but includes the catalytic function of the latter.

4. HERBICIDE RESISTANCE

Unicellular cyanobacteria such as *Synechococcus* and *Synechocystis* proved to be suitable organisms to study the mode of action of bleaching herbicides *in vivo* and *in vitro* (Sandmann and Böger 1989) and to carry out quantitative structure-activity investigations (Sandmann and Böger 1992). They were also used recently as convenient model species to select mutants resistant against bleaching herbicides and to investigate modifications of the target enzyme phytoene desaturase.

Resistance is manifested by the ability of cells to grow in the presence of comparatively high doses of herbicides. Resistance may originate from several possible changes in the organism caused by the mutation. The mutants either may have developed an exclusion mechanism for the herbicide or exhibit detoxification by enhanced degradation. In addition, a modification of the target enzyme may be the essential mechanism of resistance (Dyer et al. 1993). The latter possibility includes overproduction of the target enzyme or alterations of its amino acid sequence, leading to reduced efficiency of herbicide binding (Mullineaux 1992). In both cases resistance can be observed, e.g., in alfalfa cell suspension cultures containing overproduced glutamine synthetase which were 100-fold resistant to phosphinothricin (Donn et al. 1984), or in higher plants and algae in which the exchange of one or more amino acid(s) of the D1 protein leads to resistance against photosystem II inhibitors (Mets and Thiel 1989; Trebst 1991).

Various lines of mutants of the cyanobacterium *Synechococcus* (Linden et al. 1990; Chamovitz et al. 1993) and *Synechocystis* (Martínez-Férez et al. 1994) resistant against bleaching herbicides have been generated and selected. Several strains with different degrees of norflurazon resistance have been analyzed and the nature of the mutation has been characterized by combined *in vivo* and *in vitro* analysis of carotenogenesis in the presence of bleaching herbicides. Biochemical characterization of norflurazon interaction with phytoene desaturation showed similar inhibition properties of carotene biosynthesis *in vivo* and *in vitro* using four different mutants. This result indicates that in the mutants assayed the affinities for norflurazon binding to phytoene desaturase are decreased.

For the characterization of the cyanobacterial mutants, cellular I_{50} values have been compared to the K_i values and the dissociation constant of the enzyme-inhibitor complex measured *in vitro*. In the majority of the mutants similar K_i and I_{50} values were found which were all higher than those of the wild-type. This result excludes contribution by restricted uptake or metabolization of norflurazon in the mutants. As norflurazon is a reversibly bound inhibitor of phytoene desaturase (Sandmann et al. 1990), we conclude that affinities for norflurazon binding to phytoene desaturase are decreased in all our mutants. If protein overproduction were the reason for resistance in the mutants, the K_i values of the mutants were expected to be in the range of the value for the wild-type enzyme because K_i in the *in vitro* assays is independent of enzyme concentration.

In the four *Synechococcus* mutants selected against norflurazon the phenotype was related to specific point mutations in *pds*, the gene encoding phytoene desaturase (Chamovitz et al. 1993). Their resistance was manifested in a cell-free carotenogenic assay. This finding demonstrated that the predicted amino acid changes in the target protein alter the enzyme-herbicide interactions. Besides their strong resistance to norflurazon (this was the original selection compound), each of the mutant strains exhibited a unique cross-resistance profile to fluridone and fluorochloridone. The characteristics of these mutants are summarized in Table 1.

Except for mutant FD5 the K_i values corresponded to the *in vivo* I_{50} values in all cases, indicating that this resistance is a direct consequence of the enzyme-inhibitor interaction. In mutant FD5 resistance was only evident *in vivo*. In this case genetic analysis revealed a deletion in the promoter region that contains the putative -35 and -10 transcription-regulating elements. Resistance in this strain could be a result of overexpression of PDS, the target protein. The amount of PDS in cells of FD5, as determined with an antiserum, was at least 20-fold higher than in cells of the wild-type strain. Another *Synechococcus* mutant selected against fluridone, FRD3U, also showed resistance *in vivo*, but not in the *in vitro* assay (Sandmann and Fraser 1993). As in FD5 this may be another example for overexpression of phytoene desaturase.

More resistant *Synechococcus* mutants were selected against fluridone and another bleaching herbicide, flurtamone, including two double mutants (Sandmann and Fraser 1993). They all exhibited the same degree of resistance *in vivo* and *in vitro*. Thus far, they have not been characterized at the gene level. Their resistance pattern is quite diverse. Multiresistance for single mutants against different bleaching herbicides was found and is specified in Table 2.

Mutants resistant against norflurazon were also selected from another cyanobacterium, *Synechocystis* (Martínez-Férez et al. 1994). *In vitro* analysis of phytoene desaturation demonstrated that they all possess a resistant enzyme. The results of DNA sequencing of the mutated *pds* genes detected a single point mutation in all the mutants modifying the same amino acids at position 195 of the polypeptide chain from arginine to serine, to cysteine, or to proline. The degree of resistance

TABLE 1 Resistance Patterns of Genetically Characterized Cyanobacterial Mutants Selected Against Norflurazon or Fluridone[a]

Mutants	Amino Acid Exchange	RF[b] Against Selecting Herbicide	Cross-Resistance
A. Synechococcus PCC7942			
NFZ4	$Val^{403} \rightarrow Gly$	24	FCD, FRD
NFZ20	$Leu^{320} \rightarrow Pro$	20	FCD, FRD
NFZ42	$Arg^{195} \rightarrow Pro$	40	FCD
NFZ49	$Leu^{436} \rightarrow Arg$	70	—
FD5	Promoter[c]	24	NFZ
B. Synechocystis PCC6803			
AV1	$Arg^{195} \rightarrow Ser$	240	—
AV3	$Arg^{195} \rightarrow Cys$	188	—
AV4	$Arg^{195} \rightarrow Pro$	29	FCD, FRD

Abbreviation of bleaching herbicides: NFZ, norflurazon; FCD, fluorochloridone; FRD, fluridone.

[a] Only in the case of FD5; all others were selected against norflurazon.
[b] RF = resistance factor, based on cellular I_{50} values.
[c] In this mutant the promoter is affected, resulting in overproduction of enzyme.

Data compiled from Chamovitz et al. 1993 and Martínez-Férez et al. 1994.

determined via carotenoid formation in cells as well as by cell-free phytoene desaturation in the presence of norflurazon was found to be high, with resistance factors from 25 to 240. Cross-resistance toward other bleaching herbicides such as fluridone, flurtamone, or fluorochloridone was only moderate (Table 2). The different patterns of herbicide cross-resistance found for the mutants listed in Tables 1 and 2 indicate that the inhibitors of phytoene desaturase may be attached to the same binding region but interact with different amino acids.

There are two possible mechanisms by which the amino acid substitutions in PDS could cause herbicide resistance. It is possible that the mutation alters the amino acid which interacts directly with the herbicides in their binding sites. Alternatively, the amino acid substitutions could induce allosteric effects by disturbing the original conformation of the enzyme protein, which may also lead to deformation of the binding niche. In this case this alteration of the protein structure could also lead to a concurrent decrease of enzyme activity. For example, resistance of 5-enolpyruvylshikimate-3-phosphate synthase obtained by site-directed mutagenesis not only increased the K_i value for glyphosate but also raised the K_m value for the substrate phosphoenolpyruvate (Kishore and Shah 1988; Barry et al. 1992). A similar effect on phytoene desaturase was observed in several of the mutants. A reduction in PDS activity was found to lead to an increase of the steady-state levels of phytoene and a decrease in total colored carotenoid content. However, the *Synechococcus* AV mutants of Table 1 exhibited only a minor impairment of carotenoid formation and low amounts of accumulating phytoene. It was concluded that the targetted amino acid arginine at position 195 may be part of a specific binding site for norflurazon. In all cyanobacterial/higher plant-type phytoene desaturases investigated to date this arginine is highly conserved. It is positioned in the middle of a stretch of nine perfectly conserved amino acids, a region which is different both from the proposed carotene binding site and from the putative cofactor binding site (Pecker et al. 1992). A two-dimensional computer prediction of the helix structure indicates that Arg^{195} is positioned between two isoleucines close to the end of the helix. The other side of the helix is very hydrophobic.

TABLE 2 Cross-Resistance of *Synechococcus* Mutants Selected Against Flurtamone and Fluridone

Mutants	FTM	FRD	NFZ	FCD
FTM1E	(+)	+	—	+
FTM1X	(+)	+	n.d.	n.d.
FTM2U	(+)	+	+	+
FTM4E	(+)	+	+	+
FRD1U	+	(+)	+	—
FRD1X	+	(+)	—	—

(+) resistance with a factor of >2; parentheses indicate that the herbicide mentioned was used for selection; n.d., not determined. FRD, fluridone; FTM, flurtamone; FCD, fluorochloridone; NFZ, norflurazon.

Data compiled from Sandmann and Fraser 1993.

5. GENETIC ENGINEERING OF RESISTANT *SYNECHOCOCCUS*

Different strategies can be followed to genetically engineer herbicide resistance. Besides the introduction of genes coding for detoxifying enzymes, transformation with genes of resistant target enzymes for the corresponding herbicide is a feasible alternative (Mullineaux 1992). This approach was used successfully to obtain herbicide-resistant plants whose target enzymes are involved in amino acid biosynthesis (see Masur and Falco 1989; Dyer et al. 1993 for review). In the case of bleaching herbicides a bacterial carotenoid biosynthesis gene naturally resistant to the herbicides was introduced into cyanobacteria and tobacco plants, and functionally expressed in these hosts.

Resistant cyanobacteria were produced by inserting a foreign gene, *crtI* coding for phytoene desaturase of the bacterium *Erwinia uredovora* (Misawa et al. 1990), whose product is naturally resistant to the herbicide. This gene was integrated into the chromosome of *Synechococcus* PCC7942 and expressed successfully under the control of different promoters in this cyanobacterium (Windhövel et al. 1994). For expression of *crtI* in the heterologous host, constructs with different promoters were introduced into a special *Synechococcus* strain which carries a so-called "integration platform" in its chromosome. The location of the platform is the *metF* gene, involved in methionine biosynthesis in this organism. Provided the medium is supplemented with methionine, *metF* can be inactivated without affecting the viability of the strain. Due to the effective recombination system of *Synechococcus* PCC7942, any fragment cloned into a pBR-derived vector can integrate into this platform as the result of a "double cross-over" event.

The *Synechococcus* transformant with the best resistance against bleaching herbicides was constructed in the following way, The integration plasmid contained the *crtI* gene from the *crt* gene cluster of *E. uredovora* (Misawa et al. 1990) downstream of the *nptII* gene of transposon Tn5, including the promoter (Beck et al. 1982). The *nptII* gene confers kanamycin resistance into a wide variety of Gram-negative bacteria (Berg and Berg 1983). As there is no transcriptional stop signal downstream of *nptII* (Putnoky et al. 1983) the latter construct was expected to provide cotranscription of *nptII* and *crtI* under the control of the *nptII* promoter. This plasmid was transferred into the *Synechococcus* integration platform resulting in the transformant R2-PIM8-BG1. A schematic map of the integration is presented in Figure 3.

After the correct integration of the construct was checked by Southern hybridization, expression of the integrated *crtI* gene in the transformants was demonstrated by Western blot analysis with cell extracts of *Synechococcus* transformants. Comparison of the intensity of the immunoreaction

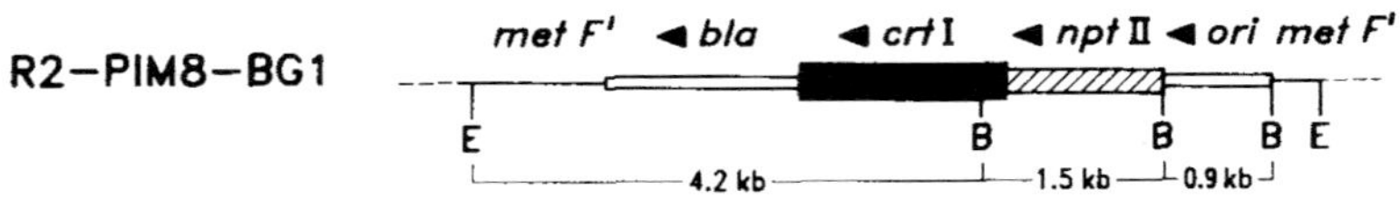

FIGURE 3 Schematic presentation of the platform region of *Synechococcus* R2-PIM8-BG1 after insertion of the *Erwinia crtI* gene construct into the *metF* gene involved in methionine biosynthesis of *Synechococcus* PCC7942 (Windhövel et al. 1994). The sizes of fragments are given that hybridize to the integration plasmid in the Southern blot analysis. Abbreviations: *bla*, gene coding for beta-lactamase, conferring ampicillin resistance; *crtI*, gene coding for phytoene desaturase of *Erwinia uredovora*; *nptII*, gene for kanamycin resistance; origin of replication.

of this protein from several transformants showed that the content of CRTI was highest in strain R2-PIM8-BG1.

Nontransformed *Synechococcus* showed substantial *in vitro* phytoene desaturation activity (Table 3). More than 20% of the radioactivity of the substrate was converted into ß-carotene during a 2-h incubation period. Labeling of intermediate carotenes was negligible. In parallel experiments the same reaction was carried out with 1 μM of norflurazon, which is 10-fold higher than the I_{50} value for inhibition of the endogenous phytoene desaturase (Sandmann et al. 1989). In the strain without the *Erwinia crtI* gene inhibition was almost complete under these conditions. For the transformant strain R2-PIM8-BG1 the conversion of phytoene to ß-carotene was higher compared to the control, as indicated by the conversion rate. The contribution of CRTI activity to the overall activity of the endogenous and newly introduced phytoene desaturase can be estimated by comparing the activity with and without norflurazon. The conversion rate for phytoene desaturation in R2-PIM8-BG1 was decreased by norflurazon from 36 to 12%, indicating that about one third of the activity can be attributed to the foreign *Erwinia* enzyme CRTI.

Analyses of herbicide resistance with transformant cultures showed that their content of colored carotenoids was unaffected up to a concentration of 10^{-6} *M* and only a slight bleaching of the culture was observed with 10^{-5} *M*. The control strain, on the other hand, was already slightly bleached at a concentration of 10^{-7} *M* and totally bleached at a concentration of 3×10^{-7} *M*. The factor of norflurazon resistance in this *Synechococcus* transformant can be estimated in the order of about 1000.

6. GENETIC ENGINEERING OF RESISTANT TOBACCO

The approach described in the previous chapter to engineer resistance against bleaching herbicides by introduction of a bacterial *crtI* gene was first realized with tobacco (Misawa et al. 1993). As the phytoene desaturase of tobacco is located in the thylakoid membranes (Linden et al. 1993), the 5′ region of the phytoene desaturase gene *crtI* from *Erwinia uredovora* was fused to the sequence for the transit peptide of a pea Rubisco small subunit, and put under the control of the cauliflower mosaic virus (CMV) 35S promoter in tobacco plants. The genetic manipulation leading to the construction of vector pYPIET4 was introduced into tobacco via *Agrobacterium* as outlined and explained in Figure 4. Several regenerated tobacco plants transformed with plasmid pYPIET4 were isolated and analyzed. The chimeric gene product was expressed and processed in the transgenic plants; the production and processing of the corresponding protein could be demonstrated by Western blotting. Immunogold localization showed targeting in the chloroplasts and location of the gene product CRTI in the thylakoids (Misawa et al. 1993).

TABLE 3 ***In Vitro*** **Phytoene Desaturase Activity of** ***Synechococcus*** **Transformants**

Strains, Conditions	Radioactivity (dpm) in Phytoene	Radioactivity (dpm) in ß-Carotene	Conversion Rate (%)[a]
Control, wild-type	16738	4602	21.6
Control + NFZ	19320	112	0.6
R2-PIM8-BG1	13497	7477	35.7
R2-PIM8-BG1 + NFZ	17637	2360	11.8

Note: [^{14}C-Phytoene was generated *in vitro* from [^{14}C]-mevalonate with extracts from *Phycomyces* mutant C5. Incubation time was 2 h. norflurazon (NFZ), 1 μ*M*, was added as indicated to inhibit the endogenous enzyme.

[a] Conversion rate, 100 x (dpm found in ß-carotene/dpm in phytoene + dpm in ß-carotene).

Functionality of the novel CRTI phytoene desaturase in the tobacco leaves was established by *in vitro* experiments. The existence of this enzyme in the transgenic tobacco plants resulted in enhanced activity for desaturation of phytoene and its transformation to ß-carotene. Thus, this bypass of the ζ-carotene pool (see Figure 2) indicated high functionality of the produced novel phytoene desaturase enzyme in the tobacco transformants.

Elevated resistance against norflurazon was observed in the transgenic tobacco plants during growth on media containing norflurazon, or on soil watered with a norflurazon solution. This resistance was confirmed to be dominantly inherited to the progeny. Further selection with a greater number of transformants was successfully carried out to obtain tobacco plants with a much higher degree of resistance against bleaching herbicides by isolating a greater number of tobacco transformants and by screening them on the media containing 3 μ*M* norflurazon. The plants shown in Color Plate 5* were either grown mixotrophically on a medium containing sucrose and 3 μ*M* fluridone (A) or in soil and watered with a 3 μ*M* norflurazon solution (B). Plants at the right of the plate are nontransformed controls. The *crtI* gene fusion construct with the transit sequence was introduced into the plants at the left. Both herbicides affect the control plant, causing the bleaching symptoms during growth when the nutrients are provided. However, when the plants rely on an intact and functioning photosynthetic apparatus, killing of the plant was observed, as the final consequence of the herbicide action (B). In contrast, the herbicide-treated transgenic plants at the left looked completely unaffected and grew with their full pigment inventory. These transgenic tobacco plants showed resistance not only to norflurazon and fluridone but also to various other bleaching herbicides (Figure 1), interfering with the higher plant-type phytoene desaturase (Misawa et al. 1994). This is the first example of a higher plant with a genetically manipulated carotenoid biosynthetic pathway, which gives rise to resistance against bleaching herbicides. The method to obtain the bleaching herbicide-resistant plants shown here is based on a novel concept, namely the introduction of a gene which has evolved naturally without an artificial selection agent, which encodes no inhibition site for a bleaching compound, and which produces a functional gene product comparable in its catalytic activity with the endogenous plant-type PDS. This strategy is an alternative used for photosynthesis inhibitors or for glyphosate. Here, generally, genes of resistant mutants or mutagenized genes were cloned, yielding enzymes with unfavorable reaction characteristics (Erickson et al. 1989; Padgette et al. 1991), leading to reduced growth and fitness of the transformants (Gressel and Segel 1990).

* Color plates follow page 84.

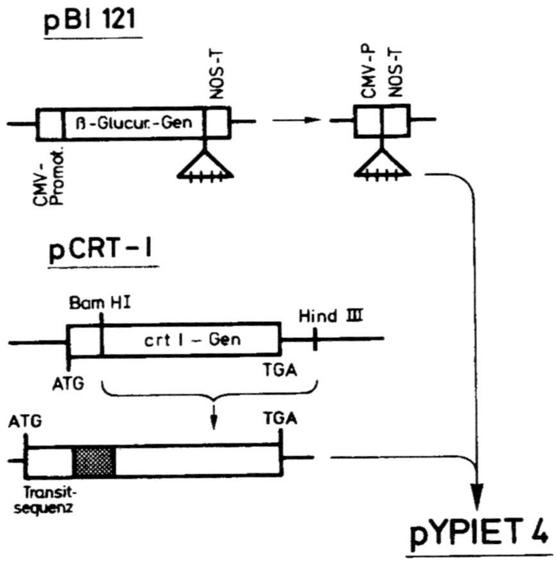

FIGURE 4 Construction of the vector pYPIET4 for transformation of tobacco which is resistant against norflurazon (Misawa et al. 1993). The transit sequence originated from the small subunit of pea ribulose-1,5-bisphosphate carboxylase. It was fused to the truncated *crtI* reading frame by insertion of a synthetic oligonucleotide (dotted section). The extended *crtI* coding region is surrounded by the cauliflower mosaic virus 35S promoter (CMV-P) and the nopalin synthase terminator (NOS-T) taken from the binary vector pBI121.

7. CONCLUDING NOTES

At the moment studies are underway to transform tobacco with a resistant (plant-type) *pds* gene coding for a phytoene desaturase with an altered amino acid sequence similar to the ones listed in Table 1 (Hirschberg, personal communication). Also, this approach resulted in tobacco plants resistant against norflurazon. Thus, there are different strategies available to genetically manipulate phytoene desaturation and make the plant resistant to bleaching herbicides. These studies on the target enzyme for bleaching herbicides were carried out so far only with tobacco as a model. The successful cloning experiments, however, demonstrate that biochemistry and genetics of carotenogenesis as well as the corresponding technology have made sufficient progress to genetically engineer resistance against bleaching herbicides into various crop plants.

Up to now, engineering of herbicide-resistant crop plants has focused on inhibitors of nitrogen and amino acid metabolism such as glufosinate, glyphosate, or inhibitors of acetolactate synthase. Some work is carried out with photosynthesis herbicides (see Mullineaux 1992 for review). Our example is the first using a target of carotenoid biosynthesis, for which a remarkable array of potent bleaching inhibitors are available. Most of them exhibit a broad weed spectrum, but their nonselectivity generally has prevented development into commercial herbicides. Transgenic resistant crops, however, may make bleaching compounds promising candidates as complementary herbicides.

REFERENCES

Babczinski, P., Blunck, M., Sandmann, G., Schmidt, R., Shiokawa, K., and Yasui, K. 1990. Substituted tetrahydropyrimidinones, a new herbicidal class of compounds inducing chlorosis by inhibition of phytoene desaturation. *Pestic. Sci.* 30: 339.

Babczinski, P., Heinemann, U., Sandmann, G., Kawamura, S., Hamada, T., Sato, R., and Sanemitsu, Y. 1992. Inhibition of carotenoid biosynthesis by interaction of 2,6-diphenylpyridine derivatives with phytoene desaturation. *J. Agric. Food Chem.* 40: 2497.

Barry, G., Kishore, G., Padgette, S., Taylor, M., Kolacz, K., Weldon, M., Re, D., Eichholtz, D., Fincher, K., and Hallas, L. 1992. Inhibitors of amino acid biosynthesis: strategies for imparting glyphosate tolerance to crop plants. In *Biosynthesis and Molecular Regulation of Amino Acids in Plants.* Ed. by Singh, B.K., Fores, H.E., and Shannon, I.C. American Society of Plant Physiologists, Rockville, MD, p. 139.

Beck, E., Ludwig, G., Auerswald, E.A., Reiss, B., and Schaller, H. 1982. Nucleotide sequence and exact localization of the neomycin phosphotransferase gene from transposon Tn5. *Gene* 19: 327.

Berg, D.E. and Berg, C.M. 1983. The prokaryotic transposable element Tn5. *Biotechnology* 1: 417.

Chamovitz, D., Pecker, I., and Hirschberg, J. 1991. The molecular basis of resistance to the herbicide norflurazon. *Plant Mol. Biol.* 16: 967.

Chamovitz, D., Sandmann, G., and Hirschberg, J. 1993. Molecular and biochemical characterization of herbicide-resistant mutants of cyanobacteria reveals that phytoene desaturation is a rate-limiting step in carotenoid biosynthesis. *J. Biol. Chem.* 268: 17348.

Cramp, M.C., Gilmour, J., Hatton, L.R., Hewett, R.H., Nolan, C.J., and Parnell, E.W. 1987. Design and synthesis of *N*-(2-4,-difluorophenyl)-2-(3-trifluoromethylphenoxy)3-pyridine-carboxamine (diflufenican), a novel pre- and early post-emergence herbicide for use in winter cereals. *Pestic. Sci.* 18: 15.

Donn, G., Tischer, E., Smith, J.A., and Goodmann, H.M. 1984. Herbicide-resistant alfalfa cells: an example of gene amplification in plants, *J. Mol. Appl. Genet.* 2: 621.

Dyer, W.E., Hess, F.D., Holt, J.S., and Duke, S.O. 1993. Potential benefits and risks of herbicide-resistant crops produced by biotechnology. In *Horticultural Reviews,* Vol. 15. Ed. by Janick, J. John Wiley & Sons, New York, chap. 9.

Erickson, J.M., Pfister, K., Rahire, M., Togasaki, R.K., Mets, L., and Rochaix, J.-D. 1989. Molecular and biophysical analysis of herbicide-resistant mutants of *Chlamydomonas reinhardtii:* structure function relationship of the photosystem II D1 polypeptide. *Plant Cell* 1: 361.

Fraser, P.D., Linden, H., Sandmann, G. 1993. Purification and reactivation of recombinant *Synechococcus* phytoene desaturase from an overexpressing strain of *E. coli. Biochem. J.* 291: 687.

Fraser, P.D., Misawa, N., Linden, H., Yamano, S., Kobayashi, K., and Sandmann, G. 1992. Expression in *E. coli*, purification and reactivation of the recombinant *Erwinia uredovora* phytoene desaturase. *J. Biol. Chem.* 267: 19891.

Gressel, J. and Segel, L.A. 1990. Modelling the effectiveness of herbicide rotations and mixtures to delay or preclude resistance. *Weed Technol.* 4: 186.

Hugueney, P., Römer, S., Kuntz, M., and Camara, B. 1992. Characterization and molecular cloning of a flavoprotein catalyzing the synthesis of phytoene and zeta-carotene in *Capsicum* chromoplasts. *Eur. J. Biochem.* 209: 399.

Kishore, G.M. and Shah, D.M. 1988. Amino acid biosynthesis inhibitors as herbicides. *Annu. Rev. Biochem.* 57: 627.

Kowalczyk-Schröder, S. and Sandmann, G. 1992. Interaction of fluridone with phytoene desaturation of *Aphanocapsa. Pestic. Biochem. Physiol.* 42: 7.

Linden, H., Lucas, M.M., de Felipe, M.R., and Sandmann, G. 1993. Immunogold localization of phytoene desaturase in higher plant chloroplasts. *Physiol. Plant.* 88: 229.

Linden, H., Misawa, N., Chamovitz, D., Pecker, I., Hirschberg, J., and Sandmann, G. 1991. Functional complementation in *Escherichia coli* of different phytoene desaturase genes and analysis of accumulated carotenes. *Z. Naturforsch.* 46c: 160.

Linden, H., Sandmann, G., Chamovitz, D., Hirschberg, J., and Böger, P. 1990. Biochemical characterization of *Synechococcus* mutants selected against the bleaching herbicide norflurazon. *Pestic. Biochem. Physiol.* 36: 46.

Martínez-Férez, I.M. and Vioque, A. 1992. Nucleotide sequence of the phytoene desaturase gene from *Synechocystis sp.* PCC6803 and characterization of a new mutation which confers resistance to the herbicide norflurazon. *Plant Mol. Biol.* 18: 981.

Martinez-Férez, I., Vioque, A., and Sandmann, G. 1994. Mutagenesis of an amino acid responsible in phytoene desaturase from *Synechocystis* for binding of the bleaching herbicide norflurazon. *Pestic. Biochem. Physiol.* 48: 185.

Masur, B. J. and Falco, S.C. 1989. The development of herbicide-resistant crops. *Annu. Rev. Plant Physiol.* 40: 441.

Mayonado, D.J., Hatzios, K.K., Orcutt, D. M., and Wilson, H.P. 1989. Evaluation of the mechanism of action of the bleaching herbicide SC-0051 by HPLC analysis. *Pestic. Biochem. Physiol.* 35: 138.

Mets, L. and Thiel, A. 1989. Biochemistry and genetic control of the photosystem-II herbicide target site. In *Target Sites of Herbicide Action.* Ed. by Böger, P. and Sandmann, G. CRC Press, Boca Raton, FL, p. 1.

Misawa, N., Nakagawa, M., Kobayashi, K., Yamano, S., Izawa, Y., Nakamura, K., and Harashima, K. 1990. Elucidation of the *Erwinia uredovora* carotenoid biosynthetic pathway by functional analysis of gene products expressed in *Escherichia coli. J. Bacteriol.* 172: 6704.

Misawa, N., Yamano, S., Linden, H., de Felipe, M.R., Lucas, M., Ikenaga, H., and Sandmann, G. 1993. Functional expression of the *Erwinia uredovora* carotenoid biosynthesis gene *crtI* in transgenic plants showing an increase of β-carotene biosynthesis activity and resistance to the bleaching herbicide norflurazon. *Plant J.* 4: 833.

Misawa, N., Masamoto, K., Hori, T., Ohtani, T., Böger, P., and Sandmann, G. 1994. Expression of an *Erwinia* phytoene desaturase gene not only confers multiple resistance to herbicides interfering with carotenoid biosynthesis but also alters xanthophyll metabolism in transgenic plants. *Plant J.* 6: 481.

Mitchell, G. and Bartlett, D.L. 1990. 2-(3-Trifluoromethylbenzyl)-3-ketomorpholine — a new phytoene desaturase inhibitor: optimisation of *in vitro* activity using computer modelling. *7th Int.Congr. Pestic. Chem. Abstr.* (Hamburg) 1: 170.

Mullineaux, P.M. 1992. Genetically engineered plants for herbicide resistance. In *Plant Genetic Manipulation for Crop Protection*. Ed. by Gatehouse, A.M.R., Hilder, V.A., and Boulter, D. C.A.B. International, Wallingford, U.K., chap. 4.

Padgette, S.R., Biest, R.D., Gasser, C.S., Eichholtz, D.A., Frazier, R.B., Hironaka, C.M., Levine, E.B., Shah, D.M., Fraley, R.T., Kishore, G.M. 1991. Site-directed mutagenesis of a conserved region of the 5-enolpyruvylshikimate-3-phosphate synthase active site. *J. Biol. Chem.* 266: 22364.

Pecker, I., Chamovitz, D., Linden, H., Sandmann, G., and Hirschberg, J. 1992. A single polypeptide catalyzing the conversion of phytoene to ζ-carotene is transcriptionally regulated during tomato fruit ripening. *Proc. Natl. Acad. Sci. U.S.A.* 89: 4962.

Putnoky, E., Kiss, G.B., Ott, I., Kondorosi, A. 1983. Tn5 carries a streptomycin resistance determinant downstream from the kanamycin resistant gene. *Mol. Gen. Genet.* 191: 288.

Sandmann, G. 1994. Carotenoid biosynthesis in microorganisms and plants. *Eur. J. Biochem.* 223: 7.

Sandmann, G. and Böger, P. 1989. Inhibition of carotenoid biosynthesis by herbicides. In *Target Sites of Herbicide Action*. Ed. by Böger, P. and Sandmann, G. CRC Press, Boca Raton, FL. p. 25.

Sandmann, G. and Böger, P. 1992. Chemical structure and activity of herbicidal inhibitors of phytoene desaturase. In *Rational Approaches to Structure, Activity, and Ecotoxicology of Agrochemicals*. Ed. by Draber, W. and Fujita, T. CRC Press, Boca Raton, p. 357.

Sandmann, G. and Fraser, P.D. 1993. Differential inhibition of phytoene desaturase from diverse origins and analysis of resistant cyanobacterial mutants. *Z. Naturforsch.* 48c: 267.

Sandmann, G., Kumita, I., and Böger, P. 1990. Untypical inhibition of phytoene desaturation by 2-(4-chloro-2-nitrobenzoyl)-5,5-dimethylcyclohexane-1,3-dione. *Pestic. Sci.* 30: 353.

Sandmann, G., Linden, H., and Böger, P. 1989. Enzyme-kinetic studies on the interaction of norflurazon with phytoene desaturase. *Z. Naturforsch.* 44c: 787.

Sandmann, G., Fraser, P.D., and Linden, H. 1992. Diversity of phytoene desaturating enzymes and corresponding genes involved in carotenoid biosynthesis of photoautotrophic prokaryotes. In *Research in Photosynthesis*, Vol. 3. Ed. by Murata, N. Kluwer Academic, Amsterdam, p. 51.

Schulz, A., Ort, O., Beyer, P., and Kleinig, H. 1993. SC-0051, a 2-benzoyl-cyclohexane-1,3-dione bleaching herbicide, is a potent inhibitor of the enzyme *p*-hydroxyphenylpyruvate dioxygenase. *FEBS Lett.* 318: 162.

Trebst, A. 1991. The molecular basis of resistance of photosystem II herbicides. In *Herbicide Resistance in Weeds and Crops*. Ed. by Caseley, J.C., Cussans, G.W., and Atkin, R.K. Butterworth-Heinemann, Oxford, p. 145.

Windhövel, U., Geiges, B., Sandmann, G., and Böger, P. 1994. Expression of *Erwinia uredovora* phytoene desaturase in *Synechococcus* PCC7942 leading to resistance against a bleaching herbicide. *Plant Physiol.* 104: 119.

Chapter 13

POSSIBLE PLEIOTROPIC EFFECTS IN HERBICIDE-RESISTANT CROPS

Carol Mallory-Smith
Charlotte V. Eberlein

CONTENTS

1. INTRODUCTION

Herbicide-resistant crops (HRCs) will find grower acceptance if there are no detrimental effects of the resistance gene(s) on plant growth, quality, and/or yield of the crop. Decreased quality or yield may be justified if there is no alternative for weed control and the weed problem is so serious that the crop cannot otherwise be grown profitably. The resistance trait must remain stable over time and in a multitude of environments. This chapter deals with crops that have been modified for resistance to specific herbicides. Resistance in this context is defined as the ability of a biotype to survive a treatment to which the wild-type was susceptible.

1-56670-045-0/96/$0.00+$.50
© 1996 by CRC Press Inc.

Herbicide resistance in crops can be achieved by: (1) altering the target site so that the herbicide no longer binds, (2) overexpressing a target enzyme so that the effect of the herbicide is overcome, or (3) detoxifying the herbicide so that it is no longer lethal to the plant (Mazur and Falco 1989). Resistant crops have been produced through traditional breeding, cell and tissue culture selection, mutation, and transformation (Dyer et al. 1993).

In order for traditional breeding to produce HRCs, genes for resistance must be available in crossing material that is compatible with a crop species. This requirement has been the limiting factor for producing HRCs with naturally occurring genes. Although resistant biotypes have been selected in over 100 weed species (LeBaron 1991), few of these species are closely related to any important crops (National Research Council 1989).

Cell and tissue culture selection and mutation breeding have produced most of the HRCs. The new technology of transgenic transformation provides promise for future production of HRCs. Genes can be moved from unrelated plant species, bacteria, viruses, and animals, and there is the possibility of producing combinations of genes and phenotypes which are outside the direct experience of traditional breeding (Dale et al. 1993). Whichever approach is taken to obtain a herbicide-resistant plant, a new genotype is produced, and this genotype carries genetic material that distinguishes it from the wild-type.

Pleiotropy is the phenomenon whereby a change at one genetic locus can bring about a variety of apparently unconnected phenotypic changes (Dawkins 1982). Pleiotropic effects of non-neutral mutations are probably the rule rather than the exception and should be expected from the complex way in which development occurs. Most non-neutral mutations presumably have several pleiotropic effects, and most mutational effects are detrimental. A gene that replaces another is usually less adaptive because it has some physiological disadvantages relative to the original gene (Haldane 1960). Therefore, it would be expected that HRCs could carry some penalty associated with the resistance trait. Reports of HRC performance vary, and although pleiotropic responses do occur, they have not been documented in all cases. When pleiotropic responses have been observed, they have not always been detrimental to field performance of the crop. This chapter reviews only reports on HRCs and not herbicide-resistant weeds, which have been reviewed extensively elsewhere (Powles and Holtum 1994).

2. TRIAZINE-RESISTANT CROPS

The first naturally occurring herbicide resistance was to the photosynthetic inhibiting triazine herbicides. In most cases, resistance is due to a point mutation in the chloroplast *psbA* gene resulting in a substitution of glycine for serine at residue 264 (Mazur and Falco 1989). This mutation prevents binding of the herbicide, but also results in reduced photosynthetic productivity. There are numerous examples of pleiotropic effects of triazine resistance in weeds and crops; only a few are discussed.

Triazine-resistant canola (*Brassica napus* L.) was produced by crossing a naturally occurring triazine-resistant *B. campestris* with *B. napus*. The resulting cross carried resistance to triazines but also resulted in a 20 to 30% yield penalty (Beversdorf et al. 1988). Triazine-resistant canola is an example of the acceptance of decreased yield in exchange for weed control. Mustard weed infestations were so serious that only triazine-resistant canola could be grown on thousands of acres in Canada (McKinlay, personal communication). Once a selective herbicide was developed, the resistant crop was no longer a viable alternative.

In a competition study using nearly isogenic lines of canola there was a decrease in fresh and dry weight, in the number and weight of the seed pods, and in seed yield of the resistant plants compared to the susceptible plants (Gressel and Ben-Sinai 1985). Triazine resistance in nearly isonuclear lines of canola led to reduced quantum yield and a decreased maximum rate of photosynthesis (Jursinic and Pearcy 1988). Hart and Stemler (1990) proposed that the decrease in efficiency of Photosystem II was due to a greater sensitivity to photoinhibition.

Although the majority of studies with triazine-resistant and -susceptible canola indicate that triazine-resistant biotypes are less fit than the wild-types, these fitness differences may be related to environmental conditions or stage of development. At 15°C resistant and susceptible plants exhibited similar carbon assimilation rates, while at 25° and 35°C the susceptible plants assimilated more carbon (Dekker and Sharkey 1992). At the reproductive stage, resistant plants assimilated more carbon than the susceptible plants, except later in the day when light was more limiting. Under high light intensity triazine-susceptible canola isoline plants produced 30 to 40% more dry weight than did the resistant isoline plants, and resource allocation differed between susceptible and resistant plants (Hart et al. 1992). However, under low light intensities little difference was found in dry weight accumulation or resource allocation between the resistant and susceptible plants. The ability to utilize a niche in which the resistant plant does not have to compete as readily with a susceptible plant may be advantageous for the survival and reproduction of a biotype, but crops do not require this type of adaptive advantage under field conditions.

Forcella (1987) evaluated the yield penalty and weed threshold associated with triazine-resistant canola in competition with wild oat. Without a herbicide application, the resistant biotype produced less biomass and 10 to 20% less yield than the susceptible biotype produced under competitive conditions where wild oat density was 0 to 128 plants per square meter. However, when the wild oat density was >32 plants per square meter applying the herbicide to the resistant plants resulted in a yield higher than the susceptible plants yielded under the same competition without a herbicide application. Thus, the detrimental effects of a herbicide-resistant gene may be overcome if the weed problem is severe enough and there is sufficient economic return.

Triazine-resistant foxtail millet [*Setaria italica* (L.) Beauv.] was produced by crossing it with its herbicide-resistant weed relative *S. viridis* (L.) Beauv. (Darmency and Pernes 1985). Under field conditions resistant plants yielded less than susceptible plants and seed harvested from field grown plants had 50% less total germination than susceptible plants (Darmency and Pernes 1989). The authors did not observe this difference in germination in greenhouse or garden experiments and suggested that the difference might have been due to delayed maturation or embryo mortality specific to the field or harvest conditions. Such differences would have profound economic consequences for the producer. Under field conditions, even a few days' delay in crop maturity can cause major production problems for growers, and embryo mortality would make the crop unsalable for seed. Because no differences were measured in early development under cool temperatures (9°C minimum and 18°C maximum) in growth chamber experiments, the authors concluded that the growth penalty that occurred at later growth stages was possibly due to higher temperatures in the field.

Because fitness differences between triazine-susceptible and -resistant biotypes can be linked directly to the mutation in the herbicide binding protein, it is unlikely that HRCs that contain this mutation will perform as well as susceptible crops. In order for triazine-resistant crops to become a viable agronomic alternative, it will be necessary to find a mutation that does not carry detrimental pleiotropic effects.

3. ACETOLACTATE SYNTHASE-INHIBITING HERBICIDES

Sulfonylurea, imidazolinone, and triazolapyrimidine herbicides bind to acetolactate synthase (ALS), also called acetohydroxyacid synthase (AHAS) (Subramanian et al. 1990). Acetolactate synthase is the first common enzyme in the biosynthetic pathways of the branched chain amino acids, leucine, valine, and isoleucine. Although the herbicides have the same site of action, there is enough difference in the binding sites of these herbicides that resistance to one herbicide does not always ensure that there will be cross-resistance to the other herbicides. This is the case both within and among herbicide families.

3.1. SULFONYLUREA HERBICIDE-RESISTANT CROPS

Commercial flax cultivars were transformed with a sulfonylurea-resistant mutant *Arabidopsis* ALS (McHughen 1989). Agronomic performance of two sulfonylurea-resistant transformed linseed flax (*Linum usitatissium* L.) lines was evaluated under field conditions in Canada (McHughen and Holm 1991). In untreated plots the resistant plants were indistinguishable from the susceptible plants. There were no differences in canopy height, biomass, seed yield, or 1000 seed weight. The resistant lines performed the same in treated and untreated plots.

Sulfonylurea-resistant tobacco (*Nicotiana tabacum* L.) also was produced through transformation using a sulfonylurea-resistant mutant ALS gene from *Arabidopsis*. Two transgenic lines of resistant tobacco, DALS15 and DALS16, were evaluated under field conditions (Brandle and Miki 1993). In the absence of sulfonylurea herbicides both of the transgenic lines were lower yielding than the untransformed control. The authors suggest that the reduced yield might not be due to the presence of the resistance trait, but might have been the result of genetic variation introduced in the tissue culture step during transformation, which indicates the need for selection or backcrossing to produce an agronomically acceptable cultivar. They also suggest that if an agronomically important gene or genes were at the site of transgene incorporation, they may have been made nonfunctional during transformation. The two lines did not respond the same to herbicide treatments. DALS16 was resistant to chlorsulfuron applied at 10 and 20 g a.i./ha. DALS15 was resistant to chlorsulfuron applied at 10 g/ha but not at 20 g/ha. Neither line was resistant to thifensulfuron plus tribenuron at either 10 or 20 g a.i./ha.

Soybeans [*Glycine max* (L.) Merr.] with sulfonylurea site-based resistance were selected from mutagenized seed (Sebastian et al. 1989). No plant growth data were provided.

Sulfonylurea herbicide-resistant lettuce (*Lactuca sativa* L.) was produced by crossing with prickly lettuce (*L. serriola* L.), a naturally occurring resistant weed relative (Mallory-Smith et al. 1993). Resistant prickly lettuce contains an altered ALS which is less sensitive to inhibition by the sulfonylurea herbicides. Resistance is due to a single point mutation at amino acid 173 in domain A, which translated to a replacement of a proline in the susceptible biotype with a histidine residue in the resistant biotype (Guttieri et al. 1992). Growth parameters of the resistant and susceptible lettuce isolines were compared. In greenhouse studies the resistant isoline accumulated biomass more quickly than did the susceptible isoline, but the two isolines produced similar total biomass after 46 days (Mallory-Smith, unpublished data). Germination at 8°, 18°, and 28°C was measured. At 8° and 18°C, small differences were measured, but at 28°C the susceptible seeds did not germinate more than 50% after 120 h, while the resistant seeds reached 100% germination within 24 h (Mallory-Smith, unpublished data). Specific activity of ALS extracted from the resistant isoline was only 46% of the specific activity of ALS extracted from the susceptible isoline (Eberlein et al. 1995). The ALS extracted from the resistant isoline was less sensitive to feedback inhibition by the branched chain amino acids, valine, leucine, and isoleucine, than was the ALS extracted from the susceptible isoline. Free branched chain amino acid concentration was higher in seeds of the resistant isoline than that in the seeds of susceptible isoline. Valine and leucine concentrations were higher in leaves of the resistant isoline than leaves of the susceptible isoline; thus, the mutation for ALS-inhibitor resistance alters the regulation of branched chain amino acid biosyntheses in the resistant isoline. These studies present a strong argument for pleiotropic effects of a mutation in the ALS gene. Even though there are several measurable consequences of the mutation, these effects need to be evaluated under field conditions to determine if they are detrimental to yield and quality.

3.2. IMIDAZOLINONE-RESISTANT CROPS

Imidazolinone-resistant corn (*Zea mays* L.) has been commercialized and grown in the U.S. since 1992. Imidazolinone resistant canola and wheat are being field tested. In the three crops, the mechanism of resistance is ALS that is less sensitive to the imidazolinone herbicides (Anderson and Georgeson 1989; Newhouse et al. 1992; Swanson et al. 1989).

Imidazolinone-resistant corn mutants were selected in callus culture, and plants were regenerated and crossed with a corn inbred line (Anderson and Georgeson 1989). There were no observable differences on plant height, flowering date, and yield between resistant and susceptible lines. However, the overall activity of ALS in the resistant lines was one half to two thirds that of the unselected line. In further studies of these resistant lines there was normal feedback regulation of the ALS enzyme and the plants displayed normal phenotypes (Newhouse et al. 1991).

Imidazolinone-resistant mutants of canola were selected from mutagenized microspores (Swanson et al. 1989). Mutagenized canola microspores were placed on media containing the imidazolinone herbicide, imazethapyr. Imazethapyr-resistant mutants were selected and plants regenerated. The mutants were similar to the susceptible parent in yield, maturity, disease resistance, and oil quality. Imidazolinone-resistant canola will be released in Canada in 1995 or 1996.

'Fidel' winter wheat (*Triticum aestivum* L.) seed were mutagenized using sodium azide (Newhouse et al. 1992). Resistant mutants were selected on media containing imazethapyr. From 120,000 seeds collected from plants grown from the mutagenized seed, four wheat plants were selected with resistance to imazethapyr. Plants displayed a normal phenotype and there were no negative effects on growth or yield. From limited yield tests no agronomic disadvantage has been associated with the mutant ALS (Krueger, personal communication). The ALS is insensitive to imidazolinone herbicides but not sulfonylurea herbicides.

Except for the difference in total ALS activity in corn, there are no reports of any pleiotropic effects in imidazolinone-resistant crops. No deleterious effects or yield reductions have been attributed to the difference in total ALS activity in corn.

3.3. TRIAZOLOPYRIMIDINE SULFONANILIDE-RESISTANT CROPS

Spontaneous mutants of tobacco (KS-43) and cotton (PS-3 and DO-2) resistant to triazolopyrimidine sulfonilide herbicides were selected in tissue culture (Subramanian et al. 1990). Acetolactate synthase from the mutants was 80 to 1000 times less sensitive to the herbicide than was ALS from the wild-types. However, ALS activity and substrate affinity were similar between the mutants and the wild-types. Acetolactate synthase from PS-3 was resistant to feedback inhibition by valine and leucine. PS-3 cultures accumulated higher levels of the branched chain amino acids as compared to the wild-type. These results are in agreement with those from studies of sulfonylurea-resistant lettuce. Feedback inhibition of ALS from KS-43 and DO-2 was not different from the wild-type ALS.

Development of ALS herbicide-resistant crops is continuing. There are pleiotropic responses related to the mutations in the enzyme that confer resistance; however, corn is the only ALS resistant crop that has been commercialized and there are no reports that under field conditions these responses are detrimental to agronomic performance.

4. GLYPHOSATE-RESISTANT CROPS

Glyphosate inhibits the enzyme 5-enolpyruvylshikimic acid-3-phosphate synthase (EPSP synthase) in the shikimic acid biosynthetic pathway that produces aromatic amino acids (Amrhein et al. 1980). Producing a resistant crop with a high enough level of resistance to provide crop safety has proved difficult. Glyphosate-resistant crops have been under development since the mid-1980s but have yet to be commercialized (Comai et al. 1985). Glyphosate-resistant crops have been produced through overproduction of the EPSP synthase, by incorporating a less sensitive EPSP synthase, and by introducing a gene that degrades glyphosate (Shah et al. 1986; Stalker et al. 1985; Barry et al. 1992). The combination of a less sensitive EPSP synthase and the degradation gene increased the level of resistance. Monsanto projects the release of glyphosate-resistant soybeans in

1996, canola in 1996 for use in Canada, cotton in 1997, and corn in 2000 (Harper, personal communication). However, no reports of field data are available.

Glyphosate-resistant petunias were regenerated from leaf disks (Shah et al. 1986). Plants survived treatment and grew to maturity. Glyphosate-resistant carrot cells accumulated fresh weight slower than the susceptible cells in the absence of the herbicide (Nafziger et al. 1984). The activity of EPSP synthase was higher and the levels of free aromatic amino acids were higher than in the susceptible cells. Under field conditions, glyphosate-resistant tomato plants yielded comparably to control plants when herbicides were not applied; however, the resistant plants did display reduced and delayed flowering (Fraley 1988).

5. SETHOXYDIM-RESISTANT CROPS

Corn tissue cultures with resistance to sethoxydim, a herbicide that inhibits the enzyme acetyl-coenzyme A carboxylase (ACCase), were selected (Parker et al. 1990). Growth of the cultures was not different in the absence of the herbicide, but plants were not regenerated. The specific activity of the ACCase from the wild-type was lower than the resistant lines, indicating that resistance was probably due to overproduction of the ACCase enzyme. The resistant cell line incorporated more acetate into nonpolar lipids than did the susceptible cell line. The authors suggested that the level of ACCase in the susceptible cell line may be a rate-limiting step in fatty acid biosynthesis and that overproduction of ACCase in the resistant line could have resulted in increased lipid synthesis.

Sethoxydim-resistant corn with an insensitive form of ACCase was tested under field conditions (Dotray et al. 1993). Yield was not reduced by any rate of sethoxydim applied at either the three- or seven-leaf stage. No differences in plant height, days to 50% silking, and rate of kernel drying between resistant and susceptible plants were measured. Germination and seedling vigor were not affected by the resistance trait.

6. BROMOXYNIL-RESISTANT CROPS

The b*xn* gene, which encodes a nitralase that detoxifies bromoxynil, was cloned from the soil bacterium *Klebsiella ozaenae* and transferred to tobacco (Stalker et al. 1988). The transgenic plants flowered and set seed normally but the plants were not grown under field conditions and other agronomic characteristics were not evaluated.

Bromoxynil-resistant potatoes were produced by transferring the *bxn* gene into the cultivar 'Lemhi Russet'. Fourteen transgenic clones were tested originally and the four best clones were field tested. In field trials the best yielding bromoxynil-resistant potato clone produced 15% less U.S. No. 1 tuber yield than the untreated, untransformed clone (Eberlein et al. 1995). It is not clear whether the difference in yield is due to the resistance gene or some other variation that may have occurred during transformation and/or regeneration because similar variation was observed in field studies of transgenic potatoes that contain constructs other than herbicide-resistance genes (Belknap et al. 1994). The transgenic plants produced fewer U.S. No. 1 tubers than nontransgenic controls. There was less variation in nontransgenic clones regenerated from tissue culture than in the transgenic lines. The authors suggested that in order to select transgenic potatoes with commercial properties several hundred individual clones with a given mutation would need to be evaluated. Dale and McPartlan (1992) report pleiotropic effects of the constructs used to produce transgenic potato plants. For three of five characters measured, the transgenic plants had lower means than plants derived from tuber-disk cultures, which indicates that the introduced genes may affect plant growth.

Approximately 10,000 acres of bromoxynil-resistant cotton was grown for seed in 1994 in Arizona (McCloskey, personal communication). There was no difference in yield between the transgenic resistant cotton and the parental line from which it was derived. There was no difference in emergence between the transgenic-resistant line and the susceptible parent line, and the transgenic line exhibited no negative growth effects. Although other crops have been transformed with the *bxn* gene there are no reports on field performance of the crops.

7. GLUFOSINATE-RESISTANT CROPS

Glufosinate inhibits glutamine synthetase, an enzyme in the biosynthetic pathway that produces the amino acid glutamine (Tachibana et al. 1986). Genes from soil microbes that degrade glufosinate have been isolated, cloned, and transferred to plants (De Block et al. 1987).

In greenhouse tests the transgenic tobacco plants were indistinguishable from nontransformed plants (Leemans et al. 1987). In a field study there was no difference in the length of the largest leaf between untreated resistant and susceptible plants. In field trials the growth and yield of glufosinate-resistant tobacco and potato lines were indistinguishable from the nontransformed, nontreated control lines (De Greef et al. 1989). The researchers did not evaluate other tuber quality characteristics, such as specific gravity or sugar content.

Glufosinate-resistant canola has been field tested (Oelck et al. 1991). Seed were harvested and tested for oil content, fatty acid composition, and glucosinolate levels. Yield was not different between resistant and susceptible plants. No detrimental effects were found in the transformed plants and the biochemical character of the plants was not changed. Glufosinate-resistant canola is projected to be produced commercially in Canada in 1995 (Anderson, personal communication).

One example of instability of a resistance trait is inactivation of a glufosinate-resistance gene in alfalfa (*Medicago sativa* L.) cell culture (Walter et al. 1992). At 25°C, 12% of the suspension cells lost the resistant trait within 150 d. After 10 d of a 37°C treatment, 95% of the cells showed no resistance. Cell division of the resistant, heat-treated culture was not affected, but the resistance trait was not expressed in subsequent generations. Polymerase chain reaction analysis showed that the resistance gene was present and intact in the cells. Whole plants were not heat treated, and therefore it is not known if inactivation under field conditions could be a problem, but temperatures in the 25° to 35°C range do occur during the growing season.

Other glufosinate-resistant crops, including sugarbeet (*Beta vulgaris* L.), alfalfa, wheat, corn, and soybeans are being produced (D'Halluin et al. 1990, 1992; Vasil et al. 1992, Anderson, personal communication). Field performance of these crops has not been reported.

8. 2-4-D-RESISTANT CROPS

The gene, *tdf*A, from the soil bacterium *Alcaligens eutrophus* JMP134, encodes an enzyme that degrades the plant growth-inhibiting herbicide 2,4-D. Resistant tobacco and cotton have been produced by transformation with the *tdfA* gene (Lyon et al. 1989, Streber and Willmitzer 1989, Bayley et al. 1992). No difference in growth behavior or final height was measured between greenhouse-grown susceptible and resistant tobacco plants, but no agronomic characteristics were evaluated. The level of resistance in cotton (*Gossypium hirsutum* L.) was high enough to prevent injury at 1.5 kg/ha, but no evaluation of growth or yield from the resistant plants was reported. Results of field studies with 2,4-D-resistant crops have not been reported.

9. DISCUSSION

Pleiotropic responses have been documented in several herbicide-resistant crops, but these responses do not always affect field performance of the crop. There does not appear to be a relationship between pleiotropic effects and the technique that was used to produce the resistant crop. A certain amount of caution should be used when evaluating the results of the field studies reported to date. The major emphasis has been to create a plant that manifests the herbicide resistance trait and many studies have evaluated the level of resistance to the herbicide in question and have not evaluated agronomic characteristics. Little data are available to compare the new herbicide-resistant lines to agronomically important cultivars. Resistant varieties will need to be evaluated by plant breeders and agronomists in field trials under diverse environments with comparison to standard varieties. Over time and in many different environments subtle differences may appear between the susceptible and resistant biotypes. In order to evaluate the pleiotropic effect of a resistance gene it is essential to compare resistant and susceptible isolines that differ only in the resistance trait. This has not been a priority in the development of herbicide-resistant crops.

REFERENCES

Amrhein, N., Deus, B., Gehrke, P., and Steinrucken, H.C. 1980. The site of inhibition of the shikimate pathway by glyphosate. *Plant Physiol.* 66: 99.

Anderson, P.C. and Georgeson, M. 1989. Herbicide-tolerant mutants of corn. *Genome* 31: 994.

Barry, G. , Kishore, G., Padgette, S., Taylor, M., Kolacz, K., Weldon, M., Re, D., Eichholtz, D., Fincher, K., and Halla, L. 1992. Inhibitors of amino acid biosynthesis: strategies for imparting glyphosate tolerance to crop plants. In *Biosynthesis and Molecular Regulation of Amino Acids in Plants.* Ed. by Singh, B.K., Flores, H.E., and Shannon, J.C. American Society of Plant Physiology, Rockville, MD, p. 139.

Bayley, C., Trolinder, N., Ray, C., Morgan, M., Quisenburry, J.E., and Ow, D.W. 1992. Engineering 2,4-D resistance into cotton. *Theor. Appl. Genet.* 83: 645.

Beversdorf, W.D., Hume, D.J., and Donnelly-Vanderloo, M.J. 1988. Agronomic performance of triazine-resistant and susceptible reciprocal spring canola hybrids. *Crop Sci.* 28: 932.

Belknap, W.R., Corsini, D., Pavek, J.J., Snyder, G.W., Rockhold, D.R., and Vayda, M.E. 1994. Field performance of transgenic Russet Burbank and Lemhi Russet potatoes. *Am. Potato J.* 71: 285.

Brandle, J.E. and Miki, B.L. 1993. Agronomic performance of sulfonylurea-resistant transgenic flue-cured tobacco grown under field conditions. *Crop Sci.* 33: 847.

Comai, L., Facciotti, D., Hiatt, W.R., Thompson, G., Rose, R.E., and Stalker, D.M. 1985. Expression in plants of a mutant aroA gene from *Salmonella typhimurium* which confers tolerance to glyphosate. *Nature* 317: 741.

Dale, P.J., Irwin, J.A., and Scheffler, J.A. 1993. The experimental and commercial release of transgenic crop plants. *Plant Breeding* 111: 1.

Dale, P.J. and McPartlan, H.C. 1992. Field performance of transgenic potato plants compared with controls regenerated from tuber discs and shoot cuttings. *Theor. Appl. Genet.* 84: 585.

Darmency, H. and Pernes, J. 1989. Agronomic performance of a triazine resistant foxtail millet (*Setaria italica* (L.) Beauv.). *Weed Res.* 29: 147

Darmency, H. and Pernes, J. 1985. Use of wild *Setaria viridis* (L.) beauv. to improve triazine resistance in cultivated *S. italica* (L.) by hybridization. *Weed Res.* 25: 175.

Dawkins, R. 1982. *The Extended Phenotype.* W.H. Freeman, Oxford.

De Block, M., Botterman, J., Vandewiele, M., Dockx, J., Thoen, C., Gossele, V., Movva, N.R., Thompson, C., Montagu, M.V., and Leemans, J. 1987. Engineering herbicide resistance in plants by expression of a detoxifying enzyme. *EMBO J.* 6: 2513.

De Greef, W., Delon, R., De Block, M., Leemans, J., and Botterman, J. 1989. Evaluation of herbicide resistance in transgenic crops under field conditions. *Bio/Technology* 7: 61.

Dekker, J.H. and Sharkey, T.D. 1992. Regulation of photosynthesis in triazine-resistant and susceptible *Brassica napus. Plant Physiol.* 98: 1069.

D'Halluin, K., Bossut, M., Bonne, E., Mazur, B., Leemans, J., and Botterman, J. 1992. Transformation of sugarbeet (*Beta vulgaris* L.) and evaluation of herbicide resistance in transgenic plants. *Bio/Technology* 10: 309.

D'Halluin, K., Botterman, J., and De Greef, W. 1990. Engineering of herbicide-resistant alfalfa and evaluation under field conditions. *Crop Sci.* 30: 866.

Dotray, P.A., Marshall, L.C., Parker, W.B., Wyse, D.L., Somers, D.A., and Gengenbach, B.G. 1993. Herbicide tolerance and weed control in sethoxydim-tolerant corn. *Weed Sci.* 41: 213.

Dyer, W.E., Hess, F.D., Holt, J.S., and Duke, S.O. 1993. Potential benefits and risks of herbicide-resistant crops produced by biotechnology. In *Horticultural Reviews*, Vol. 15. Ed. by Janick, J. John Wiley & Sons, New York, chap. 9.

Eberlein, C.V., Guttieri, M.J., and Belknap, W.R. 1994. Performance of Transgenic Potato Plants Expressing the *bxn* Gene. Abstr. No. 088, Potato Association of America Calgary.

Eberlein, C.V., Guttieri, M.J., Mallory-Smith, C.A., and Thill, D.C. 1995. In vitro and in vivo evidence for physiological effects of mutation for ALS-inhibitor resistance. *Weed Science Society of America.* 35: 64.

Forcella, F., 1987. Herbicide-resistant crops: yield penalties and weed thresholds for oilseed rape (*Brassica napus* L.). *Weed Res.* 27: 31.

Fraley, R.T. 1988. Field testing genetically engineered plants. In *Current Communications in Molecular Biology.* Ed. by Fraley, R.T., Frey, N.M., and Schell, J. Cold Spring Harbor Laboratory, Cold Spring Harbor, NY, p. 83.

Gressel, J. and Ben-Sinai, G. 1985. Low intraspecific competitive fitness in a triazine-resistant, nearly nuclear-isogenic line of *Brassica napus. Plant Sci.* 38: 29.

Guttieri, M.J., Eberlein, C.V., Mallory-Smith, C.A., Thill, D.C., and Hoffman, D.L. 1992. DNA sequence variation in domain A of the acetolactate synthase genes of herbicide resistant and susceptible weed biotypes. *Weed Sci.* 40: 670.

Haldane, J.B.S. 1960. More precise expressions for the cost of natural selection, *J. Genet.* 3: 351.

Hart, J.J., Radosevich, S.R., and Stemler, A. 1992. Influence of light intensity on growth of triazine-resistant rapeseed (*Brassica napus*). *Weed Res.* 32: 349.

Hart, J.J. and Stemler, A. 1990. High light-induced reduction and low light-enhanced recovery of photon yield in triazine-resistant *Brassica napus* L. *Plant Physiol.* 94: 1301.

Jursinic, P.A. and Pearcy, R.W. 1988. Determination of the rate limiting step for photosynthesis in a nearly isonuclear rapeseed (*Brassica napus* L.) biotype resistant to atrazine. *Plant Physiol.* 88: 1195.

LeBaron, H.M. 1991. Distribution and seriousness of herbicide-resistant weed infestations worldwide. In *Herbicide Resistance in Weeds and Crops.* Ed. by Caseley, J. C., Cussans, G W., and Atkins, R.K. Butterworth-Heinemann, London, p. 27.

Leemans, J., De Block, M., D'Halluin, Botterman, J., and De Greef, W. 1987. The use of glufosinate as a selective herbicide on genetically engineered resistant tobacco plants. *Br. Crop Protection Conf. - Weeds, 8C*-1: 867.

Lyon, B.R., Llewellyn, D.J., Huppatz, J.L. Dennis, E.S., and Peacock, W.J., 1989. Expression of a bacterial gene in transgenic tobacco plants confers resistance to the herbicide 2,4-dichlorophenoxyacetic acid. *Plant Mol. Biol.* 13: 533.

Mallory-Smith, C., Thill, D.C., and Dial, M.J. 1993. ID-BR1: sulfonylurea herbicide-resistant lettuce germplasm. *HortScience* 28: 63.

Mazur, B.J. and Falco, S.C. 1989. The development of herbicide resistant crops. *Annu. Rev. Plant Physiol. Plant Mol. Biol.* 40: 441.

McHughen, A. 1989. *Agrobacterium*-mediated transfer of chlorsulfuron resistance to commercial flax cultivars, *Plant Cell Rep.* 5: 81.

McHughen, A. and Holm, F. 1991. Herbicide resistant transgenic flax field test: agronomic performance in normal and sulfonylurea-containing soils. *Euphytica* 55: 49.

Nafziger, E.D., Widholm, J.M., Steinrucken, H.C., and Killmer, J.L. 1984. Selections and characterization of a carrot cell line tolerant to glyphosate. *Plant Physiol.* 76: 571.

National Research Council. 1989. *Field Testing Genetically Modified Organisms.* National Academy of Sciences, Washington, D.C., p. 48.

Newhouse, K., Singh, B., Shaner, D., and Stidham, M. 1991. Mutations in corn (*Zea mays* L.) conferring resistance to imidazolinone herbicides. *Theor. Appl. Genet.* 83: 65.

Newhouse, K.E., Smith, W.A., Starrett, M.A., Schaefer, T.J., and Singh, B.K. 1992. Tolerance to imidazolinone herbicides in wheat. *Plant Physiol.* 100: 882.

Oelck, M.M., Phan, C.V., Eckes, P., Donn, G., Rakow, G., and Keller, W.A. 1991. Field Resistance of Canola Transformants (*Brassica napus* L.) to Ignite (Phosphinotricin). GCIRC Int. Rapeseed Congress, Saskatchewan, Canada. p. 292.

Parker, W.B., Somers, D.A., Wyse, D.L., Keith, R.A., Burton, J.D., Gronwald, J.W., and Gengenbach, B.G. 1990. Selection and characterization of sethoxydim-tolerant maize tissue cultures. *Plant Physiol.* 92: 1220.

Powles, S.B. and Holtum, J.A.M. Eds. 1994. *Herbicide Resistance in Plants.* CRC Press, Boca Raton, FL.

Sebastian, S.A., Fader, G.M., Ulrich, J.F., Forney, D.R., and Chaleff, R.S. 1989. Semidominant soybean mutation for resistance to sulfonylurea herbicides. *Crop Sci.* 29: 1403.

Shah, D.M., Horsch, R.B., Klee, J.J., Kishore, G.M., Winter, J.A., Tumer, N.E., Hironaka, C.M., Sanders, P.R., Gasser, C.S., Aykent, S., Siegel, N.R., Rogers, S.G., and Fraley, R.T. 1986. Engineering herbicide tolerance in transgenic plants. *Science* 233: 478.

Stalker, D.M., Hiatt, W.R., and Comai, L. 1985. A single amino acid substitution in the enzyme 5-enolpyruvylshimate-3-phosphate synthase confers resistance to glyphosate. *J. Biol. Chem.* 260: 4724.

Stalker, D.M., McBride, K.E., and Malyj, L.D. 1988. Herbicide resistance in transgenic plants expressing a bacterial detoxification gene. *Science* 242: 419.

Streber, W.R. and Willmitzer, L. 1989. Transgenic tobacco plants expressing a bacterial detoxifying enzyme are resistant to 2,4-D. *Bio/Technology* 7: 811.

Subramanian, M.V., Hung, H.Y., Dias, J.M., Miner, V.W., Butler, J.H., and Jachetta, J.J. 1990. Properties of mutant acetolactate synthases resistant to triazolopyrimidine sulfonanilide. *Plant Physiol.* 94: 239.

Swanson, E.B., Herrgesell, M.J., Arnoldo, M., Sippell, D.W., and Wong, R.S.C. 1989. Microspore mutagenesis and selection: canola plants with field tolerance to the imidazolinones. *Theor. Appl. Genet.* 78: 525.

Tachibana, K., Watanabe, T., Sekizuwa, Y., and Takematsu, T. 1986. Accumulation of ammonia in plants treated with bialaphos. *J. Pestic. Sci.* 11: 33.

Vasil, V., Castillo, A.M., Fromm, M.E., and Vasil, I.K. 1992. Herbicide resistant fertile transgenic wheat plants obtained by microprojectile bombardment of regenerable embryogenic callus. *Bio/Technology* 10: 67.

Walter, C., Broer, I., Hillemann, D., and Puhler, A. 1992. Heat frequency, heat treatment-induced inactivation of the phosphinothricin resistance gene in transgenic single cell suspension cultures of *Medicago sativa*. *Mol. Gen. Genet.* 235: 189.

PART II

EFFECTS ON WEED CONTROL

Chapter **14**

THE NICHE FOR HERBICIDE-RESISTANT CROPS IN U.S. AGRICULTURE

John W. Wilcut
Harold D. Coble
Alan C. York
David W. Monks

CONTENTS

1. INTRODUCTION

Differential cultivar tolerance to herbicides in crops has been recognized for years (Barrentine et al. 1976, 1982; Graf and Ogg 1976; Harrison et al. 1985; Shaw and Wesley 1991; Stephenson et al. 1976). Until recently, however, few attempts have been made to consciously develop cultivars with improved herbicide tolerance. There was an effort in soybean [*Glycine max* (L.) Merr.] to develop cultivars with better tolerance of metribuzin [4-amino-6-(1,1-dimethylethyl)-3-(methylthio)-

1-56670-045-0/96/$0.00+$.50
© 1996 by CRC Press Inc.

1,2,4-triazin-5(4*H*)-one]. The cultivar 'Tracy M' was developed as part of this effort (Barrentine et al. 1982; Hartwig et al. 1980).

Tremendous technological advances in identifying and transferring genetic material coded for specific traits such as herbicide resistance, and regenerating plants with this genetic information from single cells, have been made in recent years (Kishore et al. 1992). Commercial products of genetic engineering are now being introduced to the consumer (Anon 1993a, b). A major research emphasis has been devoted to developing crop plants genetically engineered for herbicide tolerance (Table 1 and Table 2), and a few genetically engineered herbicide-resistant crop (HRC) cultivars are nearing commercialization. Additionally, efforts to develop crop cultivars with enhanced herbicide tolerance through conventional breeding and selection continue (Beversdorf and Kott 1987; Duke et al. 1991; Forney et al. 1989; Hartwig, 1987; Newhouse et al. 1991).

2. POTENTIAL BENEFITS OF HERBICIDE-RESISTANT CROPS

There are many potential benefits in weed management to be obtained from use of HRCs. A major benefit will be the opportunity for new strategies and/or increased flexibility in the management of problem weeds. The more weed management options available to producers, the more likely they will use multiple and integrated approaches in their weed management systems. For example, soybean producers will be able to apply glyphosate [*N*-(phosphonomethyl)glycine] postemergence (after emergence of the crop) to resistant cultivars. This normally nonselective herbicide effectively controls numerous annual and perennial grasses and broadleaf weeds. A management system utilizing glyphosate applied postemergence may allow producers to forego prophylactic applications

TABLE 1 Publicly Funded Research to Develop HRCs Through Conventional Breeding or Biotechnology

Herbicide[a]	Crop	Organization	State	Fund Source[b]	Method[c]
Acifluorfen (Blazer®)	Tomato	Univ. Illinois	IL	CSRS	A
	Numerous species	Clemson Univ.	SC	CSRS	A
Atrazine (Aatrex®)	Numerous species	Clemson Univ.	SC	CSRS	A
Bromoxynil (Buctril®)	Potato	Univ. Idaho	ID	CSRS	B
Chlorsulfuron (Glean®)	*Brassica* spp.	Univ. Arizona	AZ	CSRS	B
	Cowpea	Clemson Univ.	SC	CSRS	A
Clomazone (Command®)	Tobacco	Clemson Univ.	SC	CSRS	A
Glyphosate (Roundup®)	Field pea	Univ. Idaho	ID	CSRS	B
	Rice	Rice Exp. Stn.	LA	SAES	B
	Alfalfa	Univ. Wyoming	WY	CSRS	A
	Alfalfa	Univ. Nevada	NV	CSRS	A, B
	Poplar	USFS	WI	USFS	A
Imazapyr (Arsenal®)	Wheat	Univ. Oklahoma	OK	CSRS	B
Metolachlor (Dual®)	Rice	Rice Exp. Stn.	LA	SAES	B
Picloram (Tordon®)	Wheat	Univ. Wyoming	WY	CSRS	A
Thiadiazuron	Wheat	Univ. Illinois	IL	CSRS	A
Not given	Pea, wheat	Washington State Univ.	WA	CSRS	A
	Legumes, cucurbits	Clemson Univ.	SC	CSRS	A
	Cucurbits	Penn State Univ.	PA	CSRS	A
	Peanut, wheat	Oklahoma St. Univ.	OK	CSRS	A

[a] Trade names are in parentheses. Some of these herbicides have several trade names. Only the most widely used trade names are provided.

[b] CSRS = Cooperative State Research Service, USFS = U.S. Forest Service, SAES = State Agricultural Experiment Station.

[c] A = biotechnology, B = conventional breeding.

of soil-applied herbicides (Doss and York 1995). Grass and broadleaf weeds currently can be controlled postemergence in soybean with combinations of grass and broadleaf weed-controlling herbicides (Parker et al. 1985). With glyphosate, a single herbicide will control both grass and broadleaf weeds without the antagonism that occurs when a graminicide is applied with a broadleaf herbicide (Byrd and York 1987a; Grichar 1991; Holhouser and Coble 1990; Jordan and York 1989; Jordan et al. 1993c). The amount of herbicide used for weed management in soybean could be reduced at least 50% in many situations with glyphosate-resistant cultivars. Herbicide-resistant crops also will facilitate the addition of conservation tillage crop production practices because of more effective postemergence treatments.

The development of crop cultivars with resistance to postemergence herbicides will encourage crop producers to use economic weed threshold predictions in making their weed management decisions (Coble and Mortensen 1992; Goodman 1987). Currently, in most agronomic crops herbicides are applied preplant incorporated (herbicide is applied before planting of the crop and mixed in the soil with tillage) or preemergence (applied before crop and/or weed emergence), based on the anticipated weed population (Byrd and York 1987b; Parker et al. 1985; Wilcut et al. 1994). Furthermore, not all crops currently have registered herbicides for postmergence management of weeds. For example, there is no registered herbicide for postemergence management of annual broadleaf weeds in cotton (*Gossypium hirsutum* L.), which does not have the potential to delay maturity or reduce yields (Guthrie and York 1989; Byrd and York 1987a).

Proposed HRCs will potentially allow the use of more environmentally benign herbicides and lower use rates of herbicides than many soil-applied herbicides (Burnside 1992; Goodman 1987; Knake 1992; Mangels 1991a, b). Of major concern is the potential impact of certain currently used herbicides on the quality of groundwater and surface water (Nielson and Lee 1987). For example, alachlor {2-chloro-*N*-(2,6-diethylphenyl)-*N*-(methoxymethyl)acetamide}, atrazine {6-chloro-*N*-ethyl-*N'*-(1-methylethyl)-1,3,5-triazine-2,4-diamine}, cyanazine {2-[[4-chloro-6-(ethylamino)-1,3,5-triazin-2-yl]amino]-2-methylpropanenitrile}, and metolachlor {2-chloro-*N*-(2-ethyl-6-methylphenyl)-*N*-(2-methoxy-1-methyethyl)acetamide}, the four most commonly used herbicides in corn (*Zea mays* L.) (Bridges 1992), have been detected in groundwater and surface water (Nielson and Lee 1987). Herbicide labels have been revised to avoid groundwater contamination (Anon 1992, 1994). Mandatory production practices that avoid or mitigate herbicide contamination of groundwater are forthcoming (Anon. 1992). Development of HRCs would provide an option to use different classes of herbicides with less environmental risk, such as glyphosate or imazethapyr {2-[4,5-dihydro-4-methyl-4-(1-methylethyl)-5-oxo-1*H*-imidazol-2-yl]-5-ethyl-3-pyridinecarboxylic acid}, in environmentally sensitive areas (Duke 1988; Goodman 1987; Harris and Gagne 1991; Mangels 1991a, b).

Evolution of HR biotypes of weeds is an increasing problem (Holt 1992, Powles and Howatt 1990). One of the major factors contributing to the evolution of resistance is continued use of herbicides with the same mechanism of action (Duke 1993; Gressel and Segel 1990; York and Murdock 1993). Hence, a component in a HR management program is use of different herbicides with various modes of action. Development of HRCs would expand herbicide choices, thus allowing herbicides with a different mode of action to be used. In regions with limited crop rotation options, use of HR cultivars resistant to herbicides with different sites of action may be the only mechanism to allow for rotation of herbicide chemistry and mode of action, which is considered crucial in avoiding or delaying evolution of HR weeds (Gressel and Segel 1990).

Few herbicides are registered for small-hectarage crops such as vegetables because of the exorbitant cost of registration and the accompanying limited market potential (Burnside 1992; Knake 1992; Radosevich et al. 1992; Wyse 1992). Similarly, few herbicides likely will be registered for such crops in the future. Development of vegetable crops with resistance to an environmentally benign herbicide already registered for use on a major field crop would not only fill a great need, but also reduce the cost of registering the herbicide for use on those crops because many of the expensive environmental fate studies would already have been completed during registration on a major crop (Burnside 1992). Because of the limited hectarage, however, it is still questionable whether there would be sufficient economic incentive for the herbicide manufacturer. Interregional Research Project No. 4 (IR-4) could facilitate registering a herbicide for minor crop use if herbicide resistance was developed in such crops.

Herbicide-resistant crops could be used advantageously even where effective herbicides are available for a crop. Crop injury resulting from the carryover of herbicides applied the preceding year can be a problem (Barnes et al. 1989; Barnes and Lavy 1991; Curran et al. 1991; Jordan et al. 1993d; Knake 1992; Mills and Witt 1991). To avoid this problem, producers must either alter their rotational plans or use another herbicide program that may be less efficacious in the preceding crop. If the crop was resistant to or tolerant of the herbicide(s) used the preceding year, problems resulting from herbicide carryover would be eliminated. For example, imidazolinone herbicides applied to soybean, peanut (*Arachis hypogaea* L.), or pinto bean (*Phaseolus vulgaris* L.) sometime carryover to corn, cotton, or sugarbeet (*Beta vulgaris* L.) planted the following year (Barnes et al. 1989; Curran et al. 1991; Renner et al. 1988; Renner and Powell 1991; York and Wilcut 1993). Planting of imidazolinone-resistant crops would eliminate the problem. Imidazolinone-resistant corn is now commercially available. Another advantage is that herbicide resistance in crops would eliminate the problem of drift to sensitive crops, such as 2,4-D {(2,4-dichlorophenoxy)acetic acid} drift from rice (*Oryza sativa* L.) to cotton in the mid-south or nicosulfuron {2-[[[[(4,6-dimethoxy-2-pyrimidinyl)amino]carbonyl]amino]sulfonyl]-*N,N*-dimethyl-3-pyridinecarboxamide} and primisulfuron {2-[[[[[4,6bis(difluoromethoxy)2pyrimidinyl] amino]carbonyl]amino]sulfonyl] benzoic acid} drift from corn injuring soybean in the midwest (Bailey and Kapusta 1993).

3. POTENTIAL LIABILITIES OF HERBICIDE-RESISTANT CROPS

Although benefits to weed management and crop production from development of HRCs far outweigh the liabilities, the liabilities do exist. A major concern is the potential for HRCs to outcross with weedy plants (Radosevich et al. 1992). This potential is not likely to be a problem in crops such as corn, cotton, peanut, and soybean, which do not have wild relatives in the U.S. However, this potential problem needs to be fully addressed in crops such as sorghum (*Sorghum bicolor* L. Moench.), which may cross with johnsongrass [*Sorghum halepense* (L.) Pers.] or shattercane [*Sorghum bicolor* (L.) Moench]; rice (*Oryza sativa* L.), which may cross with red rice (*Oryza sativa* L.); wheat (*Triticum aestivum* L.), which may cross with jointed goatgrass (*Aegilops cylindrica* Host); oats (*Avena sativa* L.), which may cross with wild oats (*Avena fatua* L.); and any other potential high risk systems (Radosevich et al. 1992).

There also may be concerns with the HRC becoming a weed itself, especially in rotational crops (Goldburg 1992; Goodman 1987). This is a particularly pertinent question in cropping systems where one crop is altered to be resistant to a commonly used herbicide in the rotational crop. For example, resistance in corn to cyclohexadione herbicides such as sethoxydim (2-[1-(ethoxyimino)butyl]-5-[2-(ethylthio)propyl]-3-hydroxy-2-cyclohexen-1-one) and aryloxyphenoxypropionate herbicides such as fluazifop-P ((*R*)-2-[4-[[-(trifluoromethyl)-2-pyridinyl]oxy]phenoxy]propanoic acid) likely could lead to problems in controlling volunteer corn in soybean. Similarly, development of canola (*Brassica napus* L. and *Brassica campestris* L.) resistant to phenoxy and sulfonylurea herbicides could lead to problems with volunteer canola in small grain crops.

Problems might also develop if all crops in the rotation were altered to have resistance to the same herbicide. An example would be corn and soybean resistant to glyphosate or corn resistant to imazethapyr, which is also used in soybean on appreciable hectarage. In such cases, rotational and weed management plans would need to be carefully considered to avoid weed population shifts or evolution of HR weed biotypes (Goodman 1987).

While these are legitimate concerns, one must realize that such potential problems are not the result of development of HRCs but rather the result of poor planning of management systems (Goodman 1987, Wyse 1992). These problems could result from excessive reliance on the same herbicide or herbicides with the same mechanism of action. In HRCs, as in non-HRCs, integrated weed management strategies aimed at preventing population shifts and resistance evolution need to be followed (York 1994a).

No one herbicide controls all weeds, and many producers now recognize that several management options are needed (York 1994a). Successful weed management generally requires an integrated management system. Several herbicides are now registered on more than one crop, such as imazethapyr in corn, peanut, and soybean; flumetsulam {*N*-[2,6-difluorophenyl]-5-methyl(1,2,4)triazolo-[1,5-a]-pyrimidine 2-sulfonamide} in corn and soybean; pendimethalin (*N*-(1-ethylpropyl)-3,4-dimethyl-2,6-dinitrobenzenamine) in corn, cotton, peanut, and soybean; and sethoxydim in cotton, peanut, soybean, and numerous broadleaf vegetables (Anon 1994). Thus, potential for mismanagement and nonrotation of herbicide modes of action in crops and crop rotations already exists without HRCs (Goodman 1987; Wyse 1992).

An additional potential problem related to use of HRCs that is seldom discussed is the potential for severe crop injury resulting from drift or unintentional application of a nonselective herbicide to a non-HRC. For example, glyphosate applied to glyphosate-resistant soybean could drift to and damage other crops not resistant to glyphosate. Also, a herbicide applicator could get fields with HRCs and non-HRCs confused and apply a normally nonselective herbicide to fields not planted to HRCs. However, where these errors occur, they will be not the result of HRC development, but rather poor management and human error.

We have attempted to discuss the potential niche(s) for HR cultivars in the various commodities that follow. It may not be economically feasible for private industry to address some of these needs because of the small hectarage of some of these crops. Also, in some cases the potential for herbicide resistance to be transferred to a weedy relative of the crop may preclude the development of HR cultivars. Nevertheless, in the following sections we discuss some of the needs or voids in current weed management programs and how HRCs could help address those needs.

4. SELECTED U.S. CROPS AND HERBICIDE RESISTANCE

4.1. CANOLA

Canola is rapeseed containing <2% erucic acid and <30 μmol of glucosinolates per gram of oil-free meal. In the U.S. canola is grown primarily in the Pacific Northwest, the northern Corn Belt and central Plains, and in the southeast.

Major weed problems in canola include kochia [*Kochia scoparia* (L.) Schrad.], wild mustard [*Brassica kaber* (DC.) L.C. Wheeler], and wild radish (*Raphanus raphanistrum* L.). Trifluralin (2,6-dinitro-*N,N*-dipropyl-4-(trifluoromethyl)benzenamine) is the only herbicide with full registration for use in canola in the U.S. (Anon 1994), although some states have state registration (Section 24-C) for use of sethoxydim. Trifluralin controls annual grasses and some small-seeded annual broadleaf weeds, while sethoxydim controls annual and perennial grasses. Neither herbicide controls major broadleaf weed problems that plague canola. Hence, it cannot be produced commercially in areas heavily infested with these broadleaf weeds.

Contamination of canola by seeds of wild radish and wild mustard create problems when canola is processed into cooking oils and margarine. Seeds of these weeds increase the erucic acid content of extracted canola oil, necessitating blending this oil with low erucic acid oils to reduce the overall erucic acid level of the final product (Wall 1992). These cruciferious weeds also are a problem in canola grown for certified seed.

Triazine-tolerant canola cultivars have been developed (Beversdorf and Hume 1984, Beversdorf et al. 1990), allowing the use of triazine herbicides in Canada on these cultivars. Triazine-tolerant cultivars yield about 20% less than traditional cultivars, and consequently are not widely grown (Wall 1992).

Imidazolinone-, sulfonylurea-, glyphosate-, and glufosinate- (2-amino-4-(hydroxymethylphosphinyl)butanoic acid) resistant cultivars are being developed (Table 1 and Table 2). Imidazolinone and sulfonylurea herbicides will control several troublesome weeds including kochia, wild mustard, wild radish, chickweed (*Stellaria spp.*), henbit (*Lamium amplexcaule* L.), yellow nutsedge (*Cyperus esculentus* L.), and purple nutsedge (*Cyperus rotundus* L.) in resistant cultivars (Newhouse et al. 1991, Wilcut unpublished data). Many of these weeds are particularly troublesome during the winter and early spring months when canola is grown in the southeast.

Glyphosate or glufosinate also would control many of the aforementioned weeds, and it would provide management options for perennial broadleaf weeds. Lack of residual activity of glyphosate and glufosinate would reduce rotational crop concerns that may exist for imidazolinone, sulfonylurea, and triazine herbicides, and reduce the potential for HR development to the acetolactate synthase (ALS)-inhibiting herbicides. If HR canola cultivars are developed and the respective herbicides registered, then producers would need management plans for volunteer HR canola in rotational crops. In the southern U.S. canola is sensitive to many herbicides commonly used in rotational crops. However, depending upon the type of resistance conferred, control of HR canola could be a problem in regions where only limited rotational crop and herbicide options are available.

Canola is sensitive to many herbicides used in preceding rotational crops, including atrazine, cyanazine, chlorimuron {2-[[[[(4-chloro-6-methoxy-2-pyrimidinyl)amino]carbonyl]amino]-sulfonyl]benzoic acid}, chlorsulfuron {2-chloro-*N*-[[(4-methoxy-6-methyl-1,3,5-triazin-2-yl)amino]carbonyl]benzenesulfonamide}, diuron, [*N'*-(3,4-dichlorophenyl)-*N,N*-dimethylurea], fluometuron {*N,N*-dimethyl-*N'*-[3-(trifluoromethyl)phenyl]urea}, imazaquin {2-[4,5-dihydro-4-methyl-4-(1-methylethyl)-5-oxo-1*H*-imidazol-2-yl]-3-quinolinecarboxylic acid}, imazethapyr, linuron [*N'*-(3,4-dichlorophenyl)-*N*-methoxy-*N*-methylurea], metribuzin, nicosulfuron, norflurazon [4-chloro-5-(methylamino)-2-(3-(trifluoromethyl)phenyl)-3(2*H*)-pyridazinone], and primisulfuron. Thus, canola production is limited to areas where these residual herbicides are not used in rotational crops, limiting canola hectarage to rotations with fallow land in the southern U.S. Herbicide-resistant canola cultivars would allow production in crop rotations where these residual herbicides are commonly used. Imidazolinone-, sulfonylurea-, or triazine-resistant cultivars would allow canola to be grown on appreciable hectarage in corn and soybean rotations in the southern U.S.

4.2. CORN

There are at least 63 registrations for herbicides or herbicide packaged mixtures for weed management in field corn in the U.S. (Anon 1994). These herbicides control annual grass and broadleaf weeds when applied preplant incorporated, preemergence, or postemergence. Numerous tank mixtures also are registered for preplant incorporated, preemergence, and postemergence application. Additionally, perennial broadleaf weeds can be controlled in corn with 2,4-D and/or dicamba (3,6-dichloro-2-methoxybenzoic acid).

Most weeds can be adequately controlled with existing herbicides in U.S. corn production. However, the development of HR corn would provide more management options for certain problem weeds, aid in management where HR weed biotypes have developed, and help delay or prevent resistance development in other areas. Additionally, HR corn would allow other management options instead of triazine and chloroacetamide herbicides where risk of groundwater contamination is high or where there are restrictions on herbicide use (Nielson and Lee 1987; Anon 1992). Management plans are currently being developed in most states and will definitely affect commonly used corn herbicides (Anon 1992, 1994). Herbicide-resistant corn development may also help in cropping systems where vapor drift of 2,4-D or dicamba is a problem, particularly in areas where cotton, tobacco (*Nicotiniana tobacum* L.), or vegetables are grown.

There are several annual grasses that are extremely difficult to control with currently registered soil-applied herbicides in corn, including broadleaf signalgrass [*Brachiaria platyphylla* (Griseb.) Nash], shattercane, and Texas panicum (*Panicum texanum* Buckl.). For example, pendimethalin is the only registered preemergence herbicide in corn for Texas panicum control. Control is highly dependent on timely rainfall and is inconsistent (Johnson and Mullinix 1990). Producers are forced to cultivate and make a layby application (time when last cultivation or herbicide application can be made) of ametryn (2-ethylamino-4-isopropylamino-6-methylthio-*S*-triazine).

Recent registrations of nicosulfuron and primisulfuron for postemergence annual and perennial grass control have provided valuable new management options for corn producers (Bruce et al. 1993; Ngouajio and Hagood 1993). Prior to these registrations, corn producers had limited options for postemergence management of annual and perennial grasses with foliar-applied herbicides. Because ALS resistance has developed in other weed species, the likelihood of resistance development to nicosulfuron and primisulfuron appears to be high, particularly in areas with corn monocultures. A major concern for U.S. corn producers would be johnsongrass and/or shattercane developing nicosulfuron and/or primisulfuron resistance. Resistance to either herbicide is likely to confer cross-resistance to the other herbicide, based on the similarities in mode of action (Kendig and DeFelice 1994). No other registered herbicides adequately control these perennial grass weeds postemergence in corn. No preemergence herbicides are available, and only the thiocarbamate herbicides which require incorporation would be available for control. The thiocarbamate herbicides provide inconsistent and generally inadequate control of johnsongrass and shattercane.

Sethoxydim-tolerant corn cultivars could enhance the use of alternative weed control strategies in corn production (Dotray et al. 1993). Sethoxydim is effective for postemergence control of most annual and perennial grasses (Byrd and York 1987a, b; Holhouser and Coble 1990; Parker et al. 1985). Sethoxydim-resistant corn would allow sethoxydim to be used or rotated with nicosulfuron or primisulfuron systems, particularly in corn monocultures. Rotation of ALS-inhibiting herbicide systems (nicosulfuron and primisulfuron) with ACCase-inhibiting systems (sethoxydim) would help prevent or delay resistance development. However, control of volunteer sethoxydim-tolerant corn in rotational crops could be a significant problem, especially if there was cross-resistance to the other postemergence graminicides (Dotray et al. 1993). Postemergence graminicides (clethodim, (*E,E*)-(±)-2-[1-[[(3-chloro-2-propenyl)oxy]imino]propyl]-5-[2-(ethylthio)propyl]-3-hydroxy-2-cyclohexen-1-one; fenoxaprop, (±)-2-[4-[(6-chloro-2-benzoxazolyl)oxy]phenoxy]propanoic acid; fluazifop-P; quizalofop, (±)-2-[4-[(6-chloro-2-quinoxalinyl)oxy]phenoxy]propanoic acid; and 2-[1-(ethoxyimino)butyl]-5-[2-(ethylthio)propyl]-3-hydroxy-2-cyclohexen-1-one) currently are the only herbicide options for controlling volunteer corn in cotton, peanut, and soybean. Glyphosate with a wiper applicator can be used postemergence, but specialized equipment is required. Unfortunately, several very troublesome grass weeds have developed resistance to the ACCase inhibitors (Heap 1994; Snipes et al. 1993; Stanger and Appleby 1989; Stoltenberg and Wiederholt 1994). This resistance further reduces the advantages of sethoxydim- or other ACCase-resistant corn cultivars.

Johnsongrass also has developed resistance to the postemergence ACCase inhibitors in Mississippi (Snipes et al. 1993). Other grass species which have developed resistance to ACCase-inhibiting herbicides include several foxtail (*Setaria* spp.) species and large crabgrass [*Digitaria sanguinalis* (L.) Scop.] (Heap 1994, Stoltenberg and Wiederholt 1994). If nicosulfuron and primisulfuron are used repeatedly for johnsongrass control without integration with other management practices, resistance to ALS-inhibiting herbicides could soon develop in areas where crops and herbicide modes of action are not rotated.

Imidazolinone-resistant and imidazolinone-tolerant corn cultivars have been developed (Table 2). This research was initiated primarily because of the efficacy of imidazolinone herbicides for weeds found in the midwestern U.S. (Hart et al. 1991; Newhouse et al. 1991). Imazethapyr is widely used in soybean because it controls velvetleaf (*Abutilon theophrasti* Medik.). The imidazolinone herbicides not only control several weed species controlled by available corn herbicides, but also shattercane and johnsongrass, which were not adequately controlled until the registrations of nicosulfuron and primisulfuron (Newhouse et al. 1991; Ngouajio and Hagood 1993; Wilcut et al. 1993b). This resistance would allow the direct use of imidazolinone herbicides including imazaquin, imazapyr [(±)-2-[4,5-dihydro-4-methyl-4-(1-methylethyl)-5-oxo-1*H*-imidazol-2-yl]-3-pyridinecarboxylic acid], imazethapyr, and AC 263,222 {(±)-2-[4,5-dihydro-4-methyl-4-(1-methylethyl)-5-oxo-1*H*-imidazol-2-yl]-5-methyl-3-pyridinecarboxylic acid} in corn (Newhouse et al. 1991; Wilcut et al. 1993).

Imazaquin and imazethapyr may carry over to nonresistant corn grown in rotation with soybean (Curran et al. 1991; Johnson et al. 1993; Mills and Witt 1991; Renner et al. 1988). However, the likelihood of significant carryover has diminished because imazethapyr has replaced imazaquin in much of the midwestern Corn Belt. Additionally, corn is more tolerant of imazethapyr residues than is imazaquin (Bhalla et al. 1991; Hart et al. 1991). Imidazolinone-resistant corn would further reduce the likelihood of injurious carryover.

In the southern U.S. AC 263,222, {(±)-2-[4,5-dihydro-4-methyl-4-(1-methylethyl)-5-oxo-1*H*-imidazol-2-yl]-5-ethyl-3-pyridinecarboxylic acid}, has given good to excellent control (>80%) over many troublesome annual broadleaf weed species including Florida beggarweed [*Desmodium tortuosum* (Sw.) DC.], *Ipomoea* morningglory species, sicklepod (*Senna obtusifolia* L.), and yellow and purple nutsedge (Richburg et al. 1994b; Wilcut et al. 1994). Additionally, AC 263,222 controls seedling and rhizome johnsongrass and has residual activity on annual grasses including large and southern crabgrass [*Digitaria ciliaris* (Retz.) Koel.], broadleaf signalgrass, and Texas panicum (Wilcut et al. 1993b). Imazethapyr is not effective for Florida beggarweed or sicklepod control, and it is less active than AC 263,222 on rhizome johnsongrass.

Imazethapyr is registered in imidazolinone-resistant corn. Other herbicides should be used in management systems and crop rotations to reduce the potential for the development of resistance in weeds (Gressel and Segel 1990). Several very economical and effective herbicides are available for application in mixtures, including atrazine and 2,4-D. The potential development of shattercane and/or johnsongrass resistance to the sulfonylurea herbicides (nicosulfuron and/or primisulfuron) may also confer cross-resistance to the imidazolinone herbicides because of similarities in mode of action between the two herbicide families (Beyer et al. 1988; Shaner and Mallipudi 1991; Stidham and Singh 1991).

Considering the widespread usage of imazethapyr, imazaquin, and other ALS-inhibiting herbicides such as chlorimuron, nicosulfuron, primisulfuron, and flumetsulam, one has to question the wisdom of further increasing the already-significant hectarage being treated with ALS-inhibiting herbicides, particularly in the midwestern portion of the U.S. (Wrubel and Gressel 1994). The development of glyphosate-resistant and/or glufosinate-resistant corn would be a valuable management option for rotation with ALS-inhibiting herbicides (flumetsulam, imazethapyr, nicosulfuron, and primisulfuron) and ACCase-inhibiting herbicides (postemergence graminicides) to circumvent resistance development. The development of glyphosate- or glufosinate-resistant corn cultivars also would provide additional management options for grass and perennial broadleaf weeds.

4.3. COTTON

Weed control in cotton is more difficult than in other agronomic crops (Snipes and Mueller 1992). Weed problems often are greater in the mid-south and southeastern U.S., due partially to warm, humid growing conditions, causing rapid herbicide breakdown, weed germination and growth (Snipes and Mueller 1992b). Fewer herbicides are available for cotton as compared to other agronomic crops (Anon 1994). Due to the increased presence of weeds, the slow growth rate of cotton, and lack of consistent performance from soil-applied herbicides, cotton producers routinely rely on numerous cultivations and postemergence-directed herbicide applications to control weeds (Snipes and Mueller 1992b; York 1994b). Additionally, cotton producers do not have postemergence herbicides that can be sprayed directly over the crop (over-the-top) for annual broadleaf weed control without the risk of yield loss and/or maturity delays (Byrd and York 1987a; Guthrie and York 1989; Snipes and Mueller 1992a). Numerous studies have documented cotton yield loss from weeds, and competitive indices have been developed for the more common and troublesome annual broadleaf weeds in cotton (Coble and Byrd 1992). Furthermore, cotton is more susceptible to yield reductions from weed interference and competition than are corn or soybean (Coble and Byrd 1993; Zimdahl 1980).

DPX-PE350 {sodium 2-chloro-6-(4,5-dimethoxypyrimidin-2-ylthio)benzoate} is being developed for preemergence and postemergence over-the-top application in cotton for control of a wide spectrum of annual broadleaf weeds (Jordan et al. 1993a, b). DPX-PE350 is an ALS-inhibiting herbicide. If used frequently in cotton monocultures typical of several areas in the mid-south and southeastern U.S., weed resistance to this herbicide likely will develop. Unfortunately, the recent development of ALS-resistant cocklebur (*Xanthium strumarium* L.) in the mid-south to imazaquin has, in some locations, conferred cross-resistance to DPX-PE350 (Kendig and DeFelice 1994). Thus, development of cotton cultivars resistant to non-ALS-inhibiting over-the-top herbicides for broadleaf weed control would be a valuable new management option for cotton producers.

Two HR cotton cultivars are currently being developed with resistance to bromoxynil (3,5-dibromo-4-hydroxybenzonitrile) and glyphosate. Bromoxynil-resistant cotton has been evaluated since 1991 by university researchers across the Cotton Belt. Bromoxynil controls numerous annual broadleaf weed species that are common and difficult to control in cotton (Jordan et al. 1993; Wilcut et al. 1992, 1993a). Bromoxynil does not control grass weeds, nutsedge species, or sicklepod, and application must be very timely for *Amaranthus* spp. and prickly sida (*Sida spinosa* L.) control.

Farther behind in the development process are glyphosate-tolerant cotton cultivars. Glyphosate controls most annual broadleaf and grass species (Richburg et al. 1994a) at 0.28 to 0.42 kg a.i./ha. At these rates, glyphosate is relatively cost effective. Sicklepod and morningglory (*Ipomoea* spp.) are more tolerant of glyphosate than most other broadleaf weed species and require higher application rates and timely application. Glyphosate is also effective for perennial nutsedge and perennial broadleaf control, but higher rates and multiple applications are usually required. Both bromoxynil and glyphosate would be viable options for postemergence control of arsenical-herbicide resistant and ALS-herbicide resistant common cocklebur, and dinitroaniline-herbicide resistant Palmer amaranth (*Amaranthus palmeri* S. Wats) (Hays et al. 1993; Kendig and DeFelice 1994). Glyphosate would also be effective for control of dinitroaniline-herbicide resistant goosegrass [*Eleusine indica* (L.) Gaertn.] (Mudge et al. 1984; Murdock et al. 1993).

Bromoxynil and glyphosate can be integrated into existing weed management systems. A typical weed management system in bromoxynil- or glyphosate-resistant cotton will start with a dinitroaniline (pendimethalin or trifluralin) herbicide applied preplant incorporated and a broadleaf-controlling herbicide such as fluometuron (*N,N*-dimethyl-*N′*-[3-(trifluoromethyl)phenyl]urea) applied preemergence after planting. Rates of soil-applied herbicides for broadleaf weed control such as fluometuron could be reduced if effective postemergence over-the-top options were available. Bromoxynil or glyphosate would be applied early postemergence as a convenient replacement for commonly used early postemergence-directed herbicides, which are limited in choice and performance. Application of early postemergence-directed herbicides requires a height differential between the cotton and weeds that often is difficult to achieve (Snipes and Mueller 1992). Additionally, application of early postemergence-directed herbicides requires precision application to avoid phytotoxicity on cotton. Producers would have the option of making a second postemergence application of bromoxynil or glyphosate, or one of the currently used postemergence-directed herbicides, if additional control was needed. In fields with light infestations of weeds, producers would be able to use total postemergence management systems. The herbicide load in these fields could be reduced by at least 50% as compared to current management systems.

Lack of effective and noninjurious herbicides for postemergence over-the-top application has been one of the major factors restricting expanded production of no-tillage cotton. Herbicide-resistant cotton will be particularly useful in reduced-tillage and no-tillage systems where early postemergence-directed herbicide application currently is critical for successful weed control (Oliver et al. 1990).

Imidazolinone-resistant cotton has not been developed. Further development of imidazolinone-resistant crops likely would be unwise in light of their widespread use in many crops and the rapid evolution of resistance to these herbicides. However, resistance of cotton to lower concentrations of imidazolinone herbicides would be beneficial. Imidazolinone carryover to cotton often is observed when they are applied to soybean or peanut planted in direct rotation with cotton (Barnes et al. 1989; Barnes and Lavy 1991; Johnson and Talbert 1991; Johnson et al. 1993; York and Wilcut 1993). Potential carryover of AC 263,222 to cotton, which effectively controls a number of problem annual broadleaf and perennial sedge species in peanut, could limit the use of this herbicide in regions where peanut and cotton are commonly planted in rotation (York and Wilcut, unpublished data).

4.4. GRAIN SORGHUM

The major deficiency in sorghum weed management is lack of postemergence control options for annual and perennial grasses. Unfortunately, annual and perennial grass weeds are among the most common and troublesome weeds for sorghum producers (Dowler 1991). Sorghum is not tolerant to nicosulfuron or primisulfuron which are used postemergence to control annual and perennial grasses in corn, nor is it tolerant of cyclohexanedione or aryloxyphenoxypropionate herbicides used to control emergent annual and perennial grasses in cotton, peanut, and soybean (Anon 1994). Development of glyphosate-, glufosinate-, or sethoxydim-resistant sorghum cultivars would allow postemergence control of these grasses. However, because sorghum and weeds such as shattercane and johnsongrass may cross-pollinate, development of HR sorghum must be approached with extreme caution (Radesovich et al. 1992).

4.5. PEANUT

Peanut generally requires significant herbicide inputs for profitable production. To our knowledge, no current research is underway for the development of HR peanut. The potential for HR peanut is limited. Most weeds can be controlled with well-planned programs using currently registered herbicides (Wilcut et al. 1994). Additionally, peanut is not overly sensitive to herbicides used in rotational crops (Wilcut et al. 1994). However, development of cultivars resistant to glyphosate, glufosinate, or other herbicides effective on perennial broadleaf weeds would be helpful. Perennial broadleaf weeds such as alligatorweed [*Alternanthera philoxeroides* (Mart.) Griseb.], bigroot morningglory [*Ipomoea pandurata* (L.) G.F.W. Meyer], Carolina horsenettle (*Solanum carolinense* L.), robust horsenettle (*Solanum dimidiatum* Raf.), silverleaf nightshade (*Solanum elaeagnifolium* Cav.), and trumpetcreeper [*Campsis radicans* (L.) Seem. ex Bureau] cannot be controlled in peanut with current registered herbicides (Wilcut et al. 1994). These weeds must be controlled in corn or fallow fields rotated with peanut.

4.6. RICE

Red rice is one of the most common and troublesome weeds for rice producers in the mid-southern U.S. (Dowler 1991). Control is achieved primarily through the use of water management and other cultural practices. Rice cultivars with glufosinate or glyphosate resistance would be valuable to producers with red rice infestations or with HR weed populations (Carey and Talbert 1994). Some glufosinate-resistant cultivars have shown excellent tolerance and control of red rice with glufosinate has been good to excellent (Braverman and Linscombe 1994). Either herbicide also controls numerous other grass and broadleaf weed species. Management systems integrating resistant cultivars with cultural practices should be very effective. Care would have to be taken that herbicide resistance in HR rice would not be transferred to red rice (Radesovich et al. 1992).

4.7 SMALL GRAINS

Annual and perennial grass management options in small grain production are severely limited. Diclofop-methyl {(±)-2-[4-(2,4-dichlorophenoxy)phenoxy]propanoic acid} is effective on several annual grass weeds but does not control *Bromus* spp. including downy brome (*Bromus tectorum* L.) and jointed goatgrass (*Aegilops cylindrica* Host) (Anon 1994). In the central Great Plains, weedy grasses are dominating no-till small grain production such that small grains cannot be grown profitably under no-tillage production (Peeper, personal communication). Cultural practices including tillage and narrow row spacing are ineffective for weedy grass control (Driver et al. 1993, Justice et al. 1993; Peeper 1984; Wicks 1984). Additionally, diclofop-methyl is an ACCase-inhibiting herbicide and resistance to this mode of action is becoming widespread (Heap 1994; Snipes et al. 1993; Stanger and Appleby 1989; Stoltenberg and Appleby 1989). Other weed management problems include spread of ALS HR weed populations primarily in the central Great Plains and the Pacific Northwest.

Development of glyphosate- and/or glufosinate-resistant small grain cultivars would provide additional management options for annual and perennial grass and broadleaf weeds. Both herbicides would control many of the resistant weed species that have evolved in small grain production areas, although glyphosate is not particularly effective for kochia control. Metribuzin resistance in small grains also would aid in the control of several annual grass and broadleaf weeds (Blackshaw 1994; Justice et al. 1993; Runyan et al. 1982).

4.8. SOYBEAN

There currently are at least 70 registrations for herbicides or packaged herbicide mixtures for weed management in soybean, plus numerous mixtures (more than one herbicide is applied simultaneously) (Anon 1994). Most weeds in soybean can be adequately controlled with existing herbicides in well-planned management systems. There are no glaring deficiencies in soybean weed management such as the lack of postemergence control options for broadleaf weeds in cotton or annual and perennial grasses in sorghum. However, several efforts are underway to develop HR soybean (Table 2). Additionally, sulfonylurea-tolerant soybean has been developed (Forney et al. 1989, Mosley et al. 1993).

Sulfonylurea-tolerant soybean cultivars are more tolerant of higher rates of the sulfonylurea herbicides chlorimuron and thifensulfuron {3-[[[[(4-methoxy-6-methyl-1,3,5-triazin-2 yl)amino]carbonyl]amino]sulfonyl]-2-thiophenecarboxylic acid}. These higher rates are necessary for controlling difficult weeds, including common lambsquarters (*Chenopodium album* L.), pigweed species, *Ipomoea* morningglories, sicklepod, and velvetleaf (Forney et al. 1989; Monks et al. 1993). These cultivars would be more tolerant of an application mixture of chlorimuron plus thifensulfuron. This mixture controls the aforementioned weeds better than either herbicide applied alone (Monks et al. 1993). Additionally, this increased sulfonylurea tolerance would reduce injury potential from adverse environmental conditions, mistakes in spray calibration, overlapping of spray booms, or where two applications are necessary to control difficult broadleaf weeds such as sicklepod and velvetleaf.

Despite these advantages, we feel the potential disadvantages are significant. For example, resistance to the ALS-inhibiting herbicides continues to increase and conceivably would only worsen with more widespread usage of sulfonylurea herbicides. Additionally, increased rates of sulfonylurea herbicides, especially chlorimuron, may adversely affect sensitive rotational crops.

Most sulfonylurea herbicides have stringent restrictions for rotational crops, especially vegetables (Anon. 1994).

Glyphosate-resistant soybean would offer several significant advantages over current weed management systems. As mentioned in the cotton section, glyphosate would provide economically effective control of numerous grass and broadleaf weeds. Glyphosate-resistant soybean would offer an alternative management system for fields with cyclohexadinone, aryloxyphenoxypropionate, imidazolinone, sulfonylurea, and dinitroaniline HR weeds. Furthermore, this technology would be ideal for reduced-tillage or no-tillage production systems. The development of these cultivars could result in less herbicide use and, thus, less expensive management systems and/or provide more flexibility in application (Doss and York 1995).

4.9. VEGETABLE CROPS

Profitable vegetable production requires high labor, economic, and management inputs. One critical component for successful production is development and implementation of a good weed management program. Sicklepod, common cocklebur, Eastern black nightshade (*Solanum ptycanthum* Dun.), morning glory species, and yellow and purple nutsedges are listed among the most troublesome weeds in U.S. vegetables (Bridges 1992). In vegetables (e.g., tomato, pepper, cantaloupe, and watermelon) where plastic mulch (black, infrared-transmissible) is used, well-planned programs utilizing herbicides and hand-weeding controls most weeds (Standifer and Beste 1985). Approximately 16.1 million kg of plastic are used in vegetable production in the U.S. (Lamont, personal communication). Obviously the disposal and environmental impact of this plastic is significant. In other vegetables (e.g., sweet potato, sweet corn, snap bean, and pea) where plastic mulch is not used in production, herbicides and multiple cultivations and hand-weeding must be used to control most weeds. Over 70% of vegetable crops currently grown lack either preemergence or postemergence herbicides that give effective, broad-spectrum weed control.

There are exceptions to this statistic. Some are the vegetables sweet corn, tomato, and potato (Giaquinta 1992; Stephenson et al. 1976). Other vegetables are produced using plastic mulch (Standifer and Beste 1985). Limited herbicide registrations are directly related to the high cost of herbicide development and to the liability that the company assumes with product usage on a high-value, minor-use crop (Duvick 1992). Since 1986 there have been over 50 cancellations in herbicide registrations in vegetables, with many of these cancellations due to the loss of diphenamid, dinoseb, or chloramben (Meister 1986). The lack of effective herbicides registered in vegetables has made it extremely difficult to produce these crops in reduced tillage systems (Hoyt et al. 1994; Standifer and Beste 1985). Beets, broccoli, cucumber, greens, muskmelon, pumpkin, squash, sweet potato, and watermelon are among the vegetables that cannot be produced commercially in reduced tillage systems due to lack of registered herbicides. Development of commercial vegetable varieties with resistance to herbicides that control troublesome weeds would allow for more efficient production of vegetables in both conventional and reduced tillage systems.

Precision cultivation is an important part of a cultural system for commercial production of most vegetables (Parrish et al. 1992). To kill weeds, cultivation must be done when weeds are small, usually 5 cm tall or less, and as close to the plant row as possible without injuring the crop. Precise cultivation involves specialized cultivating equipment that must be properly adjusted. The number of cultivations required for different vegetable crops depends on the weed species present and weed population density, but generally ranges from three to five per season (Patriquin 1987; Tothe et al. 1993). This intense program of cultivation can be detrimental to crops such as potato (Strand and Rude 1992) and can lead to enhanced wind and water erosion of soil. Thus, the commercial development of HR vegetables would likely reduce the number of cultivations required to produce vegetable crops. Additionally, during seasons with high rainfall, when cultivation is often less effective or impossible, a postemergence herbicide to which the vegetable crop is resistant would be a valuable option.

TABLE 2 Representative Listing of Research on Development of HRCs by Private Industry

Herbicide[a]	Crop	Company	Method[b]	Strategy[c]
Atrazine (Aatrex®)	Canola	Allelix Crop Technol.	B	Site
Bialaphos	Corn	DeKalb Plant Genetics	A	?
	Soybean	Upjohn	A	?
Bromoxynil (Buctril®)	Tobacco	Calgene	A	Degrade
	Cotton	Calgene	A	Degrade
Glufosinate (Basta®)	Tobacco, potato, tomato	Plant Genetic Systems	A	Degrade
	Sugarbeet, oilseed rape	Plant Genetic Systems	A	Degrade
	Alfalfa	Plant Genetic Systems	A	Degrade
	Maize	DeKalb Plant Genetics	A	Degrade
Glyphosate (Roundup®)	Poplar	Calgene	A	Site
	Soybean	Monsanto Co.	A	Site
	Cotton	Monsanto Co.	A	?
	Tomato	Monsanto Co.	A	?
Imidazolinones	Canola	American Cyanamid/Allelix Crop Technol.	B	?
	Maize	American Cyanamid/Plant Science Research	C	Site
Sulfonylureas	Tobacco	DuPont	A	Site
	Canola	Allelix Crop Technol.	C	Site
	Cotton	DuPont	A	?
2,4-D	Tobacco	Schering AG	A	Degrade

[a] Trade names are in parentheses. Only the most widely used trade names are provided.
[b] A = gene transfer by biotechnololgy, B = microspore mutagenesis and selection, C = cell selection.
[c] Site = site of action, Degrade = enzyme(s) for degradation of herbicide.

Giddings, L.V. 1991. List of issued and pending permits from the Biotechnology, Biologics, and Environmental Protection Division of APHIS for field tests of transgenic organisms.

There are several herbicides being investigated for commercial development of resistance in vegetables, including sulfonylurea and imidazolinone herbicides, glufosinate, and glyphosate (Table 1 and Table 2). Vegetables with resistance to any of the aforementioned herbicides would provide valuable management options for producers and decrease management inputs for labor. The sulfonylurea and imidazolinone herbicides offer outstanding broad spectrum control of numerous species, including many of the weeds that are considered troublesome. These herbicides also have low environmental impact and low mammalian toxicity.

While the aforementioned herbicides offer valuable weed management options, they also have significant rotation restrictions for numerous vegetables that are not resistant. Many vegetable producers have intense cropping systems, often producing two or more crops per year. Their production systems must be flexible for a rapidly changing marketplace. Also, in the U.S. vegetables are commonly produced on beds and this practice increases the likelihood of injury to rotational crops from herbicide residue. Thus, unless herbicide resistance could be commercially developed in a majority of vegetable crops, use of residual herbicides that adversely impact rotational flexibility would be a major limitation in many regions. Nonresidual postemergence options that would integrate with existing cultural practices and not limit rotational crops would be more beneficial (Eberlein et al. 1992).

As previously mentioned, vegetables are high input crops and yield reductions have a greater impact in high-value crops than most agronomic crops (Standifer and Beste 1985). Thus, any attempt to commercially develop resistance in these crops that would result in lower yield or quality would limit its use. Additionally, vegetables in which herbicide resistance has been commercially developed could become weedy in sequentially planted crops. Vegetables having seed with hard seed coats or vegetative reproductive systems and having resistance to herbicides registered in sequentially planted crops would appear to have the greatest potential for becoming weedy. Vegetables that have the potential to become weedy include greens (i.e., mustard, spinach, and turnip), okra, onion, potato, southern pea, sweet corn, and tomato (Monks, Shaw, Gibbs, and Stall, personal communication). Certain cultural methods could be used to prevent these crops from surviving to the next cropping season and being weedy. One such method would be to apply excess irrigation water to fields after potato harvest in order to cause decay in potato tubers, therefore eliminating the survival of viable tubers to the next season. Thus, HR potatoes would not emerge to cause problems in corn, cotton, or peanut the next season. Another option would be to plant a competitive summer crop in the summer after temperatures are hot. Potatoes do not thrive under hot conditions.

There are benefits for developing commercial vegetable varieties having herbicide resistance. Development of HR vegetables resistant to more effective herbicides than currently registered in vegetables would be extremely advantageous. More effective herbicides would reduce the actual amount of herbicide being applied to many vegetables because of greater efficiency.

Unlike the development of insect and disease resistance in these crops, herbicide resistance in vegetable crops will require a two-step process: (1) development of herbicide resistance in a commercial vegetable variety while maintaining crop yield and quality, and (2) registration of a herbicide in a minor-use crop. Thus, development of commercial vegetable varieties having resistance to a registered herbicide may be more difficult than in agronomic crops or those with insect and disease resistance. Herbicide registration would probably be handled through Interregional Research Project No. 4. This process has reduced the difficulty in registering herbicides in these minor use crops.

5. SUMMARY

Herbicide-resistant crops would allow innovative weed management strategies for numerous crops in the U.S. If properly used, this technology would help promote and increase the hectarage of crops under IPM production, and likely would reduce the amount of herbicides being used. Herbicide-resistant crops also could be used to preclude the development of herbicide resistance and in the management of HR weeds.

The uncertainties that exist, most of the possible dangers, and potential for mismanagement already exist with conventional breeding and plant tissue culturing (Goodman 1987). Most of these pitfalls are predictable, and their impact can be minimized if managed correctly. In the past we have used herbicide safeners premixed with the herbicide or directly placed on crop seed (Stephenson 1987). Genetic engineering enables us to put biological safeners directly into the seed and allows more options for reducing the environmental impact of weed management while improving net profitability for producers.

REFERENCES

Anon. 1992. Pesticide State Management Plan Guidance for Groundwater Protection. Implementation Document for the Pesticides and Groundwater Strategy, U.S. EPA Offices of Pesticides Programs, Environmental Protection Agency, Washington, D.C.

Anon. 1993a. BST getting close to market. *Agric. Biotechnol. News* January: 1.

Anon. 1993b. Flavr Savr launch on schedule. *Biotech. Rep.* July: 1.

Anon. 1994. *Crop Protection Chemicals Reference,* 9th ed. Chemical Pharmaceutical Publishing and John Wiley & Sons, New York.

Bailey, J.A. and Kapusta, G. 1993. Soybean (*Glycine max*) tolerance to simulated drift of nicosulfuron and primisulfuron. *Weed Technol.* 7: 740.

Barnes, C.J. and Lavy, T.L. 1991. Injury and yield response of selected crops to imazaquin and norflurazon residues *Weed Technol.* 5: 598.

Barnes, C.J., Goetz, A.J., and Lavy, T.L. 1989. Effects of imazaquin residues on cotton (*Gossypium hirsutum*). *Weed Sci.* 37: 820.

Barrentine, W.L., Edwards, C.J., Jr., and Hartwig, E.E. 1976. Screening soybeans for tolerance to metribuzin. *Agron. J.*, 68: 351.

Barrentine, W.L., Hartwig, E.E., Edwards, C.J., Jr., and Kilen, T.C. 1982. Tolerance of three soybean cultivars to metribuzin. *Weed Sci.* 30: 344.

Beversdorf, W.D. and Hume, D.J. 1984. OAC Triton spring rape. *Can. J. Plant Sci.* 64: 1007.

Beversdorf, W.D. and Kott, L.S. 1987. Development of triazine resistance in crops by classical plant breeding. *Weed Sci.* 35(Suppl. 1): 9.

Beversdorf, W.D., Hume, D.J., Gostovic, P., Chu, G., Montiminy, W., and Muscat, J. 1990. OAC Triumph summer rape. *Can. J. Plant Sci.* 70: 857.

Beyer, E.M., Jr., Duffy, M.J., Hay, J.V., and Schlueter, D.D. 1988. Sulfonylureas. In *Herbicides: Chemistry, Degradation, and Mode of Action,* Vol. 3. Ed. by Kearney, P.C. and Kaufman, D.D., Marcel Dekker, New York, chap. 3.

Bhalla, P., Hackett, N., Hart, R., and Lignowski, E. 1991. Imazaquin herbicide. In *The Imidazolinone Herbicides.* Ed. by Shaner, D.L. and O'Connor. S.L. CRC Press, Boca Raton, FL, chap. 19.

Blackshaw, R.E. 1994. Downy brome (*Bromus tectorum*) control in winter wheat and winter rye. *Can. J. Plant Sci.* 74: 185.

Braverman, M.P. and Linscombe, S.D. 1994. Field evaluation of transgenic-glufosinate resistant rice. In *Proc. South. Weed Sci. Soc.*, Vol. 47. Ed. by Street, J.E. Southern Weed Science Society, Champaign, IL.

Bridges, D.C. 1992. *Crop Losses Due to Weeds in the United States - 1992.* Weed Science Society of America, Champaign, IL.

Bruce, J.A., Penner, D., and Kells, J.J. 1993. Absorption and activity of nicosulfuron and primisulfuron in quackgrass (*Elytrigia repens*) as affected by adjuvants. *Weed Sci.* 41: 218.

Burnside, O.C. 1992. Rationale for developing herbicide-resistant crops. *Weed Technol.* 6: 621.

Byrd, J.D., Jr., 1993. Report of the 1992 cotton weed loss committee, in *Proc. Beltwide Cotton Conf.,* Vol. 3. Ed. by Herber, D.J. and Richter, D.A. National Cotton Council of America, Memphis, TN, p. 1499.

Byrd, J.D., Jr. and York, A.C. 1987a. Interaction of fluometuron and MSMA with sethoxydim and fluazifop. *Weed Sci.* 35: 270.

Byrd, J.D., Jr. and York A.C. 1987b. Annual grass control in cotton (*Gossypium hirsutum*) with fluazifop, sethoxydim, and selected dinitroaniline herbicides. *Weed Sci.* 35: 388.

Carey, V.F. and Talbert, R.E. 1994. Investigation of propanil-resistant barnyardgrass (*Echinochloa crus-galli* L. Beauv.) in Arkansas. *Weed Sci. Soc. Am. Abstr.* 34: 13.

Coble, H.D. and Byrd, J.D., Jr. 1992. Interference of weeds with cotton. In *Weeds of Cotton: Characterization and Control,* Cotton Foundation Ref. Bk. Ser. No. 2. Ed. by McWhorter, C.G. and Abernathy, J.R. Cotton Foundation Publishers, Memphis, TN, chap. 4.

Coble, H.D. and Mortensen, D.A. 1992. The threshold concept and its application to weed science. *Weed Technol.* 6: 191.

Curran, W.S., Knake, E.L., and Liebl, R.A. 1991. Corn (*Zea mays*) injury following use of clomazone, chlorimuron, imazaquin, and imazethapyr. *Weed Technol.* 5: 539.

Doss, L.G. and York, A.C. 1995. Roundup Ready™ soybean response to glyphosate and acetolactate synthase inhibitor combinations. In *Proc. South. Weed Sci. Soc.*, Vol. 48. Ed. by Street, J.E. Southern Weed Science Society, Champaign, IL.

Dotray, P.A., Marshall, L.C., Parker, W.P., Wyse, D.L., Somers, D.A., and Gengenbach, B.G. 1993. Herbicide tolerance and weed control in sethoxydim-tolerant corn (*Zea mays*). *Weed Sci.* 41: 213.

Dowler, C.C. 1991. Weed survey — southern states. In *Proc. South. Weed Sci. Soc.*, Vol. 44. Ed. by Witt, W.W. Southern Weed Science Society, Champaign, IL, p. 426.

Driver, J.E., Peeper, T.F., and Koscelny, J.A. 1993. Cheat (*Bromus secalinus*) control in winter wheat (*Triticum aestivum*) with sulfonylurea herbicides. *Weed Technol.* 7: 851.

Duke, S.O. 1988. Glyphosate. In *Herbicides: Chemistry, Degradation, and Mode of Action*, Vol. 3. Ed. by Kearney, P.C. and Kaufman, D.D. Marcel Dekker, New York, chap. 1.

Duke, S.O. 1993. Mechanisms for resistance of weeds to herbicides. In *Proc. Beltwide Cotton Conf.*, Vol. 3. Ed. by Herber, D.J. and Richter, D.A. National Cotton Council of America, Memphis, TN, p. 1509.

Duke, S.O., Christy, A.L., Hess, F.D., and Holt, J.S. 1991. *Herbicide-Resistant Crops.* Council Agriculture Science Techology, Ames, IA.

Duvick, D.N. 1992. Concerns of seed company officials with herbicide-tolerant cultivars. *Weed Technol.* 6: 640.

Eberlein, C.V., Belknap, W.R., and Guttieri, M.J. 1992. Bromoxynil tolerance in transgenic potato plants expressing a detoxification gene. *Weed Sci. Abstr.* 32: 88.

Forney, D.R., Sebastian, S., Fader, G.M., and Ulrich, J. 1989. Development of sulfonylurea-resistant soybeans and associated weed control options, in *Weed Sci. Soc. Am. Abstr.* 29: 176.

Giaquinta, R.T. 1992. An industry perspective on herbicide-tolerant crops. *Weed Technol.* 6: 653.

Goldburg, R.J. 1992. Environmental concerns with the development of herbicide-tolerant plants. *Weed Technol.* 6: 647.

Goodman, R.M. 1987. Future potential, problems and practicalities of herbicide tolerant crops from genetic engineering.

Weed Sci. 35(Suppl. 1): 28.
Graf, G.T. and Ogg, A.G., Jr. 1976. Differential response of potato cultivars to metribuzin. *Weed Sci.* 24: 137.
Gressel, J. and Segel, L.A. 1990. Modelling the effectiveness of herbicide rotations and mixtures as strategies to delay or preclude resistance. *Weed Technol.* 4: 186.
Grichar, W.J. 1991. Sethoxydim and broadleaf herbicide interaction effects on annual grass control in peanuts (*Arachis hypogaea*). *Weed Technol.* 5: 321.
Guthrie, D.S. and York, A.C. 1989. Cotton (*Gossypium hirsutum*) development and yield following fluometuron postemergence applied. *Weed Technol.* 3: 501.
Harris, J.E. and Gagne, J.A. 1991. Toxicology of the imidazolinone herbicides. In *The Imidazolinone Herbicides*. Ed. by Shaner, D.L. and O'Connor, S.L. CRC Press, Boca Raton, FL, chap 14.
Harrison, H.F., Jr., Jones, A., and Dukes, P.D. 1985. Differential response of six sweet potato (*Ipomoea batatas*) cultivars to metribuzin. *Weed Sci.* 33: 730.
Hart, R., Lignowski, E., and Taylor, F. 1991. Imazethapyr herbicide. In *The Imidazolinone Herbicides*. Ed. by Shaner, D.L. and O'Connor, S.L. CRC Press, Boca Raton, FL, chap. 20.
Hartwig, E.E., Barrentine, W.L., and Edwards, C.J., Jr. 1980. Registration of Tracy-M soybeans. *Crop Sci.* 20: 285.
Hartwig, E.E. 1987. Identification and utilization of variation of herbicide tolerance in soybean (*Glycine max*) breeding. *Weed Sci.* 35(Suppl. 1): 4.
Hays, R.M., Shankle, M.W., and Shelby, P.P. 1993. Documentation and control of DSMA, MSMA resistant common cocklebur. In *Proc. Beltwide Cotton Conf.*, Vol. 3. Ed. by Herber, D.J. and Richter, D.A. National Cotton Council of America, Memphis, TN, p. 1526.
Heap, I.M. 1994. Multiple resistance to dinitroaniline and ACCase inhibiting herbicides in green foxtail [*Setaria viridis* (L.) Beauv.). *Weed Sci. Soc. Am. Abst.* 34: 56.
Holhouser, D.L. and Coble, H.D. 1990. Compatibility of sethoxydim with five postemergence broadleaf herbicides. *Weed Technol.* 4: 128.
Holt, J.S. 1992. History of identification of herbicide-resistant weeds. *Weed Technol.* 6: 615.
Hoyt, G.D., Monks, D.W., and Monaco, T.J. 1994. Conservation tillage for vegetable production. *Hort. Tech.* 4(2): 129.
Johnson, D.H. and Talbert, R.E. 1991. Effect of imazaquin residues on growth and yield of cotton. In *Proc. South. Weed Sci. Soc.*, Vol. 44. Ed. by Witt, W.W. Southern Weed Science Society, Champaign, IL, p. 409.
Johnson, D.H., Jordan, D.L., Johnson, W.G., Talbert, R.E., and Frans, R.E. 1993. Nicosulfuron, primisulfuron, imazethapyr, and DPX-350 injury to succeeding crops. *Weed Technol.* 7: 641.
Johnson, W.C., III and Mullinix, B.G., Jr., 1990. Efficacy and economic analysis of Texas panicum (*Panicum texanum*) management systems in corn (*Zea mays*). *Weed Technol.* 4: 754.
Jordan, D.L. and York, A.C. 1989. Effect of ammonium fertilizers and BCH 81508S on antagonism with sethoxydim plus bentazon mixtures. *Weed Technol.* 3: 450.
Jordan, D.L., Frans, R.E., and McClelland, M.R. 1993a. Cotton (*Gossypium hirsutum*) response to DPX-PE350 applied postemergence. *Weed Technol.* 7: 159.
Jordan, D.L., Frans, R.E., and McClelland, M.R. 1993b. Total postemergence herbicide programs in cotton (*Gossypium hirsutum*) with sethoxydim and DPX-PE350. *Weed Technol.* 7: 196.
Jordan, D.L., Smith, M.C., McClelland, M.R., and Frans, R.E. 1993c. Weed control with bromoxynil applied alone and with graminicides. *Weed Technol.* 7: 835.
Jordan, D.L., Johnson, D.H., Johnson, W.G., Kendig, J.A., Frans, R.F., and Talbert, R.E. 1993d. Carryover of DPX-PE350 to grain sorghum (*Sorghum bicolor*) and soybean (*Glycine max*) on two Arkansas soils. *Weed Technol.* 7: 645.
Justice, G.G., Peeper, T.F., Solie, J.B., and Epplin, F.M. 1993. Net returns from cheat (*Bromus secalinus*) control in winter wheat (*Triticum aesitivum*), *Weed Technol.* 7: 459.
Kendig, J.A. and DeFelice, M.S. 1994. ALS-resistant cocklebur (*Xanthium strumarium*) in Missouri. *Weed Sci. Soc. Am. Abstr.* 34: 12.
Kishore, G.M., Padgette, S.R., and Fraley, R.T. 1992. History of herbicide-tolerant crops, methods of development and current state of the art — emphases on glyphosate tolerance. *Weed Technol.* 6: 626.
Knake, E.L., 1992. Technology transfer for herbicide-tolerant weeds and herbicide-tolerant crops. *Weed Technol.* 6: 662.
Mangels, G. 1991a. Behavior of the imidazolinone herbicides in the aquatic environment. In *The Imidazolinone Herbicides*. Ed. by Shaner, D.L. and O'Connor, S.L. CRC Press, Boca Raton, FL, chap. 15.
Mangels, G. 1991b. Behavior of the imidazolinone herbicides in the soil — a review of the literature. In *The Imidazolinone Herbicides*. Ed. by Shaner, D.L. and O'Connor, S.L. CRC Press, Boca Raton, FL, chap. 16.
Meister, R.T. 1986. Vegetables. In *Weed Control Manual and Herbicide Guide*. Meister Publishing, Willoughby, OH, chap. 5.
Mills, J.A. and Witt, W.W. 1991. Dissipation of imazaquin and imazethapyr under conventional and no-tillage soybean (*Glycine max*). *Weed Technol.* 5: 586.
Monks, C.D., Wilcut, J.W., and Richburg, J.S., III. 1993. Broadleaf weed control in soybean (*Glycine max*) with chlorimuron plus acifluorfen or thifensulfuron mixtures. *Weed Technol.* 7: 317.

Mosley, C., Hatzios, K.K., and Hagood, E.S., 1993. Uptake, translocation, and metabolism of chlorimuron in soybean (*Glycine max*) and morningglory (*Ipomoea spp.*). *Weed Technol.* 7: 343.

Mudge, L.C., Gossett, B.J., and Murphy, T.R. 1984. Resistance of goosegrass (*Eleusine indica*) to dinitroaniline herbicides. *Weed Sci.* 32: 591.

Murdock, E.C., York, A.C., and Gossett, B.J. 1993. The history and present status of herbicide-resistant weed biotypes in South Carolina cotton. In *Proc. Beltwide Cotton Conf.*, Vol. 3. Ed. by Herber, D.J. and Richter, D.A. National Cotton Council of America, Memphis, TN, p. 1507.

Newhouse, K.E., Wang, T., and Anderson, P.C. 1991. Imidazolinone-resistant crops. In *The Imidazolinone Herbicides*. Ed. by Shaner, D.L. and O'Connor, S.L. CRC Press, Boca Raton, FL, chap. 10.

Ngouajio, M. and Hagood, E.S., Jr., 1993. Weed control in corn (*Zea mays*) with primisulfuron as influenced by rate, timing, and herbicide combinations. *Weed Technol.* 7: 65.

Nielsen, E.G. and Lee, L.K. 1987. *The Magnitude and Costs of Groundwater Contamination from Agricultural Chemicals*, Environmental Protection Agency, Washington, D.C.

Oliver, G.W., York, A.C., and Worsham, A.D. 1993. Weed management in no-till cotton. In *Proc. Beltwide Cotton Conf.*, Vol. 3. Ed. by Herber, D.J. and Richter, D.A. National Cotton Council of America, Memphis, TN, p. 350.

Parker, W.B., Thompson, L., Jr., and Godley, F.M. 1985. Integrating sethoxydim into soybean (*Glycine max*) weed management systems. *Weed Sci.* 33: 100.

Parrish, R.L., Bracy, R.P., and Bergeron, P.E. 1992. Precision Cultural Practices for Commercial Vegetable Production, La. Agric. Exp. Stn. Bull. No. 836, Louisiana State University, Baton Rouge.

Patriquin, D.G. 1987. Weed control in organic farming systems. In *Weed Management in Agroecosystems: Ecological Approaches*. Ed. by Altieri, M.A. and Liebman, M. CRC Press, Boca Raton, FL, chap. 17.

Peeper, T.F. 1984. Chemical and biological control of downy brome (*Bromus tectorum*) in winter wheat and alfalfa in North America. *Weed Sci.* 32(Suppl. 1): 18.

Powles, S.B. and Howat, P.D. 1990. Herbicide-resistant weeds in Australia. *Weed Technol.* 4: 178.

Radosevich, S.R., Ghersa, C.M., and Comstock, G. 1992. Concerns a weed scientist might have about herbicide-tolerant crops. *Weed Technol.* 6: 635.

Renner, K.A., Meggitt, W.F., and Penner, D. 1988. Effect of soil pH on imazaquin and imazethapyr adsorption to soil and phytotoxicity to corn (*Zea mays*). *Weed Sci.* 36: 78.

Renner, K.A. and Powell, G.E. 1991. Response of sugarbeet (*Beta vulgaris*) to herbicide residues in soil. *Weed Technol.* 5: 622.

Richburg, J.S., III, Wilcut, J.W., and Ingram, E.G. 1994a. Weed efficacy with Buctril and Roundup. In *Proc. Beltwide Cotton Conf.*, Vol. 3. Ed. by National Cotton Council of America, Memphis, TN, p. 1702.

Richburg, J.S., III, Wilcut, J.W., and Wehtje, G.R. 1994b. Toxicity of AC 263,222 to purple (*Cyperus rotundus*) and yellow nutsedge (*Cyperus esculentus*), *Weed Sci.* 42: 398.

Runyan, T.J., McNeil, W.K., and Peeper, T.F. 1982. Differential tolerance of wheat (*Triticum aestivum*) cultivars to metribuzin. *Weed Sci.* 30: 94.

Shaner, D.L. and Mallipudi, N.M. 1991. Mechanisms of selectivity of the imidazolinones. In *The Imidazolinone Herbicides*. Ed. by Shaner, D.L. and O'Connor, S.L. CRC Press, Boca Raton, FL, chap. 7.

Shaw, D.R., and Wesley, M.T. 1991. Wheat (*Triticum aestivum*) cultivar tolerance and Italian ryegrass (*Lolium multiflorum*) control with diclofop, BAY SMY 1500, and metribuzin. *Weed Technol.* 5: 776.

Snipes, C.E. and Mueller, T.C. 1992a. Influence of fluometuron and MSMA on cotton yield and fruiting characteristics. *Weed Sci.* 42: 210.

Snipes, C.E. and Mueller, T.C. 1992b. Cotton (*Gossypium hirsutum*) yield response to mechanical and chemical weed control. *Weed Sci.* 40: 249.

Snipes, C.E., Barrentine, W.L., and Smeda, R.J. 1993. Documentation of graminicide-resistant johnsongrass in cotton. In *Proc. Beltwide Cotton Conf.*, Vol. 3. Ed. by Herber, D.J. and Richter, D.A. National Cotton Councilil of America, Memphis, TN, p. 1508.

Standifer, L.C. and Beste, C.E. 1985. Weed control methods for vegetable production with limited tillage. In *Weed Control in Limited Tillage Systems*. Ed. by Wiese, A.F. Weed Science Society of America, Champaign, IL, p. 93.

Stanger, C.E. and Appleby, A.P. 1989. Italian ryegrass (*Lolium multiflorum*) accessions tolerant to diclofop. *Weed Sci.* 37: 350.

Stephenson, G.R. and Ezra, G. 1987. Chemical approaches for improving herbicide selectivity and crop tolerance. *Weed Sci.* 35(Suppl. 1): 24.

Stephenson, G.R., McLeoud, J.E., and Phatak, S.C. 1976. Differential tolerance of tomato cultivars to metribuzin. *Weed Sci.* 24: 161.

Stidham, M.A. and Singh, B.K. 1991. Imidazolinone-acetohydroxy acid synthase interactions. In *The Imidazolinone Herbicides*. Ed. by Shaner, D.L. and O'Connor, S.L. CRC Press, Boca Raton, FL, chap 6.

Stoltenberg, D.E. and Wiederholt, R.J. 1994. Resistance of large crabgrass [*Digitaria sanguinalis* (L.) Scop.] biotypes to acetyl-coenzyme A carboxylase inhibitors. In *Weed Sci. Soc. Am. Abstr.* 34: 94.

Strand, L.L. and Rude, P.A. 1992 Integrated Pest Management for Potatoes in the Western United States. University of California Division of Agriculture and Natural Resources Publ. No. 3316.

Tothe, S.J., Duncan, H.E., Monks, D.W., Sorensen, K.A., and Wilson, L.G. 1993. Potato Pest Management 1989: A Survey of Pesticide Use and Other Pest Management Practices by North Carolina Potato Producers, Bull. No. AG-497. North Carolina State University Coop. Ext. Serv., Raleigh, NC, p. 1.

Wall, D.A. 1992. Flurtamone for wild mustard (*Sinapis arvensis*) control in canola (*Brassica napans*) and (*B. campestris*). *Weed Technol.* 6: 878.

Wicks, G.A. 1984. Integrated systems for control and management of downy brome (*Bromus tectorum*) in cropland. *Weed Sci.* 32(Suppl. 1): 26.

Wilcut, J.W., Coble, H.D., Murdock, E.C., and Snipes, C.E. 1992. Weed response to bromoxynil in Georgia, North and South Carolina, and Mississippi cotton. In *Proc. Beltwide Cotton Conf.*, Vol. 3. Ed. by Herber, D.J. and Richter, D.A. National Cotton Council of America, Memphis, TN, p. 1317.

Wilcut, J.W., Eastin, E.F., and Richburg, J.S., III. 1993. Buctril systems and efficacy for transgenic cotton in Georgia. In *Proc. Beltwide Cotton Conf.*, Vol. 3. Ed. by Herber, D.J. and Richter, D.A. National Cotton Council of America, Memphis, TN, p. 1524.

Wilcut, J.W., Eastin, E.F., Richburg, J.S., III, Vencill, W.K., Walls, F.R., and Wiley, G. 1993b. Imidazolinone systems for southern weed management in resistant corn. *Weed Sci. Soc. Am. Abstr.* 33: 335.

Wilcut, J.W., York, A.C., and Wehtje, G.R. 1994. The control and interaction of weeds in peanut. *Rev. Weed Sci.* 6

Wrubel, R.P. and Gressel, J. 1994. Are herbicide mixtures useful for delaying the rapid evolution of resistance? A case study. *Weed Technol.* 8: 635.

Wyse, D.L. 1992. Future impact of crops with modified herbicide resistance. *Weed Technol.* 6: 665.

York, A.C. 1994a. Weed Management in Soybeans. Publ. No. AG-274. North Carolina State University Coop. Ext. Serv., Raleigh, NC.

York, A.C. 1994b. Weed management in cotton. In *1994 Cotton Information*, Publ. No. AG-417. North Carolina State University Coop. Ext. Serv., Raleigh, NC, chap. 10.

York, A.C. and Murdock, E.C. 1993. Management strategies for herbicide-resistant weeds. In *Proc. Beltwide Cotton Conf.*, Vol. 3. Ed. by Herber, D.J. and Richter, D.A. National Cotton Council of America, Memphis, TN, p. 1527.

York, A.C. and Wilcut, J.W. 1993. Insecticides do not affect cotton (*Gossypium hirsutum*) response to imazaquin and imazethapyr. *Weed Sci.* 41: 269.

Zimdahl, R.L. 1980. *Weed Crop Competition, A Review*. International Plant Protection Center, Oregon State University, Corvallis.

Chapter 15

THE POTENTIAL ROLES FOR HERBICIDE-RESISTANT CROPS IN WORLD AGRICULTURE

Jonathan Gressel

CONTENTS

1-56670-045-0/96/$0.00+$.50
© 1996 by CRC Press Inc.

1. INTRODUCTION — AGRICULTURE'S VARYING NEEDS

Previous chapters have described the technologies available, the results achieved, and the market niches filled by herbicide-resistant crops (HRCs) in U.S. agriculture. This chapter will discuss weed control needs that the author feels could be met, at least in part, by HRC, and further chapters try to assess whether it is economic, ecologically desirable, and safe to use HRCs to fill the needs, although it is hard discussing meeting the needs while ignoring these other issues. This review is an update and extension of longer, more comprehensive reviews by this author on similar subject matter (Gressel 1992a, 1993a, b; Gressel et al. 1994).

Weed control is in a state of flux, with many new problems, all of which can be assessed for having possible potentials uses and/or needs for HRC.

1. Many herbicides are being removed from the market and fewer new ones are being released. Can HRCs expand the use range of the remaining compounds by conferring new selectivities?
2. There are legislative demands, especially in Europe, to lower the amounts of herbicides applied to fields. Are there HRCs that will require less herbicide to adequately control weeds? Is less quantity actually better?
3. As subsidies drop, especially in Europe, it may be no longer economical to control weeds with the more expensive of the herbicides. Can we find replacements by developing HRCs for the lower-cost herbicides?
4. Weeds resistant to triazine as well as to acetolactate synthase (ALS) and acetyl coenzyme A carboxylase (ACCase) -inhibiting herbicides already infest an estimated 6 million ha. Are mistakes being made in generating HRCs resistant to the ALS and ACCase herbicides? Can we generate more HRCs for herbicides that are far less prone to the evolution of resistance? Can we use HRCs to overcome the multiple resistance problems evolving in grass weeds of wheat?
5. There are weeds that never were problems and never were controlled by selective herbicides because they had been "held down" by more competitive weeds or by agronomic procedures. Newer selective herbicides controlled the competitive weeds and these other weeds are infesting and reducing yields of crops. These include *Bromus* spp. and barleys in wheat as examples of the first type, and perennial weeds such as the nutsedges (*Cyperus* spp.) that proliferate in conservation (minimum) tillage systems. Can HRCs help us overcome such problems?
6. There were weed problems, especially in developing countries that were never adequately addressed by industry, and there are no readily usable selective herbicides for their control. This is exemplified by the parasitic broomrapes and witchweeds. Can HRCs alleviate these problems?
7. There are many ways to increase yields, including growing hybrids, using nitrogen-fixing organisms, etc. Can HRCs help facilitate making any of these processes more cost effective?

Examples are presented showing that HRCs can assist in meeting these needs. In too many of these cases, HRCs can become a "quick chemical fix" for the problem, meeting short-term but not the longer-term needs of agriculture. Many of the problems delineated above have come from an overreliance on chemistry to the point of an addictive and total chemical dependency and to the exclusion of other methods of weed control: cultivation, crop rotation, as well as an overdependence on some herbicides to the exclusion of others (lack of herbicide rotation).

This author feels strongly that HRCs can meet the critical needs of agriculture, e.g., food production to sustain humanity. This author does not believe that herbicides should be used as a sole chemical fix to the exclusion of time-proven alternative means of weed control. Herbicide-resistant crops should be a part of integrated weed management systems for sane management of our agricultural ecosystems. Indeed, while some HRCs can mitigate or correct past agroecosystem management mistakes, others can exacerbate them. Thus, generalizations cannot be made beyond the generalization that generalizations cannot be made. Examples are given below discussing how HRC imidazolinone-resistant corn may lead to greater resistance problems in corn/soybean rotations, when both crops receive ALS inhibitors, but can be the answer to controlling parasitic witchweeds. Thus, each HRC must be evaluated in the management system where it is to be used, even when discussing the same gene/crop.

The previous chapter described the specific uses/needs for HRCs in the U.S., leaving this chapter to deal with the rest of a rather varied world. Only a few examples are employed.

2. SPECIFIC NEEDS IN EUROPE

Subsidized agriculture in western Europe that rendered the use of even the most expensive herbicides as cost effective led to multiple applications of herbicides on major crops such as corn, wheat, and sugar beets. The Green Revolut for pesticide use (as distinguished from the Green Revolution in India) has led to the banning of certain major herbicides in some countries and legislation to reduce total herbicide use as well. The recently ratified GATT treaty will decrease subsidies for major agricultural commodities, leading to herbicides becoming a larger percentage of input costs, and changing use patterns are to be expected. Much land will be "set aside," often without herbicides, which can lead to new weed problems when such lands are brought back into production. Clearly, some agriculturally marginal lands will be taken out of crop production and some will go to agroforestry, in areas where weeds present a problem to young trees (as they do everywhere). An estimated 2 million ha of triazine-resistant weeds in western Europe are also an (expensive) problem with which to deal. In eastern Europe the problems are varied. Herbicide use is on the upswing throughout the area because they are so cost effective. Some areas such as Hungary, where a limited number of herbicides were overused (such as atrazine in monoculture corn), resistance is high on most of the 1 million ha of corn land, and replacement herbicides are needed.

2.1. CORN: BIOTECHNOLOGY MAY CONTRIBUTE TO THE RESISTANCE PROBLEMS

It was previously discussed that the major herbicides used in corn, atrazine and alachlor (or metolachor), are under severe pressure, and their use has been banned in many countries for purported environmental or toxicological reasons (Gressel 1992a). Additionally, an estimated >1 million ha of cornfields in eastern Europe (mainly Hungary) are infested with triazine-resistant weeds. Much of the corn cultivation areas throughout Europe are being heavily treated preemergence with atrazine or some other triazine and then postemergence with either dicamba or pyridate (Table 1, section A). Pyridate, while being more expensive than dicamba, exerts negative cross-resistance; i.e., it controls the resistant biotype of many weeds far better than the susceptible biotype (Arlt and Jüttersonke 1992; De Prado et al. 1992). This is a very strong tool for the management of triazine resistance (Gressel and Segel 1990).

The amounts of herbicides used and their costs are shown in Table 1, section A. Two groups of herbicides are actively competing to replace the triazines and chloracetamides: the inhibitors of ALS (sulfonylureas, imidazolinones, triazolopyrimidines, etc.), and the inhibitors of ACCase, the -fops (e.g., diclofop) and the -dims (e.g., sethoxydim) . These groups are prone to the evolution of resistance (see Section 4). The ease in achieving resistant crops through cell culture, pollen selection, or finding a mutant gene in *Arabidopsis* should have warned researchers that resistance could evolve in the field with equal ease (see Chapters 8, 9, and 11). The same genes should appear at the same

TABLE 1 Herbicide Regimes for Conventional and Imidazolinone-Resistant Corn where Triazine-Resistant Weeds are Problematic[a,b]

Timing	Herbicides	Rate (kg a.i./ha)	Cost/ha ($)
A. Conventional			
Preemergence	Atrazine	2.25	16
	Plus chloracetamide[a]	1.70	35
Postemergence	Dicamba[b]	0.28	9
	(or pyridate)[b]	0.7	
	Total	4.23–4.65	60
B. Imidazolinone-Resistant Corn			
Preemergence	Chloroacetamide[a]	1.70	35
	Imidazolinone	0.14	50
	Total	1.84	85

[a] The chloroacetamide (alachlor or metolachlor) is often dropped from the mixture when certain grasses are not part of the weed problem.
[b] The postemergence treatment with dicamba (or pyridate at about the same cost) is often omitted when triazine resistance is not a problem.

Adapted from Bartle 1991.

high frequencies in weeds. There were agronomic reasons to want -fop/-dim resistance in corn. *Sorghum halapense* (shattercane) is a major weed in corn for which there are no selective herbicides, and -fops and -dims would be good, but for how long? *S. halapense* has already evolved resistance to these herbicides in fields of other crops treated with these herbicides (Smeda et al. 1993).

The pesticide industry is intent on having an ALS or ACCase inhibitor for each crop in a corn rotation; excellent ones were available for soybeans, the major crop in rotation with corn. Fewer were available for corn, until biotechnology intervened. Four companies are marketing imidazolinone-resistant corn, and the same or related ALS inhibitors can be used in both crops. Indeed from both the points of view of cost and the need to use fewer chemicals, the use of the new imidazolinone-resistant corn seems very advantageous (Table 1, section B). The use of the imidazolinone regime (Table 1, section B) is thought to control *S. halepense* and *Cyperus* spp. (nutsedges) better than the conventional regime, but the conventional regime is thought to be better at controlling *Cirsium arvensis* (Canada thistle) and *Agropyron repens* (quackgrass) (Bartle 1991).

The demand to use less pesticides, when based on quantity of chemical, are specious and misleading. Demand should be based on biological activity and not on grams vs. kilograms per hectare. Atrazine and alachlor are both herbicides with long biological lives in the ecosystem, but so are the presently available imidazolinones and many of the other ALS inhibitors. While comparable data on biological life are not available, one of the reasons stated for developing imidazolinone-resistant corn is that there is often remaining phytotoxicity to the corn from the same imidazolinone used in soybeans the year before. Thus, there seems to be little biological difference between the imidazolinone and atrazine, despite the differences in actual amount (Table 1). The relatively long residual activity of atrazine is often shortened by using less while mixing with another triazine, cyanazine, but no such short-lived imidazolinone is yet available. There are shorter-lived sulfonylurea-type ALS inhibitors becoming available that may well be used to replace the imidazolinones. At least some of the imidazolinone-resistant corns being marketed are also cross-resistant to sulfonylureas. Imidazolinone-resistant *Xanthium* (cocklebur) already was found in monoculture U.S. soybean fields continuously treated with imazaquin (Barrentine 1993). The much larger areas in corn/soybean rotations have heretofore not seen continuous ALS inhibitors. Mixtures are proposed to prevent resistance problems; the mixing partner may control volunteer imidazolinone-resistant corn in soybeans and imidazolinone-resistant soybeans in corn. They will have little effect in preventing the evolution of resistance in weeds; the mixing partners do not cover the same weed spectra nor do they have the same persistence (Wrubel and Gressel 1994). For many weeds, it is as if the mixing partner is not there.

Thus, with corn, biotechnology and herbicide abandonment are depleting instead of broadening the number of target sites for herbicide action. The switch from multisite-affecting herbicides such as alachlor and a herbicide with a low frequency of resistance such as atrazine to single-site, high-resistance frequency herbicides to be used without meaningful rotations or mixtures is ominous. Despite even more widespread use than atrazine, no weed has been reported to evolve alachlor resistance in corn. In 30 years of use of atrazine alone in monoculture corn, there are >4 million ha of resistant weeds. Resistance has been appearing faster and more extensively in weeds with the ALS and ACCase inhibitors than they had with the triazine herbicides. Atrazine resistance was delayed where mixtures and rotations were used; the same quality of mixing partners is not available for the new compounds. Biotechnology should be broadening, not narrowing, the spectrum of available herbicide target sites. Still, there is the possibility that an imidazolinone-resistant corn using an imidazolinone herbicide employed in rotation with regular corn using the conventional regime, especially if used in rotation with other crops, will delay triazine and ALS resistance where they have not yet evolved. The use of imidazolinone-resistant corn where triazine-resistant weeds already exist raises the spectre of evolution of multiply resistant weeds, i.e., resistant to both triazines and ALS inhibitors. This could occur very quickly as at the sites where there are triazine-resistant weeds; the weed stands are dense and the seed banks full of weed seed. Where there is more seed to choose from, where the herbicide exerts high selection pressure (as do the imidazolinones), and where the frequency of resistance is great, resistant populations evolve more quickly than where weed seeds are sparse. Thus, it is strongly advisable to use other means to deplete triazine-resistant weed seed banks *before* using an imidazolinone.

There is a bright spot for the use of imidazolinone-resistant corn. It was predicted (Gressel 1992) and recently found (Eplee 1993; Ransom et al. 1995) that it is possible to control witchweed (*Striga* spp.) with imidazolinones in these HRC corns (see discussion in Section 5). Still, this is probably only a stop-gap measure until a better system is found. Single *Striga* plants produce tens of thousands of seeds, and there is no reason not to believe that they will rapidly evolve resistance to ALS-inhibiting herbicides.

2.2. SUGAR BEETS

Sugar beets have been widely cultivated in Europe in a heavily subsidized manner and the price of production is higher than world sugar prices. Being at least partially self-sufficient in sugar production has had emotional, defensive, and political implications, and it is not clear how much this uneconomical production will be cut back following the recent tariff agreements. Sugar beets are not the most competitive of crops, resulting in the need for chemical weed control, because at least within the rows, mechanical or manual weed control is not feasible. Conventional chemical weed control regimes consist of one preemergence and three or four postemergence herbicide treatments (Table 2), which are rather expensive. The average yield is ca. 8.5 T raw sugar per hectare, with a world sugar price of $250/T and a yield equivalent to about $2100 raw sugar per hectare. Clearly, the $250 spent on herbicide alone would be a large, variable expense if sugar was at world prices. With 2 million ha in production in the European Union alone, the 8000 T of herbicide used is a large environmental load. Despite all this herbicide use, weed control is hardly complete, as the conventional herbicide regime used still gives inadequate control of *Solanum, Polygonum,* and *Mercurialis* spp. (Bartle 1991). The possible use of transgenic glyphosate- (Chapter 4) and glufosinate- (D'Halluin et al. 1992) resistant sugar beets can clearly alleviate some of these problems. Less total herbicide is needed, with fewer treatments, at a much lower total cost (e.g., Table 2, section B). Additionally, the harder to control species mentioned above are better handled by these herbicides than by the conventional ones. The use of the transgenic crop can also allow control of volunteer forage beets as well as volunteers from older, lower-yielding sugar beet varieties that plague sugar beet growers. Conversely, because of these volunteers, farmers will be obliged to use these herbicides in and around the fields to prevent genetic transfer of the resistance genes by hybridization to the volunteer species, at least in seed production areas where the beets are allowed to flower.

Sugar beets are among the species most susceptible to the sulfonylurea herbicide chlorsulfuron. Sugar beets are 1000 times more susceptible than wheat and 100 times more susceptible than most weed species (Beyer et al. 1988). This led to severe carryover problems for sugar beets, as chlorsulfuron tends to persist for exceedingly long times. Acetolactate synthase-level resistant sugar beets were developed by tissue culture selection (Hart et al. 1993). How and whether these will be commercialized for general use or just to be used after ALS-inhibiting herbicides on other crops is an open question.

TABLE 2 Use of Glyphosate-Based Weed Control Systems with Transgenic Sugar Beets vs. A Conventional Regime in Europe

Timing	Treatment	Product	Rate (a.i.) (kg/ha)	Cost ($/ha)
A. Conventional Regime				
Preemergence	1	Metamitron	0.70	
		+Pyrazon[a]	0.43	45
Postemergence	2	Phenmedipham	0.13	
		+Ethofumesate	0.17	
		+Metamitron	0.60	53
	3	Phenmedipham	0.15	
		+Ethofumesate	0.20	
		+Metamitron	0.70	62
	(4)[b]	Same as (3)		62
	5	Fluazifop	0.06	24
Totals	5		4.19	246
B. Glyphosate Based-Regime				
Postemergence	1	Pyrazon[a]	0.86	40
	2	Glyphosate	1.08	48
	(3)[b]	Glyphosate	1.08	48
	4	Fluazifop	0.06	24
Total	4		3.08	160

[a] The common name of pyrazon is chloridazon in Europe.
[b] Treatments in parentheses are optional, based on weediness.

Collated from data in Bartle (1991) for the glyphosate and conventional regimes.

3. SPECIALTY CROPS

The chemical industry is primarily interested in developing herbicides for a few major crops. Lead compounds are modified until derivatives are found that lack toxicity toward these major crops. The companies only register such compounds for minor uses when they also have a market in major crops. No major efforts are made for minor crops, where minor is considered to be most crops other than corn, soybeans, wheat, rice, and cotton. "Minor" includes high-value vegetable and tree crops; i.e., the definition is based on potential herbicide sales and not on the value of the crop.

Thus, growers often lack good herbicides for many crops. The problem becomes more acute with some machine-harvested vegetables, where weeds can jam sensitive machinery. Hand weeding is often performed instead of hand harvesting. Many minor crops, especially vegetables, where seeds command high prices, can be economically transformed to resistance by seed companies of the public sector, if the available gene constructs discussed in earlier chapters are released by industry for use by seed companies or the public sector. Some of these high-value crops have special weed problems, as discussed in Section 5. Considering the value of the crops and the losses engendered by weeds, the costs of generating these HRCs can be recouped by taxes (on income from crop or land), value to the national economy, seed price, etc., even though the amount of actual herbicide sold is negligible.

3.1. FORESTRY

It is expected that at least part of the area taken out of crop production in Europe and elsewhere will be dedicated to agroforestry, either for short-term multiple cuttings of regenerable trees or longer-term afforestation. One of the greatest problems in forestry is weed control in tree nurseries, and for the first few years after planting, until canopy closure or after cutting of regenerable species until after canopy reclosure. Herbicides, often aerially applied, could be a viable alternative to impede weed growth, but only if the trees were resistant (Nelson and Haissig 1986). Mechanical cultivation is not economically viable for forestry, at least not in developed countries. Rapid tree growth is important for the purifying effect of trees on the environment and to make forest planting more attractive to those who must cover the costs. The trick is to find herbicides having acceptable environmental and herbicidal characters, and then introduce resistance into the trees. Many acceptable candidates seem to be available. The chemical industry is not likely to perform the necessary research and development to obtain resistant forest trees, because the market for herbicides would be restricted to one or two applications in new plantings and no applications for the following decade(s) until maturity. Indeed, the transformation of poplar with a gene for glufosinate resistance (Devillard 1992) seems to be a movement in the right direction. Poplars also have been engineered with resistance to ALS-inhibiting herbicides (Brasilerio et al. 1992). Because of the rare use of herbicides in forestry, there is little reason not to select for trees resistant to ALS-inhibiting herbicides. The widespread evolution of weeds resistant to this group will be very slow in the context of forestry.

4. WORLDWIDE NEEDS IN WHEAT

4.1. OVERCOMING AND PREVENTING HERBICIDE RESISTANCE

2,4-D and MCPA were the prime herbicides in wheat for more than 40 years, with a large degree of effectiveness, except for controlling grass weeds. Diclofop-methyl, trifluralin, chlorotoluron, and chlorsulfuron, herbicides affecting different sites of action, gave excellent results in controlling some of these grass weeds, and chlorsulfuron controlled broadleaf weeds as well. The widespread use of these newer herbicides hastened the evolution of resistance often with cross-resistance to all other wheat-selective herbicides (Gressel 1988a, b; Saari et al. 1994; Moss 1992; Powles 1993). The potential needs for HR wheat were first noted 6 years ago (Gressel 1988a, b). The needs are no longer "potential"; the situation with herbicide resistance covers millions of hectares of wheat lands in Australia alone. There is potential for this situation to worsen. Additionally, weeds such as *Bromus* spp. and wild barley species that were never controlled by selective herbicides are gaining importance in areas where the better grass-selective herbicides are still controlling the competition. All metabolically selective herbicides for which there is published information are degraded in wheat by monooxygenase(s) (Gressel 1988b). Thus, it is easy for grasses such as *Bromus* spp. to possess similar monooxygenases in advance or to evolve them as an evolutionary biochemical mimicry, as has been proposed as a mechanism used by many grass species (Gressel 1988a). The needs for HR wheat can be fully understood only after the extent of the resistance problems is realized. Thus, resistance to wheat-selective herbicides is described below.

4.1.1. Resistance to Broadleaf Herbicides in Wheat

2,4-D has been the standard for broadleaf control in wheat for 50 years. It has low selection pressure, both because of imperfect kill and because weeds in many ecosystems can germinate after it has dissipated, diluting resistance (Gressel and Segel 1982). Still, monoculture with 2,4-D in a cool climate with longer 2,4-D persistence and less late germination took its toll. 2,4-D (and MCPA) -resistant populations of *Sinapis arvensis* (wild mustard) appeared in western Canada (Heap and Morrison 1992). This resistance could not be traced to adsorption, translocation, or metabolism, and there is evidence that it may be due to a modified auxin binding site, having low affinity for phenoxy herbicides (Hall et al. 1993).

4.1.2. Resistance to Acetolactate Synthase Inhibitors

The ALS inhibitors have replaced much of the 2,4-D used, as they control most broadleaf weed species and some grass weeds. The first cases of resistance quickly appeared due to the severe selection pressure exerted by the highly persistent chlorsulfuron (Thill et al. 1991; Saari et al. 1994). It does not appear that the switch to shorter-lived sulfonylureas has changed the trend where chlorsulfuron had been used. DNA sequence analyses of resistant weeds indicated single amino acid transversions in the ALS at different sites in different resistant isolates (Guttieri et al. 1992; Guttieri and Eberlein 1993), partly explaining the high mutation frequency to resistance. Not all weeds have evolved just target-site resistance to sulfonylureas in wheat. *Lolium rigidum*, with its ubiquitous resistance to selective herbicides, has evolved different methods in different areas of Australia (Powles 1993; Burnet et al. 1991). In Western Australia, where chlorsulfuron was used repeatedly, target site resistance evolved (Christopher et al. 1992). Elsewhere in Australia, where diclofop methyl was used, the cross-resistance to chlorsulfuron, which had not been used, was due to enhanced degradation (Cotterman and Saari 1992; Christopher et al. 1992; Gill and Diggle 1993), although one subpopulation had both types of resistance (Burnet et al. 1994).

Even if all highly persistent sulfonylureas were banned from wheat, there are basic reasons to expect more cases of resistance: (1) target site resistance can evolve with the lesser persistent, postemergence sulfonylureas. This is expected first among those weeds that germinate mostly in a single flush, and all weeds come in contact with the herbicide; (2) inhibitors of acetolactate synthesis are being developed for almost all rotational crops; thus, weeds will be under yearly pressure from this group; (3) highly persistent sulfonylureas (used for roadside weed control) selected for resistant *L. multiflorum* (Taylor and Coats 1993) and *Amaranthus retroflexus* (Rubin et al. 1992), add to the roadside weeds that can migrate into wheat fields; and (4) the mixtures of sulfonylureas with other herbicides can apply selection pressure for the metabolic resistances so well documented in *L. rigidum*.

4.1.3. Resistance to Dinitroanilines

Trifluralin is widely used in wheat and wheat/rape rotations in western Canada, and resistance has evolved in *Setaria viridis* at many sites (Morrison et al. 1991). Both the single dominant gene controlling resistance (Jasieniuk et al. 1993) and the far reaching cross-resistances to other herbicides that act on microtubule formation (Smeda et al. 1992) strongly suggest a target-site resistance. These include pendimethalin, also used in wheat. Unlike a similar resistance in *Eleusine,* where the tubulin was modified, here resistance is thought to be due to a modification in a protein that stabilizes microtubular arrays (Smeda et al. 1992). A slightly greater selection pressure exerted by preplant incorporated trifluralin (in rape) rather than pre-emergence incorporated trifluralin (in wheat) is thought to have brought about the rapid resistance (Beckie and Morrison 1993). Trifluralin would probably be available for a longer time if it were restricted for grass weed control in wheat, and other grass killers were advised for rape.

Alopercurus myosuroides and *L. rigidum* have evolved cross-resistances to particular dinitroaniline herbicides under the selection pressure of other wheat-selective herbicides, as described earlier (Moss 1992; Powles 1993). An *Amaranthus* sp. has evolved resistance to trifluralin used in cotton (Gossett et al. 1992). There is no reason not to expect the same to happen in wheat.

4.1.4. Triallate Resistance in Wild Oats

Triallate is a thiocarbamate herbicide widely used to control wild oats (*Avena* spp.) in small grains. Thiocarbamates are thought to affect many targets, and thus were considered not to be prone to resistance. Nearly 20 years of continuous use yielded their due. Nearly one half the nonrandom samples of wild oats from 34 wheat fields in British Columbia were resistant to field rates of triallate, with cross-resistance to difenzoquat, another widely used wild oat herbicide used in wheat (O'Donovan et al. 1992). Difenzoquat has a different but unknown site(s) of action based on symptomology. Triallate-resistant wild oat populations also evolved throughout an isolated 1200 km^2 area in Montana dedicated to the cultivation of malting barley (Malchow et al. 1993). The rapid concurrent distribution of resistance may have been due to contaminated seed, as all seed came from the same supplier. These populations also have cross-resistance to difenzoquat.

There are only two other groups of metabolically selective herbicides for wild oats, one of the most pernicious grass weeds of wheat: various inhibitors of ACCase and isoproturon (which provides only marginal control). They are discussed below.

4.1.5. Resistance to ACCase Inhibitors

Some of these herbicides are degraded by wheat, allowing selective grass weed control. Their use is widespread, especially where more than one grass weed is problematic. Resistance to these herbicides has become almost as widely distributed as their use, especially in Australia (mainly in *L. rigidum,* but also in *Avena* spp.) (Powles and Matthews 1992). The situation at present, based on a chemical industry survey, is that >2500 wheat farms are infested with diclofop-methyl-resistant weeds that cannot be controlled by any selective herbicide (Powles 1993). In Western Australia alone there are an estimated 0.4 million ha of resistant *L. rigidum* (Gill and Diggle 1993). Industry surveys often underestimate the magnitude of resistance. The cross-resistance in *L. rigidum* to all other wheat-selective herbicides suggests that the weeds evolved a biochemical mechanism to mimic wheat. Initially the researchers could not find any diclofop-resistant biotypes with an altered target site, and they were killed by other ACCase inhibitors that are toxic to wheat. The resistant biotypes rapidly degrade other herbicides that they had never previously encountered (Powles 1993; Cotterman and Saari 1992), yet they are only slightly more efficient than the wild-type at degrading diclofop-methyl (Holtum et al. 1991). Unless this slight metabolic difference results from a large difference in a small, critical part of the tissue, it cannot adequately explain resistance. More recently, a *L. rigidum* biotype was found with resistance to sethoxydim (which is not selective for wheat) having an altered ACCase (Tardif et al. 1993).

Target site resistances seem to have evolved (based at least on cross-resistances to many -fops and -dims, including those toxic to wheat) in *Setaria* spp. in the Canadian western plains (Heap and Morrison 1993) and the U.S. Great Plains (Stoltenberg and Wiederhoot 1993; Marles et al. 1993b), *Sorghum halepense* in the U.S. (Marles et al. 1993a), *Eleusine indica* in Scotland (Marshall et al. 1993), *L. multiflorum* in the U.S. plains (Gronwald et al. 1992), and *Alopecurus myosuroides* in England (Moss 1992). Different wild oat (*Avena*) species have evolved resistance to diclofop-methyl use in the U.S., Canada, and Australia. The U.S. and Australian strains had different cross-resistances among -fops and -dims (Seefeldt et al. 1992, 1993; Mansooji et al. 1992), no target site differences could be found among some Canadian types (Devine et al. 1992), and others were not characterized for mode (Olufunmilayo et al. 1990).

The inhibitors of ACCase are the most widely used herbicides in the developed world for grass control in wheat. A plethora of awful weeds, over most of the areas used, have evolved resistance by at least one mechanism. Where there were six isolated instances in Australia 8 years ago, resistance is now ubiquitous. The cross-resistances to other herbicide groups are even more foreboding.

4.1.6. Isoproturon Resistance, A New Problem of the Developing World

"Green Revolution" dwarf wheat varieties have transformed China and India into the second and third largest producers of wheat in the world. These varieties, with their high yield indices (much grain, little straw) tripled yields, and rendered it economical to introduce fertilizers and irrigation. The self-sufficiency contributed extensively to geopolitical stability. These dwarf wheat varieties cannot compete with grass weeds, and herbicides had to be introduced. Unfortunately, in retrospect India settled on one herbicide, isoproturon, which is now used on about one half the wheat in Haryana and Punjab. Three companies produce ca. 2000 T/year. China introduced these wheats, and then the same herbicides, a few years after India. Simple calculations based on past experience allowed the prediction of resistance problems (Gressel 1993c).

Resistance to isoproturon or its analog chlorotoluron has been well documented in England, Germany, Israel and Spain (cf. Moss 1992; Menendez et al. 1993). In England it took a long time for resistance to evolve at a small number of sites. The predominant resistant strains have cross-resistance to all wheat-selective herbicides (Moss 1992). The first inklings of resistance appeared in India during the 1991/1992 growing season in *Phalaris minor,* the major grass weed in Indian wheat, but the difference between putatively resistant and susceptible biotypes was not great (Malik et al. 1992). Presumably, a generation of self- or cross-pollination helped, as in 1992/1993, field rates no longer controlled *Phalaris.* A second application was often made without effect, and the fields had to be cut for fodder. The problem is now quite severe in two districts (0.5 million ha wheat). Control was near zero at seven locations tested by the researchers in Haryana and in considerable areas in Punjab as well. The isoproturon-resistant *Phalaris* could be at least partially controlled by pendimethalin and by diclofop-methyl (Malik and Singh 1993).

A large part of the areas in England treated by isoproturon, or its analog, chlorotoluron, are now being treated with diclofop-methyl or other ACCase inhibitors. It required over a decade to obtain a few populations of *Alopecurus* resistant to isoproturon/chlorotoluron. In less than 3 years there were populations with ACCase target-site resistance (Moss 1992). Based on this and previously described experience, diclofop-methyl can provide only a temporary stop gap solution to the problem in India. The only long-term solution is having other herbicides. The best way to have other herbicides that are not prone to rapid evolution of resistance is to introduce new resistances into wheat.

4.2. MEETING THE NEEDS FOR NEW SELECTIVE HERBICIDES IN WHEAT

Metabolically selective herbicides, especially those controlling grass weeds, do not have a great future. 2,4-D may last a long while for broadleaf control, unless the sulfonylurea/2,4-D mixtures exert strong selection pressure for a joint metabolic resistance. The only commercially selective herbicide that does not have metabolic resistance is the thiocarbamate prosulfocarb, which has placement selectivity in wheat (Glasgow et al. 1987). It has not been reported whether triallate (thiocarbamate)-resistant weeds have cross-resistance to prosulfocarb.

The best herbicide answer to natural and evolved resistance is genetically engineered wheat (Gressel 1988a). The genes already are available, and their advantages have been detailed (Gressel 1992a, 1993a). The two best genes described are those for glyphosate and glufosinate resistance. Both do not seem to be highly prone to have resistance evolve to them, or when resistance does evolve, it is probably due to an overproduction of the target enzymes (cf. Hollander-Czytko et al. 1992) and the progeny would be very unfit. Both herbicides have low persistence in the environment and low mammalian toxicity. Still, plants possess pathways to metabolize glyphosate (Komossa et al. 1992) and glufosinate (Komossa and Sandermann 1992; Dröge et al. 1992), and evolution of enhanced rates of such pathways might lead to resistance. The efficiency of the glufosinate resistance gene may well have been enhanced by generating a "synthetic" gene for resistance, where the preferred codon usage was changed from that of the actinomycete source of the gene to typical plant codon usage (Donn and Eckes 1992).

A gene conferring resistance to the grass-controlling herbicide Dalapon was isolated recently from a pseudomonad and was eventually used to transform tobacco plants (Buchanan-Wollaston et al. 1992). Even though exceedingly high levels of this herbicide must be used, it is relatively inexpensive and is rapidly degraded in the environment, and the gene is being considered for use in wheat. Dalapon does not control broadleaf weeds whereas glyphosate and glufosinate control most weed species, which is a disadvantage of Dalapon.

The main problem in achieving herbicide-resistant wheat was the inability to transform wheat. A relatively low-efficiency system was finally worked out whereby embryogenic calli of specific varieties could be transformed by biolistically shooting them with pellets covered with DNA. A few stable regenerants were obtained (Vasil et al. 1992). Fortuitously, the gene coding for glufosinate resistance was used as their selectable marker, serendipitously meeting a need of agriculture. More recently, biolistic transformation of scutellar wheat cultures using superior regeneration techniques led to high efficiency, transient expression of the marker gene in a number of common varieties (Perl et al. 1992). Many of the plants seem stably transformed, as assayed through two generations (Galili et al. 1993). Successful use of basically the same system was confirmed in a larger study (Weeks et al. 1993).

Thus, it appears that the genes are available to protect wheat from its enemy weeds (with the help of herbicides), and the technologies are ready to perform the task. The previous sections show how great the needs can be. Present oversupplies of wheat in the world hover at 1 month's supply. If the transgenic wheat varieties are not readily available in 5 to 10 years, the oversupply will disappear due to resistant weeds. Finding suitable genes, genetic engineering, testing, multiplying, and registering varieties take time, and at the rate resistance is spreading, usually without quarantine and containment to dampen the fast spread. Little time is left.

4.3. NEEDS TO LOWER COST OF HYBRID WHEAT SEED PRODUCTION

Hybrid vigor is a property that often increases yield. Hybrid seed is expensive to produce and in wind-pollinated crops such as wheat, there is a problem of lack of sufficient pollen movement. The pollen-parent must be planted as close as possible to the seed parent, and the seed parent must then be harvested separately at great effort and expense. If the seed parent had a maternally inherited or a dominant resistance to a herbicide that can kill the pollen parent after anthesis, susceptible males could be interplanted with seed parents in one sowing and then be culled by herbicide (Beversdorf et al. 1987; Gressel 1988b). Mixed planting of pollen and seed parents also would be advantageous and cheaper for insect-pollinated species as well, if the pollen parent could be culled. Such hybrid seed would have three positive economic traits: cheaper production costs than conventionally produced hybrids, higher yield, and a herbicide resistance. These could all be coupled to making hybrid wheat an economically viable proposition. This resistance would be expressed only in the F1 generation if there is nuclear inheritance, or in all subsequent seed generations from seed parents if resistance is maternally inherited. If breeding is left to the private sector, the greatest advances in yield will usually derive from the crops the seed companies can continually sell, i.e., hybrids where seed must be repurchased each year. Thus, there is the added long-term potential for the appearance of continually higher yielding hybrids due to the economic incentive to breed them, and to the lowered cost of hybrid seed production.

5. THE NEED TO CONTROL PARASITIC WEEDS

Many crops are attacked by higher plant weeds that attach themselves to the growing plant. The parasites usually curtail photosynthesis in cases where there is some initially, and then they remove photosynthate, minerals, and water from the host crop (Stewart and Press 1990). The most common parasites on field crops are dodders (*Cuscuta* spp.), broomrapes (*Orobanche* spp.) (Foy et al. 1989), and witchweeds (*Striga* spp.) (Stewart 1990). Mistletoes (Loranthaceae) and other parasitic plants attack forest trees. The most devastating of these species on the field crops are the broomrapes (Foy et al. 1989) and the witchweeds (Stewart 1990). The areas infested are vast and expanding, especially in tropical and subtropical countries. For example, a survey of 180,000 km^2 in Nigeria found 70% of fields infested with witchweed seeds (Hartman and Tanimonure 1991). In the seven agroecological zones of sub-Saharan Africa witchweed species are listed as the worst pests affecting agriculture (including insects and diseases) in all except the mountainous and forested regions (Akobundu 1991). The areas affected are expanding due to agricultural practices. Hand and mechanical weeding of other species does not remove attached parasites. The lack of rotation or fallowing supports the parasites, especially with the lack of selective herbicides to control the parasites.

Agriculture in developing countries has learned to use dangerous insecticides, but not herbicides. More than 60% of Third World farmers' time is spent in weeding. The weeds usually have irreversibly depressed yields by the time weeding is performed, and the parasitic weeds cannot be removed. Witchweeds infest the grain crops of >200 million people in sub-Saharan Africa and Asia, reducing yields by 50%, and by more in drought years. Broomrapes devastate the legume and vegetable crops in much of northern Africa and elsewhere around the Mediterranean, severely limiting the local production of protein crops for the diet of the inhabitants (Foy et al. 1989). Over 30% of Egypt's broadbean cultivating areas had to be removed from production over a 10-year period due to broomrape infestations (Sauerborn 1991). Broomrapes and witchweeds presently seem to have a greater deleterious effect on the health and welfare in these Third World areas than AIDS is expected to have when it peaks (Anderson et al. 1991). Why parasitic weeds are not high on the list of the "world's worst weeds" (Holm et al. 1977) is unclear. Crop yields could be doubled without added fertilizer or irrigation if these weeds could be effectively controlled.

There is some marginal natural resistance to herbicides available in some crops, but it is often only phenotypically expressed at specific stages of growth (Foy et al. 1989; Sauerborn et al. 1989; Dawson 1990; Kotoula-Syka et al. 1991; Liu et al. 1991; Garcia-Torres and Lopez-Granados 1991). These host crops of the parasites are clearly not sufficiently resistant to herbicides for widespread field use without yield loss, especially where uncalibrated backpack sprayers are used, and there is a possibility of underdose with lack of parasitic weed control or overdose with ensuing crop damage. At times, with some metabolically resistant crops, frequent retreatments are necessary. For example, tomatoes are resistant to the sulfonylurea herbicide rimsulfuron, due to their rapid metabolism of the herbicide. This herbicide is also rapidly degraded in the soil. Rimsulfuron effectively controls broomrape for only a few weeks, so that 25 g/ha herbicide must be reapplied at 10-d intervals. This action seems to be directly through the soil, as foliar application alone (with a soil barrier) was ineffective (Kleifeld et al. 1994a).

Low rates of glyphosate have often been tried for broomrape control, especially in broadbeans and vegetables (Jacobsohn and Kelman 1980; Foy et al. 1989; Sauerborn et al. 1989). Even low rates of glyphosate suppress the production of induced phytoalexins that defend against fungal attack (Sharon et al. 1992). Thus, it is not unexpected that glyphosate-treated broadbeans are often infected with diseases such as chocolate spot, due to glyphosate-decreased natural immunity. It was proposed that target site resistance introduced into crops could allow selective control of these weeds (Gressel 1992a). Metabolic resistance to foliar contact herbicides should be of little use to control underground parasitic weeds because the herbicides are degraded before reaching the parasite. Indeed, we tested this hypothesis using transgenic tomato plants with metabolic resistance to glufosinate. There was no control of *Orobanche* with foliar treatments, even though it is known that glufosinate can be translocated and that it is toxic to *Orobanche* (cf. Gressel et al. 1994; Joel et al 1995). What is quickly metabolized cannot be translocated.

The second type of case, in which the target enzymes affected by the herbicides are modified in a manner precluding herbicide binding, is of primary interest for controlling parasites. The target site mutation must be in such a position that the target enzyme can still carry out its normal function. With target site resistance, the parent herbicides remain present in the plant in the active form and are translocated to the roots, to be lethally bioaccumulated by the root parasitic *Striga* and *Orobanche* spp., or translocated to stems, in the case of *Cuscuta* spp. The herbicide must be one that can affect the parasite. The most common target site resistance available is target site triazine resistance. It has been transferred by protoplast fusion into potatoes and rape (Gressel et al. 1990; Thomzik and Hain 1990). Alas, triazines only affect photosynthesis, which is lacking in the broomrapes attacking potatoes and rape and in most other parasites when attached to their crop hosts. Photosynthesis is not active underground, where *Orobanche* and *Striga* do their damage.

As the parasite species are all dicotyledenous, they are naturally resistant to ACCase inhibitors used for HR graminaceous crops, so their use is unnecessary (Chapter 11).

Positive preliminary results have been achieved with HRCs with two target site resistance. These are described below.

5.1. GLYPHOSATE-RESISTANT HERBICIDE-RESISTANT CROPS

Glyphosate resistance has been engineered into tobacco, tomato, rape, and soybeans (Chapter 4), all species affected by broomrapes. There also may be problems of residues of glyphosate in crops, as the herbicide is not readily degraded. Efforts have been made to find, clone, and engineer genes into plants that degrade this herbicide. There are now available transformants of some crops containing both the target site resistance and a slow metabolic resistance (Chapter 4).

Preliminary observations with transgenic oilseed rape possessing target site glyphosate resistance are that excellent control of broomrape is achieved with a 250 or 750 g/ha glyphosate treatment 5 weeks after transplanting seedlings into broomrape-infested pots (cf. Gressel et al. 1994 Joel et al 1995). Control (untransformed) rape plants were killed by the herbicide at both rates, and untreated transgenic rape plants were killed by the broomrape at the high inoculum level used. Field-yield experiments are needed to ascertain whether a single treatment is sufficient and whether late-season parasitization reduces yields/produces broomrape seed, etc. The fact that very low rates (one fourth the normal rates) can be effectively used means that herbicide costs would be very low for this herbicide, which, being off patent, is now very reasonably priced.

Transgenic crops with similar gene constructs are resistant to high field rates as well. In many areas of southern Africa the manual (usually "femanual") weeding leaves an open ecological niche for the horrible perennial *Cyperus* spp. (Akubundo 1991). High levels of glyphosate also control *Cyperus* spp., as well as *Striga* spp. Thus, there would be two advantages to such transgenic crops. At present the only cost-effective control for *Cyperus* and *Striga* is the too-commonly practiced abandonment of farms.

5.2. HERBICIDE-RESISTANT CROPS WITH MODIFIED ACETOLACTATE SYNTHASE

It has been very easy to select for resistance to the sulfonylurea, imidazolinone, and triazolopyrimidine herbicides inhibiting this key enzyme in the synthesis of branch-chained amino acids (Chapters 8 and 9). The problem is that resistance to these herbicides has also readily evolved in weeds in the field (Saari et al. 1994), primarily to the members of the herbicides in this group that have long residual activities in soil.

Chlorsulfuron effected excellent control of *Orobanche aegyptiaca* in preliminary pot tests, with transgenic tobacco having a modified ALS. Foliar application of 15 g/ha chlorsulfuron 10, 20, or 30 d after transplanting prevented development of the parasite without reducing tobacco yield (Joel 1992; Joel et al. 1995).

The use of imidazolinone herbicides with the resistant corn (Chapter 9) was predicted to affect the control of *Striga* spp. (Gressel 1992a). Such analyses were made in 1993 with the commercially available resistant corn and there was excellent control of *Striga* (Eplee 1993; Ransom et al 1995).

The transgenic sulfonylurea-resistant flax already available (McSheffrey et al. 1992) might make flax an interesting trap/catch crop for controlling broomrape beyond its present use (Kleifeld et al. 1994b).

The potential evolution of herbicide resistance in parasitic weeds is a potential problem that must be considered, especially due to the propensity for resistance to evolve to ALS inhibitors. The high natural frequency of such mutations, and the huge seed output of *Orobanche* and *Striga* spp., exacerbate the problems. Conversely, the large seed bank of susceptible *Orobanche* seed in the soil will have a delaying effect on resistance. It will be necessary to design proactive, preemptive resistance management strategies to delay evolution of resistant populations by lowering the selection pressure, or by other means. One way might be to use shorter-persistence ALS inhibitors so that there is a late season attack on the crop by the parasitic weeds, which will then set susceptible seeds, severely diluting and competing with selected resistant seeds. Seed dressing of the target-site HRC with these herbicides might affect only the parasites germinating near the seed, but allow infection of peripheral roots by susceptible parasites later in the season. This later infestation might have little effect on yield but would produce susceptible seed, diluting any resistance that might have evolved. Other strategies might be to leave internal refuges of susceptible weeds in the field. The most important and useful strategy is crop rotations including rotations with catch or trap crops, or with crops resistant to other types of herbicides, so that ALS inhibitors will never be used 2 years in row. There is good reason to believe that the evolution of resistance to EPSP synthase inhibitors is much slower than to ALS inhibitors. This is based on the epidemiological information gathered over the years and the extent of glyphosate use as compared to the use of ALS inhibitors.

5.3. DIHYDROPTERATE SYNTHASE

A bacterial mutant gene conferring resistance to sulfonamide antibiotics was modified and engineered into tobacco, conferring cross-resistance to asulam (Guerineau et al. 1990). Both the sulfonamides and asulam inhibit dihydropterate synthase in the pathway to folic acid. *Orobanche* does not have an obligate requirement for folic acid (Ben-Hod et al. 1991), and might be inhibited by asulam, which is known to be translocated. The transgenic plants are resistant to field rates, which could bring new uses for a not-widely used herbicide, maybe even to control parasitic weeds. Transgenic asulam-treated resistant tobacco plants grown in broomrape-infested soil had much less broomrape attached in preliminary experiments (Joel et al. 1995).

5.4. CELLULOSE BIOSYNTHESIS

Separate mutant *Arabidopsis* strains have been isolated that have target site resistance to isoxaben (Heim et al. 1990) and dichlobenil (Heim et al. 1991), herbicides, inhibiting cellulose biosynthesis. The different mutants have no cross-resistance between the two herbicides, suggesting that different enzymes in the pathway or distant sites on the same enzyme are affected. Presumably, efforts are underway to isolate the genes for further engineering.

Parasitic weeds must synthesize cellulose *de novo;* cellulose cannot be translocated from the host. Does that mean that resistant crop plants treated with isoxaben or dichlobenil will have the parasite controlled? This is an open question. Isoxaben and diclobenil are not translocated within the crops, at least not in those crops naturally having metabolic resistance. There is probably a requirement that the herbicide be taken up directly by the parasites from the soil. It is now known that parasites take up some organic compounds from the soil such as the herbicide rimsulfuron (see above). If the parasites do take up a cellulose biosynthesis-inhibiting herbicide from the soil, metabolic resistances in crops could be considered for parasitic weed control with those soil-applied herbicides.

5.5. PHYTOENE DESATURASE

A mutant gene has been isolated and cloned from cyanobacteria that upon transformation into other cyanobacteria conferred resistance to the pyridazinone and other herbicides inhibiting phytoene desaturase, a key enzyme in the pathway to carotene (Chapter 12). Despite widespread use of the pyridazinone norflurazon in cotton, resistant weeds have not yet evolved. Most of the phytoene desaturase inhibitors are nonselective, and engineered resistance into crops can greatly change their market potential. There are two open questions about whether such resistance genes will be useful in our context: (1) Can crops be successfully transformed with such genes, leading to HRCs? The cyanobacterial gene could be either inactive or lethal in higher plants. Despite efforts, transgenic plants with this trait have yet to be regenerated. (2) Which parasitic weeds can be controlled? Carotenes can protect against damage caused by light and are clearly present in dodders attacking the above-ground portions of plants. Will carotene-free dodder be photodegraded by sunlight? This is questionable, as carotene-free (norflurazon-treated) plant cells in suspension had to contain chlorophyll to be killed by light. Achlorophyllous, carotene-free cells were not hurt by light (Gaba et al. 1988). Carotenes are present in some broomrapes, but their function is obscure; therefore, guessing the results of inhibiting their synthesis is equally obscure.

6. NEED FOR HERBICIDE-RESISTANT NITROGEN FIXERS IN RICE

Herbicide-resistant cyanobacteria are needed to facilitate rice growth where paddy rice is cultivated with little or no nitrogen fertilizer. The mineral nitrogen comes from fixation by cyanobacteria (blue-green algae), either free living or in association with *Azolla,* a fern. Herbicides have come into use, but many, especially those inhibiting photosynthesis affect the cyanobacteria as well. Even some herbicides not affecting photosynthesis such as alachlor, and fungicides such as mancozeb also inhibit the nitrogen-fixing cyanobacteria (Vaishamapayan and Prassad 1984). Thus, there is reason to select for strains that can withstand these herbicides. Cyanobacterial mutants to the herbicides butachlor, fluchloralin, alachlor, atrazine, and propanil were found in *Gleocapsa* sp. and transformed into the nitrogen-fixing *Nostoc muscorum* (Singh et al. 1987; Modi et al. 1991). The addition of glufosinate resistance, obtained by spontaneous mutation in *Nostoc* facilitated added ammonium secretion into the medium (Modi et al. 1991).

The system will be more complicated when the *Azolla* nitrogen-fixing cyanobacterium mutualistic system is used to supply fixed nitrogen to rice. It will be necessary there that both the fern and the cyanobacterium be resistant to the herbicides used in rice. Conversely, not only do some herbicides control higher plant weeds, they control other non-nitrogen-fixing algae competing with the cyanobacteria for light and minerals, and their control could have an added advantage.

7. CONCLUDING REMARKS

There are great needs for HRCs, both in the developed and developing worlds, in specific crops and situations. While some of those needs are being met by the industries involved, many are not, e.g., the needs of wheat and those in a variety of crops to control parasitic weeds. The reasons stated by the chemical and biotechnology industries for not pursuing these weeds are economic; profits from such ventures are not perceived. As there would be great national economic profitability from the control of grass weeds in wheat or of parasitic weeds, such work must be done by the public sector. There is often willingness, but the author has witnessed cases where there is considerable reluctance by industry to supply the already isolated and used gene constructs for such purposes. Further problems would ensue when it is time to register herbicide use for such HRCs.

There is also a tendency for industry to develop HRCs bearing resistance to already widely used herbicides, increasing the chances of evolution of herbicide resistance. This is clearly already an acute problem with the single target site herbicides affecting ALS and ACCase. Despite the setting up of committees to deal with these resistance problems by the chemical industry, they seem to have no teeth, as newer resistant crops keep coming on market despite their dubious long-term possibilities of use.

Thus, there seems to be a mismatch between the needs of agriculture for certain HRCs, where there are few good alternatives, and the immediate goals of the biotechnology industries to meet the needs of the farmers. This is unfortunate, as there are many mouths to feed on this planet.

ACKNOWLEDGMENTS

The many who supplied ideas, comments, feedback, and unpublished material are gratefully thanked. None but the author are to be held responsible for the views expressed herein. The author's research on HRC for use in parasitic weed control is supported by grant no. HNE-0158-G-00-3060-00 from the trilateral (U.S./Israel/Egypt) U.S. Agency for International Development (AID) program. The author holds the Gilbert de Botton Chair of Plant Sciences.

REFERENCES

Akobundu, I.O. 1991. Weeds in human affairs is sub-Saharan Africa. *Weed Technol.* 5: 860.

Anderson, R.M., May, R.M., Boily, M.C., Garnett, G.P., and Rowley, J.P. 1991. The spread of HIV-1 in Africa: sexual contact patterns and the predicted demographic impact of AIDS. *Nature* 352: 581.

Arlt, K. and Jüttersonke, B. 1992. Zur negativen Kreizresistenz bei triazine resisten unkranten, inbesondere *Chenopodium album* L. *Z. Pflanzenkrank. Pflanzen.* 13: 483.

Barrentine, W.B. 1993. Personal communication.

Bartle, I.D.G. 1991. *Herbicide Tolerant Plants — Weed Control With the Environment in Mind.* I. C. I. (Zeneca) Seeds, Jealotts Hill, U.K.

Beckie, H.J. and Morrison, I.N. 1993. Effective kill of trifluralin-susceptible and resistant green foxtail (*Setaria viridis*). *Weed Technol.* 7: 15.

Ben-Hod, G., Losner, D., Joel, D.M., and Mayer, A.M. 1991. *In vitro* culture of *Orobanche aegyptica*. *Ann. Bot.* 68: 413.

Beversdorf, W.D., Erickson, L.R., and Grant, I. 1987. Hybridization Using Cytoplasmic Male Sterility, Cytoplasmic Herbicide Tolerance and Herbicide Tolerance from Nuclear Genes. U.S. Patent 4,658,085.

Beyer, E.M., Duffy, M.J., Hay, J.V., and Schleuter, D.D. 1988. Sulfonylurea herbicides. In *Herbicides: Chemistry Degradation and Mode of Action,* Vol. 3. Ed. by Kearney, P.C. and Kaufmann, D.D. Dekker, New York p. 117.

Brasilerio, A.C.M., Tourneur, C., Leplé, J.-C., Combes, V., and Jouanin, L. 1992. Expression of the mutant *Arabidopsis thaliana* acetolactate synthase gene confers chlorsulfuron resistance to transgenic poplar plants. *Transgenic Res.* 1: 133.

Buchanan-Wollaston, V., Snape, A., and Cannon, F. 1992. A plant selectable marker gene based on the detoxification of the herbicide dalapon. *Plant Cell Rep.* 11: 627.

Burnet, M.W.M., Christopher, J.T., Holtum, J.A.M., and Powles, S.B. 1994. Identification of two mechanisms of sulfonylurea resistance within one population of rigid ryegrass (*Lolium rigidum*) using a selective germination medium. *Weed Sci.* 42: 468.

Burnet, M.W.M., Hildebrane, O.B., Holtum, J.A.M., and Powles, S.B. 1991. Amitrole, triazine substituted urea, and metribuzin resistance in a biotype of rigid ryegrass (*Lolium rigidum*). *Weed Sci.* 39: 317.

Christopher, J.T., Powles, S.B., and Holtum, J.A.M. 1992. Resistance to acetolactate synthase inhibiting herbicides in annual ryegrass (*Lolium rigidum*) involves at least two mechanisms. *Plant Physiol.* 100: 1909.

Cotterman, J.C. and Saari, L.L. 1992. Rapid metabolic inactivation is the basis for cross-resistance to chlorsulfuron in diclofop-methyl-resistant rigid ryegrass (*Lolium rigidum*) biotype SR3/84. *Pestic. Biochem. Physiol.* 43: 182.

D'Halluin, K., Bossut, M., Bonne, E., Mazur, B., Leemans, J., and Botterman, J. 1992. Transformation of sugarbeet (*Beta vulgaris* L.) and evaluation of herbicide resistance in transgenic plants. *Bio/Technology* 10: 309.

Dawson, J.H. 1990. Newly seeded alfalfa (*Medicago sativa*) tolerates glyphosate and SC-0224 at doses that control dodder (*Cuscuta* spp.). *Weed Technol.* 4: 876.

De Prado, R., Sanchez, M., Jorin, J., and Dominguez, C. 1992. Negative cross resistance to bentazone and pyridate in atrazine-resistant *Amaranthus cruentus* and *Amaranthus hybridus* biotypes. *Pestic. Sci.* 35: 131.

Devillard, C. 1992. Transformation *in vitro* du tremble (*Populus tremula x Populus alba*) par *Agrobacterium rhizogenes* et regeneration de plantes tolerantes au basta. *C. R. Acad. Sci. Paris* 314: 291.

Devine, M.D., MacIsaac, S.A., Romano, M.L., and Hall, J.C. 1992. Investigation of the mechanism of diclofop resistance in two biotypes of *Avena fatua. Pestic. Biochem. Physiol.* 42: 88.

Donn. G. and Eckes, P. 1992. Basta-verträgliche kulturpflanzen durch übertragung eines synthetischen phosphinothricin-acetyltransferasegens. *Z. Pflanzenkrank. Pflanzen.* 13: 499.

Dröge, W., Broer, I., and Pühler, A. 1992. Transgenic plants containing the phosphinothricin-*N*-acetyltransferase gene metabolize the herbicide L-phosphinothricin (glufosinate) differently from untransformed plants. *Planta* 187: 142.

Eplee, R.E. 1993. Personal communication.

Foy, C.L., Jain, R., and Jacobsohn, R. 1989. Recent approaches for chemical control of broomrape (*Orobanche* spp.). *Rev. Weed Sci.* 4: 123.

Gaba, V., Cohen, N., Shaaltiel, Y., Ben-Amotz, A., and Gressel, J. 1988. Light-requiring acifluorfen action in the absence of bulk photosynthetic pigments. *Pestic. Biochem. Physiol.* 31: 1.

Galili, G., Perl, A., and Galun, E. 1993. Personal communication.

Garcia-Torres, L. and Lopez-Granados, F. 1991. Control of broomrape (*Orobanche crenata* Forsk.) in broad bean (*Vicia faba* L.) with imidazolinones and other herbicides. *Weed Res.* 31: 227.

Gill, G. and Diggle, A.J. 1993. Herbicide resistance in annual ryegrass in the cropping belt of western Australia — a case-study for integrated weed management. In *Integrated Weed Management for Sustainable Agriculture.* Vol. 1. Ed. by Malik, R K. Indian Weed Science Society, Hisar, p. 203.

Glasgow, J.L., Mojica, E., Baker, D.P., Tillis, H., Gore, N.R., and Kurtz, P.G. 1987. SC-0574, a new selective herbicide for use in winter cereals. *Br. Crop Protect. Conf. - Weeds* p. 27.

Gossett, B.J., Murdock, E.C., and Toler, J.E. 1992. Resistance of palmer amaranth (*Amaranthus palmeri*) to the dinitroaniline herbicides. *Weed Technol.* 6: 587.

Gressel, J. 1988a. Multiple resistances to wheat selective herbicides: new challenges to molecular biology. *Oxford Surv. Plant Mol. Cell Biol.* 5: 195.

Gressel, J. 1988b. *Wheat Herbicides: The Challenge of Emerging Resistance.* Biotechnology Affiliates, Checkendon/ Reading, U.K., p. 247.

Gressel, J. 1992a. The needs for new herbicide-resistant crops. In *Achievements and Developments in Combating Pesticide Resistance.* Ed. by Denholm, I., Devonshire, A.L., and Hollomon, D.W. Elsevier, London, p. 283.

Gressel, J. 1992b. Genetically engineered herbicide-resistant crops — a moral imperative for world food production. *Agro. Food Ind. Hitech.* 3(6): 3.

Gressel, J. 1993a. Advances in achieving the needs for biotechnologically-derived herbicide-resistant crops. *Plant Breeding Rev.* 11: 155.

Gressel, J. 1993b. Improving weed control by biotechnologically conferring herbicide resistance on crops — a progress report. In *Opportunities for Molecular Biology in Crop Production.* Ed. by Beadle, D.J., Bishop, D.H.L., Copping, L.G., Dixon, G.K., and Hollomon, D.W. BCPC/SCI, Farnham, U.K., p. 245.

Gressel, J. 1993c. Evolution of herbicide resistance in weeds: causes, prevention, and ameliorative management. In *Integrated Weed Management for Sustainable Agriculture,* Vol. 1. Ed. by Malik, R.K. Indian Weed Science Society, Hisar, p. 173.

Gressel, J. and Segel, L.A. 1982. Interrelating factors controlling the rate of appearance of resistance: the outlook for the future. In *Herbicide Resistance in Plants.* Ed. by Le Baron, H. and Gressel, J. John Wiley & Sons, New York, p. 325.

Gressel, J. and Segel, L.A. 1990. Negative cross-resistance; a possible key to atrazine resistance management: a call for whole plant data. *Z. Naturforsch.* 45C: 470.

Gressel, J., Aviv, D., and Perl, A. 1990. Method of Producing Herbicide Resistant Plant Varieties and Plants Produced Thereby. *U.S. Patent* 4,900,676.

Gressel, J., Kleifeld, Y., and Joel, D.M. 1994. Genetic engineering can help control parasitic weeds. In *Biology and Management of Orobanche.* Ed. by Pieterse, A.H., Verkleij, J.A.C., and ter Borg, S.J. Royal Tropical Inst., Amsterdam, p. 406.

Gronwald, J.W., Eberlein, C.V., Betts, K.J., Baerg, R.J., Ehlke, N.J., and Wyse, D.L. 1992. Mechanism of diclofop resistance in an Italian ryegrass (*Lolium multiflorum* Lam.) biotype. *Pestic. Biochem. Physiol.* 44: 126.

Guerineau, F., Brooks, L., Meadows, J., Lucy, A., Robinson, C., and Mullineaux, P. 1990. Sulfonamide resistance gene for plant transformation. *Plant Mol. Biol.* 15: 127.

Guttieri, M.J. and Eberlein, C.V. 1993. Diverse mutations in the acetolactate synthase genes of kochia [*Kochia scoparia* (L.) Schrad.] biotypes confer resistance to chlorsulfuron. *Weed Sci. Soc. Am. Abst.* 33: 193.

Guttieri, M.J., Eberlein, C.V., Mallory-Smith, C.A., Thill, D.C., and Hoffman, D.L. 1992. DNA sequence variation in domain A of the acetolactate synthase genes of herbicide-resistant and susceptible weed biotypes. *Weed Sci.* 40: 670.

Hall, J.C., Alam, S.M.M., and Muir, D.P. 1993. Ethylene biosynthesis following foliar application of picloram to biotypes of wild mustard (*Sinapis arvensis* L.) susceptible or resistant to auxinic herbicides. *Pestic. Biochem. Physiol.* 47: 36.

Hart, S.E. , Saunders, J.W., and Penner, D. 1993. Semidominant nature of monogenic sulfonylurea resistance in sugarbeet: cross-resistance and physiological basis of resistance. *Weed Sci.* 41: 317.

Hartman, G.L. and Tanimonure, O.A. 1991. Seed populations of *Striga* species in Nigeria. *Plant Dis.* 75: 494.

Heap, I.M. and Morrison, I.N. 1992. Resistance to auxin-type herbicides in wild mustard (*Sinapis arvensis* L.) populations in Western Canada. *Weed Sci. Soc. Am. Abst.* 32: 164.

Heap, I.M. and Morrison, I.N. 1993. Resistance to aryloxyphenoxypropionate and cyclohexanedione herbicides in green foxtail (*Setaria viridis* (L.) Beauv.). *Weed Sci. Soc. Am. Abst.* 33: 185.

Heim, D.R., Roberts, J.L., Pike, P.D., and Larrinua, I.M. 1990. A second locus *Ixr B1* in *Arabidopsis thaliana* that confers resistance to the herbicide isoxaben. *Plant Physiol.* 92: 858.

Heim, D.R., Bjelk, L.A., and Larrinua, I.M. 1991. Isolation and characterization of a dichlobenil-resistant mutant of *Arabidopsis thaliana. Weed Sci. Soc. Am. Abst.* 31: 76.

Holländer-Czytko, H., Sommer, I., and Amrhein, N. 1992. Glyphosate tolerance of cultured *Corydalis sempervirens* cells is acquired by an increased rate of transcription of 5-enolpyruvylshikimate 3-phosphate synthase as well as by a reduced turnover of the enzyme. *Plant Mol. Biol.* 20: 1029.

Holm, L., Pancho, J.V., Herberger, J.P., and Plucknett, D.L. 1977. *The World's Worst Weeds: Distribution and Biology.* University Press of Hawaii, Honolulu.

Holtum, J.A.M., Matthews, J.M., Häusler, R.E., Liljegren, D.R., and Powles, S.B. 1991. Cross-resistance to herbicides in annual ryegrass (*Lolium rigidum*). III. On the mechanism of resistance to diclofop-methyl. *Plant Physiol.* 97: 1026.

Jacobsohn, R. and Kelman, Y. 1980. Effectiveness of glyphosate in broomrape (*Orobanche* spp.) control in four crops. *Weed Sci.* 28: 692.

Jasieniuk, M., Brûlé-Babel, A.L., and Morrison, I.N. 1993. The genetics of trifluralin resistance in green foxtail (*Setaria viridis* (L.) Beauv.). *Weed Sci. Soc. Am. Abst.* 33: 181.

Joel, D.M. 1992. Control of broomrape (*Orobanche aegyptiaca* Pers.) with chlorsulfuron on a transgenic sulfonylurea-resistant crop. *Weed Sci. Soc. Am. Abst.* 32: 193.

Joel, D.M., Kleifeld, Y. Losner-Geshen, D., Herzlinger, G., and Gressel, J. 1995. Transgenic crops against parasites, Nature 374: 220.

Kleifeld, Y., Goldwasser, Y., Herzlinger, G., Golan, S. Blumenfeld, T., and Buxbaum, H. 1994a. Selective control of Orobanche in tomatoes with rimsulfuron (DPXE-9636), In *Biology and Management of Orobache.* Ed by Pieterse, A.H., Verkleij, J.A.C., and ter Borg, S.J. Royal Tropical Inst., Amsterdam, p. 561.

Kleifeld, Y., Goldwasser, Y., Herzlinger, G., Joel, D.M., and Kahanna, D. 1994b. The effect of flax (*Linum usitatissimum L.*) and other crops as trap-and-catch crops for control of Egyptian broomrape (*Orobanche aegyptiaca Pers.*). Weed Res. 34: 37.

Komossa, D., Gennity, I., and Sandermann, H., Jr. 1992. Plant metabolism of herbicides with C-P-bonds: glyphosate. *Pestic. Biochem. Physiol.* 43: 85.

Komossa, D. and Sandermann, H., Jr. 1992. Plant metabolism of herbicides with C-P-bonds: phosphinothricin. *Pestic. Biochem. Physiol.* 43: 95.

Kotoula-Syka, E. and Eleftherohorinos, I.G. 1991. *Orobanche ramosa* L. (broomrape) control in tomato (*Lycopersicon esculentum* Mill.) with chlorsulfuron, glyphosate and imazaquin. *Weed Res.* 31: 19.

Liu, Z.Q., Fer, A., and Lecocq, F.M. 1991. L'imazaquine: un herbicide prometteur pour la lutte curative contre la cuscute (*Cuscuta* spp.) dans les cultures de soja *(Glycine max). Weed Res.* 31: 33.

Malchow, W.E., Fay, P.K., Maxwell, B.D., and Dyer, W.E. 1993. Wild oat (*Avena fatua* L.) resistance to triallate in Montana. *Weed Sci. Soc. Am. Abst.* 33: 54.

Malik, R.K. and Singh, S. 1993. Evolving strategies for herbicide use in wheat: resistance and integrated weed management. In *Integrated Weed Management for Sustainable Agriculture,* Vol. 1. Ed. by Malik, R.K. Indian Weed Science Society, Hisar, p. 225.

Malik, R.K., Singh, S., and Singh, A. 1992. Effect on environment and biotypes on the potency of isoproturon against *Phalaris minor* Retz. *Weed Sci. Soc. India Abst.* 6: 14.

Mansooji, A.M., Holtum, J.A., Boutsalis, P., Matthews, J.M., and Powles, S.B. 1992. Resistance to aryloxyphenoxypropionate herbicides in two wild oat species (*Avena fatua* and *Avena sterilis* ssp. *ludoviciana*). *Weed Sci.* 40: 599.

Marles, M.A.S., Devine, M.D., and Hall, J.C. 1993a. Herbicide resistance in *Setaria viridis* conferred by a less sensitive form of acetyl coenzyme-A carboxylase. *Pestic. Biochem. Physiol.* 46: 7.

Marles, M.A.S., Devine, M.D., and Hall, J.C. 1993b. Herbicide resistance in green foxtail (*Setaria viridis* (L.) Beauv.) and johnsongrass (*Sorghum halpense* (L.) Pers.) biotypes conferred by an insensitive form of acetyl coenzyme-A carboxylase. *Weed Sci. Soc. Am. Abst.* 33: 186.

Marshall, G., Kirkwood, R.C., and Leach, G.E. 1993. Studies on the biology and control of graminicide-resistant and susceptible biotypes of goosegrass (*Eleusine indica* L. Gaertn.). *Weed Sci. Soc. Am. Abst.* 33: 184.

McSheffrey, S.A., McHughen, A., and Devine, M.D. 1992. Characterization of transgenic sulfonylurea-resistant flax (*Linum usitatissimum*). *Theor. Appl. Genet.* 84: 480.

Menendez, J., DePrado, R., Jorrin, J., and Taberner, A. 1993. Penetration, translocation and metabolism of diclofop-methyl in chlorotoluron-resistant and susceptible biotypes of *Alopecurus myosuroides*. *Brighton Crop Protect. Conf. - Weeds* p. 213.

Modi, D.R., Singh, A.K., Rao, K.S., Chakravarty, D., and Singh, H.N. 1991. Construction of multiple herbicide resistant ammonia excreting strains of cyanobacterium *Nostoc muscorum*. *Biotechnol. Lett.* 13: 793.

Morrison, I.N., Beckie, H., and Nawolsky, K. 1991. The occurrence of trifluralin resistant *Setaria viridis* (green foxtail) Western Canada. In *Herbicide Resistance in Weeds and Crops*. Ed. by Caseley, J.C., Cussans, G.W., and Atkin, R.K. Butterworth, Oxford, p. 67.

Moss, S. 1992. Herbicide resistance in the weed *Alopecurus myosuroides* (blackgrass): the current situation. In *Achievements and Developments in Combating Pesticide Resistance*. Ed. by Denholm, I., Devonshire, A.L., and Hollomon, D.W. Elsevier, London, p. 28.

Nelson, N.R. and Haissig, B.E. 1986. Herbicide stress: use of biotechnology to confer herbicide resistance to selected woody plants. In *Stress Physiology and Forest Productivity*. Ed. by Hennessey, T.C., Dougherly, P.M., Kossuth, S.V., and Johnson, J.D. Martinus Nijhoff, Boston, p. 191.

O'Donovan, J.T., Sharma, M.P., Newman, J.C., and Feddema, H. 1992. Triallate resistant wild oats (*Avena fatua* L.) are cross-resistant to difenzoquat. *Weed Sci. Soc. Am. Abst.* 32: 265.

Olufunmilayo, O.J., Hobbs, S.L.A., and Jana, S. 1990. Diclofop resistance in wild oat (*Avena fatua*). *Weed Sci.* 38: 475.

Perl, A., Kless, H., Blumenthal, A., Galili, G., and Galun, E. 1992. Improvement of plant regeneration and GUS expression in scutellar wheat calli by optimization of culture conditions and DNA-microprojectile delivery procedures. *Mol. Gen. Genet.* 235: 279.

Powles, S.B. 1993. Multiple herbicide resistance in annual ryegrass (*Lolium rigidum*), a driving force for integrated weed management. In *Integrated Weed Management for Sustainable Agriculture*, Vol. 1. Ed. by Malik, R.K. Indian Weed Science Society, Hisar, p. 189.

Powles, S.B. and Matthews, J.M. 1992. Multiple herbicide resistance in annual ryegrass (*Lolium rigidum*). A driving force for the adoption of integrated weed. In *Achievements and Developments in Combating Resistance*. Ed. by Denholm, I., Devonshire, A., and Hollomon, D. Elsevier, London, p. 75.

Ransom, J.K., Odhiambo, G.D., Kisumu, K. and Gressel, J. 1995. Seed dressing maize with imidazolinone herbicides to control *Striga hermonthica* (Del.) Benth., *Weed Sci. Soc. Amer. Abst.* 35: 15.

Rubin, B., Sibony, M., Benyamini, Y., and Danino, Y. 1992. Resistance to sulfonylurea herbicides in redroot pigweed (*Amaranthus retroflexus* L.). *Weed Sci. Soc. Am. Abst.* 32: 197.

Saari, L.L., Cotterman, J.C., and Thill, D.C. 1994. Resistance to acetolactate-synthase-inhibitor herbicides. In *Herbicide Resistance in Plants: Biology and Biochemistry*. Ed. by Powles, S.O. and Holtum, J.A.M. Lewis Publishers, Chelsea MI.

Sauerborn, J., Saxena, M.C., and Meyer, A. 1989. Broomrape control in faba bean (*Vicia faba* L.) with glyphosate and imazaquin. *Weed Res.* 29: 97.

Sauerborn, J. 1991. The economic importance of the phytoparasites *Orobanche* and *Striga*. In *Proc. 5th Int. Symp. on Parasitic Weeds*, CIMMYT, Nairobi, Kenya, p. 137.

Seefeldt, S.S., Gealy, D.R., Fuerst, E.P., Brewster, B.D., and Appleby, A.P. 1992. Investigations of diclofop-methyl resistant wild oat (*Avena fatua* L.) biotypes from Oregon. *Weed Sci. Soc. Am. Abst.* 32: 266.

Seefeldt, S.S., Gealy, D.R., Pullman, B., and Brewster, D. 1993. Diclofop-methyl resistant wild oat in the Willamette valley of Oregon: distribution, case histories and cross-resistance to other herbicides. *Weed Sci. Soc. Am. Abst.* 33: 55.

Sharon, A., Amsellem, Z., and Gressel, J. 1992. Glyphosate suppression of an elicited defense response; increased susceptibility of *Cassia obtusifolia* to a mycoherbicide. *Plant Physiol.* 98: 654.

Singh, D.T., Nirmala, K., Modi, D.R., Katiyar, S., and Singh, H.N. 1987. Genetic transfer of herbicide resistance gene(s) from *Gloeocapsa* spp. to *Nostoc muscorum*. *Mol. Gen. Genet.* 208: 436.

Smeda, R.J., Vaughn, K.C., and Morrison, I.N. 1992. A novel pattern of herbicide cross-resistance in a trifluralin-resistant biotype of green foxtail [*Setaria viridis* (L.) Beauv.]. *Pestic. Biochem. Physiol.* 42: 227.

Smeda, R.J., Barrentine, W.L., and Snipes, C.E. 1993. Johnsongrass (*Sorghum halepense* (L.) Pers.) resistance to postemergence grass herbicides. *Weed Sci. Soc. Am. Abst.* 33: 53.

Stewart, G.R. and Press, M.C. 1990. The physiology and biochemistry of parasitic angiosperms. *Annu. Rev. Plant Physiol. Plant Mol. Biol.* 41: 127.

Stewart, G. 1990. Witchweed: a parasite weed of grain crops. *Outlook Agric.* 19: 115.

Stoltenberg, D.E. and Wiederholt, R.J. 1993. Giant foxtail *(Setaria faberi)* resistance to acetyl-CoA carboxylase inhibitors. *Weed Sci. Soc. Am. Abst.* 33: 183.

Tardif, F.J., Holtum, J.A.M., and Powles, S.B. 1993. Occurrence of a herbicide resistant ACCase mutant in annual ryegrass selected by sethoxydim. *Planta* 190: 176.

Taylor, J M. and Coats, G.E. 1993. Sulfometuron resistant Italian ryegrass (*Lolium multiflorum)* in Bermudagrass rights-of-way. *Weed Sci. Soc. Am. Abst.* 33: 339.

Thill, D.C., Mallory-Smith, C.A., Saari, L.L., Cotterman, J.C., Primiani, M.M., and Saladini, J.L. 1991. Sulfonylurea herbicide resistant weeds: discovery, distribution, biology, mechanism, and management. In *Herbicide Resistance in Weeds and Crops.* Ed. by Caseley, J.C., Cussans, G.W., and Atkin, R. K. Butterworths, Oxford, p. 115.

Thomzik, J.E. and Hain, R. 1990. Introduction of metribuzin resistance into German winter oilseed rape of double-low quality. *Pflanzenschutz-Nachr. Bayer* 43: 61.

Vaishampayan, A. and Prasad, A.B. 1984. Inter-strain transfer of a pesticide-resistant marker in the N_2-fixing cyanobacterium *Nostoc muscorum. Mol. Gen. Genet.* 193: 195.

Vasil, V., Castillo, A.M., Fromm, M.E., and Vasil, I.K. 1992. Herbicide resistant fertile transgenic wheat plants obtained by microprojectile bombardment of regenerable embryogenic callus. *Bio/Technology* 10: 667.

Weeks, J.T., Anderson, O.D., and Blechl, E.E. 1993. Rapid production of multiple independent lines of fertile transgenic wheat (*Triticum aestivum). Plant Physiol.* 102: 1077.

Wrubel, R P. and Gressel, J. 1994. Are herbicide mixtures useful for delaying the rapid evolution of resistance? A case study. *Weed Technol.* 8: 635.

PART III

THE ECONOMICS

Chapter 16

SEED COMPANY PERSPECTIVES

Donald N. Duvick

CONTENTS

1. INTRODUCTION

Attempts by farmers to fight weeds with chemicals antedate seed companies, and even agrichemical companies. In the 19th century, table salt was recommended as a weed killer. It was applied to burdock stumps, following the hoe. In 1940, as knowledge of chemical–plant interactions was increasing, the desire for targeted chemical assistance in weed control was expressed clearly in a farm institute publication: "For many years farmers have been hoping for some low cost magic fluid that could be applied in small quantities to destroy the weeds without permanently injuring the

1-56670-045-0/96/$0.00+$.50
© 1996 by CRC Press Inc.

land or the accompanying crop plants. To date such a 'miracle' has not been accomplished" (Burlison et al. 1940).

In the ensuing 50 years numerous herbicides (today's name for "low cost magic fluids") have been developed and put to use. The herbicides have been designed to be "crop specific;" i.e., all cultivars of certain crop species are tolerant to a given herbicide, but other crops, as well as the target weeds, are not tolerant to the herbicide.

During the past 15 years, advances in genetic manipulation have enabled the production of specially designed herbicide-tolerant cultivars within a nontolerant species. Advances in tissue culture, mutagenesis, and molecular biology have given new power to the modification of genotypes in regard to herbicide tolerance (Dyer et al. 1993). Research-based seed companies have now become involved with weed control by means of chemicals. They have the power to broaden the use of herbicides, to extend the sales lifetime of some, or to shorten the life of others. They will be able to offer farmers new options for controlling difficult weeds, or new options for herbicide safety or cost control. This new power for seed companies has aroused interest and concern among two disparate groups of people. One group consists of the members of boards of directors, research departments, and business management teams of seed companies and agrichemical companies. A second group is composed of members of organizations devoted to the protection of the environment, and (often) to social action.

The following discussion deals with the points of view and probable courses of action of the seed companies, but does so by contrasting them with viewpoints of chemical companies and environmental groups. Statements and conclusions are based on the author's personal experience in the seed industry and on conversations with representative seed company officials.

2. DISCUSSION

2.1. THE SEED BUSINESS

Profits are the basis for the continued existence of seed companies, as they are for any business enterprise. The profits of research-based seed companies depend on their success in developing, producing, and selling superior new cultivars.

Research-based seed companies aiming for success are impelled to excel in three areas:

1. They must excel in plant breeding and its supporting fields of science; they must be able consistently to produce improved, new cultivars.
2. They must excel in the technology of seed production, producing seeds of new cultivars economically and with high levels of purity and germination.
3. They must excel in marketing and sales; they need to predict changes in market demand years in advance, to give breeders time to develop the right kinds of new cultivars; and they must know how to advertise and sell the new cultivars to the largest possible number of customers.

In the end seed company profits will depend on the difference between cost of goods sold and income from sales. Sales volume alone cannot sustain a company. Overhead costs (including costs of research and development) and operating costs (including costs of seed production and marketing) must be kept in line with income from sales. Income from sales depends not only on volume of sales but also on the price of the product.

2.2 NEW PRODUCTS

Seed companies continually test the market for potential demand for new products in their sphere of interest. When market research indicates potential demand for new products, the companies attempt to add them to their product line. Such additions can be simple, e.g., the addition of a new kind of disease tolerance to sunflower cultivars, or they can be complex and difficult, e.g., entry into breeding, production, and sales of an entire new crop species such as oilseed rape.

Breeding for herbicide tolerance is not really new in the seed business. Since the advent of herbicides, seed companies have selected passively in favor of herbicide-tolerant cultivars; that is, breeding materials and cultivars are grown under current herbicide practices, and so undergo natural selection for ability to perform well under those conditions.

Additionally, seed companies have discarded the strongly susceptible genotypes that occasionally turn up in a crop that is supposed to be uniformly tolerant to a given herbicide. Hybrid seed corn companies have discontinued newly released hybrids that turned out to be highly susceptible to widely used corn herbicides under certain environmental conditions. Soybean breeders in the southern U.S. have selected in segregating populations against susceptibility to metribuzin [4-amino-6-(1,1-dimethylethyl-3-(methyl-thio)-1,2-4-triazin-5(4H)-one]. This susceptibility traces back to a widely used donor cultivar, popular before use of metribuzin as a herbicide (Barrentine et al. 1976; Hartwig et al. 1980).

2.3. HERBICIDE TOLERANCE, MADE TO ORDER

Seed companies and others with expertise in gene manipulation in plants now can provide stronger, directed selection for tolerance to specific herbicides. Genetic transformation, selection in tissue culture, and treatment with mutagens can give rise to cultivars with new kinds of herbicide tolerance. New genes, new levels of regulating existing genes, and new ways of nullifying action of a herbicide give breeders the option of producing tailor-made tolerance to specific herbicides (Dyer et al. 1993).

Plant breeding activities thus can greatly extend the use of a given herbicide, enabling its use on many crops instead of only one or a few (or even no) crops. Breeders can add useful herbicide tolerance to minor crops hitherto neglected by herbicide companies because the potential for sales and profits was too low. They can increase the number of herbicides that can be used on a given crop, giving the farmer greater flexibility in choice and timing of herbicide application. They can breed for tolerance to the most environmentally acceptable herbicides, thus increasing their proportionate use, or conversely they can breed for tolerance to herbicides with less acceptable effects on the environment.

2.4. THREE POINTS OF VIEW

These new possibilities for breeding herbicide tolerant cultivars have excited the interest of three groups of people: environmentalists, chemical companies, and seed companies.

Environmental organizations and certain other public interest groups believe the seed companies' new powers do not bode well for a healthy environment because they enlarge the scope for use of herbicides. "[A] major objective of companies now employing the new biotechnologies is to produce a new generation of agricultural crops and forest trees that are tolerant to chemical herbicides . . . At a time when pesticide residues are being found increasingly in the food supply . . . it is both inexcusable and unacceptable that biotechnology be used to further pesticide use in agriculture . . . [H]erbicide tolerance is not compatible with sustainable agriculture. It diverts us from the paths that could lead to reduced chemical dependency in agriculture . . . It ought to be exposed for what it is: a way for the agrichemical establishment to control the direction of agricultural biotechnology ... [H]erbicide-tolerant crops are at best a distressing misstep, at worst a cynical marketing strategy." (Goldburg et al. 1990)

Herbicide companies look on the new technology with mixed feelings. They see the possibility for greatly increased sales for some herbicides, but they also can envision reduced sales of other herbicides, due to competition from the newly favored ones. Because nearly all U.S. cropland is already treated with herbicides, the increased use of any one herbicide means a reduction in use of others (Dyer et al. 1993). Put more directly, increased business for any one company must come at the expense of sales by another company. Widespread use of specially designed herbicide tolerant cultivars could introduce entirely new factors into the competitive arena. A new kind of volatility has been introduced into the herbicide business (Wheat 1993).

Seed companies have a third point of view. They look on the new possibilities as introducing an uncertain and not necessarily a brighter future. Concerns expressed by environmental organizations have placed the seed companies in a new and uncomfortable role. They have looked upon themselves as developers of environmentally beneficial products. Their business is to develop cultivars with less dependence on chemical protectants, i.e., with improved disease and insect tolerance. They know that during the past 50 years successive new cultivars have shown steady improvement in tolerance to heat and drought, to low soil fertility, and to abnormally cold, wet weather (Duvick 1984). More crop product per unit of land and other inputs means a net saving of environmental costs. Seed company officials understand that specially designed herbicide-tolerant cultivars could stimulate the use of the most environmentally acceptable herbicides. They agree with environmentalists that such changes could extend the range and volume of the use of those herbicides, but they believe that this expanded use would reduce rather than increase environmental problems because of concurrent reduction in the use of less desirable herbicides. Most seed company officials conclude, therefore, that development of herbicide-tolerant cultivars can be compatible with the development of a healthier environment.

2.5. SEED COMPANY CONCERNS, QUESTIONS

Seed company officials are not convinced, however, that the development of novel herbicide-tolerant cultivars is necessarily compatible with increasing profits for their companies. In considering their course of action regarding these potential additions to the product line, they ask the following questions:

1. Will a given herbicide still be environmentally acceptable by the time, years later, that the company's research and development programs have produced cultivars with specially bred tolerance for the herbicide?
2. Will that herbicide still be in demand in the marketplace by the time we have developed cultivars that are tolerant to it?
3. Will a given cultivar still be in demand in the marketplace by the time it has been converted to specific herbicide tolerances? A corollary to this concern is: will the technology of conversion to herbicide tolerance improve? (At present, it is expensive and slow, especially when gene transformation is a necessary first step.)
4. Will introducing herbicide-tolerance genes entail agronomically undesirable changes such as reduced yield, or different time of maturity?
5. Will improvements in alternative systems of weed control (special tillage systems, crop rotations, and microbial treatments) reduce the need for bred-in herbicide tolerances?
6. Will wholesale conversion of cultivars to tolerance for a single favored herbicide increase the incidence of herbicide-tolerant weeds?
7. Could herbicide-tolerant crops become weeds, difficult to eradicate from other crops with tolerance to the same herbicide?
8. In some crops can imparted herbicide tolerance genes pass from herbicide-tolerant cultivars to weedy relatives of the crop? If so, what would be the consequences?
9. Will government regulations hasten the demise of herbicides that today are widely used, thus encouraging use of newer, more acceptable herbicides?
10. Will patents on genetic systems for herbicide tolerance give a monopoly to some companies?

11. Will added complexities in inventory control (the need to have multiple forms of a given cultivar) detract from overall profitability?
12. Can herbicide-tolerant cultivars be sold for a premium price, because of the value they might add to a farmer's operations? If not, can they increase sales significantly? (As noted, added income can come from either higher prices or more sales volume.)
13. Will targeted herbicides be priced low enough to make the cultivar–herbicide combinations economically attractive to the farmer?
14. In sum, seed company officials are looking for an answer to the key question: will the addition of herbicide-tolerant hybrids to our product line increase, decrease, or have no effect on profitability of the company? Will added returns from sales justify the added expense of developing, producing, and selling herbicide-tolerant cultivars? Or alternatively, must we add these new products simply to protect our position in the market?

The following commentary examines these questions in greater detail, and suggests possible seed company actions on them.

2.6. ENVIRONMENTAL ACCEPTABILITY OF HERBICIDES

In recent years some of the older herbicides have been reclassified to lower levels of environmental acceptability. Because the addition of herbicide tolerance to elite breeding materials and cultivars requires several years of concentrated effort, seed companies must take care that they do not invest in research to impart tolerance to a herbicide that might be regulated out of the market before or soon after the work is completed. Predicting the future action of government regulators is, of course, difficult, but a careful study of trends will be required before a prudent seed company sets out to develop genetic tolerances to any given herbicide.

2.7. HERBICIDES: PRODUCT TURNOVER, FLUCTUATING DEMAND

Industrial research and development to produce new and improved herbicides remains strong, although increasing proportions of research and development funds are used to satisfy regulation requirements. Turnover in herbicides is continual; new products replace older ones. Although herbicide products tend to have a longer life in the market than most field crop cultivars, the odds are high that several of today's widely used herbicides will be off the market or used in only limited amounts in 10 years, the time now needed to breed, test, and market a cultivar with novel herbicide tolerance. The demand for cultivars with tolerance to an outdated herbicide will be low.

2.8. CULTIVARS: CONVERSION TIME AND EFFICIENCY

Seed company researchers are fully occupied in breeding new cultivars. Any time and effort expended on the development of novel herbicide-tolerant cultivars must be subtracted from that now spent on breeding and associated research for other traits, unless new money can be found to support the added costs of developing herbicide-tolerant cultivars. At present, the genetic transformation of the major field crops is time-consuming, uncertain, and expensive. Tissue culture and mutagen treatments also are expensive in time and effort per useful product.

Seed companies, therefore, are not likely to start very many conversion projects, especially those requiring search and transformation of new herbicide resistance genes, until more efficient procedures are in place. Only those companies with ample resources can undertake large-scale conversion projects. Even they operate under the double constraint that cultivars and herbicides seem to have increasingly short lifetimes, due to rapid replacement by improved versions. Matching up cultivar and herbicide in the hope that both will still be popular when conversion is complete becomes more difficult each year.

2.9. POSSIBILITY OF DELETERIOUS EFFECTS OF HERBICIDE TOLERANCE

In at least one example the genetic system imparting herbicide tolerance also caused reduced yields in cultivars (Foreham 1989; Grant and Beversdorf 1985; Jursinic and Pearcy 1988; Robbelen 1987). Other examples of undesirable side effects may be found. A difficult but real complication is that undesirable effects may not turn up except in certain environmental conditions or with certain background genotypes. Extensive testing is needed, therefore, to establish the efficacy of a given genetic system for herbicide tolerance. Of prime importance will be whether such anticipated problems are due to tight linkages of tolerance genes with genes of undesirable effect, or to pleiotropic effects of the tolerance genes or genetic systems. The first case eventually can be corrected by selection for recombinant genotypes; the second probably cannot be changed.

2.10. ALTERNATIVE METHODS OF WEED CONTROL

Alternative agriculture by whatever name has as one of its primary objectives the reduction or elimination of chemical weed control at no cost to profitability of the farming operation (Crosson 1989). Farmers continually are experimenting with new tillage systems; researchers are looking for new biological methods to control weeds (Calloway 1992; Gunsolus 1990). Some of the new technologies are said to be successful now; more will be in the future. To the extent that the new methods are adopted, herbicide use will decrease, and perhaps the need for novel kinds of herbicide tolerance also will decrease. A compromise goal of alternative agriculture might be to encourage the use of only the most environmentally compatible herbicides. If that point of view should prevail, engineering cultivars with tolerance to those favored herbicides would be a promising endeavor. Predicting future trends in demand for herbicides and herbicide-tolerant cultivars is obviously a difficult task.

2.11. HERBICIDE TOLERANCE IN WEEDS

Heavy and continuous use of a particular herbicide can result in inadvertent selection for populations of herbicide-tolerant weeds. This phenomenon is more likely to occur with some herbicides than with others, but nevertheless the total number of reported cases of new herbicide tolerance in weeds increases annually (Holt and LeBaron 1990; Dyer et al. 1993). In some cases resistant weed populations have developed within a few years after the introduction of a herbicide. These results demonstrate clearly that to convert all or most of the crops in a region to tolerance to a single herbicide could destroy the utility of that herbicide. From the seed company point of view, therefore, developing cultivar tolerance to just one herbicide might be a poor investment in the long term, even though it could be more economical in the short term. The weed tolerance results also indicate that it might be well to avoid conversion of cultivars to tolerance for certain classes of herbicides.

A special problem for certain crop species is the cross-fertility between the crop plants and their weedy relatives, for example, between cultivated and wild annual sunflower or between grain sorghum and shattercane. Exotic genes for herbicide tolerance, introduced into such cultivars, could easily become incorporated in populations of the weed relatives. Such tolerance could negate the effectiveness of the target herbicide, not only for use in the sunflower or sorghum cultivars (for example), but also for use in all other crops. Seed company officials understand clearly that for crops such as sorghum or sunflower, engineered herbicide tolerance is not an option.

2.12. HERBICIDE-TOLERANT CROPS AS WEEDS

Herbicide-tolerant cultivars could be a carryover problem in succeeding crops using the same herbicide. Corn could become a troublesome weed in soybeans, as is shattercane in grain sorghum. Would this mean that farmers would discriminate against such herbicide-tolerant cultivars?

2.13. NEW MARKET POTENTIALS

Even though seed company officials are concerned about the possibility of low demand for some herbicide-tolerant cultivars, they also worry that they might miss the chance to sell in a windfall market for other kinds of tolerant cultivars. Regulators might ban one or more widely used herbicides and thereby open the market for other herbicides that are not widely used at the present time. Demand for cultivars tolerant to the newly popular herbicides conceivably could escalate rapidly. To be late in entering such an important new market could be financially troublesome to seed companies.

Seed companies, therefore, generally feel they must put at least some effort into developing herbicide-tolerant cultivars, to avoid unacceptably late entry into a possibly important market. Their problem, as already stated, is in deciding which tolerances are likely to be most important 10 years and more into the future.

2.14. PATENTS AND MONOPOLIES

In some instances seed companies without internal resources to do the necessary research are entering into contractual agreements with other seed companies, or with chemical companies, to give them access to herbicide-tolerant genotypes if and when they are needed. They look on these agreements as insurance against the future. Companies letting the contracts do so in order to widen the market for their specialized proprietary products. Such agreements ensure against a monopoly by any one company. Patented products are licensed rather than held away from other users.

Monopoly control is further hindered, of course, by the fact that farmers are free to not buy herbicide-tolerant cultivars. They can continue to use standard herbicide–crop combinations, which will be supplied in proportion to the demand for them.

As noted in an earlier section of this chapter, environmental groups believe that herbicide-tolerant cultivars will allow "the agrichemical establishment to control the direction of agricultural biotechnology..." (Goldburg et al. 1990). They suggest, for example, that agrichemical companies which own or control seed companies may require them to breed and sell cultivars with special tolerance to the parent company's proprietary herbicides; "seed-herbicide packages" will be the goal.

Such controls, such goals, seem unlikely. There is no compelling reason for farmers to have a strong interest in a corporation's self-serving seed–herbicide packages. Farmers make up their own input combinations, designed to fit their unique needs and opinions. They buy cultivars based on total performance, as observed on their own farms. They will show very little interest in cultivars with novel herbicide tolerance, but lacking top yielding ability or other essential traits such as resistance to important disease or insect pests. Farmers will not be interested in a seed–herbicide package unless they are satisfied with both components in the package. A small minority of farmers with special weed problems or with an unusual herbicide carryover situation may have a genuine interest in a package containing the needed herbicide paired with a mediocre cultivar, but their purchases alone will not keep a seed company in business.

Aside from the pervasive influence of government regulations, the control of agricultural biotechnology applied to plant breeding lies in the hands of the end user — the farmer.

2.15. INVENTORY CONTROL

Overproduction and subsequent dumping of inventory play havoc with seed company profits. This problem is especially burdensome in production of seeds with low margin and/or a short shelf life, such as wheat and soybeans. Seed companies worry about the added inventory complexity that will result if they make several herbicide tolerance versions of each cultivar. They wonder if increased income from added sales of herbicide-tolerant cultivars would be canceled out by losses from increased amounts of discarded seed.

2.16. VALUE ADDED PRODUCTS

Income from sales is altered either by volume of sales or price per unit. Pricing herbicide-tolerant cultivars is worrisome for seed companies. These specialty cultivars will cost more to develop, produce, and sell than standard cultivars. It might seem that such cultivars could command an extra price, based on their added value for farmers, but as noted earlier, farmers usually buy cultivars on the basis of total performance, rather than because of specific added traits. Except for a small minority with special needs, today's farmers are not likely to place high value on added herbicide tolerance. Farmers who grow low-volume horticultural crops do have special needs, for few or no herbicides have been registered for those crops (Dyer et al. 1993). Seed companies, up to now, have not put a higher price on a new cultivar simply because of a single added agronomic trait. Pricing has been based on overall merit of the cultivar, as compared to its competition. Improvements in important agronomic traits (e.g., yield, pest resistance, drought tolerance) are advertised to the customer, but pricing of a cultivar is not overtly tied to any specific improvement. Whether to price herbicide-tolerant hybrids as unique value added products, therefore, is of concern to seed companies.

A few herbicide-tolerant corn hybrids have now been put on the market. Reports from the field are that no matter how the hybrids were priced, demand was lower than expected. Farmers looked first to general performance of the hybrids, and then to their unique herbicide tolerance qualities. These early results indicate that currently available herbicide-tolerant corn hybrids will not command a special value added price, and further, they will be purchased in sizable quantity only when their general performance is equivalent to that of the best standard hybrids.

2.17. PROFITABILITY

In the end decisions about whether to add herbicide-tolerant cultivars to the product line of seed companies will be based on estimates of profits. "Profitability ... is the major concern for seed companies. Even if adding novel kinds of herbicide tolerance to cultivars is ecologically acceptable, and even if it is useful to farmers, will it be worthwhile for seed companies?" (Duvick 1992).

Added activities may be undertaken to increase a company's reputation in the investment community, or in the broader public service arena. Such enhancements may be laudable for short-term investment purposes or for demonstrating corporate responsibility to the general public. However, they do not necessarily increase profits, and they may reduce them. For increased profitability, added activities must produce products desired by customers and that can be sold at a profit. In the case of herbicide-tolerant cultivars farmers must want them, and seed companies must be able to develop, produce, and sell them at a profit.

Farmers, too, will want (or not want) herbicide-tolerant cultivars for their own business reasons. Their decisions to buy (or not buy) will be based on the ability of the new cultivars to add to the profitability of farming. Farmers, like seed companies, will not wish to use products that harm the environment, and when they can they will prefer the use of practices that enhance the environment. A few farmers refrain from all or nearly all herbicide use because of strongly held beliefs about environmental safety and human health. Farmers, like seed companies, sometimes adopt practices or buy products because they enhance their reputation among their peers.

Other things being equal, profit enhancement is the farmer's first consideration when buying new products, including new cultivars. The charge and obligation for seed company officials, therefore, is to determine whether herbicide-tolerant cultivars will bring profits to farmers and farming.

3. PREDICTIONS AND RECOMMENDATIONS

Despite the lukewarm attitude of seed company officials toward the production and sale of herbicide-tolerant cultivars, they are moving forward in developing them (Wheat 1993). They do so cauticusly, and with concern for the attitude of the public as well as of their customers. Environmentalist arguments against herbicide-tolerant cultivars have not prevailed with the seed companies. They believe some herbicides are environmentally acceptable, and intend to impart tolerance to those herbicides. As a guideline, they look to those herbicides that have, and that they think will continue to have, government approval. Such approval implicitly indicates environmental acceptability.

It also has been noted that environmental groups are not in full agreement on the matter of herbicide use. Those whose chief concern is soil erosion seem inclined to favor the use of herbicides to enhance minimum tillage operations, if the herbicides are in the most desirable category. Use of cultivars tolerant to those herbicides therefore might also be favored.

Customer demand for herbicide-tolerant cultivars is weak to nonexistent at present, depending on the crop. As with any new technology, it will take time and experimentation for farmers and seed companies to find useful and biologically sensible combinations of farming practice and plant genotype.

Herbicide-tolerant cultivars will not be a one-shot cure for weeds. The Biotechnology Working Group was right when it decried the "silver–bullet" approach for weed control, recommending instead "a truly integrated understanding of the biological systems that make crop and animal agriculture possible in the first place" (Goldburg et al. 1990). Farmers and seed companies may differ with the Working Group regarding use of herbicide-tolerant cultivars in weed control, but they nevertheless will be well advised to take the Working Group's advice about understanding the operation and integration of biological systems in crop agriculture. Such knowledge will enable them to devise effective, economic, and biologically sound combinations of agronomic practice, chemicals, and herbicide-tolerant cultivars. Put in simpler language: broadminded common sense and patience can in the end give new tools for farming with profit and in harmony with nature.

Although seed companies will continue to deplore the added complexities of extra research (either in-house or purchased), extra inventory control, and extra effort in sales operations, they will continue to develop and sell herbicide-tolerant cultivars. In many of their operations, not just with herbicide-tolerant cultivars, seed companies now are identifying specialized small markets with increasingly great precision, looking toward sales to these groups. The hoped-for sales to the small specialty markets are intended to be in addition to, not in place of, present sales to large and relatively homogeneous markets. It seems probable that herbicide-tolerant cultivars will fit into the low-volume specialty market category, but it is not likely that they will add to total sales volume. They will simply replace nontolerant cultivars.

Research and development programs of the seed companies will become increasingly complex in order to serve the more varied markets of the future, but they also are becoming more efficient; their technology is improving. For example, molecular marker technology is now available to greatly speed up backcrossing programs to insert useful genes into elite cultivars. Conversion time will be cut in half, or even more. Unfortunately such technology is also expensive, probably too expensive for routine use. Costs are coming down and efficiencies are improving in this one example of increasingly efficient applications of technology to plant breeding needs.

4. CONCLUSION

Research-based seed companies regard herbicide-tolerant cultivars as a complication rather than a complement to their activities, but they will continue to develop and sell them in proportion to their demand by farmers. Knowledge of how best to use herbicide-tolerant cultivars and how to breed them efficiently is rudimentary but growing. The best results will come from pooling knowledge as it is gained by farmers, plant breeders, agronomists, chemists, and plant biologists with strong knowledge of ecology and systems interactions.

REFERENCES

Barrentine, W.L., Edwards, C.J., Jr., and Hartwig, E.E. 1976. Screening soybeans for tolerance to metribuzin. *Agron. J.* 68: 351.

Burlison, W.L., Hackleman, J.C., and Sherwood, L.V. 1940. *Fighting Illinois Weeds.* Illinois Farmers' Institute, Urbana, IL.

Calloway, M.B. 1992. A compendium of crop varietal tolerance to weeds. *Am. J. Alternative Agric.* 7: 169.

Crosson, P. 1989. What is alternative agriculture? *Am. J. Alternative Agric.* 4: 28.

Duvick, D.N. 1984. Progress in conventional plant breeding. In *Gene Manipulation in Plant Improvement.* Ed. by Gustafson, J.P. Plenum Press, New York, chap. 2.

Duvick, D.N. 1992. Concerns of seed company officials with herbicide-tolerant cultivars. *Weed Technol.* 6: 640.

Dyer, W.E., Hess, F.D., Holt, J.S., and Duke, S.O. 1993. Potential benefits and risks of herbicide-resistant crops produced by biotechnology. In *Horticultural Reviews,* Vol. 15. Ed. by Janick J. John Wiley & Sons, New York, pp. 367–408.

Foreham, M.A. 1989. The effect of the triazine-resistant cytoplasm in selected backcrosses and F_1 hybrids of *Brassica campestris, B. juncea* and *B. napus. Can. J. Plant Sci.* 69: 546.

Goldburg, R., Rissler, J., Shand, H., and Hassebrook, C. 1990. *Biotechnology's Bitter Harvest — Herbicide Tolerant Crops and the Threat to Sustainable Agriculture, A Report of the Biotechnology Working Group.* Environmental Defense Fund, New York.

Grant, I. and Beversdorf, W.D. 1985. Agronomic performance of triazine-resistant single-cross hybrid oilseed rape (*Brassica napus* L.). *Can. J. Plant Sci.* 65: 889.

Gunsolus, J.L. 1990. Mechanical and cultural weed control in corn and soybeans. *Am. J. Alternative Agric.* 5: 119.

Hartwig, E.E., Barrentine, W.L., and Edwards C.J., Jr. 1980. Registration of Tracy-M soybeans. *Crop Sci.* 20: 825.

Holt, J.S. and LeBaron, H.M. 1990. Significance and distribution of herbicide resistance. *Weed Technol.* 4: 141.

Jursinic, P.A. and Pearcy, R.W. 1988. Determination of the rate limiting step for photosynthesis in a nearly isonuclear rapeseed (*Brassica napus* L.) biotype resistant to atrazine. *Plant Physiol.* 88: 1195.

Robbelen, G. 1987. Measurement of yield penalty for triazine tolerance in oilseed rape, *Brassica napus* L. *Cruciferae Newsletter* 12:124.

Wheat, D. 1993. Big changes are coming to herbicide markets. In *The Bowditch Group Inc., Newsletter,* The Bowditch Group, Inc., Boston.

Chapter 17

HERBICIDE-RESISTANT CROPS: PERSPECTIVES FROM A HERBICIDE MANUFACTURER

F. Dan Hess

CONTENTS

1. INTRODUCTION

Imidazolinone-resistant corn is now on the market and several other crops resistant to these and other herbicides are slated for introduction to the market during the remainder of the 1990's. An important topic of discussion within the herbicide industry is what will the impact of these resistant crop–herbicide combinations have on weed control practices? The answer is unknown; however, most believe the availability of this technology will supplement currently used weed control practices (e.g., selective herbicides and cultivation) rather than bringing about a substantial and sustained change to agriculture. This view has been quite consistent throughout the development of herbicide-resistant crops (HRCs). In 1986, Benbrook and Moses stated, "it is not likely that the structure of agribusiness will change dramatically simply because of herbicide resistant plants." In a 1988 review of engineering herbicide resistance into plants, Botterman and Leemans concluded, "herbicide-resistant cultivars will not cause a major green revolution." Four years later, Giaquinta (1992) summarized, "it is unlikely that herbicide-tolerant crops will revolutionize weed

1-56670-045-0/96/$0.00+$.50
© 1996 by CRC Press Inc.

management." The same conclusion seems valid in 1995. Unless there is a specific need for using the resistant crop–herbicide combination, the primary decision for their use will be economic. An effective mechanism for producers to increase profit is to reduce input costs for growing the crop. Weed control is an input cost that growers are constantly trying to minimize.

For the major agronomic crops (corn, rice, cereals, soybeans, cotton, peanuts), which currently use the bulk of the herbicides sold, it is unlikely there will be an increase in herbicide use with the introduction of HRCs. The primary reason is the majority of hectares are already treated. In the U.S., for example, >95% of the corn and soybean production area currently receives at least one herbicide treatment each year (Fehr 1991; Giaquinta 1992).

The dose of herbicide applied to control weeds has been decreasing since 1985 in large part due to the high unit activity of newly introduced herbicides [e.g., acetolactate synthase (ALS) inhibitors]. The trend of developing lower use rate herbicides is strongly supported by the chemical industry and will continue regardless of whether selectivity is inherent in the crop or introduced by genetic engineering. For industry, low-dose herbicides reduce environmental impact which increases the probability of obtaining and retaining a registration, and in many cases, reduces costs which increases the ability of a company to compete in the mature herbicide market.

There are new regulatory issues related to the added genes in resistant crop plants; however, the regulations related to the use of herbicides on food and nonfood crops are in place and can effectively be applied to herbicides being registered for use on HRCs. Thus, new regulations for the herbicide component of the HRC–herbicide technology are not needed.

Even though HRCs add new options for weed control, and clear advantages for HRC technology exist (see Burnside 1992), the chemical industry must manage this new tool so that maximum benefit to agriculture, and thus to the industry, is realized. This includes using only the most environmentally sound herbicides in HRC programs and not excessively relying on a single herbicide, or a single site of action, to the degree that selection pressure results in the appearance of significant increases in tolerant and resistant weed populations (Fehr 1991).

2. IMPACT OF HERBICIDE-RESISTANT CROPS ON AGRICULTURE

There are many excellent options for controlling weeds in major agronomic crops such as rice, cereals, maize, and soybeans. Here, the adoption of HRCs by the farming community will primarily be an economic decision. The resistant crop–herbicide combination will be forced to compete with new technology consistently being brought to the market. For example, imidazolinone-resistant corn (IR) from Pioneer Hybred International and imidazolinone-tolerant corn (IT) from ICI (Zeneca) Seeds must compete for acceptance in a market in which new herbicide options will be increasing significantly during the next several years [e.g., dimethenamid (SAN 582H), acetochlor, flumetsulam (DE 498), sulcotrione (ICIA 0051), flumiclorac-pentyl (V-23031), halosulfuron (Mon 12000), prosulfuron (CGA 152005)]. One exception to HRCs being linked primarily to economics will be their use in managing weed resistance. Nearly all herbicides released by chemical company research groups since 1985 inhibit one of only three enzymes; ALS, acetyl-coenzyme A carboxylase (ACCase), and protoporphyrinogen oxidase (Protox). Two of these sites of action (ALS and ACCase) exhibit significant occurrences of weed resistance. Herbicide-resistant crops that are not ALS or ACCase based will permit the use of alternative herbicides to manage the resistant weed populations. Another exception to the selection of HRCs being solely economic will be when the herbicide–crop combination can be used to control one or more weeds not effectively controlled with other available technology [e.g., downy brome (*Bromus tectorum* L.) in wheat]. However, these opportunities are not great. For example, at a conference on HRCs held at Iowa State University in October 1990, it was estimated that only about 5% of the crop land in Iowa had weeds which were not adequately controlled by current technology (Fehr 1991). Using HRC technology for controlling difficult weeds in a specific crop may create a problem. Often, these weeds are genetically similar to the crop; therefore, in certain instances there may be potential for the resistant trait to outcross to the weed (e.g., grain sorghum to johnsongrass (*Sorghum halepense* L. Pers.). This outcrossing potential must be carefully assessed prior to the introduction of a HRC.

3. ECONOMICS OF HERBICIDE-RESISTANT CROPS

As stated above, unless there is a clear performance advantage of the HRC–herbicide combination, or there is a superior environmental profile when compared to available herbicide options, the principle driving force for adoption of this technology will be economic. There is concern by some that when these traits are available in crops, farmers will be forced to use the associated herbicides. As stated by Dunwell and Paul (1990), "it is illogical to suppose that the availability of a crop variety resistant to a particular herbicide will force the farmer to purchase supplies of the compound. He or she will make a decision on economic grounds and the herbicide regime chosen will be that which can provide effective control at the lowest possible cost." Clearly, the principle question to be asked by those considering use of this new technology is what is the cost (added seed cost plus herbicide) compared to current input costs for *equivalent* weed control? If the cost is less for equivalent weed control, then the technology will likely be adopted; if not, commercial success will be minimal or be driven by other factors such as reduced environmental impact.

Regarding economics, the most significant impact may well be for the nonselective herbicide glyphosate. If the price of glyphosate from Monsanto, and after patent expirations other generic producers, continues to decline, this herbicide will indeed be an attractive weed control strategy for agriculture. Added to this will be the availability of similar herbicides such as sulfosate (Zeneca) which no doubt will be cross-resistant. The potential reduction in herbicide costs to farmers with the availability of inexpensive generic glyphosate, and a wide range of HRCs for glyphosate, has been estimated to be hundreds of millions of dollars each year if only one application is needed for the weed control required to sustain yield (David Wheat, The Bowditch Group, Inc., Boston, personal communication). However, it is critical to be aware of, and to fully understand, the significant long-term risk in relying on one herbicide with one site of action for all weed control. This risk is discussed in Section 5.

Some are concerned that the high level of resistance being introduced into HRCs will encourage excessive use of the herbicide associated with the HRC. While this possibility will technically exist, the economics of agriculture will be a substantial deterrent. To reduce costs and increase profits, farmers are already experts at quickly determining the absolute minimum amount of herbicide needed on their land to achieve the desired level of weed control. As stated by Fehr (1991), "the value of the dollar has been and will continue to be the main deterrent to excessive application."

4. ENVIRONMENTAL EFFECTS OF HERBICIDE-RESISTANT CROPS

The claim that HRC technology will increase the use of herbicides is not correct with respect to major agronomic crops. As stated in the Introduction, nearly all of the agricultural land used to grow these crops is already treated with herbicides, and alternatives to herbicides are currently not adequate (e.g., biological agents such as mycoherbicides) or are environmentally unsound (e.g., overuse of cultivation with resulting soil erosion from wind and rain). In fact, HRC technology can permit the use of the most environmentally sound, low use rate, herbicides in crop production, which is now not possible due to sensitivity of the crop to the herbicide (Dunwell and Paul 1990).

It is important that industry be diligent in providing proper stewardship of this technology so as to maximize its benefit to agriculture. Over the long term it will not serve the best interests of agriculture, and thus the industry, to abandon known good management practices in favor of overreliance on one herbicide or one mode of action. These good management practices are rotation of herbicide modes of action, proper use of cultivation, the use of cover and smother crops where applicable, and crop rotation. Even though HRCs open new possibilities for weed control, industry must continue its search for low use rate, environmentally sound, crop-selective herbicides with new mechanisms of action. Fehr (1991) stated, "the worst thing that could happen would be a reduction by the private sector in their research for the identification of new compounds." Research on application and formulation technology must continue to maximize the effectiveness of herbicide use. Industry should also support internal and external long-term research for weed control alternatives such as natural product chemistry, allelopathic agents, and biological control.

5. UTILITY OF USING HERBICIDE-RESISTANT CROPS IN A WEED CONTROL PROGRAM

The most intense discussion within industry regarding HRCs is related to the impact of using nonselective herbicides such as glufosinate and glyphosate in a "total postemergence" herbicide program. The intent is to replace current weed control practices which involves preemergence and postemergence herbicide use as well as cultivation, and in some crops, hand weeding. The benefits of a total postemergence herbicide program were discussed by Wyse (1992). Wyse concludes total postemergence weed control allows farmers to use the "weed threshold" concept of weed management, it reduces or eliminates the need for cultivation and hand weeding (which enhances soil conservation and reduces production costs), and it allows farmers to try alternative weed control methods because postemergence herbicides are available as insurance treatments.

While the first analysis of a total postemergence weed control program may appear to be a major attribute for this technology, there are associated risks. In a program that relies only on postemergence herbicides, the timing of application will be critical to assure that competition from the developing weeds do not affect crop yield prior to being controlled. If weather conditions delay herbicide application, continued weed growth during this delay may have a significant effect on crop growth, development, and yield. If a very early application is needed in order not to risk crop yield loss, then multiple applications may be necessary if the herbicide does not have adequate residual activity.

Nonselective herbicides do not control all weeds to the same degree of efficiency; therefore, during several years of application of the same postemergence herbicide to HRCs, weeds less well controlled will dominate in the population. This shift in weed species can also occur with conventional herbicide application (Haas and Streibig 1982). There is also the issue of selecting resistant weed populations from the environment by overuse of a single herbicide target site of action (Holt 1992). Based on insecticide and fungicide experiences, it is naive to assume any herbicide is immune to the selection of resistant biotypes from populations. A justified concern of HRCs is potential outcrossing of the resistant trait from the crop to nearby weeds [e.g., from cultivated rice into red rice (*Oryza sativa* L.)]. Although not a problem in many crops, Dale (1992) published an overview of specific cases of crops being engineered for herbicide resistance with examples of sexually compatible wild species to these crops. A hidden problem with the reality of outcrossing between crop and weed is the use of herbicide resistance as a selectable marker while engineering crops for other traits (Gressel 1992). A final concern regarding the utility of HRC–herbicide combinations is the potential added difficulty in controlling the HRC if it germinates in a rotated crop where the same herbicide could be used for control (e.g., control of ACCase-resistant corn in soybeans using ACCase herbicides). The chemical industry must continue to recognize the importance of using a mix of herbicide modes of action to manage weeds and to integrate herbicide use with cultural practices (e.g., crop rotation and timely cultivation) known to minimize the impact of weeds on production agriculture. As eloquently stated by Giaquinta (1992), HRCs "are not a license to violate sound weed management practices."

Although significant potential of HRCs has been identified for minor crops (e.g., Duke et al. 1991, Dalling 1992), registration of herbicides for use on these crops will face many of the same issues already recognized for these crops. The market size is small for minor crops and may not justify registration expense even with the help of government assistance programs (e.g., the IR-4 program in the U.S.). The financial risk for the herbicide manufacturer associated with herbicide use on high-value minor crops may not justify the potential increased sales. Finally, as discussed in Section 6, minor crop registrations present special issues that must be assessed with respect to benefits to a herbicide manufacturer. For these reasons the tendency may be for the chemical industry not to register herbicides for the minor HRCs.

6. REGULATORY ISSUES RELATED TO HERBICIDE-RESISTANT CROPS

The use of herbicides on HRCs should not require any additional regulations. In the U.S., for example, there are currently precise regulations for registering a herbicide for the use on a new crop with the Environmental Protection Agency (EPA) under the Federal Insecticide, Fungicide and Rodenticide Act (FIFRA). Part of the registration process for the use of herbicides in crops involves an assessment of metabolic degradation of the herbicide in the crop. All metabolites with concentrations higher than 0.01 ppm must be identified. Then the residue level of the parent herbicide and all identified metabolites must be determined in the crop. If the crop is processed, residues must be determined in the processed form of the crop. If crop residue exists and the crop is fed to animals, residue in the animal products (meat, milk, and eggs) must also be determined.

Based on the levels of residues found in the crop, a "Tolerance" residue level for the parent herbicide, plus all identified metabolites, is assigned by the EPA. Based on the toxicology data acquired for the herbicide and its major metabolites, the EPA establishes a Reference Dose (RfD) [formerly called Acceptable Daily Intake (ADI)] for that herbicide. The RfD is set using the lowest "No Observed Adverse Effect Level," or NOAEL, found in critical animal toxicology studies (expressed in milligrams per kilograms per day) divided by a safety factor of usually 100 or more. The portion (percentage) of the total RfD utilized by each crop is based on an EPA computer program (Dietary Risk Evaluations System; DRES) that calculates the dietary intake of the crop by various segments of the human population and then multiplies this value by the Tolerance residue assigned to the crop. For the purposes of these calculations it is assumed all planted hectares are sprayed with the maximum registered rate of the herbicide. The more a particular crop is consumed as food by various segments of the population, the greater the percentage of the RfD assigned to that crop. The number of hectares grown is not a consideration. As an example, when a constant Tolerance residue value (e.g., 1 ppm) is input into the DRES computer program, a larger percentage of the assigned RfD is utilized by vegetable crops when compared to agronomic crops (Table 1). In this example nearly 30% of the RfD is assigned to tomato even though the hectares of tomatoes in the U.S. is substantially lower than field corn, which is assigned only 6.7% of the RfD. Soybeans are assigned nearly the same RfD as lettuce (6.8 and 4.5%, respectively), even though the hectares of soybeans are profoundly larger than the hectares of lettuce. Cotton products that eventually reach the food chain (seed oil and seed meal fed to animals) are few, thus cotton represents only a small portion of the RfD (0.4%). Based on the DRES analysis, there may be a reluctance by manufacturers to register a herbicide for use on herbicide-resistant vegetables. Clearly, the small land area for potential herbicide sales in vegetables such as tomato and lettuce, when compared to the sales potential for field corn, soybeans, and even cotton, would be a significant decision factor if the RfD were limiting.

When actual Tolerance residue values are used in place of a constant Tolerance residue, the portion of the assigned RfD utilized by vegetables often increases. This is because residues are usually higher in vegetables (e.g., lettuce) when compared to agronomic crops (e.g., field corn). The cause is largely due to vegetables being consumed sooner after herbicide application, which allows less time for degradation of the herbicide and its dilution during crop growth. As an example, DRES was used to calculate the percent of the assigned RfD utilized for specific crops using assigned Tolerance residues for a particular registered pesticide (Table 2). Based on the tomato consumption

TABLE 1 Amount of Total Assigned RfD (5 μg/kg/d) Used by Specific Crop Registrations

	TMRC[a] (μg/kg/d)	RfD[b] (%)
Cotton	0.021	0.4
Lettuce	0.226	4.5
Broccoli	0.049	1.0
Soybeans	0.340	6.8
Peanuts	0.070	1.4
Tomatoes	1.454	29.1
Sweet corn	0.236	4.7
Field corn	0.334	6.7
Wheat	1.411	28.2
Total	4.141	82.8

Note: A constant Tolerance residue (1 ppm) was used for each crop. The RfD used was for the total population in the 48 contiguous United States.

[a] Theoretical Maximum Residue Concentration consumed as food from individual crops.
[b] Portion of Reference Dose Represented by individual crops.

TABLE 2 Portion of Total Assigned RfD (13 μg/kg/d) for a Registered Pesticide on Specific Crops

	Tolerance residue[a] (ppm)	TMRC[b] (μg/kg/d)	RfD[c] (%)
Cotton	0.05	0.001	<0.01
Tomato	10.0	14.546	111.9
Squash	10.0	0.641	4.9
Wheat	0.05	0.071	0.5
Red meat	0.01→0.5	0.035	0.3
Dairy products	0.01	0.106	0.8
Totals		15.400	118.4

Note: The RfD used was for the total population in the 48 contiguous United States.

[a] Tolerance residue (parent plus metabolites) assigned to individual crops.
[b] Theoretical Maximum Residue Concentration consumed as food from individual crops.
[c] Portion of Reference Dose Represented by individual crops.

profile used in the DRES analysis and the assigned Tolerance residue of this pesticide on tomatoes, more than the total assigned RfD for the population in the 48 contiguous United States is utilized for this single crop.

When the cumulative RfD for all crops on which a particular herbicide is registered reaches 100%, no new crops can be added to the registration unless an existing registration is canceled. However, before any cancellations are initiated, the EPA usually recalculates the percent of the RfD utilized by each registered crop using the anticipated residue (average expected residue) in place of the Tolerance residue (the maximum legal limit of residue). Because actual residue in a crop is usually less than the maximum residues allowed by law, the percentage of the total RfD utilized deceases, which allows for additional crop registrations without altering the original safety factors built into the RfD.

7. CONCLUSIONS

Herbicide-resistant crops and their associated herbicides are now, and will continue to be, an option for managing weeds in production agriculture. Their availability supplements current weed control practices, but will not revolutionize weed management. Because few unsolved weed problems exist, the principal reason for producers to adopt HRC–herbicide combinations will be economic (i.e., the least expensive option). Industry must support this technology being used only as part of diversified and environmentally sound strategies for weed control. These strategies should include rotation of crops and herbicides with different mechanisms of action and the use of cultural practices (e.g., cultivation, cover crops, and smother crops) which minimize the impact of weeds on crops. The potential for selecting resistant weeds from populations, for outcrossing from crops to certain weeds, and for inducing species shifts toward difficult-to-control weeds must be evaluated and understood by the herbicide manufacturer prior to marketing a herbicide for HRCs. With the availability of HRC technology, industry must resist the temptation to promote overreliance on one herbicide mechanism of action. Finally, HRC technology is not the penultimate cure for weed influences on crop production, and thus must not stifle the search by industry for new and better weed control technology.

ACKNOWLEDGMENTS

The DRES analyses for RfD determinations were conducted by Mr. M. Graben, Regulatory Group, Sandoz Agro, Inc. A literature search, which served as the foundation for this publication, was conducted by Ms. M. Manion, Library and Information Services, Sandoz Agro, Inc.

REFERENCES

Benbrook, C.M. and Moses, P.B. 1986. Engineering crops to resist herbicides. *Technol. Rev.* 89: 54.

Botterman, J. and Leemans, J. 1988. Engineering of herbicide resistance in plants. In *Biotechnology and Genetic Engineering Reviews*, Vol. 6. Ed. by Russell, G.E. Intercept Ltd, Wimborne, chap. 8.

Burnside, O.C. 1992. Rationale for developing herbicide-resistant crops. *Weed Technol.* 6: 621.

Dale, P.J. 1992. Spread of engineered genes to wild relatives. *Plant Physiol.* 100: 13.

Dalling, M.J. 1992. Development of crops resistant to herbicides. In *Proc. 1st Int. Weed Control Congress.* Ed. by Combellack, J.H., Levick, K.J., Parsons, J., and Richardson, R.G. Weed Science Society of Victoria, Inc., Melbourne, Australia.

Duke, S.O., Christy, A.L., Hess, F.D., and Holt, J.S. 1991. *Herbicide-Resistant Crops, Comments from CAST No. 1991-1.* Council for Agriculture Science and Technology, Ames, IA, 24 pp.

Dunwell, J.M. and Paul, E.M. 1990. Impact of genetically modified crops in agriculture. *Outlook Agric.* 19: 103.

Fehr, E.R. 1991. Herbicide tolerance in crops. In *Agricultural Biotechnology at the Crossroads; Biological, Social and Institutional Concerns.* NABC Rep. 3. Ed. by MacDonald, J.F. National Agriculture and Biotechnology Council, Ithaca, NY, p. 179.

Giaquinta, R.T. 1992. An industry perspective on herbicide-tolerant crops. *Weed Technol.* 6: 653.

Gressel, J. 1992. Indiscriminate use of selectable markers — sowing wild oats. *Trends Biotechnol.* 10: 382.

Haas, H. and Streibig, J.C. 1982. Changing patterns of weed distribution as a result of herbicide use and other agronomic practices. In *Herbicide Resistance in Plants.* Ed. by LeBaron, H.M. and Gressel, J. John Wiley & Sons, New York, chap. 4.

Holt, J.S. 1992. History of identification of herbicide-resistant weeds. *Weed Technol.* 6: 615.

Wyse, D.D. 1992. Future impact of crops with modified herbicide resistance. *Weed Technol.* 6: 665.

Chapter 18

FARMER AND PUBLIC PERSPECTIVES OF HERBICIDE-RESISTANT CROPS

Loren W. Tauer

CONTENTS

1. INTRODUCTION

The public desires quality and safe food at a reasonable price. At the same time, they want agricultural production to meet standards of environmental stewardship, community amity, and rural aesthetics. The difficulty in attaining these goals is that vast uncertainty exists concerning correctly measuring them and determining their trade-offs under different agricultural alternatives. Individuals also place different values on the various attributes of agriculture. Thus, in much of the debate on herbicide resistance in crops, the discussion depends as much on values as it does on the possible impacts and trade-offs of herbicide resistance on the characteristics of agriculture and food. Unfortunately, our values often determine how we estimate or at least interpret scientific evidence (Shrader-Frechette 1985). Even if we claim to be objective, simply the selection of a scientific problem to investigate indicates that we feel that problem is important enough that we give up the opportunity to investigate other scientific problems. We will not be able to resolve those issues here. Instead, the author attempts to point out the concerns of the public and what we might know or what must be known before any resolution can be reached. The acceptability of herbicide-resistant crops (HRCs) will be determined in the economic and political arenas.

It is common for economists to separate the members of an economy into producers and consumers because at the most elementary level we are all consumers and a number of us are producers of a commodity or service. This is the division used in this chapter. The commodity here is more than just food, however; it includes the characteristics of the production process and all of its end products, including any externalities of pollution, or environmental and social degradation or enhancement. In most debates the public also includes those intermediaries or organizations that

1-56670-045-0/96/$0.00+$.50
© 1996 by CRC Press Inc.

are formed because of the institutional arrangements that have evolved in an economy. Not much time is spent discussing these various organizations here. They are important in helping producers and consumers reach their objectives, and often take on a life of their own, but nonetheless are only the result of consumers and producers bound by some common purpose within a specific institutional setting.* The concentration is on those preferences and objectives.

2. PREFERENCES OF FARMERS

The economic decision for a farmer concerning the use of herbicide-resistant technology is quite simple: Is herbicide resistance something that he or she can benefit from without harming the self, the family, or others? The concept of harm is comprehensive here, but not exhaustive, in that it includes, for instance, water contamination that would make water unsuitable to drink, yet does not include the fact that using HRCs may allow him or her to produce a crop at a lower price and indirectly drive other farmers out of business. As a society we have decided that price competition that may force some out of business is necessary in order to ensure that resources are used efficiently, so that maximum output can be produced with the limited resources of society. Not permitting price competition would allow for inefficient use of resources, diminishing the total output that can be produced, as recently failed economic systems in eastern Europe have demonstrated. However, we do not generally allow nonprice competition in order to eliminate competition. That would not necessarily lead to more efficient use of resources.

At the other end of the economic criteria are equity considerations. Many believe that society is better off if wealth and income are not concentrated in the hands of a few, but are more evenly distributed across individuals. Unfortunately, there are often trade-offs between efficiency and equity. The opportunity to earn more income than others provides an economic incentive for entrepreneurs, but may allow wealth to amass into a few hands, and may even ultimately threaten price competition if the wealthy use their power to curb price competition. Thus, we often use institutional restrictions to maintain some degree of equality. These include progressive income and inheritence taxes. Another such mechanism is the restrictions in various forms against corporate farming in many states.

It is hard to envision that the use of HRCs would have a detrimental impact by itself on the distribution of income (Fehr 1991). Weeds are extensively controlled now with chemical herbicides, and modifying which chemicals are used on which crops should not significantly alter farm incomes. Nonetheless, if farmers adopt herbicide-resistant technology, they do so because it provides some benefit to them. That benefit may be lower-cost weed control, increases in yield, or a reduction in yield variability. In a competitive economy all of these benefits ultimately lead to a reduction in crop prices.

What is the impact on crop prices if HRCs are introduced? Tauer and Love (1989) found that the price effect would probably be minor. They used an econometric model to simulate what impact herbicide-resistant corn would have on agricultural producers and consumers. The model they used was a temporal (10-year) partial equilibrium model that included the major crops and livestock sectors of the U.S. Therefore, although they only introduced herbicide resistance in corn, they were able to determine the impact of that introduction not only on the corn market but also on the other crops and livestock groups over each of 10 years, thus providing both short- and long-run impacts.

It is currently not known how herbicide resistance would affect herbicide usage by type and quantity, how it would affect yields, or how it would affect production costs per hectare. Because

* Organizations are also formed to change the institutions. The New Institutional Economics is described by Eggertsson (1990) and North (1990). In that literature institutions are the rules of the game and organizations are the players.

so many herbicides are involved and weed problems are unique to individual farmers/regions, the pattern of change in herbicide usage/application will probably be complex and not easily summarized by a blanket usage change, as many attempt to do. Thus, Tauer and Love (1989) ran a number of scenarios representing a number of different types of changes. Some of these scenarios were extreme cases in order to determine the boundaries of possible impacts.

Herbicides are applied to nearly all corn hectarage (Hanthorne et al. 1982), and the number of treatments per hectare averages about 1.3 (USDA, 1989). A combination of preemergence and postemergence applications are used. As the earlier chapters of this book demonstrate, work is progressing on making the corn plant resistant to many different herbicides. Because many of these are for postemergence use, some believe that herbicide applications and usage will decrease because farmers can ascertain weed infestation before making a decision to apply herbicide (although the decision to plant the herbicide-resistant corn must be made before then). This may reduce herbicide usage, although the type of herbicide may change as will the resultant impact. The actual cost may increase or decrease depending on application mode and the price of the herbicide and corn seed. Others have argued that application rates may increase, especially if the following year's crop becomes resistant to a herbicide when it was not previously resistant (Dekker and Comstock 1992). An increase in application rates will increase costs, increase yields, and probably not decrease the environmental impact. Thus, it is not clear yet whether herbicide resistance will increase or decrease herbicide use, measured either in application numbers or rates.

Costs will depend on the pricing strategies of seed and chemical companies. A technology that can increase corn yields by controlling more weeds or reduce application costs can be priced higher and still be adopted by farmers. However, the herbicide market is a mature market used on most hectarage, so capturing market share at a low price may be a profitable strategy for a seed/chemical company or coalition.

Under current weed control technology, the U.S. Corn Belt only loses about 4% of possible corn production from weed pests (NAPIAP 1985). Therefore, even assuming that herbicide-resistant corn would permit complete control of weeds and substantial adoption of that technology, the increase in corn output would be minor, with yield increases in the Corn Belt only about 251 kg/ha, on average (Tauer and Love 1989). If the herbicide-resistant corn seed and companion chemical increase the cost of producing corn by $5.27/ha, it would barely be profitable to use in the Corn Belt and probably would not be profitable in the Northern Plains of the U.S. It would be used only on hectarage with severe weed infestation that could not be controlled by other methods. Only if the herbicide resistance technology were similarly priced, so that the cost per hectare was similar to the current weed control costs, would the herbicide resistance technology be widely adopted in the Corn Belt.

The story is different outside the Corn Belt, especially in the southeast regions of the U.S. (Tauer and Love 1989). Corn yields are lower outside the Corn Belt, and part of the reason is poor weed control. Alternatives that could lead to better weed control would be profitable to use. This, in itself, presents a strategic decision for the seed companies. Although the potential need for herbicide-resistant corn outside the Corn Belt may be greater, the corn hectarage and thus seed sales are lower. Corn hectarage would increase only slightly outside the Corn Belt, even with better weed control because there are other production limitations. Should the seed companies first target the varieties for the Corn Belt or those outside the Corn Belt? That decision has regional income implications. If the technology is first made available in varieties suitable for the Corn Belt, the small increase in corn yield per hectare would reduce the corn price for farmers in all regions. If instead it is first made available in varieties grown outside the Corn Belt, producers outside the Corn Belt could gain without significantly increasing aggregate corn production or lowering corn price. Regardless which region uses the technology first, however, corn producers in the Southeast (Delta, Appalachia, Southeast) always benefit, even if aggregate corn production increases and lowers the price of corn. That is not so of producers in the Corn Belt because under most scenarios their per hectare corn profits decrease after herbicide-resistant corn is widely adopted.

The fact that farmers would use a technology that ultimately reduces their profit perplexes some people, but the reason is due to the individual decision vs. the collective impact of all farmers. Figure 1 shows possible profit per hectare for an individual farm without the introduction of herbicide-resistant corn, with adoption, and with availability but not adoption. Note that early adoption increases the profit per hectare until a large number of farmers adopt, which increases aggregate corn production, lowers corn price, reduces revenue, and thus profit per hectare. Eventually, the profit per hectare is lower than the farmer would have experienced without the introduction of herbicide-resistant corn. Should the farmer cease using the herbicide-resistant corn? The farmer could, but then profits would be even lower because other farmers would continue to use the herbicide-resistant technology.

Cast in a dynamic setting, this leads to the technology treadmill story of Cochrane (1958). New technologies of various sorts are constantly released to farmers. By being one of the first to always adopt, a farmer can actually enhance his or her income as compared to no new technological change. Even if a farmer is not the first to adopt, he or she must eventually adopt in order to remain competitive with the vast majority of farmers who do adopt. Empirical evidence comparing income distributions of farmers from countries with various rates of technological change has found that in countries with greater rates of technological change the income of well-educated farmers who are in a position to continually adopt first are much greater than poorly educated farmers (Welch 1970).* This may explain why some farmers actively support agricultural research while others resist it, although other moral, social, or economic reasons may also be responsible.

Increased corn production from better weed control affects other commodities (Tauer and Love 1989). Livestock producers benefit from lower corn prices and respond by increasing their production, leading to greater demand for protein meal and increasing the price of soybeans, at least temporarily. Soybean hectarage eventually increases, not only because of higher soybean prices, but also because corn production becomes relatively less profitable. Outside the regions where soybeans are grown, alfalfa hay output increases instead. Consumers benefit because greater quantities of

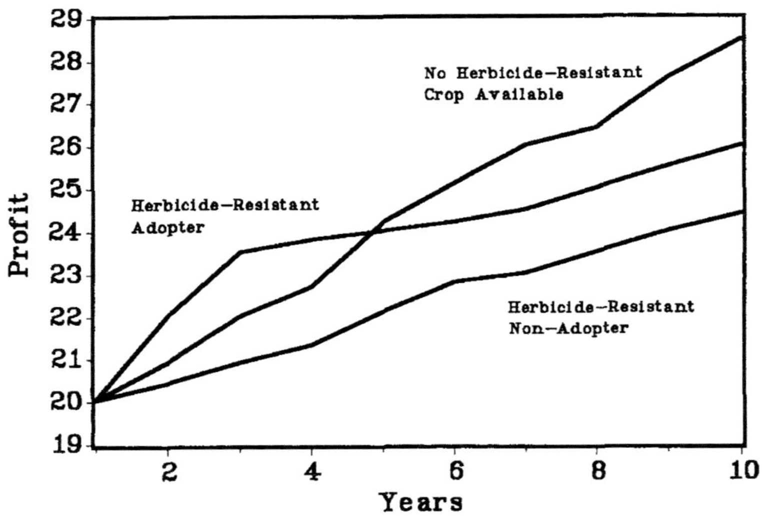

FIGURE 1 Profits of adopter, nonadopter, and nonavailability of a HRC.

* It is believed that education allows the farmer to correctly process the information necessary to successfully adopt. Characteristics other than education, such as wealth, may also be responsible.

commodities are produced at a lower price, and exports increase slightly. The net result is that total farm income falls, but consumer benefits increase, such that there is a net gain in economic welfare; society is economically better off because more can be produced with limited resources (Tauer and Love 1989).

To my knowledge, no other economic study has been completed on HRCs in the U.S., although as the earlier chapters in this book show, resistance to crops other than corn is being developed. One important crop is soybeans. Herbicide use in soybean production amounts to over $1 billion annually, nearly half of the U.S. herbicide market value (Netzer 1984), and 95% of all U.S. soybean hectarage is treated with herbicides one to two times per season (USDA 1987). Weeds still cause about 9.8% loss in annual U.S. soybean production. Love and Tauer (1989), using aggregate supply and demand curves, estimate that a biotechnology in soybean production which completely eliminates weed losses by 1995 to 1997 could increase annual production by 5.6 million Mg. The net value of this increased production accrues to consumers.

The Tauer and Love (1989) study is a standard economic analysis that does not include organizational or institutional change. Some have argued that the prospects of herbicide resistance has and will cause structural changes in the agricultural sector that might have a detrimental impact on social welfare in the food system (Comstock 1989). Tauer and Love (1989) also do not include any estimate of the possible environmental damage or benefits that herbicide resistance may have because little confident information concerning cause and effect is known. Finally, they ignore possible structural impact on rural communities. Opponents to herbicide resistance have argued that these are important issues of concern which cannot be ignored (Goldburg et al. 1990).

Doyle (1985) discussed the possible impacts on agriculture from the structural changes in the seed and chemical industries. He concluded that the ongoing developments were not conducive for a type of agriculture that he and others envisioned as ideal. Their best outcome would be an agriculture that provides farmers many different seed variety and chemical options. A large selection would permit farmers to select which is best for themselves. As seed and chemical firms consolidate, Doyle worries that the genetic diversity of seeds may be insufficient to meet the environmental challenges that nature presents. The current solution is to resort to chemical control, which leads agriculture toward a nondiversified route, with greater risks of calamity. Doyle also does not have much hope for the agricultural colleges in helping to develop a more ideal agriculture. They have surrendered much of their plant breeding programs to industry and have shifted to genetic engineering, including engineering for herbicide resistance.

Farmers need to be concerned about the market power of input suppliers and the impact that may have on their choice and cost of inputs. It is not clear, however, whether the firm consolidation that Doyle discusses is a result of firms positioning themselves to gain from any economies of biotechnology research and market gains, or whether consolidation may have occurred anyway, even without biotechnology. If that consolidation would have occurred anyway, it still needs to concern farmers, but that discussion is not dependent upon engineering HRCs. The key to ensuring that market power does not extract excessive profits from farmers is to make sure that farmers have viable alternatives. Public development and control of some effective crop varieties, the availability of off-patent chemical herbicides, and the development of nonchemical weed control methods all can provide viable alternatives to check the exercise of monopoly power. Even if these alternatives are not extensively utilized by farmers, their potential use will put constraints on the actions of input suppliers.

Farmers are concerned about the environmental and health aspects of chemical herbicides on both themselves and others. Except for a few specific chemicals, the evidence of cause and effect and damage is vague, and each farmer reacts differently to incomplete knowledge. Some are convinced that all chemicals are undesirable, and have eliminated their dependence on chemicals.

Others believe the environmental and health effects are nonexistent or insignificant as compared to the benefits of chemical herbicides. Only more information will help to resolve these uncertainties.

Farmers are also concerned about the structure of their communities. However, most farmers realize that the structure of agriculture will have little impact on rural communities. That is because except for the Great Plains of the U.S. and a few isolated pockets elsewhere, very few U.S. counties are agriculturally dependent (Ahearn et al. 1988). Even with a multiplier effect, losing farms or farm consolidation has very little impact on the social structure of the community. We are no longer a nation of farmers.

A final concern of farmers is that HRCs might accelerate the occurrence of herbicide-resistant weeds. The argument is that HRCs may lead farmers to depend upon only one herbicide or one family of herbicides, permitting weeds resistant to that herbicide to greatly expand in numbers as other weeds are eliminated. This may also happen if the weed can capture the necessary genetic material from a herbicide-resistant relative crop. The end result is that the herbicide is no longer effective in that region as the immune weeds spread.

Economists refer to this type of phenomenon as public property or problems of the Commons. Although not normally considered an asset, weeds are public property owned by all of us. What one farmer does to this public property affects not only himself but others as well. If a farmer produces a field of weeds immune to a specific herbicide, then he no longer is able to effectively use that herbicide. Worse yet, if those weeds spread to other farmers' fields (and weeds have a habit of doing that), then other farmers are also not able to effectively use that herbicide, although they were not responsible for the irresponsible act that caused the immunity. Yet, it is the company producing the herbicide that is the most affected. If resistant weeds spread, then their herbicide becomes ineffective and sales plummet. Naturally, it is in the best interest of chemical companies to ensure that a large and concentrated population of immune weeds does not develop. One strategy to prevent this is to instruct farmers not to use the same herbicide year after year. Unfortunately, it only takes a few to ignore this advice for a problem to develop, and a few will ignore it if the herbicide in question appears to be effective currently in controlling weeds. Developing crop resistance to a specific herbicide across crop species will mean this is not only a problem for monocultural farming but also for farmers rotating crops. In fact, even without engineered herbicide resistance some families of herbicides can be used on dissimilar crops. The solution for chemical companies is straightforward. Make sure that the susceptible herbicide is marketed as tank mixes with other chemicals that will keep immune weeds in check. The problem, however, is that what may be a relatively environmentally "friendly" susceptible herbicide may be combined with a herbicide that is not as environmentally friendly.

3. PREFERENCES OF CONSUMERS

The consuming public wants quality and safe food at a reasonable price. Some are not convinced that chemical weed control is necessary to achieve this end. Even if chemicals are useful in controlling weeds, others feel that chemicals are unethical to use and believe that engineering crop resistance to herbicides is unwarranted. They believe that the use of chemical herbicides may be detrimental to the environment, unsafe for farmers or workers to handle, or lead to unsafe food. Still others believe that the use of chemical herbicides will lead to the demise of the family farm with adverse effects on rural communities (Comstock 1989). The development of herbicide resistance is seen as a continuation and reinforcement of chemical use in agriculture.

The environmental concerns are discussed in other chapters, and those discussions provide insight on the environmental issues. Not discussed is whether the use of chemicals is necessary in order to ensure adequate food at a reasonable price. For even if the use of chemicals imposes some environmental risks, many would believe that those risks are necessary in order to produce food for humans.* A minority would still assert that this is an anthropocentric perspective and that humans should not be accorded more importance than other natural beings, with rights that could be violated via environmental impact.

At first glance, one would opt to conclude that chemicals are necessary simply because they are so commonly used, and were widely adopted during a period when food production per hectare increased dramatically, in part because of chemical usage. Numerous economic studies have shown that restricting or banning some or all chemical herbicides would have a draconian impact on food production (Burton and Martin 1987; Knutson et al. 1990; Swanson and Grube 1986). Those studies typically are completed by modeling the withdrawal of the chemical and substituting current available technology without the use of the chemical. Dissenters argue that this is an unfair comparison. The use of chemical herbicides evolved under an institutional structure that favored the use of those chemicals as compared to nonuse of those chemicals. They argue that the agricultural research community has put all of its research efforts into one basket, and little effort has been expended to develop nonchemical weed control technologies. If those efforts had been made at the same level as chemical control efforts during the last 50 years, then viable nonchemical controls might be available today.

If that is true, then why did society take the chemical route? The argument that small historical events can lead one technology to win out over another, even though down the road it may be an inferior technology, is suggested by Arthur (1989). Once technology develops along a particular path, given increasing economic returns in research and implementation, alternative paths and alternative technologies may be shoved aside and ignored. How the story can evolve with chemicals in agriculture is rather compelling. World War II saw an expansion of our ability to produce chemicals in mass (Adler et al. 1977). After the war, chemical companies looked for a need for that capacity and saw the fertilization (nitrogen) and pest control needs of farmers. Knowledge is a public good unless it can be bound up by property rights, in this case patents and trademarks owned by corporations. Weed control by rotation, tillage, and other management practices was not extensively investigated because that developed knowledge could not be readily embedded into products that could be controlled and sold. Instead, efforts were developed in finding chemical compounds that could be patented. Many would not find fault with businesses pursuing this direction in research as long as it was moral, but public institutions, such as the agriculture colleges, should pursue research programs that are in the best interest of the public regardless of any private gain (Comstock 1989). Because those colleges pursued chemical herbicide programs, would it not be true then that route was the most socially optimal? Not necessarily if we acknowledge the historical event thesis of Arthur (1989). Mechanical weed control research may have been the norm until the discovery and release of many effective chemical controls during the 1960s, starting with atrazine in 1958. The fact that these worked so well at reasonable costs leads to the abandonment of much alternative weed control research.

Are chemicals really needed for weed control (Madden 1988; Fox et al. 1991)? There is no doubt that they work. That they worked, and did so relatively easily, explains why they were adopted by most farmers rather quickly. Integrated weed management (without herbicides) may work also, but not as simply and as relatively easily as chemicals. Integrated pest management requires knowledge and entails management costs that are much greater than applying a chemical herbicide. Not all farmers have the necessary management skills to implement an integrated weed control strategy, but most have the ability to successfully apply

* A pervasive statement in various forms, the originality of which I could not determine, is that chemical control may cause a few cancers, but these chemicals have allowed the bountiful production of low-cost food with nutritional value that have saved even more lives.

a chemical herbicide.* Even if a farmer has the managerial ability to successfully implement a nonchemical weed control scheme, is that the best use of limited time? Might not time be better spent on marketing or financial decisions? The answer to that is probably yes, unless there are externalities involved in chemical weed control that are not included in production costs. Some counter that there are environmental costs associated with chemical weed control that are not included in the production costs of individual farmers, but are part of the social cost of producing food. If these costs were recognized, then they argue that the chemical route of producing food is a greater cost than integrated weed management, even with its larger management cost component.

The task is to determine the cost of agricultural chemical pollution and then internalize those costs to the firm by institutional change. As other chapters of this book show, we are still in the process of determining the extent and cost of chemical pollution. We have much to discover. At the same time, institutions are evolving to reflect the cost of any pollution onto those who produce that pollution. The cost of potential lawsuits is shaping decisions even where specific laws and regulations restricting chemical use are not in place (Olexa 1991). In other sectors, the right to pollute is being traded as a commodity (Baumol and Oates 1988). In some instances public assistance and funding is provided to lower the cost of alternative control mechanisms, swinging the economics away from chemical control. Even some in the chemical industry applaud these efforts, but suggest that some chemical usage still will be socially acceptable if it entails new chemicals that are effective, target specific, and have negligible toxicity. As the chapters in this book on the various herbicides demonstrate, our developing knowledge of how herbicides work is a big step forward from the research method of searching, screening, and testing chemical compounds. This knowledge may allow advances in the biological control of weeds. Others realize that herbicides are only one method of controlling weeds, and although advances in herbicide control and use can improve social welfare, advances in crop varieties, machinery, and tillage and management are also important.

Finally, will HRCs unduly impact the structure of rural communities? Probably not. The use of chemical herbicides is pervasive in agriculture, and introducing different combinations of crops/herbicides than what are currently available should not alter structure. To argue that chemical herbicides may have affected rural structure is another issue, but to argue that herbicide resistance will further affect structure would be a tenuous argument.** The technology is too similar to what farmers are currently using, the same players are involved, and as stated earlier, the price effects probably will be minor.

4. CONCLUSIONS

The author hopes that this discussion has presented the relevant issues and concerns of the public on herbicide-resistant technologies. An attempt was made to point out the concerns of farmers and consumers and what information is necessary to resolve these issues. However, any astute observer of the debate on HRCs realizes very quickly that it is merely one battle of a larger war. That war, of course, is the struggle for the future structure of food production, agriculture, and rural America. Like many wars, this may rage on for many years because both sides believe that they have truth and right behind them. This struggle will lead to institutional change as the various factions exercise their economic and political might. It is hoped that the decision reached will be the best for

* If greater skills are required to successfully utilize pest management without chemicals compared to chemical control, and those skills are not evenly distributed across farms, nonchemical control may lead to greater farm income inequality than chemical technology. Like guns in the old west, chemicals may be the "equalizer."

** Just because rural communities evolved to a form thought less desirable by some as chemical herbicides were adopted does not necessarily mean that chemical herbicides were responsible.

society, although North (1990) has repeatedly indicated that the end result may be stable but not necessarily efficient. Maybe discussions like this book will lead to an equilibrium that does not deviate significantly from the efficient norm.

ACKNOWLEDGMENTS

The author thanks Robert Kalter and Gerald White for their comments.

REFERENCES

Adler, E.F., Wright, W.L., and Klingman, G.C. 1977. Development of the American herbicide industry. In *Pesticide Chemistry in the Twentieth Century*. Ed. by Plimmer, J.R. American Chemical Society, Washington, D.C.

Ahearn, M., Bentley, S., and Carlin, T. 1988. Farming-Dependent Counties and the Financial Well-Being of Farm Operator Households. Agric. Info. Bull. No. 544, U.S. Department of Agriculture, Washington, D.C.

Arthur, W.B. 1989. Competing technologies, increasing returns, and lock-in by historical events. *Econ. J.* 99: 116.

Baumol, W.J. and Oates, W.E. 1988. *The Theory of Environmental Policy, 2nd ed.* Cambridge University Press, New York.

Burton, R.O. and Martin, M.A. 1987. Restrictions on herbicide use: an analysis of the economic impacts on U.S. agriculture. *N. Cent. J. Agric. Econ.* 9: 181.

Cochrane, W.W. 1958. *Farm Prices: Myth and Reality*. University of Minnesota Press, Minneapolis.

Comstock, G. 1989. Genetically engineered herbicide resistance. I. *J. Agric. Ethics*. 2: 263.

Dekker, J. and Comstock, G. 1992. Ethical and environmental considerations in the release of herbicide resistant crops. *Agric. Hum. Values*, Summer: 31.

Doyle, J. 1985. *Altered Harvest: Agriculture, Genetics, and the Fate of the World's Food Supply*. Viking, New York.

Eggertsson, T. 1990. *Economic Behavior and Institutions*. Cambridge University Press, Cambridge, U.K.

Fehr, W.R. 1991. Herbicide tolerance in crops. In *Agricultural Biotechnology at the Crossroads, Biological, Social and Institutional Concerns*. Rep. No. 3. Ed. by MacDonald, J.F., National Agricultural Biotechnology Council, Ithaca, NY.

Fox, G., Weersink, A., Sarwar, G., Duff, S., and Deen, B. 1991. Comparative economics of alternative agricultural production systems: a review, *Northeast J. Agric. Resource Econ.* 20(1): 124.

Goldburg, R., Rissler, J., Shand, H., and Hassebrook, C. 1990. Biotechnology's Better Harvest: Herbicide-Tolerant Crops and the Threat to Sustainable Agriculture, The Biotechnology Working Group, Washington, D.C.

Hanthorne, M., Osteen, C., McDowell, R., and Roberson, L. 1982. 1980 Pesticide Use on Field Corn in the Major Producing States. Staff Rep. No. AGES820202. U.S. Department of Agriculture, Washington, D.C.

Knutson, R.D., Taylor, C.R., Penson, J.B., Jr., and Smith, E.G. 1990. Economic impacts of reduced chemical use. In *Choices*. American Agricultural Economics Association.

Love, J.M. and Tauer, L.W. 1989. Regional Factors Affecting the Impact of Biotechnology in U.S. Crop Production. A.E. Ext. 89-6. Cornell University, Ithaca, NY.

Madden, P. 1988. Low-input/sustainable agricultural research and education — challenges to the agricultural economics profession. *Am. J. Agric. Econ.* 70(5): 1167.

National Agricultural Pesticide Impact Assessment Program (NAPIAP). 1985. Pesticide Assessment of Field Corn and Soybeans: Corn Belt States, Staff Rep. No. AGES85024A. U.S. Department of Agriculture, Washington, D.C. (five other regions available).

Netzer, W.J. 1984. Engineering herbicide tolerance: when is it worthwhile? *Bio/Technology* 2: 939.

North, D.C. 1990. *Institutions, Institutional Change and Economic Performance*. Cambridge University Press, Cambridge, U.K.

Olexa, M.T. 1991. Contaminated collateral and lender liability: CERCLA and the new age banker. *Am. J. Agric. Econ.* 73(5): 1388.

Shrader-Frechette, K.S. 1985. *Science Policy, Ethics, and Economics Methodology*. D. Reidel, Dordrecht.

Swanson, E.R. and Grube, A.H. 1986. Economic impact of the cancellation of the use of Trifluralin on soybeans: a comparison of selected estimation models. *N. Cent. J. Agric. Econ.* 8(1): 143.

Tauer, L.W. and Love, J. 1989. The potential economic impact of herbicide-resistant corn in the U.S.A. *J. Prod. Agric.* 2(3): 202.

U.S. Department of Agriculture. 1987. Agricultural Resources: Inputs, Situation and Outlook Report. USDA, Washington, D.C.

U.S. Department of Agriculture, Economic Research Service. 1989. Agricultural Resources: Outlook and Situation. AR-13, USDA-ERS, Washington, D.C.

Welch, F. 1970. Education in production. *J. Political Econ.* 78: 35.

PART IV

ENVIRONMENTAL CONCERNS

Chapter 19

CROP RESISTANCE TO HERBICIDES: EFFECTS ON SOIL AND WATER QUALITY

Thomas B. Moorman
Kyle E. Keller

CONTENTS

1. INTRODUCTION

Research since 1985 has heightened scientific and public awareness of the potential for pesticide movement from agricultural fields to surface and groundwater supplies. We have also seen the development and commercial introduction of many new herbicides and the coincidental development of herbicide-resistant crops (HRCs). Some analyses indicate that HRCs may have economic advantages in some situations (Hayenga et al. 1992). The impact of HRCs on soil and water quality is likely to result from the changes in the patterns of herbicide use and the environmental impact of those herbicides. This chapter focuses on the environmental fate of herbicides which could receive

1-56670-045-0/96/$0.00+$.50
© 1996 by CRC Press Inc.

increased usage from the adoption of HRC varieties. The environmental fate of related compounds is also considered, as these compounds may become important as new HRCs are developed.

2. PROCESSES CONTROLLING HERBICIDE FATE IN THE ENVIRONMENT

The key processes controlling the fate of herbicides in the environment include retention in the soil (adsorption, desorption, and strong binding); the chemical and biological degradation in air, soil, and water; and the transport processes in air and soil (volatilization, runoff, and leaching). Predicting the distribution of herbicide residues in the environment requires an integration of these processes, often with some requirement for site-specific knowledge. Other factors that can contribute to the distribution of the herbicide in the environment include the method and timing of application, crop production practices such as type of tillage, timing of herbicide applications, and soil and landscape properties. Finally, the impact that a particular herbicide has in the environment will depend upon the distribution of residues in the environment and their toxicity.

2.1. ADSORPTION, DESORPTION, AND BOUND RESIDUES

The adsorption of organic chemicals, including herbicides, to soils is a complex chemical process between the herbicide in solution and the soil surface. Adsorbed herbicide is herbicide that is reversibly bound to the soil surface. If leaching, volatilization, or degradation reduces the solution concentration of the herbicide, then a portion of the adsorbed herbicide may reenter the soil solution in a process termed desorption. The chemical reactions and mechanisms responsible for herbicide sorption include ion exchange, van der Waals forces, hydrogen bonding, charge transfer, and hydrophobic partitioning. The chemical structure and properties (e.g., charge, water solubility) of the herbicide, the amount and structure of the soil organic matter, and the type of clay all determine which chemical mechanisms predominate in the binding of a particular herbicide. Anionic herbicides are generally less strongly sorbed than are neutral or cationic compounds.

The extent and kinetics of sorption reactions are typically assessed by batch equilibrium assays where aqueous herbicide solution is reacted with soil. Sorption and desorption reactions are rapid and reversible, reaching a near-equilibrium condition within 24 to 48 h. The resulting partition of herbicide between soils and water is most simply expressed as a K_d (Equation 1), where C_s is the concentration of herbicide sorbed to soil and C_w is the solution phase concentration. Desorption can be determined using similar methods, starting with herbicide adsorbed to the soil. The fraction of herbicide sorbed to the soil can vary with the concentration of herbicide in the system. To overcome this problem, data from batch equilibrium tests performed at several concentrations are often fit to the Freundlich isotherm (Equation 2), in which K_f is an index of the adsorption capacity, and $1/n$ is a constant controlling the degree of curvature in the isotherm. Another index of sorption is the K_{oc}, which is the partition coefficient (K_d) normalized for organic carbon fraction (OC, Equation 3). The K_{oc} is used to predict the sorption of nonionic herbicides over a broad range of soils, because these compounds are principally sorbed by the soil organic matter.

$$C_s = C_w K_d \quad \text{or} \quad K_d = C_s/C_w \tag{1}$$

$$C_s = K_f C_w^{(1/n)} \tag{2}$$

$$K_{OC} = K_d/\text{OC} \tag{3}$$

Adsorption coefficients have often been used as predictors of herbicidal activity in soil. Weber and Peter (1982) reported that bioactivity of herbicides was inversely correlated with their sorption to a series of soils. This is often reflected in the different rates of herbicides used for soils of different organic matter contents. The adsorption process is explicitly considered in all major simulation models that are used to describe pesticide leaching in soils. In general, these models require a K_d or an estimate of K_d from K_{oc} to calculate the solution-phase herbicide concentration, which is available for transport. Evidence suggests that model performance can be improved by considering nonlinear adsorption (Equation 2, $n \neq 1$) and rate-limiting sorption kinetics (Jury and Ghodrati 1989). Estimates of sorption are also required to effectively predict herbicide losses in runoff water and sediment (Leonard 1988). Although factors other than sorption affect the magnitude of herbicide losses in runoff, the proportions lost in water and sediment are strongly controlled by the sorption and desorption processes.

Soil organic matter also forms strong covalent bonds or complexes with herbicides and their degradation products which are termed bound or fixed residues. Establishment of covalent chemical bonds is also termed chemisorption (Hassett and Banwart 1989). Bound residues can account for 20 to 70% of the original herbicide and are generally unavailable for leaching, degradation, or plant uptake (Duke et al. 1991; Khan 1991). Both biological and nonbiological mechanisms are involved in bound residue formation (Bollag 1991; Khan 1991).

2.2. DEGRADATION AND PERSISTENCE IN THE ENVIRONMENT

The abiotic and biotic degradation processes in agricultural soils, waters, sediments, and air ultimately remove herbicides from the environment. Degradation in soils is a critical process because the soil receives the greatest load of herbicide. Abiotic processes include hydrolysis under basic or acidic conditions or oxidation and reduction reactions (Beland et al. 1974; Burkhard and Guth 1981). Photochemical degradation on the soil surface and plant leaves in the atmosphere and in water is an important process for some herbicides. The overall dissipation (loss from soil) process in soils is often described using an approximation of first-order kinetics:

$$C = C_0 e^{-kt} \tag{4}$$

where C is the concentration of herbicide at some time (t), C_0 is the amount present at $t = 0$, and k is the first-order degradation rate constant. While the use of first-order kinetics has been successful in describing the initial dissipation of herbicides in soils, many other studies have shown deviations from this pattern (e.g., Moorman and Harper 1989).

Soil microorganisms conduct a wide array of enzymatic reactions on pesticides (Moorman 1994). In soil these reactions may be conducted by a multispecies community of microorganisms or by individual microorganisms, with each microorganism performing one or more degradative reactions. The complete degradation of some herbicides, such as 2,4-D, to mineral constituents (H_2O, Cl, and CO_2) is accomplished by individual soil microorganisms. Studies with ^{14}C-labeled 2,4-D show that some of the ^{14}C is respired as $^{14}CO_2$ (20 to 60 % of added ^{14}C), incorporated within the soil biomass (10 to 20%), and another portion (30 to 70%) remains in the soil as bound residue (Scott et al. 1983.) Other herbicides, such as metribuzin, are transformed with little mineralization (Moorman and Harper 1989). Herbicide degradation in soil may also be accomplished by a consortia or community of soil microorganisms. The degradation products from both abiotic and biotic reactions are typically less phytotoxic than the parent (Duke et al. 1991), although some herbicide residues retain considerable toxicity to plants or other organisms (Duke et al. 1991; Pothuluri et al. 1991).

The microorganisms responsible for herbicide degradation in field soils are affected by oxygen concentration, temperature, moisture, bioavailability of the herbicide, soil depth, and the microbial response to the herbicide (Moorman 1994). These factors account for the wide variation in the rates of herbicide dissipation that are observed under laboratory or field conditions. Repeated applications of some herbicides induce changes in the populations and activity of the soil microorganisms resulting in enhanced or accelerated degradation of the herbicide (Roeth 1986). In some soils herbicide degradation rates are increased sufficiently to reduce herbicidal efficacy. Herbicides that appear most susceptible to this phenomenon are the carbamothioate and the phenoxy classes, although other structural classes are also affected (Roeth 1986). Herbicide degradation rates decline in the soil profile due to reduced microbial biomass and activity in these deeper strata. For instance, Mueller et al. (1992) reported that the half-life of fluometuron increased from 18 d in the 0 to 7.5 cm depth to 147 d in the 90 to 120 cm depth. Microbial biomass declined from 262 to 113 mg carbon/kg soil as depth increased in this soil profile. Similar reductions in degradation rate with increasing depth have been observed for alachlor (Pothuluri 1990), metribuzin (Moorman and Harper 1989), and atrazine (Kruger et al. 1993). The increased persistence of herbicides in subsurface soils contributes to further leaching of herbicides that have entered the subsurface soil horizons.

2.3. RUNOFF AND LEACHING

Herbicide residues have been detected in surface water (Thurman et al. 1992) and groundwater (Burkart and Kolpin 1993) throughout the U.S. Midwest and elsewhere. Runoff and leaching of herbicides and the resulting contamination of water resources are complex processes that are dependent upon pesticide physical and chemical properties, soil properties and processes, climate, and farming practices. The detection of herbicides in groundwater is related to depth of groundwater, groundwater recharge rates, and herbicide use at both the watershed and regional scales (Burkart and Kolpin 1993; Pionke et al. 1988). More recently, there recognition is increasing that drainage tiles are potentially important sources of herbicides in surface water (Gaynor et al. 1992; Jayachandran et al. 1994). Soil factors that contribute to increased leaching are low sorptive capacity and slow herbicide degradation. In general, herbicides with K_{oc} values of <100 and half-lives >50 d are potentially mobile in soils. However, the actual risk of groundwater contamination is highly influenced by site-specific soil and hydrologic factors. Furthermore, the results of field or column leaching studies need to be evaluated in terms of health advisory levels and application rates.

Herbicide losses in runoff are affected by surface topography, pesticide sorption, time after application, and rainfall intensity (Wauchope 1978). Losses from individual fields are generally <5% of the applied amount, with the greatest losses occurring during runoff events within a few days after application. Herbicides that are more tightly sorbed are less susceptible to loss during runoff events. Conservation tillage practices tend to reduce pesticide loss in both the water and sediment components of runoff (e.g., Seta et al. 1993). In the midwestern U.S. there is some evidence that runoff losses of herbicide are less with postemergence applications than with preemergence applications (Pantone et al. 1992). This finding appears to be due to greater interception by the plant canopy, drier soil conditions and fewer intense rainstorms producing runoff when postemergence applications are made.

2.4. VOLATILIZATION

Volatilization and drift are the principal mechanisms for herbicide entry into the atmosphere. Volatilization refers to loss of herbicide from soil surfaces, crop residues, or live vegetation, whereas drift refers to movement of herbicide in aerosols created during spray applications. The volatilization process has been examined in closed laboratory systems and under field conditions under a variety of conditions. Taylor and Spencer (1990) compiled literature reports of the amounts of herbicide volatilized under field conditions. As expected, the amounts vary widely, even for the same herbicide. For instance, volatilization of trifluralin ranged from 3.4% in 90 d to 87% in 50 h. Factors that increase volatilization from soil include high temperatures, moderate to high soil moisture, and low sorption to soil (Taylor and Spencer 1990). Volatilization losses of the insecticides fonofos and chlorpyrifos were considerably greater than atrazine following surface application to soil (Whang et al. 1993). The relative losses of these three pesticides corresponded to their relative differences in vapor pressure and Henry's constant.

3. ENVIRONMENTAL FATE OF HERBICIDES USED WITH RESISTANT CROPS

3.1. ATRAZINE

Atrazine is a widely used herbicide in many crops, although it is used principally for control of broadleaved weeds in corn. Atrazine is a symmetric nitrogen-containing heterocyclic triazine that is a weak base (pKa = 1.7), which becomes positively charged with decreasing soil pH. Atrazine is adsorbed with moderate strength to soil, with K_{OC} values generally ranging between 47 and 291 (Montgomery 1993). Atrazine adsorption is principally related to organic matter content in soils, where it is adsorbed by hydrogen bonding. Adsorption coefficients normally increase with decreasing pH, reflecting increased protonation and cationic behavior of atrazine under acidic conditions (Burkhard and Guth 1981). Atrazine is adsorbed to varying extents to clay minerals (Laird et al. 1992). The adsorption to clays was inversely related to surface charge density, suggesting atrazine was adsorbed as a neutral molecule at the pH values in this study. The dealkylated metabolites of atrazine, deethylatrazine (DEA), and deisopropylatrazine (DIA) are adsorbed to soil to a lesser extent than atrazine (Brouwer et al. 1990). Based upon adsorption coefficients and soil thin-layer chromatography (TLC) measurements, atrazine is considered to be moderately mobile in soil.

Atrazine degradation in soil is an example of the complexity in biological and chemical reactions that contribute to atrazine fate in the environment. Abiotic dechlorination of atrazine results in the formation of hydroxyatrazine (Figure 1). Armstrong et al. (1967) compared hydrolysis rates over a range of pH levels in solution and in soil. Hydrolysis reaction half-lives increased as soil pH increased. At pH 3.9 hydrolysis rates were increased by almost tenfold by the addition of sterile soil into the system, which suggested catalytic effects of the soil. Later studies of sorption and hydrolysis kinetics suggested that the majority of adsorption sites were catalytically active in the formation of hydroxyatrazine (Gamble and Khan 1990). Hydroxyatrazine is more strongly sorbed than atrazine due to its greater basicity and low water solubility. Microorganisms dealkylate and dechlorinate atrazine to produce DEA, DIA, desethyl-DIA, and hydroxyatrazine (Bekhi et al. 1993). These compounds are further dealkylated and dechlorinated prior to ring cleavage, with microbial utilization of atrazine as a nitrogen source (Cook 1987). In neutral and high pH soils biodegradation may be the primary degradation process in soil.

Atrazine contamination of ground- and surface water has been linked to agricultural usage of this herbicide. The incidence of atrazine detection in groundwater at both regional and watershed scales is linked to both usage and groundwater vulnerability (Burkart and Kolpin 1993; Pionke et al. 1988). Atrazine and degradation products have also been detected in tile drainage waters at several midwestern (U.S.) sites (Buhler et al. 1993; Gaynor et al. 1992; Jayachandran et al. 1994). Concentrations in tile drainage water emanating from fields treated with atrazine range from 0.5 to 3 μg/l, although Gaynor et al. (1992) reported mean concentrations above 6 μg/l. Studies following the movement of atrazine in soil are inconsistent. Bowman (1989) found only traces of atrazine beyond 50 cm in field soil columns using a low organic matter sand. Buhler et al. (1993) found atrazine residues ranging from nondetectable to 6 μg/kg in field soils at a depth of 105 to 150 cm. Several mechanisms may account for the movement of small amounts of atrazine deep into the subsurface environment, even in soils with moderate organic carbon contents. Movement of herbicides through macropores may allow rainfall events to transport herbicide deep into the subsoil. Zins et al. (1991) showed that root channels could act as macropores which increased leaching of atrazine.

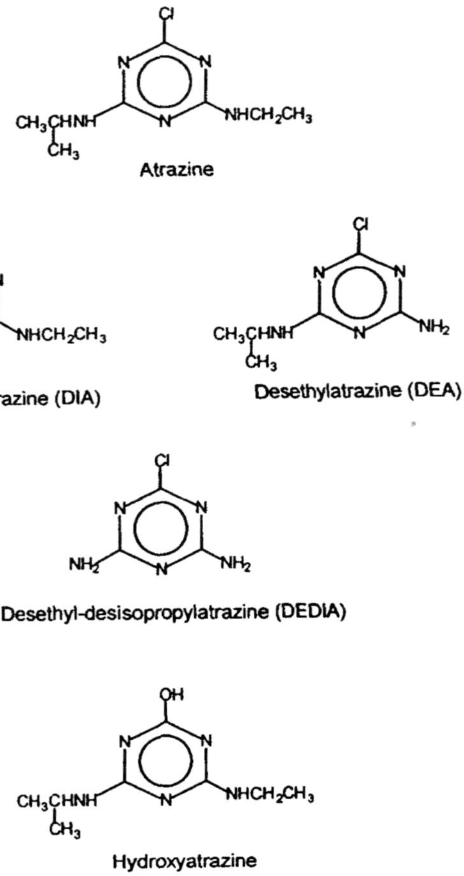

FIGURE 1 Principal products from atrazine degradation in soil. Atrazine is dealkylated and dechlorinated by soil microorganisms to form DEA, DIA, hydroxatrazine and DEDIA. Alternatively chemical hydrolysis produces hydroxyatrazine, which can then undergo dealkylation to produce the dechlorinated analogs of DEA and DIA. Microorganisms open the triazine ring of DEDIA after dechlorination in several enzymatic steps to form CO_2 and NH_3.

In the subsurface environment slower microbial degradation and reduced adsorption increase the potential for subsequent movement. Klint et al. (1993) spiked sandy aquifer sediments with atrazine and observed no degradation in experiments lasting 174 d.

3.2 BROMOXYNIL

Bromoxynil is an ionic, highly soluble herbicide applied as a salt or ester with a pK_a of 4.20 (Cessna and Grover 1978). Above this pH this herbicide will be increasingly anionic and mobile in soil. Bromoxynil is degraded rapidly in soil (Smith 1971). Isolated bacteria, *Klebsiella pneumoniae,* and *Flavobacterium* sp. degrade bromoxynil in defined cultures. *K. pneumoniae* (McBride et al. 1986) hydrolyzed the cyano group of bromoxynil, resulting in 3,5-dibromo-4-hydroxybenzoate (Figure 2). The cyano group served as a nitrogen source to the bacterium. In contrast, the *Flavobacterium* sp. cleaves the cyano group to produce cyanide and 2,6-dibromohydroquinone, which is eventually mineralized (Topp et al. 1992).

Bromoxynil has been detected in stream runoff samples and shallow groundwater, although concentrations were <1 μg/l (Waite et al. 1992). No bromoxynil was detected in pond water in the same study. The absence of bromoxynil in pond water may reflect active biodegradation or

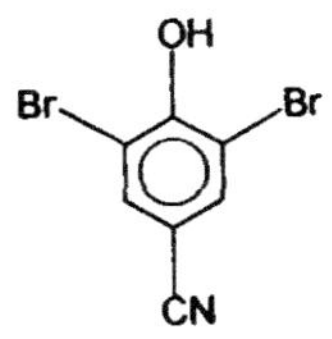

Bromoxynil

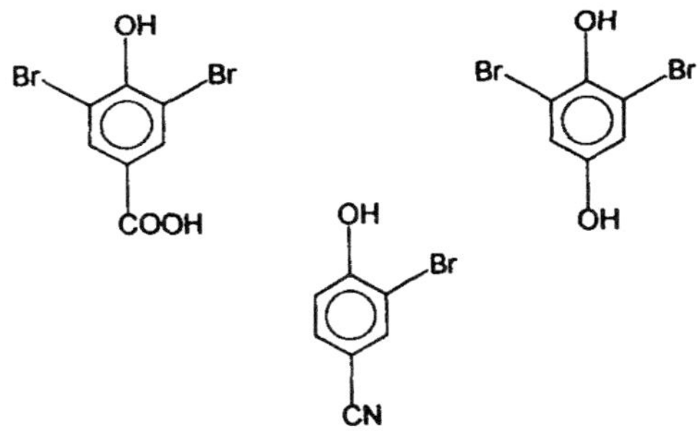

FIGURE 2 The herbicide bromoxynil (3,5-dibromo-4-hydroxybenzonitrile) and the principal degradation products 3,5-dibromo-4-hydroxybenzoate, 3-bromo-4-hydroxybenzonitrile, and 2,6-dibromohydroquinone (left to right).

photolysis in this environment. Laboratory studies indicate that bromoxynil is rapidly degraded in aqueous media, with the production of dehalogenated metabolites (Kochany et al. 1990). In studies in which bromoxynil was applied to pond and marsh areas, bromoxynil degraded rapidly at pond surfaces with half-lives of <3 h (Muir et al. 1991). At lower water depths, the half-life averaged 9 days. Bromoxynil breakdown products included 3-bromo-4-hydroxybenzonitrile. Low concentrations of bromoxynil (octanoate ester and phenol forms) and the dehalogenated metabolite were found in pond sediments up to 120 days after application. These studies show that bromoxynil is degraded in surface waters fairly rapidly, although the studies of Muir et al. (1991) indicate that small concentrations (100 μg/l) can impact aquatic ecosystems.

3.3. GLYPHOSATE AND GLUFOSINATE

Glyphosate and glufosinate are nonselective, ionic, highly soluble herbicides. Both have relatively little herbicidal activity in soil. Glyphosate is strongly adsorbed to soil, clay minerals, and sediments (Zaranyika and Nyandoro 1993). One mechanism accounting for the strong adsorption of glyphosate and the metabolite aminomethylphosphonic acid (AMP) is the formation of ligand complexes with exchangeable metals adsorbed to clays (McBride and Kung 1989, McBride 1991). Thin layer chromatography using soil indicated that glyphosate is relatively immobile (Rueppel et al. 1977).

Glyphosate is degraded by a wide variety of soil microorganisms by two different pathways. Glyphosate degrades to form AMP in soil (Rueppel 1977) and in microbial cultures (Balthazor and Hallas 1986). Aminomethylphosnic acid is also degraded, producing inorganic phosphate, glyoxylate, and methylamine (Figure 3). Alternatively, the phospate group is hydrolyzed from glyphosate to form sarcosine, which is then demethylated to form glycine (Liu et al. 1991; Pipke et al. 1988). In another *Pseudomonas* isolate a carbon–nitrogen bond is broken to produce glycine and methylphosphonate (Jacob et al. 1985). The ability to metabolize glyphosate is widely distributed among soil bacteria. Species of *Arthrobacter* (Pipke et al. 1988), *Agrobacterium* (Wackett et al. 1987), *Actinomyces*, *Bacillus* (Quinn et al. 1989), *Flavobacterium* (Balthazor and Hallas 1986), *Rhizobium* (Liu et al. 1991), and *Pseudomonas* (Jacob et al. 1985; Quinn et al. 1989) have been

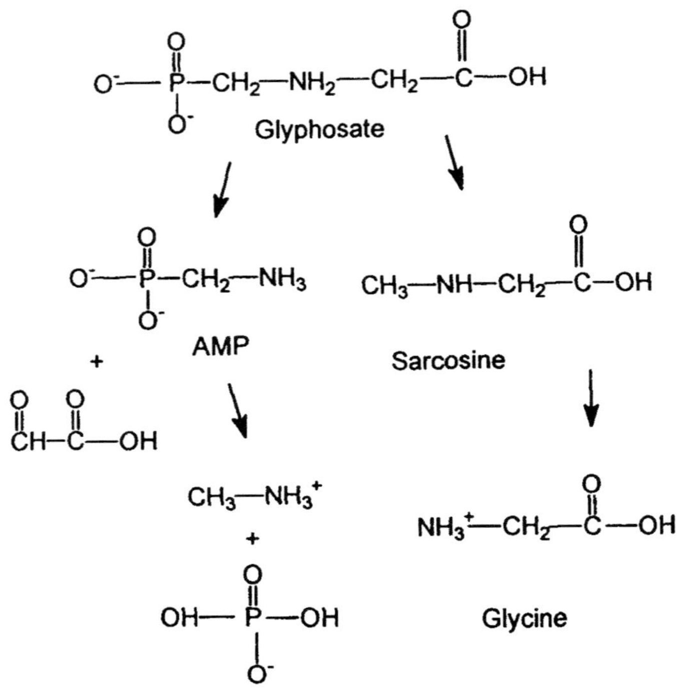

FIGURE 3 Pathways of glyphosate breakdown by soil microorganisms.

isolated which utilize glyphosate as a source of phosphorus. The activity of *Pseudomonas* and *Bacillus* strains were not inhibited by inorganic phosphate in the growth medium and strains which degraded glyphosate did not degrade glufosinate (Quinn et al. 1989). Jacob et al. (1985) also showed that the carbon in glyphosate becomes incorporated into the microbial biomass, which is consistent with microbial mineralization of ^{14}C-glyphosate (Balthazor and Hallas 1986).

Glyphosate degradation in soil is similar to that obtained with isolated microorganisms, with the exception that AMP is the most prevalent metabolite recovered from soil (Rueppel et al. 1977). Aminomethylphosphonic acid concentrations were equal to or greater than glyphosate concentrations 8 months to 1 year after application in soil at several sites in Sweden, suggesting that AMP is more persistent than glyphosate (Torstensson et al. 1989). Smith and Aubin (1993) estimated glyphosate half-lives in soil to be <40 d, based on mineralization studies with three agricultural soils. Microorganisms mineralize varying amounts of ^{14}C-glyphosate in soil (Heinonen-Tanski 1989; Rueppel et al. 1977). Differences in mineralization rates may be due to variation in the microbial activity in soil or the bioavailability of glyphosate sorbed to the soil. In some soils degradation kinetics are characterized by rapid initial rates of degradation followed by slower rates of degradation (Nomura and Hilton 1977). Similar results were obtained in sediment and river water microcosms (Zaranyika and Nyandoro 1993). A kinetic model describing glyphosate persistence in these microcosms was based upon rapid degradation of solution phase glyphosate followed by slower degradation of glyphosate bound to colloidal or sediment particles.

The rapid biodegradation and strong sorption of glyphosate indicate that this compound is relatively immobile in soils. Jacoby et al. (1992) reported only 7 of 247 groundwater samples contained glyphosate and none of these concentrations exceeded the MCL. However, glyphosate can be transported from plant and soil surfaces in runoff to surface water. Amounts up to 1.8% of the applied glyphosate were lost from small watersheds in storms immediately following application (Edwards et al. 1980). In most years of the study losses were <1% of applied herbicide. Presumably, any glyphosate entering surface waters will be sorbed to sediments or degraded by the aquatic microorganisms (Zaranyika and Nyandoro 1993). Small amounts of glyphosate may persist in soils for long periods of time, particularly under cold conditions, but the hazard to rotational crops appears to be minimal at this time.

Glufosinate dissipates rapidly in soils (Figure 4) to form 3-methylphosphinylpropanoic acid (MPPA-3) and 2-methylphosphinylacetic acid (MPPA-2) (Behrendt et al. 1990; Gallina and Stephenson 1992). There was no evidence of a direct attack on the carbon–phosphorus bond of glufosinate as the initial step in degradation. Between 50 and 80% of the added glufosinate was degraded in agricultural soils in the first 7 d post-treatment (Gallina and Stephenson 1992; Smith 1989). 3-Methylphosphinylpropanoic acid was the principal metabolite formed in these studies, accounting for as much as 56% of the original glufosinate. Glufosinate is weakly adsorbed to soil with K_d values of 0.2 to 0.4 at initial concentrations of 5 mg/l, as calculated from the data of Gallina and Stephenson (1992), which suggests that it would be mobile in soils. We are not aware of studies monitoring glufosinate in surface or groundwater.

3.4. IMIDAZOLINONE HERBICIDES

Imazaquin and imazethapyr represent a relatively new herbicide class known as the imidazolinones. These compounds are applied pre- or postemergence at low rates ranging from 0.03 to 0.15 kg a.i./ha. Both herbicides are capable of reacting chemically either as an acid or as a base. In most agricultural soils these herbicides would be found predominantly in the anionic form due to their acidic pk_a being <4 (Goetz et al. 1986; Renner et al. 1988). In extreme acidic conditions both compounds are protonated due to a basic pk_a near 2 (Stougaard et al. 1990). The water solubility of these compounds is generally high: 60 mg/l solubility for imazaquin and 1400 mg/l for imazethapyr at pH 7.0. However, imazethapyr is more strongly sorbed than imazaquin (Che et al. 1992; Loux et al. 1989; Renner et al. 1988; Stougaard et al. 1990).

Due to the amphoteric nature of the imidazolinone herbicides, sorption increases as soil pH decreases (Goetz et al. 1986; Loux et al. 1989; Mangels 1991; Renner et al. 1988). In general, sorption also increases with clay content and type, and more importantly, organic matter content (Che et al. 1992; Loux et al. 1989). In agricultural soils with pH values >5.5 imidazolinone

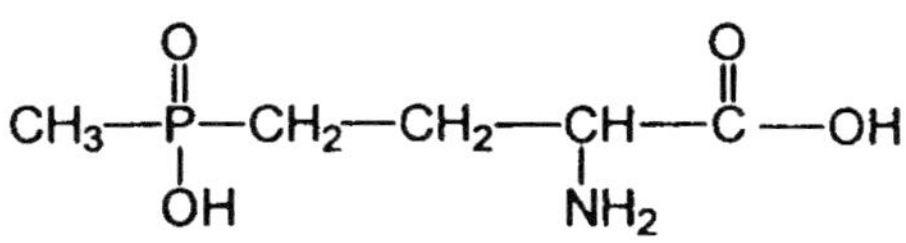

Glufosinate

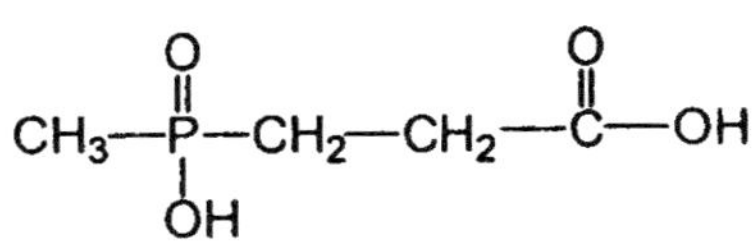

MPPA-3

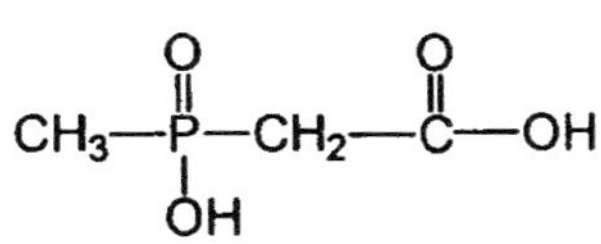

MPPA-2

FIGURE 4 The herbicide glufosinate and its principal degradation products in soil.

herbicides are weakly sorbed, with K_f values generally <1.5 for imazaquin and <2.5 for imazethapyr (Loux et al. 1989). At this and larger pH values the predominantly anionic herbicide molecules are repelled by the negatively charged clay and organic surfaces (Loux et al. 1989; Stougaard et al. 1990). However, in lower pH soils there is some sorption of the imidazolinones as anions to pH-dependent kaolinite and aluminum and iron hydrous oxides (Che et al. 1992; Goetz et al. 1986; Loux et al. 1989). The amount of positive charge exhibited by these minerals increases as pH decreases (Bohn et al. 1979). Furthermore, in acid conditions the basic moiety may acquire a positive charge, resulting in cationic binding and subsequent increased sorption (Renner et al. 1988; Stougaard et al. 1990).

Imidazolinone herbicides are degraded in soil by microorganisms (Basham and Lavy 1987; Cantwell et al. 1989; Goetz et al. 1990). Under laboratory conditions microorganisms converted 9 to 44% of the applied ring-labeled ^{14}C-imazaquin and ^{14}C-imazethapyr, respectively, to $^{14}CO_2$ after 12 weeks (Cantwell et al. 1989). In the same study 3 to 16% of the applied herbicides were transformed to extractable metabolites during the 12-week period. There was no transformation or mineralization of these herbicides in sterilized soils. In other soils small amounts of ^{14}C-imazaquin (Basham and Lavy 1987) and ^{14}C-imazethapyr (Goetz et al. 1990) were mineralized by microorganisms over longer time periods. Goetz et al. (1990) also reported that a significant fraction of ^{14}C-imazethapyr added to soil was converted to uncharacterized metabolites. The small amounts of mineralized herbicide with production of metabolites suggest co-metabolic degradation, whereby the microorganisms degrade the herbicide without obtaining metabolic energy or carbon for growth.

Imidazolinone herbicides vary in their persistence. Imazethapyr half-lives at constant temperature and moisture were 18 to 46 d at 25° and 35°C (Goetz et al. 1990). Both laboratory and field studies show that greater degradation occurs in soils that adsorb less herbicide (Cantwell et al. 1989; Goetz et al. 1990). Field applications can result in detectable residues for considerable periods following application. These residues can affect rotational crops, depending upon the application rate and the sensitivity of the rotational crop (Mangels 1991). The effects of imazethapyr and imazaquin on rotational crops suggests that increased use of these herbicides in conjunction with HRCs poses some risk, depending upon the specific crops, soils, and climatic factors (Barnes et al. 1989; Johnson et al. 1993).

Time of exposure, soil moisture, and soil type influence photodecomposition of imidazolinones. Using ring-labeled ^{14}C-imazaquin applied to soil-coated glass slides, Basham and Lavy (1987) measured significant losses (25 to 50%) of ^{14}C during exposure to sunlight or ultraviolet light. They concluded that time of exposure was more important than temperature in photodecomposition. Similar studies (Goetz et al. 1990) using ^{14}C-imazethapyr showed that photodecomposition increased as exposure to sunlight increased, accounting for up to 8% of the herbicide. Curran et al. (1992) discovered that soil type and water content affected photodegradation of imidazolinones. After 48 h, there was no significant loss of imazaquin or imazethapyr on dry sand or silty clay loam, as compared to 45% photodegradation of imazaquin and 52% for imazethapyr on moist media. Differences in photodegradation were due to greater sorption of the herbicides in the dry silty clay loam, which apparently reduced exposure of the herbicide to light. Nevertheless, photodecomposition should be minimal when the imidazolinone is immediately incorporated or rainfall occurs soon after application to move the herbicide beneath the soil surface.

Limited information exists pertaining to imidazolinone movement in soil. In an Arkansas field leaching study with 8 cm of irrigation over a 21-d period nearly 97% of the ^{14}C-imazaquin remained in the upper 10 cm of soil (Basham et al. 1987). Companion studies, however, suggest that if high rainfall continues for a long period of time, some leaching would occur. Using soil TLC with five soils varying in clay content, organic matter, and pH, Goetz et al. (1986) observed that imazaquin was more mobile than metribuzin and atrazine, which is similar to their relative strength of sorption to soil. Using the same technique with a silty clay loam, silt loam, and sandy loam, each at three pH levels, Stougaard et al. (1990) noted that imazethapyr was less mobile than imazaquin at pH 5, 6, and 7, respectively. The herbicides were the most mobile in the sandy loam at pH 7. Although sorption measurements and soil column leaching experiments suggest that imazaquin and imazethapyr may leach under high-intensity rainfall, we are not aware of any studies that have found these herbicides below the top meter of soil or in groundwater. Mangels (1991) suggested that plant withdrawl of soil water would be a factor limiting the movement of these herbicides, which are principally applied postemergence.

In a 2-year runoff study with five different soybean cropping systems Shaw et al. (1992) found losses of imazaquin as high as 10% in 1987 and 0.13% in 1988. Loss of imazaquin was highest in double-crop soybean following incorporation of the stubble of the preceding year's wheat crop. This occurrence was related to higher runoff. Even though imazethapyr was not in the runoff study, imazethapyr would be expected to be less susceptible to runoff loss than imazaquin due to its greater sorption to soil (Che et al. 1992; Loux et al. 1989; Renner et al. 1988; Stougaard et al. 1990).

Relatively little is known about imidazolinone volatilization from soil. Laboratory results indicate that the imidazolinones are slightly volatile or nonvolatile. Basham and Lavy (1987) found 23% of the imazaquin applied to soil-coated glass slides maintained in darkness at 42°C may have volatilized. Dissimilarly, volatilization of imazethapyr from soil-coated glass slides was <2% of the applied herbicide (Goetz et al. 1990).

3.5. PHENOXY HERBICIDES

The phenoxyacetic acid herbicides, including 2,4-D, MCPA, and dichlorprop, are applied postemergence as esters or salts, which rapidly degrade to the herbicidal acids. The pK_a of these compounds is generally below 3.5 (Cessna and Grover 1978), which indicates that they will exist as anions in soil. Consequently, these compounds are weakly sorbed to soils. Helweg (1987) reported K_d values for MCPA ranging from 0.7 to 1.0. 2,4-D is also weakly adsorbed in soils, with K_f values of 0.13 to 1.30 (Amonette and O'Connor 1980; Grover 1973). Sorption was greater in the soils with lower pH and higher organic carbon contents. Ester formulations and surfactants had little effect on sorption.

These herbicides dissipate rapidly due to metabolism by soil microorganisms in many different soils. 2,4-D is metabolized by microorganisms, with 2,4-dichlorophenol and 2,4-dichlorocatechol as intermediates (Moorman 1994). The half-lives of 2,4-D and 2,4-DB were <7 d, while dichlorprop had half-lives of 8 to 12 days in soils incubated at 20°C (Smith 1978). Helweg (1987) reported Q_{10} values of 3.3 and 2.9 for MCPA in the 0° to 29°C range, indicating that degradation rates are strongly dependent on temperature. In soils that are repeatedly treated with 2,4-D microbial populations adapt and are able to metabolize 2,4-D and MCPA at accelerated rates (Kirkland and Fryer 1972; Smith et al. 1989; Torstensson et al. 1975). A similar phenomenon was reported for mecoprop in groundwater sediments (Klint et al. 1993). Initially, mecroprop degradation followed after a 40-d lag period, but after the initial adaptation mecoprop was degraded rapidly without a lag. The basis for adaptation to phenoxy herbicides is probably a combination of physiological changes, such as the induction of efficient degradative enzymes, and the growth of the herbicide-degrading population (Moorman 1994).

The low adsorption of 2,4-D to soil indicates that it is potentially mobile in soil and it has been reported as a groundwater contaminant (Jacoby 1992), although the magnitude of contamination is apparently considerably less than for atrazine. In shallow groundwater 2,4-D was present in 10% of the samples, with a mean concentration of <1 ppb (Waite et al. 1992). Losses of 2,4-D in runoff occurred principally in the water phase, with losses greatest after precipitation events near the time of application (White et al. 1976). Total losses were <0.5% of that applied. In contrast, 2,4-D applied to wheat in the fall was more susceptible to loss in runoff, with total losses in spring averaging 4.1% of the applied amount (Nicholaichuk and Grover 1983). The greater runoff may be attributed to slower degradation in the fall and early spring, resulting in more 2,4-D being available for runoff.

3.6. ARYLOXYPHENOXY HERBICIDES

This group of herbicides is used postemergence and is distinguished by the structural features of an phenoxy alkanoic acid group linked to an additional aromatic structure. Members of this group include the oxyphenoxy group, in which the aromatic rings are joined by an ether linkage (e.g., diclofop), and the benzoyl-*N*-phenoxy (e.g., flamprop), quinoxalinyloxyphenoxy (e.g., quiazalofop), and pyridinyloxyphenoxy acids (Duke and Kenyon 1988). Most of these herbicides are formulated as the esters of the acid forms, which reduces water solubility slightly as compared to the acid form. Despite their low water solubilities, fluazifop-butyl and haloxyfop-methyl were not strongly sorbed to a variety of soils and showed some mobility in soil columns (Rick et al. 1987). Jacoby et al. (1992) reported the presence of fluazifop in groundwater, although the sampling was limited in scope.

The degradation of these herbicides begins with the hydrolysis of the ester to release the acid form. The speed of this reaction is variable, with the half-life of flamprop-isopropyl ranging from 4 to 20 weeks, while the half-lives of diclofop-methyl and fluazifop-butyl are <1 week in soil (Hitchings and Roberts 1979; Martens 1978; Nègre et al. 1988). The resulting acidic metabolites decrease after formation, with as much as 40 d required for 50% degradation of diclofop acid (Martens 1978). Under field conditions, half-lives are more difficult to determine due to the washoff of the ester form from foliage to soils. Bosio (1982) observed increases in the soil concentration of benzoylprop-ethyl as the herbicide was washed from the foliage to the soil. The estimated half-lives, combining the ester and acid forms, ranged from 4 to 16 weeks, but there was relatively little residue carryover to the next growing season.

The role of microorganisms in the degradation of these herbicides was confirmed by the isolation of a stable, mixed-species culture of microorganisms which degraded fluazifop-butyl to fluazifop, which in turn was degraded to 4-(5-trifluoromethyl-2-pyridyl)oxyphenol (Nègre et al. 1993). This metabolite was then cleaved at the diphenyl-ether linkage. However, in studies with radiolabeled herbicides added to soil, microorganisms mineralized <35% of the ring components of diclofop-methyl, flamprop-isopropyl, and haloxyfop-methyl (Hitchings and Roberts 1979; Martens et al. 1978; Smith 1985). Significant amounts of these herbicides were converted into bound residues in these studies.

3.7. SULFONYLUREA HERBICIDES

Sulfonylureas are a major, new class of herbicides that exhibit unprecedented herbicidal activity. Generally, these herbicides are applied postemergence at low rates ranging from 0.002 to 0.075 kg a.i./ha. From this class, chlorsulfuron, chlorimuron, tribenuron, and sulfometuron have undergone testing in developing HRCs (CAST 1991; Chaleff and Ray 1984; Hatmaker 1994; Newhouse et al. 1990). Due to the extensive literature on chlorsulfuron and chlorimuron, the fate and behavior of sulfonylureas in soil are characterized primarily from these two compounds.

Sulfonylureas are weak acids, with pK_a values of 3 to 5 for the sulfonamide nitrogen (Hay 1990). Water solubilities of sulfonylurea herbicides increase substantially as pH increases. Like the imidazolinones, the anionic form of sulfonylureas will predominate in the soil solution of most soils at neutral and higher soil pH values. Consequently, adsorption is reduced due to repulsive interactions with negatively charged clay and organic surfaces (Borggaard and Streibig 1988; Goetz et al. 1989; Mersie and Foy 1985; Shea 1986). Sulfonylurea adsorption increases as pH decreases (Goetz et al. 1989; Mersie and Foy 1985; Shea 1986) and organic matter increases (Mersie and Foy 1986; Walker et al. 1989). However, anionic sulfonylureas adsorb to pH-dependent iron compounds, aluminum oxides (Borggaard and Streibig 1988; Goetz et al. 1989), and anionic exchange resins (Shea 1986). Undoubtedly, sorption to the iron and aluminum oxide occurs in the low pH soils commonly found in the southeastern U.S. Minimal chlorsulfuron adsorption to silicate clays has been observed (Borggaard and Streibig 1988; Shea 1986). Typical K_d values for chlorsulfuron are <1.0 in soils with pH >5.5 (Walker et al. 1989; Walker and Welch 1989).

Degradation of sulfonylureas in soil occurs primarily by chemical hydrolysis and biodegradation, and to a smaller extent, photolysis. Chemical hydrolysis, a pH-sensitive process, occurs through cleavage of the sulfonylurea bridge to yield the corresponding sulfonamide and heterocyclic amine (Beyer et al. 1987). In a laboratory study Joshi et al. (1985) noted that the rate of chlorsulfuron breakdown due to chemical hydrolysis was 15 times faster in acid soil (pH 5.9) than alkaline soil (pH 8.0). As soil pH decreases, the neutral sulfonylurea bridge predominates. This neutral sulfonylurea bridge is at least 250 to 1000 times more susceptible to hydrolysis than the anionic form (Brown 1990). After 10 weeks, Joshi et al. (1985) also noted as much as 80% and as little as 10% of the chlorsulfuron degraded by chemical hydrolysis at soil pH values of 5.9 and 8.0, respectively. Chemical hydrolysis also appears to be important in the degradation of chlorimuron in air-dry soil (Fuesler and Hanafey 1990).

Microorganisms play an important role in the degradation of sulfonylureas in soil. Joshi et al. (1985) noted that degradation of chlorsulfuron was significantly faster in the nonsterile soils than in sterilized soils. In the same study after a 10-week period microorganisms degraded 79% of the herbicide in the acid soil (pH 5.9) and 91% in the alkaline soil (pH 8.0). Conversely, using the same herbicide, Thirunarayanan et al. (1985) and Fredrickson and Shea (1986) observed reduced microbial degradation when soil pH was >7. Microorganisms also mineralize the phenyl ring-labeled ^{14}C-sulfonylurea or metabolites to yield $^{14}CO_2$ (Anderson and Dulka 1985; Miller 1992). Conditions that promote microbial activity also promote sulfonylurea degradation. Increased degradation of chlorsulfuron (Beyer et al. 1987; Thirunarayanan et al. 1985) and chlorimuron (Fuesler and Hanafey 1990) was observed with increasing temperature. Similarly, increased degradation of chlorsulfuron (Thirunarayanan et al. 1985) and chlorimuron (Fuesler and Hanafey 1990) were measured as soil moisture increased. Furthermore, microbial biomass influences sulfonylurea degradation. Degradation rates of chlorsulfuron and metsulfuron-methyl decreased with increasing soil depth (Walker et al. 1989). Within the soil profiles, degradation rates were positively correlated with microbial biomass and inversely correlated with soil pH. Similarly, Miller (1992) noted longer half-lives for primisulfuron as soil depth increased.

Most studies of sulfonylurea herbicide photodegradation processes have been conducted in aqueous solutions and on glass surfaces. Beyer et al. (1988) reported that chlorimuron applied to soil-coated glass plates degraded more than twice as fast as the herbicide kept in the dark. In the same study there was no photolysis of metsulfuron. Schneiders et al. (1993) noted no photodegradation of rimsulfuron applied to soil-coated glass slides. In aqueous solutions Thomas and Harrison (1990) noted that 63% of the chlorimuron and 30% of the metsulfuron photodegraded after 8 d. Surfactants in the aqueous solutions greatly enhanced photolysis. These researchers theorized that surfactants partitioned the herbicides in a more favorable medium for photolysis or acted as photosensitizers. Venkatesh et al. (1993) also noted increased photolysis of chlorimuron in aqueous solutions by the addition of photosensitizers. They concluded that soluble and naturally occurring organic compounds may enhance photodegradation of chlorimuron in surface waters.

Sulfonylurea herbicides show a wide range of persistence in both laboratory and field experiments, depending upon soil pH, temperature, and soil moisture conditions. Half-lives in surface soils incubated under uniform temperature and moisture conditions from eight locations ranged from 21 to 74 d for chlorsulfuron and from 23 to 73 d for metsulfuron-methyl (Walker et al. 1989). Half-lives of 3 to 140 d were reported for chlorsulfuron in other soils (Beyer et al. 1988). Less than 10 d were required for 50% degradation of chlorimuron in soils under a variety of temperature and moisture conditions (Fuesler and Hanafey 1990). Persistence of sulfometuron-methyl in field lysimeters was greatest in Colorado and Saskatchewan soils, which had the highest soil pH and coolest climates as compared to the other sites (Anderson and Dulka 1985). In cool, alkaline soils chlorsulfuron can be extremely persistent, with half-lives exceeding 100 d (Smith 1986). Under field conditions between 3 and 16% of the previous year's application can remain in soil (Smith and Hsiao 1985), which can affect rotational crops several years after application (e.g., Moyer et al. 1990). Beyer et al. (1988)

argued that the problems with the carryover of chlorsulfuron to rotational crops was due to the extreme sensitivity of these crops. The impacts of sulfonylurea herbicides used in conjunction with herbicide-resistant crops would be greatly influenced by the specific characteristics of the sulfonylurea herbicide, the soil characteristics, and the degree of sensitivity of the rotational crops.

The movement of sulfonylurea herbicides, particularly chlorsulfuron, in soil has been studied. Using soil TLC with four soils, Beyer et al. (1987) reported that chlorsulfuron and metsulfuron were the most mobile of six sulfonylurea herbicides. This study and others (Fredrickson and Shea 1986; Mersie and Foy 1986) showed that in most instances, sulfonylurea mobility increases with increasing soil pH and decreasing organic matter. Additional factors that influence mobility in the field include other chemical and physical properties of the soil and climatic conditions (Helling and Gish 1986; Weber 1991; Weber and Miller 1989). These relationships are quite evident in the following field studies.

In undisturbed field lysimeters (1-m length) leached with 45 cm water very little chlorsulfuron and metsulfuron methyl were detected in leachate after 7 months (Bergstrom 1990). Concentrations ranged from below detection to 21 ng/l in one soil and up to 19 ng/l in another. The amounts leached were calculated to be <1% of the applied herbicide. Beckie and McKercher (1990) used intact soil columns to show movement of chlorsulfuron to depths of 24 cm, as determined by bioassay methods. Movement was greatest in the alkaline soils. In other lysimeters of 1 m in length ^{14}C-primisulfuron was shown to be very mobile in one soil and immobile in another (Keller 1992; Warren and Weber 1993). More than 20% of the ^{14}C-primisulfuron and/or metabolites accumulated in leachate from a coarse-textured loamy sand after 127 d.

Other field studies have shown chlorsulfuron, triasulfuron, and metsulfuron to be relatively mobile, with herbicides detected at depths of 30 to 50 cm (Blacklow and Pheloung 1992; Walker and Welch 1989). Blacklow and Pheloung (1992) observed the greatest mobility in a high pH soil. In general, some sulfonylureas are classified as highly mobile under certain soil and rainfall conditions (Beyer et al. 1987, 1988). Even though sulfonlyurea herbicides have been detected in lower depths of the soil profile, upward movement is possible under conditions of high evapotranspiration. Upward movement of chlorsulfuron and triasulfuron has been observed in soil columns (Mahnken and Weber 1988; Nilsson 1985). Upward movement of sulfonlyurea herbicides may reduce groundwater and surface water concerns because the herbicide(s) may be degraded in the microbially active upper portion of the soil profile, may be sorbed to soil organic matter and become unavailable, or may be taken up by the crop. It is possible that upward movement of these herbicides may contribute to carryover concerns for the following year's crop.

There is limited information concerning sulfonylurea herbicides in runoff waters. A recent runoff study in North Carolina showed chlorimuron concentrations ranged from 5 to 40 μg/l from a conventional and no-till soybean field in a piedmont and coastal plain soil (Afyuni, personal communication). Higher concentrations of chlorimuron in runoff occurred in both tillage systems 5 min into a simulated rainfall as compared to 20 min. Concentrations of chlorimuron in runoff were greater in conventional tillage as compared to no-till in both locations.

To our knowledge, there are few studies concerning sulfonlyurea herbicides in surface waters. If present in surface waters, the persistence of these compounds appears to be quite variable. Thompson et al. (1992), using *in situ* enclosures in a mixed-wood/boreal forest lake in Ontario, Canada, observed concentration-dependent dissipation of metsulfuron-methyl. Slow rates (DT_{50} > 84 d) were noted at 10^3 μg/l concentrations and rapid rates (DT_{50} = 29 d) at a more environmentally realistic concentration (10 μg/l).

The sulfonylureas are considered nonvolatile with vapor pressures ranging from 1.3×10^{-10} to 5.5×10^{-16} mmHg at 25°C (Beyer et al. 1988). However, Fletcher et al. (1993) reported concerns over sulfonylurea drift in south-central Washington State. It appears that cherry (*Prunus avium* L.) growers attributed low cherry yields during the late 1980s to sulfonylurea drift, because low yields coincided with the first extensive use of sulfonlyurea herbicides in that portion of the state. To simulate potential herbicide drift, Fletcher et al. (1993) applied chlorsulfuron directly to cherry branches with various developmental stages. Chlosulfuron was applied at 0, 0.001, 0.002, 0.01, and 0.1 of the recommended tank mixture for use on small grains in Washington, Oregon, and California. They concluded that cherry trees were highly sensitive to chlorsulfuron and that small quantities of the herbicide, such as those transported during drift, may change plant reproduction without altering vegetative growth. The amount of herbicide transported in drift depends upon several factors, including method of application, herbicide volatility, temperature and wind speed, and herbicide formulation.

4. CONCLUSIONS

The environmental impact of HRCs depends upon the extent that these varieties are used and the particular properties of the herbicide. Herbicide losses from the site of application are dependent upon a number of interactive processes: degradation, sorption, volatilization, leaching, and runoff. The predominant processes for a particular herbicide are dependent on the characteristics unique to that herbicide (e.g., high vapor pressure or a slow degradation rate) and the interaction of processes with site-specific soil and climate factors. Another factor that may affect the impact of herbicides associated with HRCs is the projected pattern of use, including the number of applications and their timing.

Despite the complexity of these processes, certain conclusions can be drawn concerning the herbicides which appear to have potential use in conjunction with HRCs. First, most of the herbicides used in this technology will be transported in runoff during storms that closely follow herbicide application. In some regions postemergence applications may reduce losses in runoff as compared to preemergence applications due to the reduced storm frequency later in the growing season and canopy interception of the herbicide. The adoption of HRCs will have uncertain effects on the movement of herbicides into groundwater. Herbicides which are applied postemergence will benefit from plant transpiration, which will reduce the water available for leaching herbicides. Some herbicides, such as the imidazolinone and sulfonylurea herbicides have the advantages of low single application rates and relatively low mammalian toxicity. However, these compounds do appear to have some potential mobility in soil, and multiple applications of herbicides to resistant crops will increase the opportunity for movement. Results of model simulations show that there is a considerable margin of safety between model predictions of sulfonylurea and imidazolinone (chlorimuron and imazaquin) concentrations in groundwater and the projected health advisory level (Hoag and Hornsby 1992). The wide variability in persistence of sulfonylurea and imidazolinone herbicides in soil would seem to offer some options for choosing herbicides that have minimal leaching potential. Other herbicides such as bromoxynil and atrazine have already been detected as groundwater contaminants. The actual environmental impact of bromoxynil- and atrazine-resistant crops would depend upon specific crops, herbicides, soils, and herbicide usage. Impacts resulting from discontinuing the use of one herbicide in favor of a herbicide and a HRC should be considered in the design of a weed control program.

REFERENCES

Amonette, J. and O'Connor, G.A. 1980. Nonionic surfactant effects on adsorption and degradation of 2,4-D. *Soil Sci. Soc. Am. J.* 44: 540.

Anderson, J.J. and Dulka, J.J. 1985. Environmental fate of sulfometuron-methyl in aerobic soils. *J. Agric. Food Chem.* 33: 596.

Armstrong, D.E., Chesters, G., and Harris, R.F. 1967. Atrazine hydrolysis in soil. *Soil Sci. Soc. Am. Proc.* 31: 61.

Balthazor, T.M. and Hallas, L.E. 1986. Glyphosate-degrading microorganisms from industrial activated sludge. *Appl. Environ. Microbiol.* 51: 432.

Barnes, C.J., Goetz, A.J., and Lavy, T.L. 1989. Effects of imazaquin residues on cotton (*Gossypium hirsutum*). *Weed Sci.* 37: 820.

Basham, G.W. and Lavy, T.L. 1987. Microbial and photolytic dissipation of imazaquin in soil. *Weed Sci.* 35: 865.

Basham, G., Lavy, T.L., Oliver, L.R., and Scott, H.D. 1987. Imazaquin persistence and mobility in three Arkansas soils. *Weed Sci.* 35: 576.

Beckie, H.J. and McKercher, R.B. 1990. Mobility of two sulfonylurea herbicides in soil. *J. Agric. Food Chem.* 38: 310.

Behki, R., Topp, E., Dick, W., and Germon, P. 1993. Metabolism of the herbicide atrazine by *Rhodococcus* strains. *Appl. Environ. Microbiol.* 59: 1955.

Behrendt, H., Matthies, M., Guildemeister, H., and Gorlitz, G. 1990. Leaching and transformation of glufosinate-ammonium and its main metabolite in a layered soil column. *Environ. Toxicol. Chem.* 9: 541.

Beland, F.A., Farwell, S.O., and Greer, R.D. 1974. Anaerobic degradation of 1,1,1,2-tetrachloro-2,2-bis(p-chlorophenyl)ethane (DTE). *J. Agric. Food Chem.* 24: 753.

Bergstrom, L. 1990. Leaching of chlorsulfuron and metsulfuron methyl in three Swedish soils measured in field lysimeters. *J. Environ. Qual.* 19: 701.

Beyer, E.M., Brown, H.M., and Duffy, M.J. 1987. Sulfonylurea herbicide soil relations. *Proc. Br. Crop Prot. Conf.-Weeds* 2: 531.

Beyer, E.M., Jr., Duffy, M.J., Hay, J.V., and Schlueter, D.D. 1988. Sulfonylureas. In *Herbicides: Chemistry, Degradation, and Mode of Action*, Vol. 3. Ed. by Kearney, P.C. and Kaufman, D.D. Marcel Dekker, New York, p. 117.

Blacklow, W.M. and Pheloung, P.C. 1992. Sulfonylurea herbicides applied to acidic sandy soils: movement, persistence and activity within the growing season. *Aust. J. Agric. Res.* 43: 1157.

Bohn, H.L., McNeal, B.L., and O'Conner, G.A. 1979. *Soil Chemistry.* John Wiley & Sons, New York.

Bollag, J.-M. 1991. Enzymatic binding of pesticide degradation products to soil organic matter and their possible release. In *Pesticide Transformation Products: Fate and Significance in the Environment.* Ed. by Somasundaram, L. and Coats, J.R. American Chemical Society, Washington, D.C., p. 122.

Borggaard, O.K. and Streibig, J.C. 1988. Chlorsulfuron adsorption by humic acid, iron oxides, and montmorillonite. *Weed Sci.* 36: 530.

Bosio, P.G., Elgar, K.E., Mathews, B.L., Woodbridge, A.P., and Wright, A.N. 1982. Depletion of residues of benzoylprop-ethyl in soil. *Pestic. Sci.* 13: 1.

Bowman, B.T. 1989. Mobility and persistence of the herbicides atrazine, metolachlor and terbuthylazine in Plainfield sand determined using field lysimeters. *Environ. Toxicol. Chem.* 8: 485.

Brouwer, W.W.M., Boesten, J.J.T.I., and Siegers, W.G. 1990. Adsorption of transformation products of atrazine by soil. *Weed Res.* 30: 123.

Brown, H.M. 1990. Mode of action, crop selectivity, and soil relations of the sulfonylurea herbicides. *Pestic. Sci.* 29: 263.

Buhler, D.D., Randall, G.W., Koskinen, W.C., and Wyse, D.L. 1993. Atrazine and alachlor losses from subsurface tile drainage of a clay loam soil. *J. Environ. Qual.* 22: 583.

Burkart, M.R. and Kolpin, D.W. 1993. Hydrologic and land-use factors associated with herbicides and nitrate in near-surface aquifers. *J. Environ. Qual.* 22: 646.

Burkhard, N. and Guth, J.A. 1981. Chemical hydrolysis of 2-chloro-4,6-bis (alkylamino)-1,3,5-triazine herbicides and their breakdown in soil under the influence of adsorption. *Pestic. Sci.* 12: 45.

Cantwell, J.R., Liebl, R.A., and Slife, F.W. 1989. Biodegradation characteristics of imazaquin and imazethapyr. *Weed Sci.* 37: 815.

Cessna, A.J. and Grover, R. 1978. Spectrophotometric determination of dissociation constants of selected acidic herbicides. *J. Agric. Food Chem.* 26: 289.

Chaleff, R.S. and Ray, T.B. 1984. Herbicide-resistant mutants from tobacco cell cultures. *Science* 223: 1148.

Che, M., Loux, M.M., Traina, S.J., and Logan, T.J. 1992. Effect of pH on sorption and desorption of imazaquin and imazethapyr on clays and humic acid. *J. Environ. Qual.* 21: 698.

Cook, A.M. 1987. Biodegradation of *s*-triazine xenobiotics, *FEMS Microbiol. Rev.* 46: 93.

Council for Agricultural Science and Technology. 1991. Herbicide-resistant crops. In *Comments from CAST No. 1991-1,* CAST, Ames, IA.

Curran, W.S., Loux, M.M., Liebl, R.A., and Simmons, F.W. 1992. Photolysis of imidazolinone herbicides in aqueous solution and on soil. *Weed Sci.* 40: 143.

Duke, S.O. and Kenyon, W.H. 1988. Polycyclic alkanoic acids. In *Herbicides: Chemistry, Degradation, and Mode of Action,* Vol. 3. Ed. by Kearney, P.C. and Kaufman, D.D. Marcel Dekker, New York, p. 71.

Duke, S.O., Moorman, T.B., and Bryson, C.T. 1991. Phytotoxicity of pesticide degradation products. In *Pesticide Transformation Products: Fate and Significance in the Environment.* Ed. by Somasundaram, L. and Coats, J.R. American Chemical Society, Washington, D.C., p. 188.

Edwards, W.M., Triplett, G.B., and Kramer, R M. 1980. A watershed study of glyphosate transport in runoff. *J. Environ. Qual.* 9: 661.

Fletcher, J.S., Pfleeger, T.G., and Ratsch, H.C. 1993. Potential environmental risks associated with the new sulfonylurea herbicides. *Environ. Sci. Technol.* 27: 2250.

Fredrickson, D.R. and Shea, P.J. 1986. Effect of soil pH on degradation, movement, and plant uptake of chlorsulfuron. *Weed Sci.* 34: 328.

Fuesler, T P. and Hanafey, M.K. 1990. Effect of moisture on chlorimuron degradation in soil. *Weed Sci.* 38: 256.

Gallina, M.A. and Stephenson, G.R. 1992. Dissipation of [^{14}C]glufosinate ammonium in two Ontario soils. *J. Agric. Food Chem.* 40: 165.

Gamble, D.S. and Khan, S.U. 1990. Atrazine in organic soil: chemical speciation during heterogeneous catalysis. *J. Agric. Food Chem.* 38: 297.

Gaynor, J.D., McTavish, D.C., and Findlay, W.I. 1992. Surface and subsurface transport of atrazine and alachlor from a Brookston clay loam under continuous corn production. *Arch. Environ. Contam. Toxicol.* 23: 240.

Goetz, A.J., Lavy, T.L., and Gbur, E.E., Jr. 1990. Degradation and field persistence of imazethapyr. *Weed Sci.* 38: 421.

Goetz, A.J., Walker, R.H., Wehtje, G., and Hajek, B.F. 1989. Sorption and mobility of chlorimuron in Alabama soils. *Weed Sci.* 37: 428.

Goetz, A.J., Wehtje, G., Walker, R.H., and Hajek, B. 1986. Soil solution and mobility characterization of imazaquin. *Weed Sci.* 34: 788.

Grover, R. 1973. The adsorptive behavior of acid and ester forms of 2,4-D on soils. *Weed Res.* 13: 51.

Hassett, J.J. and Banwart, W.L. 1989. The sorption of nonpolar organics by soils and sediments. In *Reactions and Movement of Organic Chemicals in Soils*. Ed. by Sawhney, B.L. and Brown, K. Soil Science Society of America, Madison, WI, p. 31.

Hatmaker, D. 1994. List of Biotechnology Permits from the Biotechnology, Biologics, and Environmental Protection Division of APHIS. Hyattsville, MD.

Hay, J.V. 1990. Chemistry of sulfonylurea herbicides. *Pestic. Sci.* 29: 247.

Hayenga, M., Thompson, L.C., Chase, C., and Kaaria, S. 1992. Economic and environmental implications of herbicide-tolerant corn and processing tomatoes. *J. Soil Water Conserv.* 47: 411.

Heinonen-Tanski, H. 1989. The effect of temperature and liming on the degradation of glyphosate in two arctic forest soils. *Soil Biol. Biochem.* 21: 313.

Helling, C.S. and Gish, T.J. 1986. Soil characteristics affecting pesticide movement into ground water. In *Evaluation of Pesticides in Ground Water*. ACS Symp. Ser. No. 315. Ed. by Garner, W.Y., Honeycutt, R.C., and Higg, H.N. American Chemical Society, Washington, D.C., p. 14

Helweg, A. 1987. Degradation and adsorption of ^{14}C-MCPA in soil-influence of concentration, temperature, and moisture content on degradation. *Weed Res.* 27: 287.

Hitchings, E.J. and Roberts, T.R. 1979. Degradation of the herbicide flamprop-isopropyl in soil under laboratory conditions. *Pestic. Sci.* 10: 1.

Hoag, D.L. and Hornsby, A.G. 1992. Coupling groundwater contamination with economic returns when applying farm pesticides. *J. Environ. Qual.* 21: 579.

Jacob, G.S., Schaefer, J., Stejskal, E.O., and McKay, R.A. 1985. Solid-state NMR determination of glyphosate metabolism in a *Pseudomonas* sp. *J. Biol. Chem.* 260: 5899.

Jacoby, H., Hoheisel, C., Karrie, J., Lees, S., Davies-Hilliard, L., Hannon, P., Bingham, R., Behl, E., Wells, D., and Waldman, E. 1992. Pesticides in Ground Water Database, Rep. EPA 734-12-92-001. U.S. Environmental Protection Agency, Washington, D.C.

Jayachandran, K., Steinheimer, T.R., Somasundaram, L., Moorman, T.B., Kanwar, R.S., and Coats, J.R. 1994. Occurrence of atrazine and degradates as contaminants of subsurface drainage and shallow groundwater. *J. Environ. Qual.* 23: 311.

Johnson, D.H., Jordan, D.L., Johnson, W.G., Talbert, R.E., and Frans, R.E. 1993. Nicosulfuron, primusulfuron, imazethapyr, and DPX-PE350 injury to succeeding crops. *Weed Technol.* 7: 641.

Joshi, M.M., Brown, H.M., and Romesser, J.A. 1985. Degradation of chlorsulfuron by soil microorganisms. *Weed Sci.* 33: 888.

Jury, W.A. and Ghodrati, M. 1989. Overview of organic chemical environmental fate and transport modeling approaches. In *Reactions and Movement of Organic Chemicals in Soils*. Ed. by Sawhney, B.L. and Brown, K. Soil Science Society of America, Madison, WI, p. 271.

Keller, K.E. 1992. Movement and Dissipation of Atrazine, Metolachlor, and Primisulfuron in Field Lysimeters. Ph.D. thesis, North Carolina State University, Raleigh.

Khan, S.U. 1991. Bound (nonextractable) pesticide degradation products in soils. In *Pesticide Transformation Products: Fate and Significance in the Environment*. Ed. by Somasundaram, L. and Coats, J.R. American Chemical Society, Washington, D.C., p. 108.

Kirkland, K. and Fryer, J.D. 1972. Degradation of several herbicides in a soil previously treated with MCPA. *Weed Res.* 12: 90.

Klint, M., Arvin, E., and Jensen, B.K. 1993. Degradation of the herbicides mecoprop and atrazine in unpolluted sandy aquifers. *J. Environ. Qual.* 22: 262.

Kochany, J., Choudhry, G.G., and Webster, G.R.B. 1990. Environmental photochemistry of the herbicide bromoxynil in aqueous solution containing soil fulvic acids. *Int. J. Environ. Anal. Chem.* 39: 59.

Kruger, E.L., Somasundaram, L., Kanwar, R.S., and Coats, J.R. 1993. Persistence and degradation of [^{14}C]atrazine and [^{14}C]deisopropylatrazine as affected by soil depth and moisture conditions. *Environ. Toxicol. Chem.* 12: 1959.

Laird, D.A., Barriuso, E., Dowdy, R.H., and Koskinen, W.C. 1992. Adsorption of atrazine on smectites. *Soil Sci. Soc. Am. J.* 56: 62.

Leonard, R.A. 1988. Herbicides in surface waters. In *Environmental Chemistry of Herbicides*, Vol. 1. Ed. by Grover, R. CRC Press, Boca Raton, FL, p. 45.

Liu, C.-M., McLean, P.A., Sookdeo, C.C., and Cannon, F.C. 1991. Degradation of the herbicide glyphosate by members of the family *Rhizobiaceae*. *Appl. Environ. Microbiol.* 57: 1799.

Loux, M.M., Liebl, R.A., and Slife, F.W. 1989. Adsorption of imazaquin and imazethapyr on soils, sediments, and selected adsorbents. *Weed Sci.* 37: 712.

Mahnken, G.E. and Weber, J.B. 1988. Capillary movement of triasulfuron and chlorsulfuron in Rion sandy loam soil. *Proc. South. Weed Sci. Soc.* 41: 332.

Mangels, G. 1991. Behavior of the imidazolinone herbicides in soil — a review of the literature. In *The Imidazolinone Herbicides*. Ed. by Shaner, D.L. and O'Connor, S.L. CRC Press, Boca Raton, FL, p. 191.

Martens, R. 1978. Degradation of the herbicide [^{14}C]-diclofop-methyl in soil under different conditions. Pestic. Sci. 9: 127.

McBride, K.E., Kenny, J.W., and Stalker, D.M. 1986. Metabolism of the herbicide bromoxynil by *Klebsiella pneumoniae* subsp. *ozaenae*. *Appl. Environ. Microbiol.* 52: 325.

McBride, M.B. 1991. Electron spin resonance study of copper ion complexation by glyphosate and related ligands. *Soil Sci. Soc. Am. J.* 55: 979.

McBride, M.B. and Kung, K.-H. 1989. Complexation of glyphosate and related ligands with iron (III). *Soil Sci. Soc. Am. J.* 53: 1668.

Mersie, W. and Foy, C.L. 1985. Phytotoxicity and adsorption of chlorsulfuron as affected by soil properties, *Weed Sci.*, 33: 564.

Mersie, W. and Foy, C.L. 1986. Adsorption, desorption, and mobility of chlorsulfuron in soils. *J. Agric. Food Chem.* 34: 89.

Miller, J.L. 1992. Degradation of Atrazine, Metolachlor, and Primisulfuron in Soil from Four Depths in a Dothan Loamy Sand. M.S. thesis, North Carolina State University, Raleigh.

Montgomery, J.H. 1993. *Agrochemicals Desk Reference: Environmental Data*. Lewis Publishers, Boca Raton, FL, p. 31.

Moorman, T.B. 1994. Pesticide degradation by soil microorganisms: environmental, ecological, and management effects. In *Soil Biology: Effects on Soil Quality*. Ed. by Hatfield, J.L. and Stewart, B.A. Lewis Publishers, Boca Raton, FL, p. 121.

Moorman, T.B. and Harper, S.S. 1989. Transformation and mineralization of metribuzin in surface and subsurface horizons of a Mississippi Delta soil. *J. Environ. Qual.* 18: 302.

Moyer, J.R., Esau, R., and Kozub, G.C. 1990. Chlorsulfuron persistence and response of nine rotational crops in alkaline soils of southern Alberta. *Weed Technol.* 4: 543.

Mueller, T.C., Moorman, T.B., and Snipes, C.E. 1992. Effect of concentration, sorption, and microbial biomass on degradation of the herbicide fluometuron in surface and subsurface soils. *J. Agric. Food Chem.* 40: 2517.

Muir, D.C.G., Kenny, D.F., Grift, N.P., Robinson, R.D., Titman, R.D., and Murkin, H.R. 1991. Fate and acute toxicity of bromoxynil esters in an experimental prairie wetland. *Environ. Toxicol. Chem.* 10: 395.

Nègre, M., Gennari, M., Andreoni, V., Ambrosoli, and Celi, L. 1993. Microbial metabolism of fluazifop-butyl. *J. Environ. Sci. Health B* 28: 545.

Nègre, M., Gennari, M., Cignetti, A., Zanini, E. 1988. Degradation of fluazifop-butyl in soil and aqueous systems. *J. Agric. Food Chem.* 36: 1319.

Newhouse, K.E., Shaner, D.L., Wang, T., and Fincher, R. 1990. Genetic modification of crop responses to imidazolinone herbicides. In *Managing Resistance to Agrochemicals: From Fundamental Research to Practical Strategies*, ACS Symp. Ser. No. 421. Ed. by Green, M.B., LeBaron, H.M., and Moberg, W.K., American Chemical Society, Washington, D.C., p. 474.

Nicholaichuk, W. and Grover, R. 1983. Loss of fall-applied 2,4-D in spring runoff from a small agricultural watershed. *J. Environ. Qual.* 12: 412.

Nilsson, H. 1985. Capillary transport of a mobile herbicide: a column experiment with chlorsulfuron. *Weeds and Weed Control, 26th Swedish Weed Conf.* 1: 209.

Nomura, N.S. and Hilton, H.W. 1977. The adsorption and degradation of glyphosate in five Hawaiian sugarcane soils. *Weed Res.* 17: 113.

Pantone, D.J., Young, R.A., Buhler, D.D., Eberlein, C.V., Koskinen, W.C., and Forcella, F. 1992. Water quality impacts associated with pre- and postemergence applications of atrazine in maize. *J. Environ. Qual.* 21: 567.

Pionke, H.B., Glotfelty, D.E., Lucas, A.D., and Urban, J.B. 1988. Pesticide contamination of groundwaters in the Mahantango Creek watershed. *J. Environ. Qual.* 17: 76.

Pipke, R. and Amrhein, N. 1988. Degradation of the phosphonate herbicide glyphosate by *Arthrobacter atrocyaneus* ATCC 13752, *Appl. Environ. Microbiol.*, 54: 1293.

Pothuluri, J.V., Hinson, J.A., and Cernaglia, C.A. 1991. Propanil: toxicological characteristics, metabolism, and biodegradation potential. *J. Environ. Qual.* 20: 330.

Pothuluri, J.V., Moorman, T.B., Obenhuber, D.C., and Wauchope, R.D. 1990. Aerobic and anaerobic degradation of alachlor in samples from a surface to groundwater profile. *J. Environ. Qual.* 19: 525.

Quinn, J.P., Peden, J.M.M., and Dick, R.E. 1989. Carbon-phosphorus bond cleavage by Gram-positive and Gram-negative soil bacteria. *Appl. Microbiol. Biotechnol.* 31: 283.

Renner, K.A., Meggitt, W.F., and Penner, D. 1988. Effect of soil pH on imazaquin and imazethapyr adsorption to soil and phytotoxicity to corn (*Zea mays*). *Weed Sci.* 36: 78.

Rick. S.R., Slife, F.W., and Banwart, W.L. 1987. Adsorption of selective grass herbicides by soils and sediments. *Weed Sci.* 35: 282.

Roeth, F.W. 1986. Enhanced degradation in soil with repeat application. *Rev. Weed Sci.* 2: 45.

Rueppel, M L., Brightwell, B.B., Schaeffer, J., and Marvel, J.T. 1977. Metabolism and degradation of glyphosate in soil and water. *J. Agric. Food Chem.* 25: 517.

Schneiders, G.E., Koeppe, M.K., Naidu, M.V., Horne, P., Brown, A.M., and Mucha, C.F. 1993. Fate of rimsulfuron in the environment. *J. Agric. Food Chem.* 41: 2404.

Seta, A.K., Blevins, R.L., Frye, W.W., and Barfield, B.J. 1993. Reducing soil erosion and agricultural chemical losses with conservation tillage. *J. Environ. Qual.* 22: 661.

Shaw, D.R., Smith, C.A., and Hairston, J.E. 1992. Impact of rainfall and tillage systems on off-site herbicide movement. *Commun. Soil Sci. Plant Anal.* 23: 1843.

Shea, P.J. 1986. Chlorsulfuron dissociation and adsorption on selected adsorbents and soils. *Weed Sci.* 34: 474.

Smith, A.E. 1971. Degradation of bromoxynil in Regina heavy clay. *Weed Res.* 11: 276.

Smith, A. E. 1978. Relative persistence of di- and tri-chlorophenoxyalkanoic acid herbicides in Saskatchewan soils, *Weed Res.*, 18: 275.

Smith, A.E. 1985. Transformation and persistence of the herbicide [^{14}C]haloxyfop-methyl in soil under laboratory conditions. *J. Agric. Food Chem.* 33: 972.

Smith, A.E. 1986. Persistence of the herbicides [^{14}C]chlorsulfuron and [^{14}C] metsulfuron in prairie soils under laboratory conditions. *Bull. Environ. Contam. Toxicol.* 37: 608.

Smith, A.E. 1989. Transformation of the herbicide [^{14}C]glufosinate in soils. *J. Agric. Food Chem.* 37: 267.

Smith, A.E. and Aubin, A.J. 1993. Degradation of ^{14}C-glyphosate in Saskatchewan soils. *Bull. Environ. Contam. Toxicol.* 50: 499.

Smith, A.E., Aubin, A.J., and Biederbeck, V.O. 1989. Effects of long-term 2,4-D and MCPA field applications on soil residues and their rates of breakdown. *J. Environ. Qual.* 18: 299.

Smith, A.E. and Hsiao, A.I. 1985. Transformation and persistence of chlorsulfuron in prairie field soils. *Weed Sci.* 33: 555.

Stott, D.E., Martin, J.P., Focht, D.D., and Haider, K. 1983. Biodegradation, stabilization in humus, and incorporation into soil biomass of 2,4-D and chlorocatechol carbons. *Soil Sci. Soc. Am. J.* 47: 66.

Stougaard, R.N., Shea, P.J., and Martin, A.R. 1990. Effect of soil type and pH on adsorption, mobility, and efficacy of imazaquin and imazethapyr. *Weed Sci.* 38: 67.

Taylor, A.W. and Spencer, W.F. 1990. Volatilization and vapor transport processes. In *Pesticides in the Soil Environment: Processes, Impacts and Modeling*. Ed. by Cheng, H H. Soil Science Society of America, Madison, WI, p. 213.

Thirunarayanan, K., Zimdahl, R.L., and Smika, D.E. 1985. Chlorsulfuron adsorption and degradation in soil. *Weed Sci.* 33: 558.

Thomas, S.M. and Harrison, S.K. 1990. Surfactant-altered rates of chlorimuron and metsulfuron photolysis in sunlight. *Weed Sci.* 38: 602.

Thompson, D.G., MacDonald, L.M., and Staznik, B. 1992. Persistence of hexazinone and metsulfuron-methyl in a mixed-wood/boreal forest lake, *J. Agric. Food Chem.* 40: 1444.

Thurman, E.M., Goolsby, D.A., Meyer, M.T., Mill, M.S., Pomes, M.L., and Kolpin, D.W. 1992. A reconnaissance study of herbicides and their metabolites in surface water of the midwestern United States using immunoassay and gas chromatography/mass spectrometry. *Environ. Sci. Technol.* 26: 2440.

Topp, E., Xun, L. and Orser, C.S. 1992. Biodegradation of the herbicide bromoxynil (3,5-dibromo-4-hydroxybenzonitrile) by purified pentachlorophenol hydroxylase and whole cells of *Flavobacterium* sp. strain ATCC 39723 is accompanied by cyanogenesis. *Appl. Environ. Microbiol.* 58: 502.

Torstensson, N.T.L., Lundgren, L.N., and Stenström, J. 1989. Influence of climatic and edaphic factors on persistence of glyphosate and 2,4-D in forest soils. *Ecotoxicol. Environ. Safety* 18: 230.

Torstensson, N.T.L., Stark, J., and Göransson, B. 1975. The effect of repeated applications of 2,4-D and MCPA on their breakdown in soil. *Weed Res.* 15: 159.

Venkatesh, R., Harrison, S.K., and Loux, M.M. 1993. Photolysis of aqueous chlorimuron and imazaquin in the presence of phenolic acids and riboflavin. *Weed Sci.* 41: 454.

Wackett, L.P., Shames, S.L., Venditti, C.P., and Walsh, C.T. 1987. Bacterial carbon phosphorus lyase: products, rates, and regulation of phosphonic and phosphinic acid metabolism. *J. Bacteriol.* 169: 710.

Waite, D.T., Grover, R., Westcott, N.D., Somerstad, H., and Kerr, L. 1992. Pesticides in groundwater, surface water and spring runoff in a small Saskatchewan watershed. *Environ. Toxicol. Chem.* 11: 741.

Walker, A., Cotterill, E. G., and Welch, S. J. 1989. Adsorption and degradation of chlorsulfuron and metsulfuron-methyl in soils from different depths, *Weed Res.*, 29: 281.

Walker, A. and Welch, S.J. 1989. The relative movement and persistence in soil of chlorsulfuron, metsulfuron-methyl and triasulfuron. *Weed Res.* 29: 375.

Warren, R.L. and Weber, J.B. 1993. Reducing risk of groundwater contamination with a simplified pesticide/soil ranking system. *Proc. South. Weed Sci. Soc.* 46: 336.

Wauchope, R.D. 1978. The pesticide content of water draining from agricultural fields — a review. *J. Environ. Qual.* 4: 459.

Weber, J.B. 1991. Fate and behaviour of herbicides in soils. *Appl. Plant Sci.* 5: 28.

Weber, J.B. and Peter, C.J. 1982. Adsorption, bioactivity and evaluation of soil tests for alachlor, acetochlor, and metolachlor. *Weed Sci.* 30: 14.

Weber, J.B. and Miller, C.T. 1989. Organic chemical movement over and through the soil. In *Reactions and Movement of Organic Chemicals in Soils.* Ed. by Sawhney, B.L. and Brown, K. Soil Science Society of America, Madison, WI, p. 305.

Whang, J.M., Schomberg, C.J., Glotfelty, D.E., and Taylor, A.W. 1993. Volatilization of fonofos, chlorpyrifos, and atrazine from conventional and no-till surface soils in the field. *J. Environ. Qual.* 22: 173.

White, A.W., Asmussen, L.E., Hauser, E.W., and Turnbull, J.W. 1976. Loss of 2,4-D in runoff from plots receiving simulated rainfall and from a small agricultural watershed. *J. Environ. Qual.* 5: 487.

Zaranyika, M.F. and Nyandoro, M.G. 1993. Degradation of glyphosate in the aquatic environment: an enzymatic kinetic model that takes into account microbial degradation of both free and colloidal (or sediment) particle adsorbed glyphosate. *J. Agric. Food Chem.* 41: 838.

Zins, A.B., Wyse, D.L., and Koskinen, W.C. 1991. Effect of alfalfa (*Medicago sativa*) roots on movement of atrazine and alachlor through soil. *Weed Sci.* 39: 262.

Chapter 20

MOVEMENT OF CROP TRANSGENES INTO WILD PLANTS

Kathleen H. Keeler
Charles E. Turner
Margaret R. Bolick

CONTENTS

1-56670-045-0/96/$0.00+$.50
© 1996 by CRC Press Inc.

1. INTRODUCTION

The technology of genetic engineering is introducing a wide variety of traits into crop plants. One of the first to be field tested is herbicide resistance. Field trials show that herbicide resistance can be a dramatically effective agricultural tool, allowing the crops to survive while general-purpose herbicides of low toxicity and persistence remove or significantly reduce competing weeds (Chapters 14 and 15). Despite these benefits, problems with this technology can be anticipated. Some of these are considered elsewhere in this book. In this chapter we discuss the problems of herbicide- resistant weeds (HRWs) and how these might arise from the widespread development and employment of herbicide-resistant crops (HRCs).

Concern has been expressed over the inadvertent development of HRWs that are harder to control than present weeds (Hauptli et al. 1985; Keeler and Turner 1991; Dyer et al. 1993). This can occur either by the HRC itself becoming a weed where it is not specifically cultivated, or by existing weeds receiving herbicide-resistant genes (HRGs) from a HRC via gene flow, leading to greater difficulty controlling these weeds. In this chapter we mention possible problems from increased weediness attained by HRCs, but focus particularly on the role of gene flow to weedy wild relatives as a potential problem (Keeler and Turner 1991; Dyer et al. 1993), because that is a matter of far greater concern. Other possible weed problems are discussed elsewhere in this book and in Keeler and Turner (1991).

Escape of herbicide resistance should not be regarded as a problem in and of itself. Escape is a problem only if this leads to undesirable consequences. Problems are unlikely in habitats where herbicides are not used, because in the absence of herbicide applications, herbicide-resistant plants will have no particular selective advantage. In heavily herbicided environments, however, herbicide resistance will favor the survival of HRWs, potentially creating a new set of weed problems.

Not all consequences are equally serious. While the most desirable outcome is a crop that survives herbicide application and makes absolutely no contribution to weed populations, many other alternatives are tolerable. For example, a transgenic weed that is resistant to the herbicide of choice but is controllable by alternate methods, e.g., mechanically, does not greatly decrease the usefulness of the resistant crop. Even when a HRW spreads, if it is less costly to control than the plants killed by the herbicide, there is a net gain. While we probably cannot evaluate and weigh these alternatives quantitatively, they are part of all decisions to develop or market a crop, and part of the U.S. Environmental Protection Agency's regulation of plant pesticides under the Federal Insecticide, Fungicide and Rodenticide Act (FIFRA). Every action, including inaction, represents a decision, so all parties are best served by well-considered decisions and actions.

Precise predictions are not possible. We have only incomplete information at present, but even if we knew much more about the interaction of weeds with HRCs, some predictions would fail because the system is dynamic: Ecosystems are continually changing and the resident organisms evolving, so that a novel genotype–environment interaction is always possible. In addition, while many HRGs occur in nature, some transgene–plant combinations are, or will be, unique. Changing environments and organismal evolution could confound any prediction.

Herbicide resistance confers survival in the presence of substances that otherwise cause severe reduction of plant growth or death. For the purpose of evaluating environmental consequences, the mode of action is less important than the functional phenotype. If the genes produce a herbicide-resistant phenotype, the application of that herbicide will produce strong selection favoring HRWs, including hybrids that received resistance genes from the crop.

We assume for the purposes of this chapter that the only effect of the herbicide resistance gene (and the facilitating DNA introduced as part of the construct) is herbicide resistance, and that it is the only way in which weeds receiving the introduced gene will differ from other local weeds. Thus, we do not consider problems that might arise when there are pleiotropic effects of the inserted genes or genes that alter the phenotype in several ways (e.g., a combined herbicide/virus/pest resistance package). However, accurate risk assessment must include these complex effects (Dyer et al. 1993; see also Chapters 25 and 26). To further simplify our analysis here, our focus is on the U.S.

2. HERBICIDE-RESISTANT WEEDS

2.1. WEED DEFINITIONS

Herbicide-resistant crops have the potential to lead to "worse weeds." "Weed" is a familiar term about which there is much debate as to definition and meaning. The basic, working definition of weed seems to be "a plant in the wrong place," a plant whose presence is objectionable to someone. Trying to cast this into an objective definition, however, has proven quite difficult (e.g., Salisbury 1961; Bucholtz 1967; Perrins et al. 1992a, b). Basically, the concept of a weed is an anthropocentric construction, and thus essentially subjective in nature. Plants may be weeds for quite diverse reasons. They may interfere with human activities because of rapid growth or because, although slow growing, they are hard to remove, or be neither of the above, but are unpalatable or poisonous to livestock. When considering possible weeds of herbicide-resistant cultivated plants, this is a very important point. If biotechnology produces a plant that interferes with someone or something, it has produced a weed, regardless of the taxonomic identity of the species.

2.2. WEED SEVERITY

Plants which are never unwanted are never weeds. There must be species in this category, but they are few. At the other extreme of this spectrum are plants that are unwelcome in most places in which they occur, and that are therefore considered weeds almost universally. For example, cocklebur (*Xanthium strumarium*), a spiny weed of 11 crops in 28 countries with foliage toxic to livestock, is likely to have very few proponents (Holm et al. 1977).

By the definition above, major and especially minor weeds abound. However, no one is going to spend much time or money removing minor weeds. The weeds of chief interest here are species with major economic or environmental impact. For this sort of weed it is easy to find agreement that they should be controlled. In the economic sphere a functional approach to determining that a weed presents a serious problem can be found in its economic impact. Holm et al. (1977) rated the "world's worst weeds" by determining which weeds were problems in the most crops in the most countries, the record being 52 crops in 92 countries (purple nutsedge, *Carex rotundata*). Plants creating problems to multiple crops in multiple countries are clearly serious weeds. Thus, a gradient exists from plants that never bother anyone to interference that dramatically reduces the productivity of major areas of cropland or rangeland. It is most critical to not genetically enhance the plants at the most damaging end of the weed spectrum.

2.3. LINES OF EVIDENCE

To predict the environmental impact of a transgenic crop requires a number of assumptions, none of which actually apply. These assumptions include:

1. the plants will always be responsibly handled
2. the plants will not evolve
3. land use patterns, e.g., agricultural and pasturage systems, will not change significantly
4. human uses of plants will remain the same
5. human evaluation of the worth of different species will remain stable
6. the distribution of people and the way they use the land (e.g., urban/rural) will be stable

Given these assumptions, one way to predict outcome is to gather information based on our historical experience with crops, weeds, and herbicide resistance to determine the relative probabilities of problems.

Science works best from well-controlled, experimental tests of ideas. For transgenic herbicide resistance plants, these do not currently exist, the PROSAMO experiment being a notable exception (Crawley et al. 1993; see also Miller and Gunary 1993). In the absence of adequate experimental evidence we look to other sources of information. All of these must be evaluated for what they can and cannot tell us about transgenic plants.

2.3.1. Cultivated Plants

The record of the cultivated plants into which the genes are inserted provide insight into the behavior of the same plant with herbicide resistance. This is true because only a few genes are being introduced into a large genome and breeders will try to make a transgenic plant that is as much like successful varieties as possible. While more extensive transgenic constructs may be present in the crops of the future, by then we will have field data on our present transgenics to use to evaluate those crops.

2.3.2. Herbicide Resistance Genes

Rather than being just an area of theoretical concern, there have however, been real problems in the field from the evolution of weeds resistant to herbicides. For example, sulfonylurea-resistant populations of prickly lettuce (*Lactuca serriola*), kochia (*Kochia scoparia*), and Russian thistle (*Salsola* sp.) are reducing the utility of sulfonylurea herbicides in and around winter wheat and along roadsides in the northwestern U.S. (Mallory-Smith et al. 1990; Alocer-Ruthling et al. 1992; Thompson et al. 1994; Stallings et al. 1994; Saari et al. 1994; also Chapter 8).

2.3.3. Environment

There is no consensus about the invasibility of environments. Species sometimes invade even intact and apparently healthy natural ecosystems, but stressed or disturbed systems are particularly vulnerable (e.g., croplands, roadsides, and overgrazed rangeland) (Mooney and Drake 1986). Fortunately, many disturbed sites, such as field edges, are easily managed for weed control.

Many of the events which can lead to the escape of transgenes are low probability events. Thus, watching for them at any particular site is likely to fail, but across a large area, they are more likely to occur. Their impact, however, depends on a complex set of local events.

Events with a general low probability can be greatly enhanced by natural selection. Thus, the emergence of herbicide-resistant phenotypes, while still improbable in any particular location, is significantly enhanced by the selection pressure created by the application of herbicides in the area. Except for the occasional case of cross-resistance, this is herbicide specific. Glyphosate resistance can be expected where glyphosate is applied, imidizolinone resistance, not glyphosate resistance, where imidizolinone is applied. These patterns increase our ability to monitor effectively for the emergence of HRWs.

2.3.4. Lessons from Introduced Species

Experience with introduced species, naturalized species from a different part of the world, has been used frequently to provide information about the behavior of transgenics (e.g., Sharples 1983, Keeler 1985, Tiedje et al. 1989). This has the following justifications:

1. Transgenics are perceived as potentially having novel interactions in an ecosystem, similar to the novel interactions of introduced species.
2. Population growth of naturalized transgenics may parallel the growth of introduced species from low initial numbers to outbreak populations
3. Introduced species produced some of the best data sets on change in number over time and invasion of new environments.
4. Control of introduced species provide some of the clearest studies of the factors that determine plant success.

However, information from introduced species must be treated carefully because these plants are outside their native range and have escaped adapted parasites, diseases, and predators. Transgenics may or may not be outside their range and free of specialist natural enemies. A given HRW is likely to be afflicted by the same consumer pests as those attacking the HRC relative. Furthermore, because they are virtually identical to a crop, most of their herbicide interactions can be readily predicted by the crop to which the HRGs were added. In contrast, an introduced species may have less predictable impact.

2.3.5 Predictions of Experts

Any plant that most weed scientists agree is likely to be a serious weed is just that. Predictive power for this category is excellent. This is based largely on knowledge of which plants are weedy somewhere in the world: the best predictor of whether a plant is potentially weedy here is that it is weedy somewhere else. There are lists of plants that are considered potential threats to U.S. agriculture (Zimdahl 1983), and noxious weed lists of weeds legally barred from entry to parts of the world where they do not currently occur (Foy et al. 1983). These plants are likely to produce further weed problems if provided with herbicide resistance; some already have it.

At the other extreme are plants for which almost no one would anticipate the development of weed problems. This group is probably also sound predictively. Note, however, that one would experience greater unhappiness if surprised by a weed from this group than if one of the noxious weeds fails to expand when introduced to a new site. A failure to predict weediness is more serious than a prediction of weediness that fails.

There remain many species for which there is no consensus on their potential for weediness (e.g., Perrins et al. 1992a, b). For these plants, experimental studies are essential. Expansion of our understanding of the genetics and ecology of weediness should have the most impact on this group. For the present, release of HRCs in this category must be treated on a case-by-case basis.

3. GENERATION OF HERBICIDE-RESISTANT WEEDS

As noted above, herbicide resistance may contribute to weed problems either by increased weediness of the crop or by escape of the resistance genes to compatible weedy relatives. The trait is then problematic or does not depend on whether herbicides are a potent selective factor in the environment of the resistant plants. We consider these problems below.

3.1. HERBICIDE-RESISTANT CROPS AS WEEDS

This alternative is rarely considered in detail because most of the major crops (e.g., corn, soybeans, wheat) are rarely, if ever, seen as weeds, and are readily controlled if they escape (Harlan 1965; and see also Table 1). These assumptions about the behavior of crops are less clearly valid in aseasonal climates, for crops recently domesticated, and for cultivated plants not grown as annual row crops. There are, unfortunately, many plants deliberately introduced as ornamentals, forage plants, or for soil stabilization (i.e., outside a row crop environment) that have become agggressive weeds. Some well-known examples include bermudagrass (*Cynodon dactylon*), Scotch broom (*Cytisus scoparius*), lantana (*Lantana camara*), Dalmatian toadflax (*Linaria dalmatica*), Japanese honeysuckle (*Lonicera japonica*), kudzu (*Pueraria montana* var. *lobata*), multiflora rose (*Rosa multiflora*), and Brazilian peppertree (*Schinus terebinthifolius*) (Foy et al. 1983; Turner 1988).

3.2. ESCAPE OF HERBICIDE RESISTANCE GENES TO WILD RELATIVES

We focus chiefly on weedy wild relatives because compatible wild plants that are already weedy constitute the greatest weed threats. If HRGs move from crops to wild relatives in the area, there is immediate potential for a serious weed problem because (1) herbicide resistance in the crop is likely to be accompanied by application of the herbicide, generating selection for the resistance, and (2) if the wild relatives were effective weeds before becoming herbicide resistant, they will be even more difficult to control as HRWs. This has the potential to significantly reduce the utility of the herbicide resistance trait and the herbicide.

With only rare exceptions, all our crops have wild relatives somewhere in the world, so escape of the transgenes is a strong possibility somewhere. The necessary conditions for the movement of an HRG from an HRC to a wild plant through pollen flow is as follows: dispersal of viable herbicide-resistant pollen from the HRC; interfertility between the HRC and a wild relative, leading to viable hybrids; and opportunity to cross, which is the presence of crop and weed in sufficiently close proximity that viable pollen reaches the compatible wild population.

3.2.1 Dispersal of Viable Pollen

Transgenes can only reach weed populations if carried to weeds on viable pollen. If the crop produces no pollen or no viable pollen, there will be no gene flow. Whatever the abundance and distribution of their compatible weedy relatives, such crops will not produce HRW populations.

This covers most, but not quite all, conditions. It is conceivable that an abandoned HRC plant could be fertilized by pollen from wild relatives and serve as the pistillate (female) parent for a hybrid seed. The hybrid seed, germinating under favorable conditions, might subsequently found (establish) a new, resistant population. For this to occur, the pistillate parent HRC would have to escape harvesting, and its hybrid seeds would have to survive to germinate, grow, and reproduce. In a cropland environment the conditions for this are sufficiently restrictive that we will not consider it further.

The potential for gene flow via pollen depends on three primary factors: (1) the amount of pollen produced, (2) the longevity of the pollen, and (3) the means of pollen transport. Secondarily it may be affected by selection on any genes expressed in the recipient gametophyte.

The amount of pollen produced per plant varies over several orders of magnitude, from tens of grains in some water-pollinated or cleistogamous (closed-flowered) species, to billions of grains in some wind-pollinated species (Stanley and Linskens 1974; Faegri and Iverson 1975; Frankel and Galun 1977). In general, wind-pollinated species (Table 2) produce the greatest amounts of pollen, followed by outcrossing species with animal pollinators (Table 3), and finally species that are facultative or obligate inbreeders (Table 4) producing the least pollen (Faegri and Iverson 1975; Cruden 1977). Studies of different varieties of crop species show genetic variation in the amount of pollen produced (Oberle and Goertzen 1952; Beri and Anand 1971; cited in Frankel and Galun

TABLE 1 Weediness of 60 Herbaceous U.S. Crop Plants[a]

Crop Name	Weedy?	Location and Severity if Weedy[b,c]
Apiaceae/Umbelliferae		
Apium graveolens, celery	No	
Daucus carota, carrot	Yes	S: 1 Af, 1 CAm, USA; C: 8, incl. Can; X: 22 other countries
Pastinaca sativa, parsnip	Yes	C: Arg, Can, Sov; X: 2 SAm, NZ, USA
Petroselinum sp. parsley	No	
Asteraceae/Compositae		
Carthamus tinctorius, safflower	Yes	X: Turkey
Cichorium endivia, endive, escarole	Yes	S: Egypt
Cynara scolymus, artichoke	No	*Note: C. cardunculus* S: Arg, Aus, USA; hybridizes freely with *C. cardunculus;* some taxonomists consider them conspecifics
Helianthus annuus, sunflower	Yes	S: Mx, USA; P: Arg
H. tuberosus, sunchoke, Jerusalem artichoke	Yes	X: USA
Lactuca sativa, lettuce	No	
Brassicaceae/Cruciferae		
Brassica rapa var. *rapa* oilseed rape/turnip	Yes	S: Can, 2 SAm, 3 CAm; P: 3 Eu, 4 SAm, 1 CAm, 1Af, NZ; C: 10 countries, HA 2 Eu, Aus Jpn 2Af; X: 16 other countries
B. napus, oilseed rape, rutabaga, swede	Yes	P: USA
B. oleracea, broccoli, Brussels sprouts, cabbage, cauliflower, collards, kale, kohlrabi	No	Its apparent progenitor, *B. maritima,* is not considered weedy either
Raphanus sativus, radish	Yes	P: Arg
Rorippa nasturtium-aquaticum (*Nasturtium officinale*), watercress	Yes	C: 3 A, 1 Eu, USA
Chenopodiaceae		
Beta vulgaris, beet, sugarbeet	Yes	S: Egypt; X: USA
Spinacia oleracea, spinach	No	
Convolvulaceae		
Ipomoea batatas, sweet potato	Yes	C: Taiwan; X: 1 A, 1 CAm
Cucurbitaceae		
Citrullus lanatus, watermelon	Yes	C: 2 Oc

TABLE 1 (continued) Weediness of 60 Herbaceous U.S. Crop Plants[a]

Crop Name	Weedy?	Location and Severity if Weedy[b,c]
Cucumis melo, melon	Yes	P: 1 SAm
Curcumis sativus, cucumber	Yes	C: Aus; X: 2 A
Cucurbita maxima, squash	No	
Curcurbita moschata, squash	No	
Cucurbita pepo, pumpkin, summer squash	Yes	X: 1 Car, 1 Oc
Euphorbiaceae		
Ricinus communis, castor, castor bean	Yes	P: 1 Af, USA; C: 1 A, 1 Oc, 2 CAm, 4 Af, HA; X: 36 other countries
Fabaceae/Leguminosae		
Arachis hypogaea, peanut	Yes	C: Taiwan
Cicer arietinum, chickpea, garbanzo	No	Unknown in wild
Glycine max, soybean	No	
Lens culinaris (L. esculenta), lentil	No	
Medicago sativa, alfalfa	Yes	P: Aus; C: 1 Eu, 2 A; X: 2 A, 1 Oc, USA
Phaseolus lunatus, lima bean	Yes	X: USA
Phaseolus vulgaris, French bean	No	
Pisum sativum, pea	No	Unknown in wild
Vicia faba, broad bean	Yes	P: 1 Af; C: 1 A
Liliaceae		
Allium ampeloprasm, leek	Yes	C:1 Eu, 1 A
Allium cepa, onion	Yes	C: 1 Eu
Allium sativum, garlic	Yes	X: CAm
Asparagus officinalis, asparagus	Yes	C: 1 Eu; X: 1 Oc
Linaceae		
Linum usitatissimum, flax	Yes	P:USA, X: 1 SAm, 2 A, Aus
Malvaceae		
Gossypium hirsutum, cotton	No	
Abelmoschus esculentus (= *Hibiscus esculentus*), okra	Yes	C: 1 Af; X: 1 Af, 2 A

TABLE 1 (continued) Weediness of 60 Herbaceous U.S. Crop Plants[a]

Crop Name	Weedy?	Location and Severity if Weedy[b,c]
Poaceae/Gramineae		
Avena sativa, oats	Yes	P: Arg
Hordeum vulgare, barley	Yes	C: USA
Oryza sativa, rice	Yes	S: 2 SAm, 2 As, 1 CAm, 1 Af, USA; P: 1 Af, 1 SAm; C: 1 Oc, 1A; X: 17 other countries; taxonomic debates put red rice (variously *O. rufipogon* and *O. sativa* var. *rufipogon*) in and out of *O. sativa;* in any event, they hybridize
Panicum miliaceum, millet	Yes	P: 1 A, USA; X: Med
Pennisetum glaucum, pearl millet	No	
Saccharum officinarum, sugarcane	Yes	C: Taiwan
Secale cereale, rye	Yes	P: USA; C: 1 SAm, 1 Eu, 1 A; X: 1A, USA; weedy annual *Secale* sp. are considered *S. cereale* by some authors (Simmonds 1976)
Sorghum bicolor, sorghum	Yes	S: USA (shattercane is a weedy race of the crop)
Triticum aestivum, bread wheat	Yes	C: USA; X: Nepal
Triticum turgidum, durum wheat	No	
Zea mays, maize, corn	Yes	C: USA
Polygonaceae		
Rheum rhaponticum, rhubarb	No	
Solanaceae		
Capsicum annuum, bell pepper, chili pepper, green pepper, paprika, red pepper, sweet pepper	Yes	X: Aus, 1 A
Capsicum annuum var. *frutescens,* chili pepper, pepper, tabasco	Yes	X: 1 A, 1 CAm, 1 Oc, 1 Af
Lycopersicon esculentum, tomato	Yes	C: 1 A; X: 1 CAm, USA; the weedy tomato is *L. esculentum* var. *ceratiforme, the* cherry tomato
Nicotiana tabacum, tobacco	Yes	C: HA
Solanum melongena, aubergine, eggplant	Yes	X: 1 A
Solanum tuberosum, potato, white potato	Yes	S: USA; C: 1 Eu

[a] Major sources: Crockett 1977; Holm et al. 1979; Schery 1972; Bridges 1992.

[b] Locations: NAm = North America, CAm = Central America, SAm = South America, A = Asia, Af = Africa, Eu = Europe, Oc = Oceania, USA = United States, Sov = former Soviet Union, Can = Canada, HA = Hawaii, Med = countries around the Mediterranean Sea, Mx = Mexico, Arg = Argentina, Aus = Australia, NZ = New Zealand, Jpn = Japan, Car = countries around the Caribbean Sea.

[c] Severity of weed problem: S = serious, P = principal, C = common, X = present and weedy, but importance uncertain (after Holm et al. 1979), S = troublesome weed of several crops and states, P = troublesome weed of a few or one crop or state, C = common weed, as listed in Bridges (1992).

1977). Superimposed on this genetic variation is ecological variation that results from differences in year-to-year conditions. Soloman (1979) reported that standard deviations are normally 80% of mean production values. There is also site-to-site variation that influences the overall vigor of the plant, and plant density effects that result in isolated plants producing more pollen than those growing in more dense stands (Faegri and Iverson 1975).

As pollen completes development in the anther, it loses water and becomes dormant (Dumas et al. 1986; Hoekstra 1992; Hoekstra et al. 1992). Successful germination and fertilization depend on controlled rehydration to break dormancy. The degree of dehydration that induces dormancy is a variable that is a predictor of pollen longevity. Most pollen is shed in a highly desiccated state with a water content of between 4 and 20% (Platt-Aloia et al. 1986; Luza and Polito 1987). Some species, such as those in the Poaceae and Cucurbitaceae, have pollen that have a much higher water content (>40%) and a higher susceptibility to death by dehydration (Kerhoas et al. 1986, 1989; Hoekstra 1992). The life span of these pollen grains is measured in hours rather than days or weeks.

A second generalization applicable to predicting pollen life span is that pollen that undergoes mitosis to form the sperm before it is released from the anther (trinucleate pollen) has a faster metabolism and shorter life span than pollen that divides to produce the sperm after germination on the stigma (binucleate pollen) (Brewbaker 1967). Trinucleate pollen is found in about 30% of angiosperms (Hoekstra and Bruinsma 1978) (see Tables 2, 3, and 4). In contrast to the grasses (trinucleate, higher water content), fruit tree pollen (binucleate, lower water content) often survives for 1 week or more, and pine and date pollen are reported to survive for as long as 1 year when stored indoors under "natural conditions" (Frankel and Galun 1977). Under true field conditions, most pollen probably remains viable for 1 d or less (Soloman 1979; Hoekstra 1992).

The functional longevity of pollen also can be limited by pollinators. For example, ants and honey bees both secrete substances that inhibit pollen germination, so that pollen that has been in contact with an ant or in the cobiculas of bees for >24 h rarely germinates (Faegri and van der Pijl 1979, Hull and Beattie 1988). Pollen

TABLE 2 Outbreeding, Wind-Pollinated Economic Plants

Scientific Name	Common Name
Beta vulgaris	Beet (3*n*)[a,c]
Ceratonia siliqua	Carob
Ricinus communis	Castor
Cocos nucifera	Coco palm (2*n*)[b]
Corylus spp.	Hazelnut (2*n*)[b]
Humulus lupulus	Hop (2*n*)[b]
Morus alba, M. nigra	Mulberry (2*n*)[b]
Zea mays	Maize (3*n*)[a,c]
Olea europaea	Olive (2*n*)[b]
Pennisetum glaucum	Pearl millet (3*n*)[a,c]
Carya illinoinesis	Pecan
Pistacia vera	Pistachio
Populus spp.	Poplar (2*n*)[b]
Secale cereale	Rye (3*n*)[a,c]
Spinacea oleracea	Spinach (3*n*)[a,c]
Saccharum robustum, S. spontaneum, S. sinense S. officinarum	Sugarcane[a]
Juglans regia	Walnut

[a] Family known to have trinucleate pollen.
[b] 2*n*, Genus or species known to have binucleate pollen.
[c] 3*n*, Genus or species known to have trinucleate pollen.

From Brewbaker 1967; Frankel and Galun 1977.

TABLE 3 Outbreeding, Animal-Pollinated Economic Plants

Scientific Name	Common Name
Allium porrum	Leek (2*n*)[a]
A. schoenoprasum	Chives (2*n*)[a]
A. cepa	Onion (2*n*)[a]
Ananas comosus	Pineapple (2*n*)[a]
Asparagus officinalis	Asparagus
Brassica oleracea	Cabbage, kale (3*n*)[b,c]
B. napus	Swedes, rapes (3*n*)[b,c]
B. rapa var. *rapa*	Turnip (3*n*)[b,c]
Carthamus tinctorius	Safflower[c]
Castanea spp.	Chestnut (2*n*)[a]
Citrullus lanatus	Watermelon
Cucumis melo	Melon (2*n*)[a]
C. sativus	Cucumber (2*n*)[a]
Cucurbita spp.	Squash (2*n*)[a]
Daucus carota	Carrot[c]
Diospyros kaki	Persimmon (2*n*)[a]
Fragaria ananassa	Strawberry (2*n*)[a]
Helianthus annuus	Sunflower (3*n*)[b,c]
H. tuberosus	Jerusalem artichoke (3*n*)[b,c]
Ipomoea batatas	Sweet potato (3*n*)[b,c]
Malus pumila	Apple (2*n*)[a]
Medicago sativa	Alfalfa (2*n*)[a]
Mentha piperita	Peppermint
Persea americana	Avocado (2*n*)[a]
Petroselium sp.	Parsley[c]
Prunus spp.	Cherry (2*n*)[a]
Prunus	Plum (2*n*)[a]
P. dulcis	Almond (2*n*)[a]
Pyrus communis	Pear
Raphanus sativus	Radish (3*n*)[b,c]
Rosa spp.	Rose (2*n*)[a]
Rubus spp.	Raspberry
Theobroma cacao	Cacao (2n)[a]

[a] (2*n*) Genus or species known to have binucleate pollen.
[b] (3*n*) Genus or species known to have trinucleate pollen.
[c] Family known to have trinucleate pollen.

From Brewbaker 1967; Frankel and Galun 1977.

TABLE 4 Normally Inbreeding Economic Plants

Scientific Name	Common Name
Abelmoschus esculentus	Okra
Apium graveolens	Celery[a]
Arachis hypogaea	Peanut
Avena spp.	Oat (3*n*)[a,c]
Capsicum annuum	Pepper (2*n*)[b]
Cicer arietinum	Chick pea
Citrus spp.	Citrus (2*n*)[b]
Glycine max	Soybean
Gossypium arboreum, G. barbadense, G. herbaceum, G. hirsutum	Cotton (2*n*)[b]

TABLE 4 (continued) Normally Inbreeding Economic Plants

Scientific Name	Common Name
Hordeum vulgare	Barley[a]
Lactuca sativa	Lettuce (3*n*)[a,c]
Lens culinaris	Lentil
Linum usitatissimum	Flax (3*n*)[a,c]
Lycopersicon esculentum	Tomato (2*n*)[b]
Nicotiana rusticua, N. tabacum	Tobacco (2*n*)[b]
Oryza sativa	Rice[a]
Panicum miliaceum	Proso millet[a]
Pastinaca sativa	Parsnip[a]
Phaseolus lunatus	Lima bean (2*n*)[b]
Pisum sativum	Pea (2*n*)[b]
Prunus persica	Peach/nectarine (2*n*)[b]
P. armeniaca	Apricot (2*n*)[b]
Sesamum orientale	Sesame (2*n*)[b]
Solanum melongena	Eggplant (2*n*)[b]
S. tuberosum	Potato (2*n*)[b]
Sorghum bicolor	Sorghum[a]
Triticosecale	Triticale[a]
Triticum spp.	Wheat (3*n*)[a,c]
Vicia faba	Broad bean (2*n*)[b]
Vitis vinifera	Grape (2*n*)[b]

[a] Family known to have tinucleate pollen.
[b] 2N, Genus or species known to have binucleate pollen.
[c] 3N, Genus or species known to have trinucleate pollen.

From Brewbaker 1967; Frankel and Galun 1977.

survival is also adversely affected by high temperature, the presence of ultraviolet radiation, and repeated cycles of wetting and drying.

In wind- (Table 2) and animal-mediated (Table 3) pollination pollen dispersal shows a leptokurtic distribution (Stanley and Linskens 1974). The average distances that viable pollen is transported by different vectors has been estimated as a few hundred meters for wasps and flies; several hundreds of meters for bees, birds, and small mammals; and, depending on wind speed and the height of the flower, from a few dozens of meters to several kilometers for airborne pollen (Stanley and Linskens 1974; Pacini 1992).

Crops vary as to whether pollination is required. Many species are raised for foliage (e.g., lettuce, cabbage) or for root parts (e.g., potatoes, carrots), so pollen production need only occur in seed producers' fields. Most cultivars are currently fertile, but it should be possible to breed for male sterility as an HRG containment mechanism.

Much more difficult in terms of gene containment are crops for which seeds or fruits are the product. These include the melons and squashes, but also tomatoes, fruit and nut trees, and the cereal grains. The size of the pollen cloud produced by wind-pollinated plants and the distances flown by common pollinators are large enough that effective removal of wild relatives that spontaeously occur within such a large pollen dispersal radius may be impractical. All other factors being equal, outcrossing HRCs producing viable pollen have a higher likelihood of producing herbicide-resistant hybrids with compatible wild relatives.

In terms of the containment of HRC pollens, this means that the mode of pollination, pollen longevity, and crop type combine to influence the probability of gene escape. For some, e.g. *Brassica* mustards and pines *(Pinus* spp.), escape of HRGs from a HRC are a virtual certainty. For others, e.g., celery or onions, this is much less probable.

3.2.2. Interfertility

Interfertility is required for hybridization, no matter how far the viable pollen disperses. Herbicide-resistant crops that are incompatible with any wild species can be disregarded as potential sources for escape of HRGs. The data on compatibility relations probably exist for most crops, but are not readily accessible. This information is critical for an analysis of and strategies for the containment of HRGs. We hope this discussion will induce researchers who know of interfertility data with wild relatives to assemble the data, evaluate it, and make it accessible to others.

A very few crop species have no close wild relatives, e.g., castor (*Ricinus communis*) (Simmonds 1976), so that their HRGs would remain contained within the HRC. A few other crops have no known compatible wild relatives, e.g., watercress (*Rorippa nasturtium-aquaticum*) and broad bean (*Vicia faba*). These too will not release herbicide resistance to wild populations.

The most common situation, however, is that the crop is a member of a genus comprised of other species, some compatible, some incompatible, and others for which interfertility is not known. Often, but certainly not always, the congeners about which compatibility with the crop are unknown consist of species that are relatively uncommom or at least not weedy. Thus, some interfertility data gaps are to be expected and should not necessarily constitute a major stumbling block to risk analysis and containment.

For some crops, the probability of transgene escape is 1.0. These crops have abundant weedy wild relatives with which they freely exchange genes, for example, sorghum and johnsongrass, rice and red rice, oats and wild oats (see below). The development and field deployment of herbicide-resistant plants of these crops is tantamount to also providing the gene to these weeds.

In general, one can expect compatible wild relatives to be other members of the same genus. However, in some notable examples crops hybridize with species in other genera. The wheat group is a dramatic example. Species of *Triticum* (wheats), *Secale* (ryes), *Hordeum* (barleys), and *Aegilops* (goat grasses) can hybridize in nature (e.g., Simmonds 1976). Similarly, *Zea* (maize) crosses with wild species of *Teosinte* (teosinte) (Doebley 1990). These are relatively stable taxonomic units with gene flow between them. In other cases evolving taxonomic understanding may be responsible for expanding and contracting genus boundaries. Groups known to us in which exchange beyond the same genus should be evaluated are *Brassica* (e.g., with *Sinapis*) and *Abelmoschus* (okra) with *Hibiscus*. Other cases may well exist or be created by future nomenclatural changes.

3.2.3. Proximity

The opportunity for hybridization is created by the proximity of the wild, compatible relative to within the distance of maximum viable pollen flow of the crop, which might be termed "reproductive sympatry." It is important to bear in mind that proximity is a dynamic rather than static variable. Crops are continually planted in new as well as old areas, and weeds colonize new areas, creating new infestations as well as spreading outward from existing infestation foci. Due to the high degree of long distance transport of people and their goods, the dispersal and naturalization of weeds to new, distant areas is now commonplace.

4. IMPACT

The mere existence of a HRW is not necessarily a problem. The HRW becomes a problem only where the herbicide to which it is resistant is applied. Under our simplifying assumption that the HRG has no other pleiotropic effects, environments where the herbicide is applied are the only places where the resistance gene is advantageous and where HRW spread would be favored.

The argument is sometimes made that the added HRGs should confer a genetic load, setting transgenic herbicide resistance or any herbicide-resistant plants at a disadvantage in nonherbicided environments, leading to their demise. Indeed, some studies have shown HRWs, e.g., triazine-resistant weeds, to be less fit outside the herbicide environment (Conard and Radosevich 1979; Warwick and Black 1981; Holt and Radosevich 1983; Jacobs et al. 1988). However, other studies indicate that this is not necessarily the case with resistance to other groups of herbicides, e.g., the sulfonylureas (Alocer-Ruthling et al. 1992; Thompson et al. 1994).

Crawley et al. (1993) point out the critical importance of genotype by environmental interactions in determining fitness, so one should be very careful about the genetic load assumption. In particular, many plants appear to successfully carry tremendous loads of nontranscribed DNA that are replicated each generation, or, in the case of polyploids, multiple apparently functional copies of every locus (e.g., Bennett 1987). In a large genome the additional DNA and the production of an additional protein may not constitute a significant cost. Likewise, plant species, especially weeds, are found in a diversity of environments, from those rich in water and nutrients to very resource-poor environments. While lineages in poor conditions may be under strong selection for efficiency, those in rich environments may show considerable inertia, carrying irrelevant genes for long periods of time. Herbicide resistance transgenes certainly constitute a potential cost to the plant, potentially reducing its fitness. Whether the cost is sufficient to cause selection against plants carrying nonadvantageous transgenes in nature is probably a function of complex local conditions and should not necessarily be counted on in weed control strategies.

4.1. BIOLOGICAL IMPACT

Biological problems may arise from the movement of HRGs into wild populations. Where the herbicide is applied, HRGs are expected to be strongly selected for, and the population should evolve to contain primarily or entirely plants carrying the HRGs. This strong selection should have a number of side effects. Alleles linked to the advantageous alleles should be carried to high frequency as well, via hitchhiking effects (e.g., Futuyma 1986). Total genetic diversity should decline, at least initially, because linkage disequilibrium should cause replacement of the existing variation with alleles associated with herbicide resistance alleles, not as diverse or certainly different from those of the wild population.

Where the herbicide is not applied, HRGs should confer no fitness advantage. If they are neutral or nearly neutral in the absence of herbicide, they can be expected to simply add to the genetic variation in the population. Sheer addition of genetic variation may be considered desirable (e.g., Templeton 1986) because it gives the population more evolutionary flexibility. If, however, massive numbers of genes from the crop come into an adjacent wild population, they could swamp the native population, so that every seed is a hybrid, greatly reducing effective variation and compromising the ability of the population to respond evolutionarily.

While theoretical genetic models for the fate of particular new neutral or nearly neutral alleles show them being rather rapidly lost due to chance events (e.g., Crow and Kimura 1970), this should not be applied to HRGs, except as they occur by random mutation. If large numbers of hybrids that carry the HRG are produced, then one begins with a relatively high frequency of the herbicide resistance alleles in the wild population and, in the absence of selection against the herbicide resistance allele, it should drift in the wild population for a very long time (Figure 1). Functionally, one can assume that once introduced into wild population from a large planting of an HRC, the HRGs will be permanently present in the population of wild relatives at the site. If the wild relatives are weeds, the herbicide can never again be applied without the rapid production of HRWs.

If the HRG is slightly deleterious in the absence of herbicide, for whatever reason, then the frequency will drop steadily each generation, eventually reaching zero (Figure 2). However, unless the gene is strongly deleterious, many generations are required.

An alternative outcome is that so many herbicide resistance alleles of lower fitness are added to the wild population that they lower the fitness of the wild population, for example, by reducing the mean number of seeds per plant, enough to endanger the survival of the wild population. If the

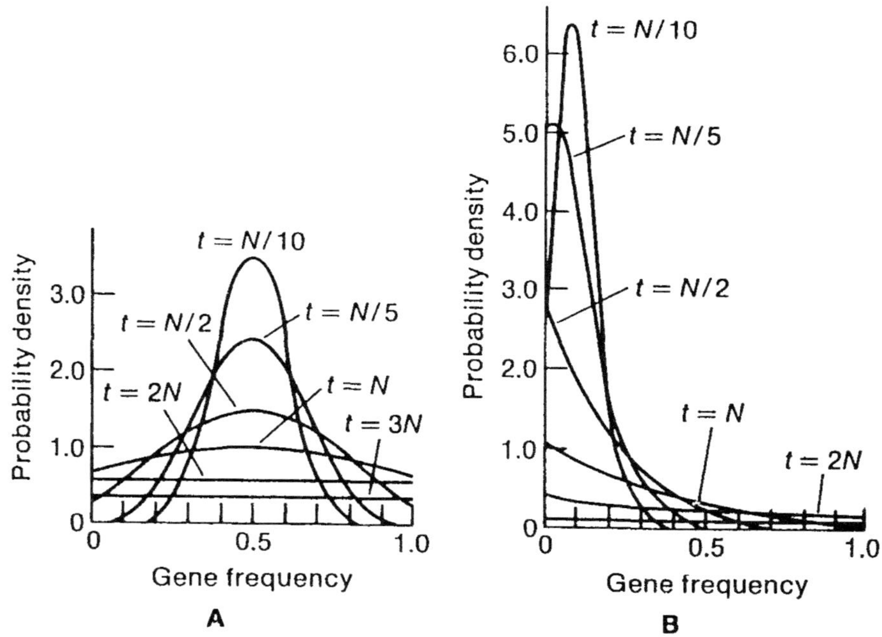

FIGURE 1 Removal of herbicide resistance genes will be very slow in the absence of selection against them. (A) Probability distribution of allele frequencies when the initial allele frequency is 0.5 (as might happen in the hybrid seed of wild relatives next to a field). t = time in generations, N is the effective population size. (B) Same, with initial allele frequency of 0.1. Proportion of populations fixed at 0 or 1 is not shown. (After Futuyma 1986.)

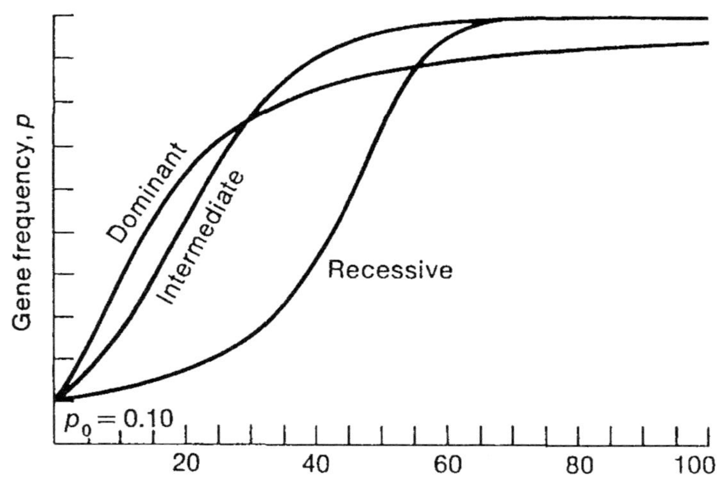

FIGURE 2 Dynamics of allele frequency change. Elimination of slightly unfit herbicide resistance genes from wild populations will take many generations. Drawn is the rise of a wild-type allele that is more fit than the transgene. Because transgenes are expected to be dominant, the right-hand curve (wild-type allele recessive) is the most applicable curve. Fitnesses used were 1, 1, 0.8 for wild-type dominant case; 1, 0.9, 0.8 for co-dominance; and 1, 0.8, 0.8 for wild-type recessive. (After Futuyma and Young 1986.)

wild population is a noxious weed, this is a beneficial side effect of using the HRC. If the wild population represents an important genetic resource, e.g., of wild germplasm for breeding or a rare native species, then this poses a potentially serious risk and action should be taken to prevent it.

Thus, genetic impact depends on how many genes are received by the wild population, whether those are advantageous, neutral, or disadvantageous, and how we value the population of wild relatives.

4.2. ECONOMIC IMPACT

The economic consequences of HRWs are potentially immense. The chief economic impact of weeds is in agriculture, where they reduce crop yield and quality by competing for limited resources. Agricultural weeds also reduce harvest efficiency and have other negative economic impact. Agricultural weed problems are universal and continuous. The agricultural losses due to weeds, including control costs, total about $15.2 billion annually in the U.S. (Chandler 1991). While machine and hand cultivation are used extensively to eliminate weeds, herbicides are the primary weapon, constituting about 62% of all pesticides applied in the U.S. (Szmedra 1991).

One challenge in the development of herbicides for weed control has been to find herbicides that do not also damage the crop. Reducing a complex story to its simplest parts, with genetic engineering, a HRC can be used so that the herbicide of choice can be applied to control the weeds, in some cases literally saving the crop. The emergence of HRWs threatens this technology specifically, and the utility of herbicides generally. Transgenic HRWs have not evolved, but naturally occurring HRWs have repeatedly threatened canola production in western Canada (Jensen and Bandeen 1979; Jana and Naylor 1982; LeBaron 1984; Beversdorf 1987), and are on the verge of threatening winter wheat production in the northwestern U.S. (Mallory-Smith et al. 1990; Stallings et al. 1994; Thompson et al. 1994).

4.3. CROP/WEED SYSTEMS OF CONCERN

The likelihood of escape of herbicide resistance into weed populations and consequent increased weed control problems varies greatly among different crops. Specifically, some species apparently do not reproduce sexually, have no relatives in North America, and are not compatible with relatives anywhere (e.g., rhubarb), therefore, weed problems from escape of herbicide resistance can be essentially disregarded. At the other extreme are crops that outcross during crop production, have seriously weedy close relatives widely distributed in North America, and hybridize freely with those relatives (e.g., sorghum). A few are native species (e.g., sunflower), whose relatives are not only weedy and compatible, but represent an indigeous gene pool many would prefer not to contaminate with introduced genes.

Table 5 is an attempt to summarize readily available data on the distribution of the weedy wild relatives of 60 important crop plants. While it contains a number of data gaps, it outlines the nature of the missing data and the level of our ignorance. Of the 60 crops, only 11 do not have congeners that are considered weeds somewhere in the world. Fifteen more do not have weedy wild congeners reported for the U.S. That leaves the majority, 34 (57%), with weedy wild relatives reported for the U.S. Hybridization with wild relatives is clearly known for only 15 (25%) of the crops.

From this analysis we can see (1) that pertinent reproductive and distributional data are apparently lacking for many crops; (2) that large geographic areas, such as countries, are very unsatisfactory ways to look for proximity of a weed; and (3) that information about abundance and distribution is constantly changing and needs continual updating. It shows, most importantly, the need to generate a database on the pollination system, interfertility relationships, and distribution of wild relatives of cultivated plants in order that informed decisions about the risks of escape of genes such as those for herbicide resistance can be made. The tables in this chapter are intended as a beginning in this effort.

TABLE 5 Distribution of Weedy Wild Relatives of 60 Herbaceous Crop Plants and Potential for Hybridization between Crop and Wild Relatives

Species	No. of Congeneric Species Reported as Weeds	If So, How Weedy, Where?[a,b]	Compatibility of Crop and Relatives?
Apiaceae			
Apium graveolens	4	*A. leptophyllum* S: HA; P: 1 SAm, 1 Af;C: 5; X: 6 more) 2 spp. C in 3 countries, 1 X	Hybrids?
Daucus carota	9	6 C in 8 countries	Hybrids?
Pastinaca sativa	0		Does not hybridize with wild congeners
Petroselinum sativum	1	*P. segetum* C: Eu	Does not hybridize with wild congeners
Asteraceae			
Carthamus tinctorius	8	*C. lanatus* S: Aus, C 2 countries, X 5 *C. oxyacantha* S: 2A, X 2 countries; 6 more C–X weeds	Not compatible with weedy wild relatives, will hybridize with wild relatives that are not considered weedy
Cichorium endivia	3	*C. intybus* S: 1 SAm, 3 Eu 1 A; P: 2, Eu 1 A, 2 Med; C: 9 incl. USA; X 12 *C. pumilum* P: Isr; C 3 spp.	Hybrids?
Cynara scolymus	3	*C. cardunculus* S: Arg, Aus; P: 1 Eu, X 4 2 C	Free hybridization between *C. scolymus* and *C. cardunculus*
Helianthus annuus	6	S: *H. ciliaris* USA; X: *H. californicus, ciliaris, grosseserratus, maximiliani, petiolaris,* and *tuberosus*	1 hybridizes freely with *H. annuus (H. petiolaris)*
H. tuberosus	6	Only *H. annuus* (c.f.) S: Mex; P: Arg, USA	Does not hybridize with any weedy congeners
Lactuca sativa	19	*L. capensis* P:1 Af; *L. floridana* C: USA; *L. scariola* P: 1 Med, *L. serriola* P: USA; C: 1 Af, Aus, Can, Med 1 SAm; *L. taraxacifolia* P: 1 Af	Hybridizes with some wild relatives (*L. serriola, L. virosa, L. saligna;* C: 1 A, Aus)
Brassicaceae			
Brassica campestris	13	*B. juncea* P: Can; C: Arg, Aus; X: 1 Oc, 1 CAm, USA; *B. kaber* P: Can; C: 1 Med, USA; *B. rapa* P: SAm; C: 1 Af, 2 Eu, USA; *B. tournefortii* P: Aus; C 1 Med; X 1 Oc; *B. nigra* (=*Sinapis nigra*) S USA, C: 10 countries, X: 7 countries	*B. campestris* hybridizes with *B. nigra* and *B. napus*
B. napus	14	See *B. campestris*	Hybridizes readily with *B. campestris* (McNaughton 1978)
B. oleracea	14	See *B. campestris*	Hybridizes with European wild congeners that are not considered weedy
Rorippa nasturtium-aquatica	1	*Rorippa* sp. P: A; C: 6 in; 8 X (incl. 2 in USA) 2 *Nasturtium* spp. X: 1 country each	Watercress is incompatible with wild relatives
Raphanus sativus	2	*R. raphanistrum* S: A, Af, Eu, Oc, SAm, USA; *R. microcarpus* C: Eu	Hybridizes with *R. raphanistrum, R. maritimus,* and *R. landra*

TABLE 5 (continued) Distribution of Weedy Wild Relatives of 60 Herbaceous Crop Plants and Potential for Hybridization between Crop and Wild Relatives

Species	No. of Congeneric Species Reported as Weeds	If So, How Weedy, Where?[a,b]	Compatibility of Crop and Relatives?
Chenopodiaceae			
Beta vulgaris	0	No weedy congeners; incompatible with wild congeners	
Spinacia oleracea	0	No weedy congeners; incompatible with wild congeners	
Convolvulaceae			
Ipomoea batatas	57	*I. cordofana* P: Af; *I. hederacea* S: USA; *I. lacunosa* S: USA; *I. pandurata* S: USA; *I. purpurea* S: USA; *I. tiliacea* P: SAm; *I. triloba* P: Oc; S: CAm, HA, USA; *I. wrightii* P: USA	*I. batatas* is not very fertile, but is compatible with *I. triloba* and others of the *I. batatas* complex
Cucurbitaceae			
Citrullus lanatus	1	C: Aus	Incompatible with weedy relatives
Cucumis melo	8	*C. myriocarpus* P: Aus; *C. anguria* P: USA	Hybridizes with wild relatives that are not considered weeds
C. sativus	8	See *Cucumis melo*; also *C. melo* P: SAm	Hybridizes with *C. hardwickii* in Asia, which is not considered very weedy (Simmonds 1976)
Cucurbita maxima	4	*C. texana* P: USA; *C. pepo* X: USA	Hybridizes with nonweedy congeners in SAm
C. moschata	4	*C. texana* P: USA; *C. pepo* X: USA	Hybridizes with nonweedy congeners in SAm
C. pepo	3	See *C. maxima*	Hybridizes with nonweedy congeners in SAm
Euphorbiaceae			
Ricinus communis	0	No other species in genus	No compatible relatives
Fabaceae			
Arachis hypogaea	0	No weedy congeners	Hybridizes with nonweedy congeners, including *A. monticola*
Cicer arietinum	0	No weedy congeners	No hybrids with other members of this genus of 39 southern Asian species
Glycine max	2	*G. soya* C: Jpn, another sp. X: Af	Hybridizes with *G. soya* to produce weedy *G. gracilis*
Lens culinaris	0	No *Lens* sp. are weedy	Interfertile with *L. orientalis*, which is not considered weedy (Hancock 1992)
Medicago sativa	20	*M. lupulina* P: A; C: USA; *M. polymorpha* C: USA; 10 other spp. C (total of 17 countries)	Hybridizes with *M. sativa* var. *falcata*, to make "*M. varia*" populations
Phaseolus lunatus	7	*P. lathyroides* P: Aus; C: HA, A; *P. trilobus* P: A	Hybrids?

TABLE 5 (continued) Distribution of Weedy Wild Relatives of 60 Herbaceous Crop Plants and Potential for Hybridization between Crop and Wild Relatives

Species	No. of Congeneric Species Reported as Weeds	If So, How Weedy, Where?[a,b]	Compatibility of Crop and Relatives?
P. vulgaris	7	See *P. lunatus*	Hybrids occur, although in general there is little geographical overlap between wild and cultivated species; the hybrids are hard to classify, leading to taxonomic problems defining the species.
Pisum sativum	1	*P. elatius* C: NAf, Por	*P. arvense* and *P. elatius* (the latter is weedy)
Vicia faba	25	*V. sativa* S: A, Eu USA; P: A, Oc; C, X: 27 countries; *V. cracca* P: 1 Eu, USA; *V. hirsuta* P: A; *V. narbonensis* P: NAf; *V. villosa* P: USA; 13 more spp. C somewhere in the world	Does not hybridize with wild congeners (Hancock 1992)
Liliaceae			
Allium ampeloprasm	13	*A. macrostemon* P: A; *A. nigrum* P: NAf; *A. vineale* P: Aus, Tur, USA; *A. canadense* P: USA; 4 spp. C somewhere	
A. cepa	13	See *C. ampeloprasm*	
A. sativum	13	See *C. ampeloprasm*	
Asparagus officinalis	1	1 *A. lucidus* C: A	Long history in cultivation, wild relatives in southern Asia
Linaceae			
Linum usitatissimum	5	*L. peyroni* C: NAf	Forms fertile hybrids with *L. africanum* and *L. angustifolium*, and 5 others, of which *L. angustifolium* is considered a weed
Malvaceae			
Gossypium hirsutum	0	No *Gossypium* spp. are weeds [the one report (*G. tomentosum*, Holt et al. 1977) is for the rare and endangered Hawaiian endemic cotton so this appears to be an error)	Hybridizes with wild congeners that are not weedy
Abelmoschus esculentus	0	No *Abelmoschus* spp. are weedy	Okra is incompatible with its wild relatives
Poaceae			
Avena sativa	10	*A. barbata* P: Eu X: USA; *A. byzantina* P: Arg, Af; *A. fatua* S: Arg, Aus, Can, Eu, SAf, USA; P: CAm, SAm; C, X: 34 more countries; *A. sterilis* spp. *ludoviciona* S: Aus, Eu; P: Af, A; *A. sterilis* S: Aus, Med; *A. strigosa* S: Af; 2 others C in 3 countries	Hybridizes with very weedy *A. fatua* despite ploidy differences (Thomas and Jones 1976)

TABLE 5 (continued) Distribution of Weedy Wild Relatives of 60 Herbaceous Crop Plants and Potential for Hybridization between Crop and Wild Relatives

Species	No. of Congeneric Species Reported as Weeds	If So, How Weedy, Where?[a,b]	Compatibility of Crop and Relatives?
Hordeum vulgare	13	*H. jubatum* S: AL; C: Can, USA; *H. murinum* S: Aus, Oc; P: Med; C, X 24 countriesu, including USA (X); *H. leporinum* P: Aus; X: Arg, USA; 7 others are C in 8 countries (*H. pusillum* USA)	Chiefly hybridizes with *H. spontaneum* which is sometimes considered part of *H. vulgare*
Oryza sativa	6	*O. punctata* S: Af; *O. barthii* P: Af; *O. perennis* P: 1 Af; 1 other is C in 1 country	Taxonomic debates put red rice (variously *O. rufipogon* and *O. sativa* var. *rufipogon*) in and out of the same species as cultivated rice.
Panicum miliaceum	59	*P. fasciculatum* S: CAm, SAm; P: SAm; C: As; X: USA; *P. maximum* S: Aus, SAm, CAm, Af, HA; P: 1 Car, SAm, Af, CAm; C: 3 countries; X: 22, incl. USA; *P. repens* S: Af, Eu, HA, Oc, As; P: 2 Oc Af, C: 5 countries, incl. USA	Hybrids?
Pennisetum glaucum	13	*P. japonicum* P: Jpn; *P. macrourum* P: Aus; *P. pedicellatum* S: Af, As; P: Aus; *P. polystachyon* S: As; P: As; *P. purpureum* S: SAm, Af; C: HA, Oc, CAm; X: 14 countries, incl. USA	Hybrids?
Saccharum officinarum	4	*S. behghalense* P: As; *S. spontaneum* S: As, Oc; P: As, CAm; X: 20 countries; 2 spp. X in 2 countries	All *Saccharum* sp. are interfertile (Hancock 1992)
Secale cereale	1	*S. montanum* X: Tur	Weedy annual *Secale* sp. are close enough to rye to be considered *S. cereale* by some authors (Simmonds 1976)
Sorghum bicolor	11	*S. bicolor* ssp. *arundinacuum* P: SAm; X: Af; *S. halepense* P: Arg, Aus, SAm, CAm, Oc, Eu, HA, As, Med, USA; S: Oc, SAm, Af, Eu, CAm; C: 18 other countries; *S. bicolor* spp. *arundinacuum* S: Af; P: Af, USA; C: 3 other countries; *S. vulgare*, *S. bicolor* ssp. *bicolor* P: As, SAm; C: 2 countries, 2 spp. C in 3 countries 5 spp. X (2 in USA)	*S. bicolor* (2x) hybridizes freely with the serious weed *S. halepense* (4x) (Ellstrand and Hoffman 1990)
Triticum aestivum	1	*T. ramosum* X: Eu, As	Hybridizes infrequently with *Aegilops (Triticum) cylindrica*, jointed goatgrass (a P weed in incl. USA)
T. turgidum	2	*T. aestivum* X: As; *T. ramosum* X: Eu, As	Hybrids form with *Secale*, *Hordeum*
Zea mays	0	No members of *Zea* are weeds	Hybridizes with teosinte (*Zea* spp.), but teosintes not considered weedy

TABLE 5 (continued) Distribution of Weedy Wild Relatives of 60 Herbaceous Crop Plants and Potential for Hybridization between Crop and Wild Relatives

Species	No. of Congeneric Species Reported as Weeds	If So, How Weedy, Where?[a,b]	Compatibility of Crop and Relatives?
Polygonaceae			
Rheum rhaponiticum	0	No weeds in *Rheum*	Propagated asexually by dividing clumps, seeds infertile; does not hybridize with congeners
Solanaceae			
Capsicum annuum	2	*C. frutescens* X: As, CAm, Oc, Med; *C. baccatum* X: Med	Hybridizes with congeners that are not considered weeds
C. annum var. frutescens	2	See *C. annuum*	Hybridizes with congeners that are not considered weeds
Lycopersicon esculentum	2	2 spp. X: both in Per	Yes?
Nicotiana tabacum	7	*N. glauca* C: Aus, HA, Af; X: SAm, Oc, USA; *N. longiflora* C: Arg; *N. suaveolus* C: Aus; 4 spp. X in 1 country each (*N. trigonophylla* USA)	Cultivated tobacco is the hybrid of two semiweeds of SAm
Solanum melongena	66	*S. alatum* P: Med; *S. americanum* P: USA; *S. auriculatum* P: Aus; *S. carolinense* P: USA; C: Can; *S. dubium* P: Af; *S. dulcamara* C: USA; *. elaeagnifolium* S: Af, USA; P: Aus; C: Arg, Af; X: SAm, CAm, USA; *S. glaucophyllum* P: Arg; *S. gracilius* P: SAm; *S. grossedentatum* P: Af; *S. hystrix* P: Aus; *S. nigrum* S: A, Aus, HA, Med, Oc, USA; P: Af, Can, CAm, Eu, SAm; C 14 other countries, incl. USA; X: 32 others); *S. nodiflorum* S: HA; C: Af; *S. ptycanthum* P: USA; *S. rostratum* S: Mx, USA; C: 1 Af, Aus; *S. saccaroides* P: USA; *S. torvum* S: Oc; P: A, Af, Aus, Oc; C: 3 other countries, incl. USA; *S. triflorum* P: Aus, USA; *S. tuberosum* P: USA; *S. villosum* S: Med; C: 1 Af; X: USA; *S. xanthocarpum* P: 1 A, As; 23 spp. C weeds somewhere	Hybridize?
S. tuberosum	66	See *S. melongena*	Hybridizes with relatives in the subsection *Potato* which are not considered weeds (Hancock 1992)

[a] Locations: NAm = North America, CAm = Central America, SAm = South America, A = Asia, Af = Africa, Eu = Europe, Oc = Oceania, USA = United States, Sov = former Soviet Union, AL = Alaska, Per = Peru, Can = Canada, HA = Hawaii, Med = countries around the Mediterranean Sea, Mx = Mexico, Arg = Argentina, Aus = Australia, NZ = New Zealand, Jpn = Japan, Car = countries around the Caribbean Ocean, Isr = Israel, NAf + North Africa, Por = Portugal, Tur = Turkey, SAf = South Africa.

[b] Severity of weed problem: S = serious, P = principal, C = common, X = present and weedy, but importance uncertain (after Holm et al. 1979), S = troublesome weed of several crops and states, P = troublesome weed of a few or one crop or state, C = common weed, as listed in Bridges (1992).

[c] Major sources: Crockett 1977; Holm et al. 1979; Schery 1972; Bridges 1992.

We briefly discuss below examples of crops for which problems with HRG escape are probable in the U.S. These highlight a challenge as well as a warning; significant ingenuity will be required to provide crops with HRGs or other potentially beneficial genes that do not exacerbate weediness in weedy, compatible wild relatives.

4.3.1. Oats

Oats (*Avena sativa*) are valued for the relatively high protein and fat content of the seeds, and are chiefly a crop of north temperate areas such as North America, northern Europe, and Russia (Langer and Hill 1982). Wild oat, *Avena fatua*, is grown as a crop in Scotland, but is a serious agricultural weed elsewhere in the world (Thomas and Jones 1976; Jana and Naylor 1982). Wild oat is weedy in grainfields, rangelands, and other environments in the U.S. and southern Canada (Reed and Hughes 1970; Bridges 1992). Although *A. fatua* is hexaploid and *A. sativa* is diploid, they freely hybridize (Thomas and Jones 1976).

4.3.2. Canola

Canola (*Brassica napus* and *B. rapa;* rapeseed) is an important oilseed crop in North America and elsewhere. The crop is sometimes itself weedy as an escape from cultivation (Rollins 1981). The canola industry of the northwestern U.S. and southwestern Canada has already experienced serious problems with spontaneous development of herbicide-resistant wild mustards (*Brassica* spp.) interfering with crop production (Beversdorf 1987; Chapter 21). This experience perhaps most clearly emphasizes the potential of new HRWs to cause problems. *B. napus* and *B. rapa* are themselves weedy, and along with *B. nigra* are widespead weeds of orchards and cultivated fields, and along ditchbanks and roadsides in North America (Reed and Hughes 1970; Rollins 1981; Whitson et al. 1992; Hickman 1993). Canola outcrosses in order to produce seed. Pollination has been well studied, and it is clear that gene flow from canola to wild *Brassica* spp. is "virtually inevitable" (e.g., papers in McCammon and Dwyer 1990).

4.3.3. Artichoke

Artichoke (*Cynara scolymus*) is an important specialty crop in California. Artichoke thistle or cardoon (*C. cardunculus*), is an aggressive colonizer of grasslands and a difficult to control weed of northern San Francisco Bay area counties and southern coastal California. The northern coastal infested region is close to the primary artichoke growing area immediately to the south. The crop and the weed are highly interfertile, and are sometimes considered to comprise a species complex (Thomsen et al. 1986).

4.3.4. Sunflower

Sunflower (*Helianthus annuus*) is grown as an oilseed crop and edible seed crop in the north-central U.S., and much of the grower seed production is done in California. Sunflower is a cultivated race of the native North American annual, *H. annuus*. Wild sunflowers, including *H. annuus* itself, are widespread weeds throughout the U.S. in grainfields and other cultivated fields, roadsides, and waste areas (Reed and Hughes 1970). Other native annual, weedy sunflowers include *H. bolanderi* and *H. petiolaris* (Robbins et al. 1970; Bridges 1992; Whitson et al. 1992). Gene flow between cultivated and wild annual sunflowers is probably common (Heiser et al. 1969; Heiser 1976). Wider crosses with perennial congeners have been made, for example, with *H. tuberosus*, but these are typically not as fertile (Heiser 1976).

4.3.5. Lettuce

Lettuce (*Lactuca sativa*) is a major leafy crop in North America. *L. sativa* is a diploid ($2n = 18$), primarily self-fertilized herbaceous annual. It is probably derived from prickly lettuce, *L. serriola* (Schery 1972). Prickly lettuce is a major weed throughout much of the U.S. in irrigated fields, orchards, and along roadsides (Reed and Hughes 1970; Bridges 1992; Whitson et al. 1992). Lettuce and prickly lettuce have the same number of chromosomes ($2n = 18$) and are interfertile (Mallory-Smith et al. 1993). The spontaneous evolution of resistance to sulfonylurea herbicides is a growing concern in the northwestern U.S. (Mallory-Smith et al. 1990; Alocer-Ruthling et al. 1992). Sulfonylurea resistance has been successfully transferred by hand pollination from resistant prickly lettuce to Bibb lettuce (Mallory-Smith et al. 1990, 1993).

4.3.6. Rice

Rice (*Oryza sativa*), an aquatic grass, is a major grain crop grown in flooded fields in California and the south-central states. Red rice is a diiferent, weedy biotype of rice, distinguished by a pigmented pericarp and other morphological and physiological differences (Langevin et al. 1990). Red rice is a serious weed of rice fields, irrigation/drainage canals, and other wet areas in the south-central U.S. (Smith et al. 1977). Measures of hybridization rates between sympatric populations of rice and red rice found that hybrid seed set in red rice varied from 1 to 52% in sampled fields of six rice cultivars (Langevin et al. 1990). Langevin et al. (1990) also report that the hybrids generally were taller and had larger flag leaves than either cultivated rice or red rice, and the hybrids produced more tillers than red rice.

4.3.7. Radish

Radish (*Raphanus sativus*) is a root crop widely grown in North America. Wild radish is currently considered conspecific with the crop radish, but is undoubtedly a different, interfertile biotype. Wild radish generally lacks the well-developed fleshy taproot of cultivated radish. Wild radish is a widespread weed in California, infesting cereal grains (where it is particularly troublesome) and other cultivated crops, roadsides, and waste areas (Robbins et al. 1970; Bridges 1992; Whitson et al. 1992). In a study of radish–wild radish hybridization Klinger et al. (1991) documented gene flow from radish to surrounding experimental plantings of wild radish, and detected gene flow to the most distant plots (1000 m) of wild radish. Klinger et al. (1992) found that hybridization rates declined with increasing linear distance between radish and wild radish, and also that there was a complex effect of recipient (wild radish) population size. In a field plot experiment, the hybrids had greater fruit and seed production and equalled wild radish in the other reproductive characters measured (Klinger and Ellstrand 1994).

4.3.8. Sorghum

Sorghum (*Sorghum bicolor*) is a diploid ($2n = 20$) annual grown for grain (chiefly for animal food), silage, and syrup in the U.S. (Schery 1972). Sorghum hybridizes freely with johnsongrass, *S. halepense*, a rhizomatous, perennial tetraploid ($2n = 40$) (Schery 1972; Baker 1974). Johnsongrass was intentionally introduced to North America as a forage grass (Foy et al. 1983), but has become a troublesome weed of cultivated fields, ditch banks, and waste areas throughout the warmer parts of the U.S. (Reed and Hughes 1970; Bridges 1992). It it listed as the sixth worst weed in the world (Holm et al. 1977).

Other crop–weed systems also may be of concern. Wheat, for example, as grown in the U.S., is primarily the hexaploid *Triticum aestivum*. There are no weedy species of *Triticum* in the U.S., and in general no species of *Triticum* has naturalized at all. However, the wheat group hybridizes freely across generic lines, so that in fact *Triticum* is compatible with *Aegilops, Hordeum,* and *Secale*. This means that HRGs could go to other crops, to wild *Hordeum* or *Secale* spp., at least in principle, because there are some weeds in these genera. Furthermore, *Aegilops cylindracea*, jointed goat grass, is a serious weed of wheat fields and partially compatible with *T. aestivum*. Not many hybrids form, and although they probably do not send many genes back to the jointed goat-grass populations (Donald and Ogg 1991), exchange with weed populations is definitely possible.

Furthermore, in the high plains of Colorado, Nebraska, the Dakotas, and western Kansas, where irrigation is impossible, wheat is grown on a fallow–wheat–fallow–wheat rotation, in which one of the worst weed problems is carryover wheat from previous years. Herbicide tolerance may be a blessing in the first year and a curse in subsequent years.

5. MANAGEMENT

Despite the great potential and increasing importance of other weed control options (Turner et al. 1992) and unwanted environmental side effects of some herbicides, herbicides constitute a very important means of weed control. The escape of herbicide resistance genes to wild, weedy plants could cause more severe weed problems, and presents a very real threat to the efficacy of herbicides as a weed control option. Therefore, management strategies that prevent, or reduce the likelihood and frequency of HRG escape through containment methods are advisable, as are mitigation plans in the event of HRG escape to wild plants.

5.1. PREVENTION/CONTAINMENT

A policy of avoidance of engineering herbicide resistance into high-risk crop–weed systems is an absolutely effective means of avoiding the transgene escape problem altogether. For now, we recommend this "abstinence" policy for the aforementioned eight crops (oats, canola, artichoke, sunflower, lettuce, rice, radish, sorghum) of special concern in the U.S. We also urge researchers to develop compatibility barriers between these crops and their presently compatible weedy relatives, so as to be able to safely take advantage of transgenic advances in the future (see below).

Containment includes the use of various genetic, ecological, and cultural methods to reduce the likelihood of HRG escape. They vary in their probable effectiveness, and none are as absolutely effective as the prevention method above. All are unproven for this specific application.

5.1.1. Genetic

Genetic containment methods include "suicide genes," interfertility barriers, and male sterility. The use of suicide genes, genes whose expression is autocidal upon escape (hybridization), amounts to an ultimate, although untried, transgene containment method in plants. The event triggering the autocidal expression upon HRG escape could be some part of the cytological process of hybridization/recombination itself, or as suggested by Ellstrand and Hoffman (1990), an inevitable part of the weed's (but not the crop's) environment. They suggest, for example, an autocidal trait expressed in seeds and triggered by exposure to cold temperatures. This could be effective for temperate crops. Crop plants could be modified to reduce their interfertility with wild relative by various methods, for example, by producing one or more unpaired chromosomes (aneuplody) in the genome, which would prevent successful hybridization (Keeler and Turner 1991). Where the crop product is not a seed or fruit requiring fertilization, male-sterile cultivars could be developed to greatly reduce the likelihood of HRG escape (Ellstrand and Hoffman 1990; Keeler and Turner 1991).

5.1.2. Ecological

The most effective ecological containment method is spatial isolation, where HRCs are isolated by distance from compatible wild relatives. For soybeans *Glycine max*, the wild relatives are currently confined to Asia. Soybeans grown in the United States are well isolated from these wild relatives. Transoceanic isolation, of course, is a drastic example of containment by spatial isolation. On a local scale, spatial isolation will probably be effective only for self-fertilizing crops, and should not be relied upon for outcrossing crops (Klinger et al. 1992). An isolation distance of 500 m should suffice for most selfers (George 1985; Ellstand and Hoffman 1990). This method is dependent on an accurate knowledge of pollen dispersal, wild plant distribution at the local scale, and interfertility relationships. The success of this method also can be affected by such complex factors as the behavior of biotic pollinators, wind movement, and the nature of intervening vegetation. One also should be mindful of the dynamic nature over time of the spatial distributions of crop plantings and wild plants, especially for weeds with excellent colonizing abilities. Klinger et al. (1992) point out the near impossibility of weed eradication even on a local scale.

All other ecological containment strategies are even less reliable than spatial isolation. For example, disjunct flowering phenologies can constitute another ecological containment method, but its effectiveness is reduced by phenological outliers and unusual seasons.

5.1.3. Cultural

Simple cultural practices can sometimes contribute to HRG containment (Ellstrand and Hoffman 1990). Where the crop product is vegetative, complete harvesting of the crop before flowering will preclude pollen production. The planting of nontransgenic plants of the crop surrounding the transgenic crop may serve to capture most of the otherwise outwardly bound transgenic pollen, although the complexities of pollination render this a method that is not necessarily dependable by itself for containing the transgenic pollen of outcrossing crops.

Wrubel et al. (1992) and Miller and Gunary (1993) are critical of the data available assessing the risk of escape of transgenes. Production of large amounts of inadequate science makes it hard to validly evaluate the lessons learned from our heavily regulated system of testing transgenics in the environment. We urge that both basic and applied scientists with environmental training address the problems of risk asssessment as critically as possible in order to set environmental release on a sound scientific basis. There will be more serious loss of public confidence if, having claimed we tested for safety, we have a problem, than if no tests were made. Regulatory agencies can assist in this process by being quite explicit about how common an event must be before it must be prevented and by approving mitigation plans rather than requiring prevention of rare events.

5.2. MITIGATION

Mitigative methods impede the survival and spread of transgenic HRWs. The types of crop and weed control (herbicide rotations and mixtures and mechanical cultivation) rotation strategies discussed by Gessel and Segel (1990) and Saari et al. (1994) to reduce the likelihood of spontaneous evolution of HRWs also should be effective in reducing the likelihood of survival and spread of transgenic HRWs. If utilized routinely as a management strategy, the effect would be to create a general landscape unfavorable to HRWs, whether transgenic or nontransgenic in origin. This is an important reason to avoid engineering multiple herbicide resistance into crops. In terms of mitigating the effects of a specific HRG escape event, speed of detection is obviously important. Readily detectable marker genes linked to HRGs could facilitate the speedy detection and mitigative response to HRG escape to wild plants.

REFERENCES

Alcocer-Ruthling, M., Thill, D.C., and Shafi, B. 1992. Differential competitiveness of sulfonylurea resistant and susceptible prickly lettuce (*Lactuca serriola*). *Weed Technol.* 6: 303.

Alcocer-Ruthling, M., Thill, D.C., and Mallory-Smith, C. 1992. Monitoring the occurrence of sulfonylurea-resistant prickly lettuce (*Lactuca serriola*). *Weed Technol.* 6: 437.

Baker, H.G. 1974. The evolution of weeds. *Annu. Rev. Ecol. Systematics* 5: 1.

Bennett, M.D. 1987. Variation in genomic form in plans and its ecological implications. *New Phytol.* 106(Suppl.): 177.

Beversdorf, W.D. 1987. Classical approaches to the development of herbicide tolerance in crop cultivars. In *Biotechnology in Agriculture and Chemistry*. Ed. by LeBaron, H.M., Mumma, R.O., Honeycutt, R.C., and Duesing, J.H. Amerrican Chemical Society, Washington, D.C., chap. 8.

Brewbaker, J.L. 1967. The distribution and phylogenetic significance of binucleate and trinucleate pollen grains in the angiosperms. *Am. J. Bot.* 54: 1069.

Bridges, D.C., Ed. 1992. *Crop Losses Due to Weeds in the United States — 1992.* Weed Science Society, Champaign, IL.

Bucholtz, K.P. 1967. Report of the terminology committee. *Weed* 15: 388.

Chandler, J.M. 1991. Estimated losses of crops to weeds. In *Handbook of Pest Management in Agriculture,* Vol. 1. Ed. by Pimentel, D. Boca Raton, FL, p. 53.

Conard, S.G. and Radosevich, S.R. 1979. Ecological fitness of *Senecio vulgaris* and *Amaranthus retroflexus* biotypes susceptible and resistant to atrazine. *J. Appl. Ecol.* 16: 171.

Crawley, M.J., Halle, R.S., Rees, M., Kohn, D., and Buxton, J. 1993. Ecology of transgenic oilseed rape in natural habitats. *Nature* 363: 620.

Crow, J.F. and Kimura, M. 1970. *An Introduction to Population Genetics Theory.* Harper & Row, New York.

Cruden, R.W. 1977. Pollen-ovule ratios: a conservative indicator of breeding systems in flowering plants. *Evolution* 31: 32.

Doebley, J. 1990. Molecular evidence for gene flow among *Zea* species. *Bioscience* 40: 443.

Donald, W. and Ogg, A. 1991. Biology and control of jointed goatgrass (*Aegilops cylindrica*), a review. *Weed Technol.* 5: 3.

Dumas, C., Kerhoas, C., Gay, G., and Gaude, T. 1986. Water content, membrane state and pollen physiology. In *Biotechnology and Ecology of Pollen.* Ed. by Mulcahy, D.L., Mulcahy, G.B., and Ottaviano, E. Springer-Verlag, New York, p. 333.

Dyer, W.E., Hess, F.D., Holt, J.S., and Duke, S.O. 1993. Potential benefits and risks of herbicide-resistant crops produced by biotechnology. *Hort. Rev.* 15: 367.

Ellstrand, N.C. 1988. Pollen as a vehicle for the escape of engineered genes? *Trends Biotechnol.* 6: 530.

Ellstrand, N.C., Devlin, B., and Marshall, D.L. 1989. Gene flow by pollen into small populations: data from experimental and natural stands of wild radish. *Proc. Natl. Acad. Sci. U.S.A.* 86: 9044.

Ellstrand, N.C. and Hoffman, C.A. 1990. Hybridization as an avenue of escape for engineered genes. *BioScience* 40: 438.

Elton, C.S. 1958. *The Ecology of Invasions by Animals and Plants.* Chapman & Hall, London.

Faegri, K. and Iverson, J. 1975. *Textbook of Pollen Analysis,* 3rd ed. Hafner Press, New York.

Faegri, K. and van der Pijl, L. 1979. *The Principles of Pollination Ecology.* Pergamon Press, Oxford.

Foy, C.L., Forney, D.R., and Cooley, W.E. 1983. History of weed introductions. In *Exotic Plant Pests and North American Agriculture.* Ed. by Wilson, C.L. and Graham, C.L. Academic Press, New York, chap. 4.

Frankel, R. and Galun, E. 1977. *Pollination Mechanisms, Reproduction and Plant Breeding.* Springer-Verlag, Berlin.

Futuyuma, D.J. 1986. *Evolutionary Biology,* 2nd ed. Sinauer Associates, Sunderland, MA.

George, R.A.T. 1985. *Vegetable Seed Production.* Longmans Press, New York.

Gessel, J. and Segel, L.A. 1990. Herbicide rotations and mixtures: effective strategies to delay resistance. In *Managing Resistance to Agrochemicals.* Ed. by Green, M.B., LeBaron, H.M., and Moberg, W.K. American Chemical Society, Washington, D.C., chap. 30.

Hancock, J.F. 1992. *Plant Evolution and the Origin of Crop Species.* Prentice-Hall, Englewood Cliffs, NJ.

Hancock, J.F. and Luby, J.J. 1993. Genetic resources at our doorstep: the wild strawberries. *BioScience* 43: 141.

Harlan, J.R. The possible role of weed races in the evolution of cultivated plants. *Euphytica* 14: 173.

Hauptli, H., Newell, N., and Goodman, R.M. 1985. Genetically engineered plants: environmental issues. *Bio/Technology* 3: 437.

Heiser, C.B., Smith, D.M., Clevenger, S.B., and Martin, W.C. 1969. The North American sunflowers (*Helianthus*). *Mem. Torrey Bot. Club* 22: 1.

Heiser, C.B. 1976. Sunflowers (*Helianthus* - Compositae). In *Evolution of Crop Plants.* Ed. by Simmonds, N.W. Longmans Press, London, chap. 13.

Hickman, J.C. Ed. 1993. *The Jepson Manual: Higher Plants of California.* University of California Press, Berkeley.

Hoekstra, F.A. 1992. Stress effects on the male gametophyte. In *Sexual Plant Reproduction.* Ed. by Cresti, M. and Tiezzi, A. Springer-Verlag, Heidelberg, p. 193.

Hoekstra, F.A. and Bruinsma, J. 1978. Reduced independence of the male gametophyte in angiosperm evolution. *Ann. Bot.* 42: 759.

Hoekstra, F.A., Crowe, J.H., Crowe, L.M., and van Bilsen, D.G.L. 1992. Membrane behavior and stress tolerance in pollen. In *Angiosperm Pollen and Ovules.* Ed. by Ottaviano, E., Mulcahy, D.L., Sari Gorla, M., and Mulcahy, G.B. Springer-Verlag, Heidelberg, p. 177.

Holm, L.G., Pancho, J.V., Herberger, J.P., and Plunknett, D.L. 1979. *A Geographical Atlas of World Weeds.* Wiley, New York.

Holm, L.G., Plunknett, D.L., Pancho, J.V., and Herberger, J.P. 1977. *The World's Worst Weeds.* University of Hawaii Press, Honolulu.

Holt, J.S. and Radosevich, S.R. 1983. Differential growth of two common groundsel (*Senecio vulgaris*) biotypes. *Weed Sci.* 31: 112.

Hull, D.A. and Beattie, A.J. 1988. Adverse effects on pollen exposed to *Atta texana* and other North American ants: implications for ant pollination. *Oecologia* 75: 153.

Jacobs, B.F., Duesing, J.H., Antonovics, J., and Paterson, D.T. 1988. Growth performance of triazine-resistant and -susceptible biotypes of *Solanum nigrum* over a range of temperatures. *Can. J. Bot.* 66: 847.

Jana, S. and Naylor, J.M. 1982. Adaptation for herbicide tolerance in populations of *Avena fatua*. *Can. J. Bot.* 60: 1611.

Jensen, K.I.N. and Bandeen, J.D. 1979. Triazine resistance in annual weeds. In *Maize*, Ciba Geigy Tech. Monogr. Ed. by Haflinger, E. Ciba Geigy, p. 55.

Jones, M.D. and Newell, L.C. 1946. Pollination cycles and pollen dispersal in relation to grass improvement. *Univ. Neb. Coll. Agric. Res. Bull.* No. 148.

Keeler, K.H. 1985. Implications of weed genetics and ecology for the deliberate release of genetically-engineered crop plants. *Recombinant DNA Tech. Bull.* 8: 165.

Keeler, K.H. 1989. Can genetically engineered crops become weeds? *Bio/Technology* 7: 1134.

Keeler, K.H. and Turner, C.E. 1991. Management of transgenic plants in the environment. In *Risk Assessment in Genetic Engineering*. Ed. by Levin, M.A. and Strauss, H.S. McGraw-Hill, New York, chap. 9.

Kerhoas, C., Gay, G., Duplan, J.C., and Dumas, C. 1986. Water content evolution in *Cucurbita pepo* during aging: a NMR study. In *Biotechnology and Ecology of Pollen*. Ed. by Mulcahy, D.L., Mulcahy, G.B., and Ottaviano, E. Springer-Verlag, New York, p. 502.

Kerhoas, C., Gay, G., Duplan, J.C., and Dumas, C. 1989. Viability of *Cucurbita pepo* pollen: biophysical and structural data. *Planta* 179: 165.

Kirkpatrick, K.J. and Wilson, H.D. 1989. Interspecific gene flow in *Cucurbita: C. texensus* vs. *C. pepo*. *Am. J. Bot.* 75: 519.

Klinger, T., Elam, D.R., and Ellstrand, N.C. 1991. Radish as a model system for the study of engineered gene escape rates via crop-weed mating. *Conserv. Biol.* 5: 531.

Klinger, T., Arriola, P.E., and Ellstrand, N.C. 1992. Crop-weed hybridization in radish (*Raphanus sativus*): effects of distance and population size. *Am. J. Bot.* 79: 1431.

Klinger, T. and Ellstrand, N.C. 1994. Engineered genes in wild populations: fitness of weed-crop hybrids of *Raphanus sativus*. *Ecol. Appl.* 4: 117.

Langer, R.H.M. and Hill, G.D. 1982. *Agricultural Plants*. Cambridge University Press, Cambridge, U.K.

Langevin S.A., Clay, K., and Grace, H.B. 1990. The incidence and effects of hybridization between cultivated rice and its related weed, red rice (*Oryza sativa* L.). *Evolution* 44: 1000.

LeBaron, H.M. 1984. Herbicide resistance in plants — future needs. *Weeds Today* 4: 2.

Luza, J.G. and Polito, V.S. 1987. Effects of desiccation and controlled rehydration on germination *in vitro* of pollen of walnut (*Juglans* spp.). *Plant, Cell Environ.* 10: 487.

Mallory-Smith, C.A., Thill, D.C., and Dial, M.J. 1993. ID-BR1: sulfonylurea herbicide-resistant lettuce germplasm. *HortScience* 28: 63.

Mallory-Smith, C.A., Thill, D.C., and Dial, M.J. 1990. Identification of sulfonylurea herbicide-resistant prickly lettuce (*Lactuca serriola*). *Weed Technol.* 4: 163.

Mallory-Smith, C.A., Thill, D.C., Dial, M.J., and Zemetra, R.S. 1990. Inheritance of sulfonylurea herbicide resistance in *Lactuca* spp. *Weed Technol.* 4: 787.

McCammon, S.L. and Dwyer, S.G., Eds. 1990. *Proceedings, Workshop on Safeguards for Planned Introduction of Transgenic Oil Seed Crucifers*. U.S. Department of Agriculture/Washington, D.C.

Miller, H.I. and Gunary, D. 1993. Serious flaws in the horizontal approach to biotechnology risk assessment. *Science* 262: 1500.

Mooney, H.A. and Drake, J.A. 1986. *Ecological Effects of Introduced Species in North America and Hawaii*. Springer-Verlag, New York.

Pacini, E. 1992. Transport mechanisms of pollen — a short review. In *Sexual Plant Reproduction*. Ed. by Cresti, M. and Tiezzi, E. Springer-Verlag, Heidelberg, p. 69.

Perrins, J., Williamson, M., and Fitter, A. 1992. A survey of differing views of weed classification: implications for regulation of introductions. *Biol. Conserv.* 60: 47.

Perrins, J., Williamson, M., and Fitter, A. 1992. Do annual weeds have predictable characters? *Acta Oecol.* 13: 517.

Platt-Aloia, K.A., Lord, E.M., DeMason, D.A., and Thomson, W.W., 1986. Freeze fracture observations on membranes of dry and dehydrated pollen from *Collomia, Phoenix*, and *Zea*. *Planta* 168: 291.

Proctor, M. and Yeo, P. 1972. *The Pollination of Flowers*. Taplinger Publishing, New York.

Reed, C.F. and Hughes, R.O. 1970. *Selected Weeds of the United States*, U.S. Department of Agriculture/Agricultural Research Service, Agric. Handb. No. 366. U.S. Government Printing Office, Washington, D.C.

Robbins, W.W., Bellue, M.K., and Ball, W.S. 1970. *Weeds of California*. California Department of Agriculture, Sacramento.

Rollins, R.C. 1981. Weeds of the Cruciferae (Brassicaceae) in North America. *J. Arnold Arboretum* 62: 517.

Saari, L.L., Cotterman, J.C., and Thill, D.C. 1994. Resistance to acetolactate synthase inhibiting herbicides. In *Herbicide Resistance in Plants: Ecology and Mechanisms*. Ed. by Powles, S.B. and Holtum, J.A. CRC Press, Boca Raton, FL, chap. 4.

Salisbury, E. 1961. *Weeds and Aliens*. Collins, London.

Schery, R.W. 1972. *Plants for Man*. Prentice-Hall, Englewood Cliffs, NJ.

Sharples, F.E. 1983. Spread of organisms, with novel genotypes: thoughts from an ecological perspective. *Recombinant DNA Tech. Bull.* 6: 43.

Simmonds, N.W., Ed. 1976. *Evolution of Crop Plants*, Longmans Press, London.

Smith, R.J., Jr., Flinchum, W.T., and Seaman, D.E. 1977. *Weed Control in U.S. Rice Production*. U.S. Department of Agriculture/Agricultural Research Service, Agric. Handb. No. 497, U.S. Government Printing Office, Washington, D.C.

Soloman, A.M. 1979. Pollen. In *Aerobiology: The Ecological Systems Approach*. Ed. by Edmonds, R.L. Dowden, Hutchinson & Ross, Stroudsburg, PA, p. 41.

Stallings, G.P., Thill, D.C., and Mallory-Smith, C.A. 1994. Sulfonylurea-resistant Russian thistle (*Salsola iberica*) survey in Washington state. *Weed Technol.* 8: 258.

Stanley, R.G. and Linskens, H.F. 1974. *Pollen: Biology, Biochemistry, Management*. Springer-Verlag, New York.

Szmedra, P.I. 1991. Pesticide use in agriculture. In *Handbook of Pest Management in Agriculture*. Vol. 1. Ed. by Pimentel, D. Boca Raton, FL.

Templeton, A.R. 1986. Coadaptation and outbreeding depression. In *Conservation Biology: The Science of Scarcity and Diversity*. Ed. by Soule, M. Sinauer Associates, Sunderland. MA, p. 105.

Thomas, H. and Jones, H. 1976. Origins and identification of weed species of *Avena*. In *Wild Oats in World Agriculture*. Ed. by Jones, D.P. Agricultural Research Council, London, p. 1.

Thomsen, C.D., Barbe, G.D., Williams, W.A., and George, M.R. 1986. Escaped artichokes are troublesome pests. *Calif. Agric.* 40(3-4): 7.

Thompson, C.R., Thill, D.C., and Shafi, B. 1994. Growth and competitiveness of sulfonylurea-resistant and -susceptible kochia (*Kochia scoparia*). *Weed Sci.* 42: 172.

Tiedje, J.M., Colwell, R.K., Grossman, Y.L., Hodson, R.E., Lenski, R.E., Mack, R.N., and Regal, P.J. 1989. The release of genetically engineered organisms: a perspective from the Ecological Society of America. *Ecology* 70: 297.

Turner, C.E. 1988. Ecology of invasions by weeds. In *Weed Management in Agroecosystems: Ecological Approaches*. Ed. by Altieri, M.A. and Liebman, M. CRC Press, Boca Raton, FL, chap. 4.

Turner, C.E., Anderson, L.W., Foley, P., Goeden, R.D., Lanini, W.T., Lindow, S.E., and Qualset, C.O. 1992. Biological approaches to weed management. In *Beyond Pesticides: Biological Approaches to Pest Management in California*. Ed. by Madden, J.P. and Schroth, M.N. University of California, Berkeley, chap. 2.

Warwick, S.I. and Black, L. 1981. The relative competitiveness of atrazine susceptible and resistant populations of *Chenopodium album* and *C. strictum*. *Can. J. Bot.* 59: 689.

Whitson, T.D., Burrill, L.C., Dewey, S.A., Cudney, D.W., Nelson, B.E., Lee, R.D., and Parker, R. 1992. *Weeds of the West*. Western Society of Weed Science, Newark, CA.

Wilson, H.D. 1990. Gene flow in squash species. *BioScience* 40: 449.

Zamir, D. and Gadish, I. 1987. Pollen selection for low temperature adaptation in tomato. *Theor. Appl. Gen* 74: 545.

Zimdahl, R L. 1983. Where are the principal exotic weed pests? In *Exotic Plant Pests and North American Agriculture*. Ed. by Wilson, C.L. and Graham, C.L. Academic Press, New York, chap. 7.

Chapter 21

MANAGING THE SPREAD OF HERBICIDE RESISTANCE

Donald C. Thill

CONTENTS

1. INTRODUCTION

Herbicides have been used to control weeds selectively in crops that were naturally tolerant to specific herbicides since the mid-1940s. The deliberate selection or production of crops resistant to specific herbicides has occurred only recently (reviewed by Mazur and Falco 1989). Herbicide-resistant crops (HRCs) may be produced either by breeding for resistance through selection or by using biotechnology to generate HRCs (Dyer et al. 1993). Using biotechnology to produce genetically altered plants, including HRCs, has caused considerable discussion regarding the environmental impact of small-scale field testing and eventual large-scale commercial production of these crops in the field (Regulatory Considerations: Genetically-Engineered Plants 1987; Field Testing Genetically Modified Organisms 1989; MacKenzie and Henry 1990). Enhanced weediness of HRCs from biotechnology is a major environmental issue. Specific concerns include the ability to control volunteer HRCs in agroecosystems, the invasive potential of these crops into natural ecosystems, gene flow from HRCs to weedy relatives, and increased selection for herbicide-resistant weeds.

Interestingly, potential weediness of crops naturally tolerant to herbicides and HRCs developed using traditional breeding methods never has been an environmental issue and never has been regulated. Also, there are no reported cases of transfer of genes responsible for naturally occurring herbicide tolerance or selected herbicide resistance in crops to related weed species (Dyer et al. 1993). However, in at least two cases germplasm from herbicide-resistant weeds has been transferred to related crop species by traditional breeding methods (Beversdorf 1987; Mallory-Smith et al. 1993). In one case herbicide resistance was shown to be inherited as a single, incompletely dominant trait (Mallory-Smith et al. 1993). These conventionally altered crops are not subject to

1-56670-045-0/96/$0.00+$.50
© 1996 by CRC Press Inc.

specific regulations governing their use in the field, while field testing and commercial production of herbicide-resistant crops from biotechnology are regulated.

> "There seems to be a view that the ecology of genetically engineered organisms is somehow different — more inscrutable perhaps, but certainly more dangerous — and that the intentional release of genetically engineered organisms poses more of a threat to the balance of nature that other kinds of organisms bred by man. This view is mistaken. While there are risks associated with the introduction of any novel organisms into a habitat, the ecology of genetically engineered organisms is exactly the same as the ecology of any other living thing." (Crawley 1990)

In this chapter managing herbicide resistance caused directly or indirectly by HRCs is discussed without regard to the method used to produce the resistant crop (Rees et al. 1990). Potential weediness and possible spread of volunteer and escaped HRC plants in agricultural and natural ecosystems is discussed. Also discussed are increased weediness and spread caused by gene flow from HRCs to related weed species and selection for herbicide-resistant weeds through frequent, repeated use of herbicides with the same mode of action as those used on HRCs.

2. HERBICIDE-RESISTANT CROP PLANTS AS WEEDS

Will HRCs become weeds? Unfortunately, there is no single answer to this question and it needs to be examined on a case-by-case basis (Ellstrand and Hoffman 1990, Keeler 1989). It ultimately depends on the relationship of the HRC with the environment (Field Testing Genetically Modified Organisms 1989). Important factors include the ability of the HRC to colonize, establish, and spread after it is introduced into or invades an ecosystem (Rees et al. 1990). Invasion is defined as the successful founding of a colony in a region where it did not previously exist, followed by rapid expansion of the range of the invading species (Barrett and Richardson 1986). There are, however, no general attributes that characterize all invaders. Characteristics of "ideal" invaders include a long list of physiological, demographic, and genetic factors (Baker 1965, 1974; Bazzaz 1986; Barrett and Richardson 1986). The susceptibility of the new habitat to invasion also is important. The ecology of biological invasion has been reviewed extensively and is not covered in this chapter (see, for example, Groves and Burdon 1986; Mooney and Drake 1986).

Herbicide-resistant crops will colonize arable and nonarable lands because seeds will be sown in large numbers of fields and will be spilled along rights-of-way during transport. Seeds also may be disseminated naturally by birds, animals, wind, and water. Humans likely will disseminate HRC seeds even further with tillage and harvest equipment, hay and other animal feed, and various travel activities. However, uncertainty exists regarding the ability of these populations to persist from one generation to the next and to spread.

Generally, domesticated plants cannot survive and reproduce unless aided by humans, although different degrees of survivability exist among various crop plants. According to Afnor (as cited in Williamson 1991), "since the start of agriculture, no exponential or anarchic spread of cultivated plants has been observed." Williamson (1991) counters that this is not true. He states that there have been infrequent but serious problems caused by agricultural crops. Holm et al. (1977) states, "Those who are not acquainted with the distribution across the world of our worst weeds often seem confused that a plant which is an important weed in one area may be a valuable crop in another place." Alfalfa (*Medicago sativa* L.), sorghum (*Sorghum* Moench sp.), and sunflower (*Helianthus annuus* L.) are examples of major crops that can become weedy, while there is little risk of weediness associated with growing crops such as corn (*Zea mays* L.), wheat (*Triticum aestivum* L.), soybean (*Glycine max* (L.) Merr.), cotton (*Gossypium hirsutum* L.), oat (*Avena sativa* L.), rice (*Oryza* L. sp.), barley (*Hordeum vulgare* L.), dry bean (*Phaseolus* L. sp.), potato (*Solanum tuberosum* L.), and tobacco (*Nicotiana tabacum* L.) (Regulatory Considerations: Genetically-Engineered Plants 1987).

Herbicide-resistant crop cultivars are less or equally fit compared to existing crop cultivars (Mallory-Smith 1995), and in the absence of the herbicide selection pressure likely will be equally or less invasive as compared to existing cultivars. This assumes that weedy pleiotropic traits are not associated with the herbicide resistance trait, which appears to be the case (Downey 1987). Additionally, weediness is a trait associated with many genes and the introduction of one or a few genes for herbicide resistance into existing cultivars is unlikely to cause the crop to become a weed (Keeler and Turner 1991), especially in the absence of the herbicide selection pressure. Thus, the invasive potential of a HRC in any ecosystem should be no greater than existing crop cultivars when the specific herbicide is not applied.

As with many other crops, HRC plants that infest subsequently planted crops are considered weeds and their control will be no more difficult than existing volunteer crop plants if the specific herbicide(s) to which the crop is resistant is avoided. Difficulty may arise if herbicides traditionally used to control weeds in the following crop have the same mode of action as the specific herbicide(s) used on the resistant crop. For example, glyphosate-resistant wheat plants that infest summer-fallow lands could not be controlled with glyphosate. Alternative strategies such as tillage or use of effective herbicides with different modes of action will control the volunteer wheat. Other examples were presented by Dyer et al. 1993.

Crops that carry two or more different herbicide resistance traits may require special management considerations. For example, wheat resistant to glyphosate and the cyclohexanedione herbicides cannot be controlled in a rotational crop of glyphosate-resistant canola (*Brassica napus* L.) with glyphosate or sethoxydim (and likely other cyclohexanedione herbicides). Cross-resistance to other families of herbicides with the same mode of action (cyclohexanediones and aryloxypropanoics) may make control more difficult.

Overall, domesticated crop plants rarely, if ever, have successfully invaded disturbed or natural ecosystems because they lack specific traits to survive from one generation to the next without human intervention. Herbicide-resistant crop plants will have, at best, an adaptive advantage compared to existing crops only in the presence of the specific herbicide to which the crop is resistant. Thus, the invasiveness of HRC plants in the absence of a specific herbicide should be no greater than existing crop plants in both disturbed and natural ecosystems. Volunteer HRCs infesting rotational crops and other disturbed ecosystems can be managed effectively if cultural strategies or herbicides with different modes of action are available.

3. GENE FLOW TO WEED RELATIVES

Hybridization between crops and their weedy relatives was reviewed previously (*Field Testing Genetically-Modified Organisms* 1989; Darmency et al. 1991, Dyer et al. 1993, Ellstrand and Hoffman 1990). It is clear that certain crops will hybridize with close weedy relatives. Examples include hybridization between cultivated rice and red rice (*Oryza rufipogon* Griff), sorghum and johnsongrass (*Sorghum halepense* (L.) Pers.), and oat and wild oat (*Avena fatua* L.) (as reviewed by Dyer et al. 1993). Isolation distances for the production of genetically pure seed for many species of field and vegetable crops have been reported and range from 18 to over 1600 m, depending on type of pollination (Association of Official Seed Certification Agencies 1971). For many crop plants, such as corn, wheat, soybean, alfalfa, cotton, barley, dry bean, and tobacco, there is no perceived risk of lateral gene transfer (Regulatory Considerations: Genetically-Engineered Plants 1987). Also, cytoplasmically controlled traits, such as triazine-herbicide resistance, cannot be inherited via pollen and pose little risk of gene transfer (Tilney-Bassett 1975).

Introgression between HRC plants and weedy relatives is considered highly probable (Ellstrand 1988, Keeler and Turner 1991) because herbicides (specific selection pressure) will be applied that may maximize selection for weed progeny resulting from the outcrossing event (Dyer et al. 1993). This may be true for crops and weeds that freely interbreed, but will not be the case for many crops grown in the U.S. because few are native to North America and they are not grown in close association with weedy relatives (Field Testing Genetically-Modified Organisms 1989). Internationally, the risk of introgression may be greater where crops are grown near their center of origin and in proximity to closely related weeds. In these cases it will be necessary to develop and implement effective integrated weed management plans to mitigate the effect of herbicide-resistant weeds formed through hybridization with HRC plants. One strategy may be to not grow HRCs in these "high risk" areas.

The likelihood of the occurrence of introgression may be reduced greatly if the weedy relative originally is susceptible to the herbicide applied to the HRC. This generally is assumed to be the case because the weed lacks the specific trait(s) conferring resistance. Few weeds will remain following the herbicide application and this will reduce greatly the probability of a successful outcrossing event. The ontogeny of weeds that escape herbicide treatment may be delayed compared to the HRC plants and flowering may not be synchronized, thus preventing hybridization, especially in crops with determinate growth.

Integrated weed management programs must be adopted to control any herbicide-resistant weeds produced through gene transfer from the resistant crop during the previous growing season. Control strategies include tillage, competitive crops, and herbicides with different modes of action (Thill et al. 1994). At most, crops resistant to the same herbicide should be grown only once every three growing seasons. This will minimize the opportunity for repeated application of the same herbicide and will help deplete herbicide-resistant weed seeds in the soil when effective integrated weed management strategies are used to control weeds (see Section 4).

Gene transfer also can occur between HRC and herbicide-susceptible crop cultivars, weeds growing in areas adjacent to fields containing HRCs, and between volunteer HRC plants and weedy relatives growing outside field boundaries. These herbicide-resistant weeds may spread to new areas, including fields where HRCs are or will be grown. Specific control strategies to minimize introgression should be implemented and must include control of volunteer HRC plants growing in association with weedy relatives, both inside and outside field borders. Gene flow from HRC to herbicide-susceptible crop cultivars also should be minimized so that the frequency of the resistance trait in the environment does not increase, which could increase the potential for introgression with weedy relatives. Likewise, cultivars of the same crop with different herbicide resistance traits that are paternally inherited should not be grown near each other to avoid gene flow between cultivars. For example, canola cultivars resistant to glufosinate and glyphosate herbicides may hybridize and form progeny resistant to both herbicides. These plants would not be controlled with either herbicide, making weed control more difficult. Surveys should be conducted and landscape maps prepared to show the location of fields where HRC and herbicide-susceptible crop plants are grown and the location of weedy relatives that may hybridize with HRCs.

Based on the previous discussion, it is apparent that some gene transfer may occur between HRC plants and their weedy relatives. It also is assumed that the hybrid will reproduce progeny that express the resistance trait, even though hybrids sometimes are sterile and produce no seed (Bing et al. 1991). Will this affect weediness? The discussion now becomes one dealing with herbicide-resistant weeds because the resistance trait has been introduced into the weed population and probably is present at some level greater than the natural mutation rate. In the presence of the specific herbicide selection pressure the resistant weed biotype has a significantly greater adaptive advantage as compared to the susceptible weed biotype, and will become the predominate biotype following repeated application of the herbicide (Gressel and Segel 1982). Biological factors also affect the evolution of resistance (in the absence or presence of the herbicide) and include fitness, gene flow, and inheritance (Maxwell et al. 1990).

A biologically based model used to simulate changes in herbicide resistance in weeds over time (Maxwell et al. 1990) showed that the proportion of resistant biotypes in the population increased slowly when the resistant biotype had a fitness advantage compared to the susceptible biotype, even in the absence of the herbicide selection pressure (Maxwell et al. 1993). This likely will be the case with some acetolactate synthase inhibitor-herbicide resistant weeds in certain environments and cropping systems (Saari et al. 1994). For example, chlorsulfuron-resistant kochia [*Kochia scoparia* (L.) Schrad.] seeds germinated faster at 8°C than susceptible biotypes (Thompson et al. 1994). This difference in germination may provide a competitive advantage for chlorsulfuron-resistant kochia in environments where crops are seeded in early spring when soil temperatures are cool. Faster germination at cool soil temperatures also could be a disadvantage when crops are seeded later after soils warm because early-germinating cohorts could be controlled with tillage or herbicides before the crop is seeded. The proportion of resistant weeds decreased when they were less fit compared to the susceptible biotype (Gressel and Segel 1982). This likely will occur for triazine-resistant weeds (Holt and Radosevich 1983; Conrad and Radosevich 1979; Jacobs et al. 1988) and, perhaps, HRC–weed hybrids, if the introduced gene(s) has a negative ecological impact on the weed (Ellstrand and Hoffman 1990).

Introgression can occur between some crops and weeds, but the potential for gene transfer is unknown for many other crops and their weedy relatives. Thus, further research is required to determine the probability of introgression between most new HRCs and their weedy relatives, especially if weedy relatives are present in the farming region where HRCs are grown. This information should be available before HRCs are grown commercially.

The proportion of resistant to susceptible weeds in subsequent generations will depend on initial population frequencies, herbicide selection pressure, and biological factors such as fitness difference between resistant and susceptible biotypes, gene flow, and inheritance of the resistance trait. Spread of herbicide-resistant weeds will be a function of species-specific dispersal mechanisms (wind, water, forced expulsion of seed, animals, human activity), further gene flow between resistant and susceptible biotypes of the same weed species (Stallings et al. 1995a), and gene transfer between other closely related weed species. Resistance is expected to spread rapidly, especially in weeds such as sulfonylurea-herbicide-resistant kochia and Russian thistle (*Salsola iberica* Sennen and Pau) that have specialized dispersal mechanisms (Stallings et al. 1995a, b).

4. SELECTION OF HERBICIDE-RESISTANT WEEDS

The evolution of weed resistance to herbicides is primarily a function of the number of years of herbicide treatment (or application per season) and the original proportion of resistant to susceptible biotypes in the field prior to the initial herbicide application (Gressel and Segel 1982). Selection pressure, fitness, and the average life span of the species in the soil seedbank also are important. Repeated usage of the same herbicide or herbicides with the same mode of action will select for weed biotypes that are resistant to the herbicide. This will happen equally in crops naturally tolerant to herbicides and HRCs. The properties and use pattern of the herbicide(s) in a cropping system and the intrinsic properties of the weed are factors important in the selection of herbicide-resistant weed biotypes. The crop is less important. Thus, growers must use the same integrated weed management strategies that delay or prevent selection of herbicide-resistant weeds regardless of whether a HRC or naturally tolerant crop is used in the cropping system.

Effective integrated weed management strategies should be based on preventing the movement of weed seeds into the soil (or vegetative reproductive parts). This can be achieved in two ways: (1) by preventing immigration of seed into the field from external sources and (2) by reducing or eliminating seed production by weeds already in the field. Immigration of most weed seed into a field can be prevented by planting clean seed, cleaning harvest and tillage equipment between fields, covering grain transported in trucks, and controlling weed infestations along roadsides, fence rows, and waste areas. Many factors influence how effectively weeds are managed in a field, and how much weed seed is returned to the soil. These include weed control practices such as herbicide use, tillage, stubble burning and rogueing, as well as agronomic and ecological factors such as crops and crop cultivars sown, type of crop rotation, relative weed and crop density, relative times of emergence of the weed and crop, crop row spacing, fertility level, and fertilizer placement (Thill et al. 1994). The probability of selecting for herbicide-resistant weed biotypes will be reduced greatly if effective integrated weed management strategies are used in all phases of a crop production system.

5. SUMMARY

Weediness of HRC and existing crop cultivars should be similar in the absence of a specific herbicide selection pressure. There are no reported cases of HRCs being more fit than existing cultivars. In fact, sometimes they are less fit (Mallory-Smith 1995). Control of volunteer HRC and nonresistant crop plants also will be the same, except that herbicide choice will be more limited with the HRC. Also, it may be necessary to control volunteer HRC plants in both arable and nonarable sites to prevent introgression with weedy relatives and existing susceptible crop cultivars.

The probability of introgression between each HRC and weedy relatives should be known before the crop is grown commercially, especially if weedy relatives are present in the farming region. If gene transfer can occur, specific management strategies will be required to mitigate introgression.

The properties and use pattern of a herbicide in a cropping system and the intrinsic properties of the weed are factors important in the selection of herbicide-resistant weeds, not whether the crop is resistant or tolerant to a herbicide. The risk of selecting herbicide-resistant weeds in a HRC is no greater than using a selective herbicide in a naturally tolerant crop. In either case repeated use of the same herbicide or herbicides with the same mode of action will eventually select for herbicide-resistant weed biotypes. This can be prevented or greatly delayed by using effective integrated weed management strategies in all parts of the crop production system. Most importantly, avoid using the same herbicide or herbicides with the same site of action routinely in any cropping system.

REFERENCES

Association of Official Seed Certification Agencies. 1971. *AOSCA Certification Handbook*, p. 23.

Baker, H.G. 1965. Characteristics and modes of origin of weeds. In *The Genetics of Colonizing Species*. Ed. by Baker, H.G. and Stebbins, G.L. Academic Press, London, p. 141.

Baker, H.G. 1974. The evolution of weeds. *Annu. Rev. Ecol. Syst.* 5: 1.

Barrett, S.C.H. and Richardson, B.J. 1986. Genetics attributes of invading species. In *Ecology of Biological Invasion*. Ed. by Groves, R.H. and Burdon, J.J. Cambridge University Press, Cambridge, U.K., p. 21.

Bazzaz, F.A. 1986. Life history of colonizing plants: some demographic, genetic, and physiological features. In *Ecology of Biological Invasions of North America and Hawaii*. Ed. by Mooney, H.A. and Drake, J.A. Springer-Verlag, New York, p. 96.

Beversdorf, W.D. 1987. Classical Approaches to the Development of Herbicide Tolerance in Crop Cultivars, ACS Symp. Ser. No. 334. *American Chemical Society,* Washington, D.C., p. 108,.

Bing, D.J., Downey, R.K., and Rakow, G.F.W. 1991. Potential of gene transfer among oilseed *Brassica* and their weedy relatives. *Proc. 8th Int'l Rapeseed Congress*, Saskatoon, Canada, p. 1022.

Conard, S.G. and Radosevich, S.R. 1979. Ecological fitness of *Senecio vulagaris* and *Amaranthus retroflexus* biotypes susceptible or resistant to atrazine. *J. Appl. Ecol.* 16: 171.

Crawley, M.J. 1990. The ecology of genetically engineered organisms: assessing the environmental risks. In *Introduction of Genetically Modified Organisms into the Environment*. Ed. by Mooney, H.A. and Bernardi, G. John Wiley & Sons, New York, p. 133.

Darmency, H., Till, I., and Reboud, X. 1991. Escape of herbicide resistance gene from transgenic crops. In *Herbicide Resistance in Weeds and Crops*. Ed. by Casely, J.C., Cussans, G.W., and Atkin, R.K. Butterworth-Heinemann, Oxford, p. 437.

Downey, R.K. 1987. Case histories of modified crop introductions. In *Regulatory Consideration: Genetically-Engineered Plants,* Workshop Summary. Boyce Thompson Institute for Plant Research, Cornell University, p. 71.

Dyer, W.E., Hess, F.D., Holt, J.S., and Duke, S.O. 1993. Potential benefits and risks of herbicide-resistant crops produced by biotechnology. *Hort. Rev.* 15: 367.

Ellstrand, N.C. 1988. Pollen as a vehicle for the escape of engineered genes. *TREE* 3: S30.

Ellstrand, N.C. and Hoffman, C.A. 1990. Hybridization as an avenue of escape for engineered genes. Strategies for risk reduction. *BioScience* 40: 438.

Field Testing Genetically Modified Organisms — Framework for Decisions. 1989. National Academy Press, Washington D.C., chap. 3, 4, 5, 6.

Gressel, J. and Segel, L.A. 1982. Interrelating factors controlling the rate of appearance of resistance: the outlook for the future. In *Herbicide Resistance in Plants.* Ed. by LeBaron, H.M. and Gressel, J. John Wiley & Sons, New York, p. 325.

Groves, R.H. and Burdon, J.J. 1986. *Ecology of Biological Invasions.* Cambridge University Press, Cambridge, U.K., p. 166.

Holm, L.G., Plucknett, D.L., Pancho, J.V., and Herberger, J.P. 1977. *The World's Worst Weeds — Distribution and Biology.* University of Hawaii Press, Honolulu, p. 609.

Holt, J.S. and Radosevich, S.R. 1983. Differential growth of two common groundsel (*Senecio vulgaris*) biotypes. *Weed Sci.* 31: 112.

Jacobs, B.F., Duesing, J.H., Antonovics, J., and Patterson, D.T. 1988. Growth performance of triazine-resistant and -susceptible biotypes of *Solanum nigrum* over a range of temperatures. *Can. J. Bot.* 66: 847.

Keeler, K.H. 1989. Can genetically engineered crops become weeds? *BioTechnology* 7: 134.

Keeler, K.H. and Turner, G.E. 1991. Management of transgenic plants in the environment. In *Risk Assessment in Genetic Engineering.* Ed. by Levin, M.A. and Strauss, H.S. McGraw-Hill, New York, p. 189.

MacKenzie, D.R. and Henry, S.C. Eds. 1990. *The Biosafety Results of Field Tests of Genetically Modified Plants and Microorganisms, Proc. Kiawah Island Conference, November 27 to 30.* Agricultural Research Institute, Bethesda, MD, p. 303.

Mallory-Smith, C.A. 1995. Potential pleiotropic difficulties with herbicide resistant crops. In *Herbicide-Resistant Crops: Agricultural, Economic, Environmental, Regulatory, and Technological Aspects.* Ed. by Duke, S.O. Lewis Publishers, Boca Raton, FL, chap. 13.

Mallory-Smith, C.A., Thill, D.C., and Dial, M.J. 1993. ID-BR1: sulfonylurea herbicide resistant lettuce germplasm. *HortScience* 28: 63.

Maxwell, B.D., Roush, M.L., and Radosevich, S.R. 1990. Predicting the evolution and dynamics of herbicide resistance in weed populations. *Weed Technol.* 4: 2.

Maxwell, B.D., Thill, D.C., Fay, P.K., Dyer, W., and Westra, P. 1993. Simulation of chlorsulfuron resistance evolution in kochia populations. In *Proc. West. Soc. Weed Sci.* 46: 128.

Mazur, B.J. and Falco, S.C. 1989. The development of herbicide resistant crops. *Annu. Rev. Plant Physiol. Plant Mol. Biol.* 40: 441.

Mooney, H.A. and Drake, J.A. 1986. *Ecology of Biological Invasions of North America and Hawaii.* Springer-Verlag, New York, p. 321.

Rees, M., Kohn, D., Hails, R., Crawley, M., and Malcolm, S. 1990. An ecological perspective to risk assessment. In *The Biosafety Results of Field Tests of Genetically Modified Plants and Microorganisms, Proc. Kiawah Island Conference, November 27 to 30.* Ed. by MacKenzie, D.R. and Henry, S.C. Agricultural Research Institute, Bethesda, MD, p. 9.

Regulatory Considerations: Genetically-Engineered Plants, Workshop Summary. 1987. Boyce Thompson Institute for Plant Research, Cornell University, Ithaca, NY, 106.

Saari, L.L., Cotterman, J.C., and Thill, D.C. 1994. Resistance to acetolactate synthase herbicides. In *Herbicide Resistance in Plants: Biology and Biochemistry.* Ed. by Powles, S. and Holtum, J. Lewis Publishers, Boca Raton, FL, p. 83.

Stallings, G.P., Thill, D.C., Mallory-Smith, C.A., and Lass, L.W. 1995a. Plant movement and seed dispersal of Russian thistle (*Salsola iberica*). *Weed Sci.* 43: 175.

Stallings, G.P., Thill, D.C., Mallory-Smith, C.A., and Shafii, B. 1995b. Pollen-mediated gene flow of sulfonylurea-resistant kochia (*Kochia scoparia*). *Weed Sci.* 43: 95.

Thill, D.C., O'Donovan, J.T., and Mallory-Smith, C.A. 1994. Integrated weed management strategies for delaying herbicide resistance in wild oat (*Avena fatua*), *Phytoprotection,* 75: 161.

Thompson, C.R., Thill, D.C., and Shafii, B. 1994. Germination characteristics of sulfonylurea resistant and susceptible kochia (*Kochia scoparia*). *Weed Sci.* 42:

Tilney-Basset, R.A.E. 1975. The genetics of plastid variegation. In *Genetics and Biogenesis of Mitochondria and Chloroplasts.* Ed. by Birky, C.W., Jr., Perlman, P.S., and Byers, R.J. Ohio State University Press, Columbus, p. 268.

Williamson, M. 1991. Assessment of the hazards from genetically-engineered plants: the work of the advisory committee on genetic manipulation intentional introduction sub-committee. In *Herbicide Resistance in Weeds and Crops.* Ed. by Caseley, J.C., Cussans, G.W., and Atkin, R.K. Butterworth-Heinemann, Oxford, p. 375.

PART V

REGULATION

Chapter 22

HERBICIDE-TOLERANT PLANTS DEVELOPED THROUGH BIOTECHNOLOGY: REGULATORY CONSIDERATIONS IN THE UNITED STATES

Diane B. Re
Stephen G. Rogers
Terry B. Stone
Frank S. Serdy

CONTENTS

1. INTRODUCTION

The development of herbicide-tolerant plants through biotechnology is continuing to advance steadily, with the commercialization of several products expected in the mid to late 1990s. Reviewing the products that are currently being field tested provides insight into the progress in this area. To date, herbicide-tolerant plants account for 28% of the field tests approved by the United States Department of Agriculture (USDA) (Foudin 1994a). Tolerance to herbicides such as bromoxynil, glufosinate, glyphosate, imidazolinones, and sulfonylureas account for the majority of the herbicide-tolerant plants being evaluated in the field today. Herbicide-tolerant crops that have been field tested in the U.S. include arabidopsis, corn, cotton, peanut, potato, rapeseed, rice, soybean, sugarbeet, tobacco, tomato, and wheat (USDA/APHIS/BBEP 1994).

In order for these products to be commercialized, an appropriate science-based regulatory framework must be in place. This framework must ensure the public of the food, feed, and environmental safety of these products without creating a duplicative and excessive regulatory obstacle for those producing these products. Anticipating the need for regulatory oversight of the

1-56670-045-0/96/$0.00+$.50
© 1996 by CRC Press Inc.

science of biotechnology and its products, the Office of Science and Technology Policy (OSTP) issued the "Coordinated Framework for Regulation of Biotechnology" on June 26, 1986 (Office of Science and Technology Policy 1986). This coordinated framework assigned regulatory responsibility of biotechnology to existing federal agencies under existing proven statutes based upon the opinion that no new agency or legislation was required. The following paragraphs discuss the elements of the current regulatory framework.

2. UNITED STATES DEPARTMENT OF AGRICULTURE (USDA)

USDA regulations are focused on the organism or the product and any risks posed by a specific use of that organism or product, rather than on the process used in production. The USDA has a statutory mandate to protect U.S. agriculture against the introduction and dissemination of plant pests. Under the authority granted by the Federal Plant Pest Act (FPPA) (Federal Plant Pest Act 1988) and the Plant Quarantine Act (PQA) (Plant Quarantine Act 1988), the Animal and Plant Health Inspection Service (APHIS) regulates the movement into and through the U.S. of plants, plant products, plant pests, and any product or article that may contain a plant pest at the time of movement (Joyce and Payne 1992).

The USDA issued regulations on June 16, 1987 (7CFR§340) covering the testing and commercialization of genetically modified plants that are plant pests or that present the potential for plant pest risk (regulated article) (APHIS/USDA 1987, 7CFR§340 1992). As of June 1994, APHIS has approved over 1091 permits for field trials of regulated articles, with over 2962 actual field test sites involving 43 different plant species in 41 states and Puerto Rico (Foudin 1994b).

Permits are required from the USDA for the importation, interstate movement, or environmental release of genetically modified plants that may be considered regulated articles. The permit application must contain sufficient information on the crop plant, the nature of the genetic modification, and the protocol for conducting the field trial to allow for an evaluation of any potential plant pest risk or environmental effects that may result from the field trial. The USDA has 120 days to approve an environmental release. During this time period, a notice of receipt is published in the *Federal Register,* a copy of the application is sent to the state(s) in which the test is being conducted (the state has 30 d to comment and add supplemental conditions if deemed necessary), and an environmental assessment (EA) is written. The permit is issued and published in the *Federal Register* if and when an EA results in a finding of "no plant pest risk" and of "no significant impact on the environment."

Due to increased experience and scientific expertise on the products of biotechnology and on the potential for interactions with the environment, APHIS amended the regulations pertaining to the introduction of certain genetically modified organisms and products to provide for a notification process for the introduction of certain plants, and to provide for a petition process allowing for a determination that certain plants are no longer regulated articles (APHIS/USDA 1987, 7CFR§340 1992).

Regulated articles which meet the requirements and the performance standards specified in 7CFR§340.3 are eligible for introduction under the notification procedure. Corn, cotton, potato, soybean, tobacco, and tomato are the six crops that are currently eligible for introduction under this procedure. APHIS has the authority to expand this list to include any additional plant species that they determine may be introduced safely. If the appropriate information is included in the notifications, APHIS will provide acknowledgment of the notifications within 10 days for interstate movement, within 30 days for importation, and within 30 days for an environmental release. APHIS has continued to keep the states informed and notifies the appropriate state regulatory officials within 5 days of receipt of a notification.

An applicant can apply for a petition for determination of nonregulated status as outlined in 7CFR§340.6. The petition must include the following information: (1) description of the biology of the nonmodified recipient plant, (2) relevant experimental data and publications, (3) a detailed description of the differences in genotype between the regulated article and the nonmodified recipient organism, and (4) a detailed description of the phenotype of the regulated article.

After filing the completed petition, APHIS will publish a notice in the *Federal Register* requesting public comment. This notice specifies that public comments will be accepted for 60 days commencing with the date of the notice, and any written comments received will become part of the petition file. APHIS has 180 days from receipt of a complete petition to respond to the applicant. The response will either approve the petition in whole or in part, or deny the petition. The decision will be placed in the public petition file in the offices of APHIS and a notice of availability published in the *Federal Register.* As of September 1994, APHIS has approved two petitions under these regulations, BXN™ Cotton (APHIS/USDA 1994a) and Roundup Ready™ Soybean (APHIS/USDA 1994b). Several other petitions are currently under review and more approvals are anticipated. Previously, APHIS approved the Flavr Savr™ Tomato through interpretive rulemaking (APHIS/USDA 1992).

3. ENVIRONMENTAL PROTECTION AGENCY (EPA)

The role that the EPA plays in the regulation of herbicide-tolerant plants is limited to approval of the herbicide that is to be applied to the plant. EPA will have no direct role in the regulation of the plant itself, although it may collaborate with the USDA on particular environmental considerations.

EPA will regulate the herbicide to be used on the genetically modified herbicide-tolerant plants in the same way it now regulates the use of herbicides on conventionally bred crop plants. Under the Federal Insecticide, Fungicide, and Rodenticide Act (FIFRA), EPA has the authority to regulate the development, sale, distribution, use, storage, and disposal of pesticides (FIFRA 1988). Extensive guidance on the regulatory requirements are outlined in 40 CFR Part 158. Data are generally required in the following areas:

- Product chemistry
- Hazard evaluation; humans and domestic animals
- Hazard evaluation; wildlife and aquatic organisms
- Hazard evaluation; nontarget plants and insects
- Environmental fate
- Residue chemistry
- Spray drift evaluation

Data requirements are based upon the characteristics of the herbicide, the proposed use, and the proposed site. If the herbicide was not previously registered for use, then all of the above data would be required. If the herbicide is registered for other uses, then only the data required to support the new or modified use would be required. The EPA will grant tolerances to cover any residues of the herbicide in the crop under Section 408 and 409 of the Federal Food Drug and Cosmetic Act.

Should testing of the herbicide on the genetically modified plant exceed 10 acres, the EPA may require that the applicant obtain an Experimental Use Permit (EUP) to allow the expanded testing of the herbicide (EUP 1992). An EUP can be requested either so that none of the food or feed items produced by the plant are allowed to enter commerce (crop destruct) or in conjunction with a request for a temporary tolerance to allow the food and feed items to be marketed. The type and amount of data required to support the EUP is contingent upon whether the applicant wishes to destroy or market the food and feed items produced.

4. FOOD AND DRUG ADMINISTRATION (FDA)

The FDA is charged with maintaining a safe and wholesome food supply and derives its authority for its actions from the FFDCA (FFDCA 1988). Traditionally, the safety of foods has been regulated through the postmarket authority under Section 402a (21CFR§402(a)(1)). A new plant variety developed by selection and plant breeding is not subject to a premarket safety review and approval, as is obtained for a food or feed additive under Section 409 (21CFR§409). Only in rare instances in which a specific concern was identified for a food from a new plant variety has FDA exercised its oversight of plants and plant products developed by traditional breeding.

The Food and Drug Administration set forth its current policy for oversight of foods in its May 29, 1992 "Statement of Policy: Foods Derived from New Plant Varieties" (FDA 1992, Kessler et al. 1992). The FDA considers that foods from traditionally produced varieties and those from genetic modifications such as somaclonal variation or recombinant DNA methods are within the scope of the regulation. As for all food products, the responsibility for safety of these products resides with the producer, whether the food is derived from a product of traditional plant breeding or from a product of genetic engineering. Because the introduction of new traits either by introgression from a wild species or by genetic engineering may raise safety questions, the FDA states, "Producers can informally consult with FDA prior to marketing to ensure that the safety and regulatory status of a new food is properly resolved." The Policy defines subject areas for such "consultations" and provides decision tree diagrams with questions that producers should answer concerning various aspects of the starting plant variety, the source of new genetic material, the method used to obtain the new variety, information on any new protein produced in the new variety, and information on any changes in the composition of the food derived from the new variety. Presently consultations are being conducted on over a dozen new varieties, including several that have been developed to be herbicide tolerant. When appropriate, companies are conducting safety assessments and consulting with FDA on the results.

FDA has not prescribed a specific outcome of the consultation process. In most cases the outcome will be that all potential issues have been resolved to FDA satisfaction and that the new variety is not significantly altered when compared to varieties of the crop currently sold. In rare cases in which there are substantial changes to the food or when broader issues need to be considered, the outcome could be the need for a food additive petition.

5. STATE INVOLVEMENT

The majority of individual states have deferred to the authority of the USDA to regulate the interstate movement and environmental release of genetically modified plants. The USDA notifies each state prior to approving an interstate movement or environmental release permit or notification. The states then have the option to approve or deny the permit, request additional information, or impose supplemental conditions in addition to those required by the USDA.

In a few instances, notably North Carolina (Genetically Engineered Organisms Act 1989) and Minnesota (Minnesota Statute 1992a, b), additional regulatory programs were established to oversee the field release of genetically modified plants. These states currently require a permit (and in some instances an environmental assessment) separate from that provided and performed by the USDA. A number of states, including Hawaii, Illinois, and Wisconsin, have passed statutes that require notification to the state of any federal permit applications (Hawaii Revised Statute 1991, Illinois

Annual Statute 1992, Wisconsin Statute 1991). These regulations are changing, however. Since 1993 North Carolina, Minnesota, and Wisconsin have proposed amendments to their regulations that would eliminate or significantly reduce the information required for obtaining state permits.

These changes coincide with the establishment of the notification procedure by the USDA, as described above, and with a sustained history of performing genetically modified plant field trials without harm to the environment or nontarget organisms. Public concern over the environmental release of genetically modified plants has also diminished, as evidenced by the decline in public comment and an increased awareness of the benefits of these products to the environment, growers, and consumers. State regulation is expected to continue to decrease with increased confidence in and reliance on the increasingly more experienced and proven federal regulatory structure to oversee the development and commercialization of these products.

6. CONCLUSIONS

A critical step toward the successful implementation of plant biotechnology is the establishment of appropriate regulatory oversight for the development and ultimate commercialization of genetically modified plants. Such oversight assures the public that products developed from biotechnology are environmentally sound and are safe for use and consumption, without placing unnecessary regulatory constraints on the biotechnology community. In addition, appropriate regulatory oversight and approvals by state and federal authorities promotes public awareness and acceptance of genetically modified plant products.

REFERENCES

Animal and Plant Health Inspection Service (APHIS)/U.S. Department of Agriculture (USDA). 1987. Introduction of organisms and products altered or produced through genetic engineering which are plant pests or which there is reason to believe are plant pests. *Fed. Reg.* 52: 22892–22915.

Animal and Plant Health Inspection Service (APHIS)/U.S. Department of Agriculture (USDA). 1992. Interpretive ruling on Calgene, Inc., petition for determination of regulatory status of FLAVR SAVR™ tomato. *Fed. Reg.* 57: 47608–47616.

Animal and Plant Health Inspection Service (APHIS)/U.S. Department of Agriculture (USDA). 1994a. Availability of determination of nonregulated status of Calgene, Inc., genetically engineered cotton lines. *Fed. Reg.* 59: 8452–8453.

Animal and Plant Health Inspection Service (APHIS)/U.S. Department of Agriculture (USDA). 1994b. Availability of determination of nonregulated status of Monsanto Co., genetically engineered soybean line. *Fed. Reg.* 59: 26781–26782.

Experimental Use Permits (EUP). 1992. 40CFR§172.

Federal Food, Drug, and Cosmetics Act. 1988. 21USC§§301-392.

Federal Insecticide, Fungicide, and Rodenticide Act. 1988. 7USC§§136-136y.

Federal Plant Pest Act. 1988. 7USC§§150—150jj.

Food and Drug Administration. 1992. Statement of policy: foods derived from new plant varieties; notice. *Fed. Reg.* 57: 22984–23005.

Foudin, A. 1994a. Public communication. National Plant Board Meeting, Chicago, IL, August 7, 1994.

Foudin, A. 1994b. Personal communication, September.

Genetically Engineered Organisms Act of 1989. 1991. N.C. General Statute §§106.765—106.777. Supplement.

Hawaii Revised Statute. 1991. §321. Supplement.

Illinois Annual Statute. 1992. Ch. 111 1/2, para. 7601 (Smith-Hurd Supplement).

Introduction of Organisms and Products Altered or Produced Through Genetic Engineering Which Are Plant Pets or Which There Is Reason to Believe Are Plant Pests. 1992. 7CFR§340.

Joyce, C A. and Payne, J. H. 1992. Pest-resistant plants developed through biotechnology: regulatory considerations. *Adv. Eng. Pestic.* p. 337–350.

Kessler, D.A., Taylor, M.R., Maryanski, J.H., Flamm, L.E., and Kahl, E.L. 1992. The safety of foods developed by biotechnology. *Science* 256: 1747.

Minnesota Statute. 1992a. §§116C.91—116C.95. Supplement.

Minnesota Statute. 1992b. §§18F.01—18F.12. Supplement.

Office of Science and Technology Policy. 1986. Coordinated framework for the regulation of biotechnology. *Fed. Reg.* 51: 23301–23350.
Plant Quarantine Act. 1988. 7USC§§151–164a, 166–167.
USDA/APHIS/BBEP. 1994. Environmental Release Standard Report Regarding Permits and Notifications. May 27, 1994.
Wisconsin Statute. 1991. §146.60. Supplement.

Chapter 23

THE REGULATION OF HERBICIDE-RESISTANT CROPS IN EUROPE

Stephen Waters

CONTENTS

1-56670-045-0/96/$0.00+$.50
© 1996 by CRC Press Inc.

1. INTRODUCTION

All of the major agricultural crops in use today are inherently herbicide resistant (HR), providing farmers with the flexibility of applying selective postemergence herbicides to control weeds appearing within the crop. Historically, the regulation of these naturally herbicide resistant crops in Europe has been confined to the certification of varieties, not for their herbicide resistance, but for other attributes, such as their morphology, yield and pest resistance. On the other hand, postemergence selective herbicides are strictly regulated to ensure their efficacy and selectivity, as well as their safety for humans, livestock, and the environment. Through classical breeding and selection, as well as through the use of newer breeding techniques, such as mutagenesis, breeders have been successful for many years in developing varieties more resistant to or resistant to more herbicides, providing farmers with even greater flexibility in managing weed problems (Beversdorf et al. 1980; Newhouse et al. 1991; Howard and Baszczynski 1992). With the developments in genetic engineering techniques it is now possible to arrive at the same result by introducing specific herbicide resistance genes into crops. This technique offers the potential advantage of being able to render varieties resistant to previously nonselective, broad-spectrum herbicides, with favorable environmental and toxicological profiles. The fact that these transgenic herbicide-tolerant crops have been developed through genetic engineering, however, categorizes them as requiring supplementary regulatory oversight, including further assurances on human and environmental safety than has been required hitherto for HR varieties.

The task of describing the regulations governing the use of transgenic HR plants in Europe is a complex one, because of the evolving nature of the regulatory situation. Not only are there no such HR varieties commercially available, but there is also as yet no clearly defined pathway for obtaining regulatory approval for the commercial use of transgenic HR varieties and their products. The information provided in this chapter is a review of the important developments in the establishment of regulations relevant to the use of transgenic HR plants in Europe prior to July 1994.

2. BIOTECHNOLOGY REGULATION IN EUROPE

Of the several hundred legislative proposals targeted for implementation by the European Commission prior to the establishment of the European Economic Community (EEC) on January 1, 1993, more than 1 dozen concern biotechnology, and most of these concern transgenic plants (Hodgson 1992). Several proposals, including the majority of those relevant to transgenic plants, are still in draft form within the Commission, or are in the process of review by the European Parliament, the Council of Ministers, or within working groups and advisory committees of these institutions.

Of the 23 Directorate Generals (DGs) of the Commission, three in particular have been involved in drafting legislation concerning the regulation of biotechnology: DGIII, responsible for Industry and Internal Affairs; DGVI, responsible for Agriculture; and DGXI, responsible for Environment. The Directives emanating from these DGs relevant to the regulation of transgenic HR plants are discussed in detail in the following sections. Other draft legislation which may impact freedom to operate with transgenic plants, including the Directive on the Legal Protection of Biotechnological Inventions (Anon 1993a) and the Plant Variety Rights Regulation (Anon 1993b), are not discussed here.

The fact that no single DG within the Commission has overall responsibility for biotechnology, combined with the disparate views of the different DGs involved in biotechnology, led to the establishment of the Biotechnology Coordination Committee toward the end of 1990. Signs of legislative coordination within the Commission are now beginning to appear. There are vociferous claims that excessive regulation, compared to the U.S. in particular, is stifling the competitivity of the European biotechnology industries, however. These claims stem from the horizontal nature of the EU legislation, which focuses on process rather than product, and which has contributed much to the regulatory and public focus on risks rather than benefits. Comparisons between the regulation of biotechnology in the EU and in the U.S. have been published elsewhere (Anon 1993c; Brauer and Schlumberger 1993).

3. REGULATION OF RESEARCH AND DEVELOPMENT WORK WITH TRANSGENIC HERBICIDE-RESISTANT PLANTS

The research and development work with transgenic HR plants involves a series of routine steps, many of which were well developed in the past by breeders and agronomists in their efforts to discover new varieties and new herbicides. These steps may be classified into three phases, according to where they are performed:

1. Laboratory: includes the production of genetic transformants and their molecular characterization
2. Glasshouse (greenhouse) or growth room: includes assessments of herbicide resistance (selectivity) and segregation patterns
3. Field: includes assessments of herbicide performance in weed management; assessments of plant morphology, yield, and quality; and backcrossing, selection and development of new varieties

Research and development activities with transgenic HR plants were initially regulated under national regulations or guidelines, and more recently according to European Community directives: the directives 90/219/EEC on the contained use of genetically modified microorganisms (Anon 1990a), which includes laboratories, growth rooms, and glasshouses, and the directive 90/220/EEC on the deliberate release of genetically modified organisms (GMOs) into the environment (Anon 1990b), which covers field (and certain glasshouse) experiments. The 90/220/EEC directive is divided into two major parts: Part B applies to research and development work, while Part C concerns "placing on the market."

The lead DG in the drafting and implementation of 90/219/EEC and 90/220/EEC is DGXI, the Commission's Environmental division. Since their conception, there has been considerable opposition to these directives from industry, from parts of the Commission itself, and from some Member States. This opposition has been based largely on the horizontal nature of the directives, which regulate all activities involving the use of GMOs in the areas of pharmaceuticals, veterinary products, additives, foods, feeds, seeds, etc., under the same regulatory regime, whereas in the past these products have been regulated vertically or sectorally. Many additional concerns, including excessive bureaucracy, dual approval requirements, lack of confidentiality, and international competitiveness, were voiced at the outset and still continue to be sources of concern. Despite these concerns, the two directives were adopted by the Council of Ministers on April 23, 1990, with the deadline for community-wide implementation set for October 23, 1991.

3.1. MEMBER STATE IMPLEMENTATION OF DIRECTIVES 90/219/EEC AND 90/220/EEC

Both directives require the appointment of a national competent authority (NCA), whose responsibilities include review and approval of requests to work with GMOs, inspection, and coordination of national activities at the European level. In many Member States the NCA for directives 90/219/EEC and 90/220/EEC are one and the same, regulating activities ranging from human gene therapy to the laboratory production of transgenic HR plants to the release of recombinant animal vaccines. The competent Ministry is in many cases the Ministry of Environment (e.g., The Netherlands, Denmark), the Ministry of Public Health (e.g., Italy), or a combination of these Ministries (e.g., Germany, U.K.). In only two countries, France and Belgium, is the Ministry of Agriculture (jointly with the Ministry of Environment and/or the Ministry of Research) responsible for the agricultural aspects of the authorization to field test genetically modified plants. In some cases the Ministries are assisted by advisory committees, such as the Advisory Committee on Releases to the Environment (ACRE) in the U.K., the Commission on Biomolecular Genetics (CGB) in France, the Central Committee for Biological Safety (ZKBS) in Germany, and the Provisional Committee on Genetic Modification (VCOGEM) in The Netherlands. The constitution of these committees varies, but generally includes representation from the scientific community, national ministries, consumer associations, environmental organizations, ecologists, trade unions, and industrial representation. In France and the U.K. different committees advise on contained use (CGG in France and ACGM in the U.K.) and deliberate release (CGB and ACRE, respectively). Other countries, including Denmark, Italy, and Belgium, currently have no such committees.

While all countries have nominated competent authorities for the application of directives 90/219/EEC and 90/220/EEC, by mid-1994 implementation appears to have been completed in 8 of the 12 Member States (not yet fully implemented in Belgium, Greece, Ireland, and Luxembourg). The approaches to implementation of the Directives have varied; Germany introduced an entirely new law in October 1990, Denmark revised its 1986 biotechnology law to be in compliance with the directives, while the U.K. and The Netherlands amended existing environmental protection laws. This led to a diverse approach to regulation.

The directive 90/219/EEC was conceived to provide measures to protect human health and the environment from genetically modified microorganisms. Being adopted under Article 130s of the Treaty establishing the EC, the objective of which is protection of human health and the environment, Member States are able to impose more stringent controls. Many Member States, including France, the U.K., Germany, and Denmark, have extended the directive 90/219/EEC to include the regulation of work with genetically modified plants under contained conditions, including glasshouses. The directive 90/220/EEC, on the other hand, adopted under Article 100a of the Treaty, whose primary objective is to ensure the free movement of goods in the community, may not be translated into regulations more stringent than the agreed directive.

The directive 90/219/EEC categorizes microorganisms into two risk categories; group I (lower risk) and group II (higher risk), and activities are divided into those for research purposes and small scale (type A) and production purposes and large scale (type B). National authorities who have adapted the directive 90/219/EEC to the contained use of genetically modified plants have generally applied the provisions for group I type A activities. The first use of installations (laboratories, glasshouses) must be notified and records of activities must be kept. In Germany yet more stringent requirements existed, requiring authorizations for individual experiments. Subsequent amendments to the German Gene Law introduced in December 1993 relaxed the regulation of the contained use of such low-risk GMOs (Anon 1993d).

3.2. FIELD TESTING OF HERBICIDE-RESISTANT PLANTS

Research and development work with genetically modified plants under noncontained ("deliberate release") conditions is regulated under Part B of the 90/220/EEC directive. Of the 928 approvals for field tests of transgenic plants worldwide up to the end of 1993, it is estimated that 342 authorizations were for field tests with herbicide-resistant crops (HRCs) (Green Industry Biotechnology Platform database on field testing of transgenic plants worldwide, unpublished). About 45% of the trials with HR plants (151) were conducted in Europe, the majority in France (77), Belgium (36), and the U.K. (21), while only 5 trials (no HR tests) were approved in Germany up until the end of 1993. Prior to the implementation of the 90/220/EEC directive in October 1991,

91 field tests with transgenic HR plants were conducted in Europe under national regulations. The number of authorizations for field tests with transgenic HR plants in Europe decreased from 36 in 1991 to 30 field tests in both 1992 and 1993. In the rest of the world the comparable numbers were 35, 53, and 59. The comparison is distorted by the fact that in North America, from 1991 to 1993, numerous multisite authorizations were granted, whereas in Europe such authorizations were less common. In 1993, for example, 81 approvals were given for field tests with transgenic plants in Europe (EU and EFTA countries), representing some 200 test sites. The corresponding 1993 figures in North America (U.S. and Canada) were 172 authorizations covering 1250 test sites.

3.3 INFORMATION REQUIREMENTS FOR APPLICATIONS TO CONDUCT RESEARCH AND DEVELOPMENT FIELD TESTS

Prior to the adoption of the 90/220/EEC directive in 1991, Member States faced with requests to conduct field tests with transgenic plants (France, Belgium, U.K., Denmark, The Netherlands) developed their own information requirements and review procedures. Despite this, the types of information required by competent authorities for review and approval of field tests with transgenic HR plants did not differ greatly between EU Member States, although the emphasis placed on the different information requirements varied. France, for instance, placed particular emphasis on the molecular characterization of transformed plants, whereas other Member States do not require as much information on this aspect.

While national review procedures have not been significantly modified with the introduction of 90/220/EEC, the information requirements are becoming increasingly similar, although the differences in national emphasis persist. The types of information which form the basis of the risk assessment are listed in detail in Annex II of the 90/220/EEC directive and are not discussed here. The fact that the Annex II applies to the field testing of microorganisms, plants, and animals has introduced further inconsistencies between Member States, because only relevant questions need be addressed. This source of inconsistency was redressed recently by the adoption of a modified Annex II specifically for genetically modified higher plants (Anon 1994a). Differences also exist in the degree of detail expected by regulators, differences in interpretation of the scientific literature (e.g, the likelihood of outcrossing from oilseed rape to wild relatives), as well as in the overall interpretation of what constitutes risk and acceptable risk. The nine different EU languages are a further complication. The idealistic notion of a single field test application and authorization valid for all EU Member States appears to be a long way off.

3.4. FEES FOR CONDUCTING RESEARCH AND DEVELOPMENT TESTS WITH TRANSGENIC PLANTS

While the majority of EU Member States do not request fees, a number of Member States have developed a scheme of charges for applications to conduct field tests with transgenic crops. Currently in France the only payments required are those necessary to cover the costs of the independent expert's review and presentation of the application to field test. In the U.K. the fee schedule was introduced along with the transposition of the 90/220/EEC directive (Anon 1992a). The fee scale is based on the costs of administrating applications, including ministerial review, review by ACRE, and inspection. The fees range from 600 ECU ($US670) for streamlined notification to 2400 ECU for full notification. The reduced fees apply to various types of trials, including repeat notifications, trials with plants containing certain marker genes, trials with plant species having "specific characteristics" such as tobacco or corn, or trials with certain plants in which specific management practices are followed. With the recent introduction of fast-track procedures it is likely that the fee scheme will be revised (Anon 1994b). In Germany and Denmark the fees appear to cover more than administrative expenses. In Denmark fees for first-time tests are 20,000 ECU with reduced fees for continuing projects, determined in relation to the administrative effort (on the order of 6500 ECU). In Germany the scale of fees ranges from 2500 ECU for research conducted by public institutes, and for experiments for which no commercial value is anticipated, up to 80,000 ECU plus expenses. It is estimated that for the first German field tests of agronomically important crops in 1993, where 3 day public hearings were required, the costs for regulatory approval alone were estimated to be on the order of 100,000 ECU. Such high costs have clearly contributed to the low level of field testing in these countries and have acted as a clear discouragement to innovation. The requirement for public hearings in Germany has now been withdrawn from the new German Gene Law (Anon 1993d).

3.5 PROCEDURES FOR THE REVIEW AND APPROVAL OF TRANSGENIC HERBICIDE-RESISTANT FIELD TESTS

The directive 90/220/EEC details the general procedure for obtaining the authorization to conduct a field test, or so-called "deliberate release," with genetically modified plants. The notification is submitted to the NCA in the Member State where the test is to be carried out, and the NCA should respond to the notifier within 90 days of the notification. In addition to the 90 days are the periods required for providing any additional data requested by the NCA, as well as the time required for any public consultation.

There are, however, many variations of this general procedure. In some countries the technical review is carried out by advisory committees, who advise the responsible Ministry or Ministries; in other countries the scientific review is conducted by individuals within the competent Ministries. While in most countries the first step is submission of the notification to the NCA, in others preliminary reviews are necessary with the NCA or its advisory committee prior to submission. In the U.K. certain details of field tests, including the objective and location of the trial, must be published in local newspapers and various national authorities must be notified following submission of the notification (Anon 1993e). Although not required by the directive, The Netherlands, Denmark, and Germany elected to include a public consultation process in the approval procedure. In the former cases this consultation is in writing, while in Germany consultation was initially by way of a public hearing (Anon 1990c). The requirement for public hearings was withdrawn in the amendment of the German Gene Law in December 1993 (Anon 1993d), although public comment may now be submitted in writing. While the policy of public consultation appears to have been useful in Denmark, where a number of trials have now been conducted without incident, the public procedures in Germany and The Netherlands have not prevented systematic legal challenges, invasion of experimental sites, and destruction of plant material (Cremers 1992; Anon 1993f).

The EU authorization procedure also involves the submission and circulation of the Summary Notification Information Format (SNIF) along with the completed notification (Anon 1991a). The objective of the SNIF is to exchange information between Member States, as well as to allow all Member States to comment on notifications. While the other Member States have no right of veto, their comments may be taken into account by the Member State of notification when considering applications to conduct field tests. A 30 day waiting period is built into the 90 day review process to allow for review and comment by other Member States.

3.6 CONDITIONS FOR THE CONSENT TO FIELD TEST TRANSGENIC HERBICIDE-RESISTANT PLANTS

Authorizations to conduct field tests with genetically modified plants are routinely subject to a series of restrictions, the nature of which vary from Member State to Member State. The extent of these restrictions is related to the level of experience with the crop and trait, but also to the "biotechnology climate" in each Member State. While the number and the degree of these restrictions has tended to diminish as experience has developed within an individual Member State, in many instances the experience gained in other Member States, even with similar environments, is not taken into account. Common requirements to address the potential environmental risks of field experiments include the use of isolation distances appropriate to each crop, in order to minimize the likelihood of outcrossing to related plants; monitoring of the trial for at least one season following the field experiment, in some cases leaving the land fallow; and prevention of flowering on crops where this is practical, e.g., sugarbeet. In Member States in which a sizeable number of field tests have now been authorized, many of the early restrictions imposed because of limited experience have now been lifted or relaxed, for instance, for a number of crops the requirement for manual planting of seedlings pregerminated in paper pots, the need to bag or remove flowers (e.g., oilseed rape or potato), the destruction of waste material by autoclaving or burning, and so on. However, some Member States, particularly those with little or no hands-on experience in field testing of transgenic plants, still apply relatively stringent restrictions to crops or genes for which considerable positive experience has been accumulated elsewhere in the EU.

3.7 ENVIRONMENTAL SAFETY ISSUES FOR REGULATORY AUTHORITIES

The prospect of the commercial introduction of transgenic HRCs raised many environmental questions early on in the research and development process. It may be argued that some of these questions are premature for small-scale, isolated field experiments, but addressing the issues at an early stage does ensure continuity in the environmental safety assessment from research and development through to commercial introduction. Some of the environmental safety questions raised are crop and trait specific, but many are of a general nature, and not unique to the EU. There is considerable commonality between the issues being addressed in the EU and in other areas of the world. Such questions include those of potential weediness of transgenic plants (invasiveness, persistence, volunteers), gene stability, transfer of genes to weedy or wild relatives by outcrossing, and so on. The environmental safety assessment for transgenic HR plants extends to include the specific question of weeds developing herbicide resistance to the broader question of the impact of transgenic HRCs on the amount and quality of the herbicides to be used in the future.

Questions on the environmental safety of transgenic plants have raised much public debate. Opponents of biotechnology and some ecologists have argued that transgenic plants will behave as exotic species, risking the disturbance of the "natural" equilibrium to the point of becoming invasive. Breeders, alternately, have argued that transgenic plants are not exotic, but merely improvements in the plant species grown and used over decades, and in some cases centuries (Anon 1987). The results of more than 1000 individual experiments conducted to date have supported the breeders' claims that there are no intrinsic dangers from transgenic plants. They have behaved as any other new plant improved through classical breeding techniques. The much publicized PROSAMO study (Crawley et al. 1993) in the U.K. confirmed this conclusion. Nevertheless, each crop has its own characteristics, and each new transgenic trait, at least over the next few years, will be characterized in sufficient detail to ensure that the improved crop poses no new risks to the environment.

It is clear that a number of Member States are nevertheless uncertain about the impact of the introduction of transgenic HRCs, and a number of countries have instigated specific studies in this area. In Germany a study conducted by the Scientific Centre for Social Research in Berlin (WZB) and financed by the Ministry of Technology was initiated in early 1991 with the aim of bringing together specialists from all spheres to discuss the impending introduction of transgenic HR plants. The areas considered were broad, including patenting, law, ethics, agronomy, environmental science, weed science, integrated pest management, soil erosion, genetic resources, Third World impact, toxicology, pesticide registration, and varietal registration. Representation was from many quarters, including academia, breeders, the chemical and biotechnology industries, agronomists, government, environmental groups, and biological farmers. Unfortunately, all environmental pressure groups withdrew from the exercise during the closing session (Loehr 1993; Ronzheimer 1993), apparently dissatisfied with the process and with the conclusion that genetic engineering approaches to plant improvement do not pose more risks than conventional breeding, and therefore should not require additional regulation (Kiper 1993). The results of this detailed and lengthy (more than 2 years) review process were published recently (van den Daele et. al. 1994). In parallel with the WZB study the German Biological Office for Agriculture and Forestry conducted its own study and concluded that "breeding for herbicide resistance is a gradual advancement of chemical weed control" (Dietz et al. 1993).

4. REGULATORY REQUIREMENTS FOR THE COMMERCIAL INTRODUCTION OF TRANSGENIC HERBICIDE-RESISTANT PLANTS

Those HR traits that have been developed most extensively in Europe have been glyphosate resistance in oilseed rape and sugarbeet; glufosinate resistance in oilseed rape, sugarbeet and corn; and bromoxynil resistance in tobacco and oilseed rape. By mid-1994, there were two requests for authorization to introduce transgenic crops containing HR genes onto the European market, according to the EU directive 90/220/EEC. The first application, submitted in France by SEITA and Rhône-Poulenc, was for the commercial use of transgenic bromoxynil-resistant tobacco seeds. Approval to market the seed was given under Article 21 of the directive 90/220/EEC, because Member State support for the approval was not unanimous (Anon 1994c, d). Approval means that the tobacco plants may now be field tested without notification, but the tobacco is not expected to be marketed for several years, as bromoxynil has yet to be registered for use on tobacco.

The second request, submitted by Plant Genetic Systems in the U.K., was for authorization to market hybrid oilseed rape produced using genetically engineered sterility and fertility restoration systems. The hybrid production system involves the use of a glufosinate-resistance gene in at least one of the parental lines, and commercially available hybrids will be completely resistant to glufosinate if the glufosinate resistance gene is present in both parental lines. The submission was first reviewed and approved by ACRE, and at the time of this writing was still under review by the competent authorities in the other EU Member States (Anon 1994e, f).

With the exception of tobacco, which has no food uses, and for which there are no variety certification requirements, transgenic HRCs and their corresponding herbicides are likely to require four regulatory clearances prior to their commercial introduction:

1. Environmental safety of the crop, according to part C of the 90/220/EEC Deliberate Release directive (Anon 1990b)
2. Food safety of the crop, according to the proposed (although not yet agreed upon) Novel Food regulation from DGIII (Anon 1992b, g)
3. Varietal certification, according to either existing national or EU requirements (Anon 1993g)
4. Herbicide safety and efficacy, according to the EU directive 91/414 Placing of Plant Protection Products on the Market, introduced by DGVI (Anon 1991b)

While the herbicide approval requirements are specific to the active ingredient and its formulation, the information concerning varietal certification, food safety, and environmental safety is interrelated and crop specific. For this reason, in developing new legislation (novel food, varietal approval) the Commission is working on the "one-door, one key" principle. In the case of transgenic HRCs, the proposition is to conduct the food safety and environmental safety assessments within the varietal approval process, so that only a single regulatory submission will be required. The aim of this approach is to avoid regulatory overlap and to ensure product-based regulatory review.

4.1. ENVIRONMENTAL SAFETY

Prior to any adoption of EU legislation concerning variety approval and food safety, the environmental safety aspects of "placing on the market" of transgenic HR varieties will continue to be assessed according to the procedure described in part C of Directive 90/220/EEC (Anon 1990b). The information required for research and development work (Annex II) must also be supplied with notifications for placing on the market, along with the SNIF for placing on the market (Anon 1992c), as well as the additional information described in Annex III of the Directive. The supplementary information is product specific, including expected use of the product, storage, packaging, and labeling information.

The approval procedure involves an initial review and consent by the Member State of notification, where the product is placed on the market for the first time. The initial review period is limited to 90 days plus additional time for any further requests for information by the competent authority. Following its consent the initial Member State forwards the dossier and any conditions of consent to the Commission, which forwards it immediately to other Member States for their review and agreement. Other Member States have 60 days in which to raise any objections. If such objections cannot be resolved between the objecting Member State and the Member State of notification, a Community decision procedure is then invoked.

4.2. FOOD SAFETY

In 1990 DGIII initiated the drafting of the Novel Food and Novel Food Ingredients Regulation, which was first published in the *Official Journal of the European Communities* in July 1992 (Anon 1992b). The initial concept of novel food was not restricted to foods derived from transgenic plants, but rather encompassed all foods which are new in the diet, or which are likely to be consumed in larger quantities than they were previously. Because of the vagueness of the definition of novel foods there has been much debate about what should be included within the scope of the Novel Food and Novel Food Ingredients regulation. The many suggestions have included products of genetic engineering not covered by existing product regulations (i.e., exclude food additives produced with the aid of genetic engineering); all (and only) products of genetic engineering; and all products of genetic engineering, in addition to other novel foods (such as mycoproteins) and novel processes (such as ohmic heating). Whatever the outcome of the debate on the scope of the regulation, it would appear that at least some of the products of transgenic plants are likely to be regulated as novel foods. This may include foods which are substantially equivalent (Anon 1992d) to foods currently in use, although they contain only one or a few new proteins, in addition to the thousands that are already present. In the extreme case the regulation may affect highly processed products, such as oil derived from a transgenic oilseed rape variety, or sugar from a transgenic sugarbeet variety. No doubt, as the first transgenic varieties are introduced, the regulatory requirements will begin to be more clearly defined.

The initial Commission proposal on Novel Foods and Novel Food Ingredients suggested that the labeling of novel foods and food ingredients be decided on a case-by-case basis. This has turned out to be the most contentious element of the Commission proposal; opponents of biotechnology and some consumer organizations have called for compulsory, systematic, process-based labeling of products developed using any genetically modified organism. The report from the Committee on the Environment, Public Health and Consumer Protection to the European Parliament (Roth-Behrendt 1993) proposed that such foodstuffs should be labeled "produced with the aid of genetic technology," in order to "safeguard consumers' rights to information, transparency, and freedom of choice when making purchases." The food industry (Anon 1993h), plant breeders (Anon 1994h), and some regulators (Kahn 1993; Anon 1993i, j) are opposed to the idea of such broad-based labeling, arguing that products of biotechnology are not inherently less safe than existing products and that process-based labels may be more misleading than informative. While there is clearly a need to inform consumers about the methods used in food production, labeling does not achieve this in a meaningful way.

Labeling also raises many practical difficulties. The commodity crops, such as rice, corn, soybean, and wheat, cannot be identity preserved under existing infrastructures. Many foods will be derived from mixtures of commodity crops, both transgenic and nontransgenic, and in addition, most of these commodity crops are traded internationally in enormous volumes. Also fresh, identity-preserved foods such as tomatoes may be quite easily labeled, but the labeling of tomato paste prepared from a blend of processed tomato varieties, or of more complex food products such as pizzas, raises many questions about how process-based labeling can be implemented in a consistent and controllable manner.

Following its publication in the *Official Journal,* the European Parliament referred the Commission novel food proposal to the Committee on the Environment, Public Health and Consumer Protection as the Committee responsible. Its report (Roth-Behrendt 1993) was tabled in July 1993, proposing significant amendments, many of which were unacceptable to the Commission. The draft regulation and the proposed amendments have been debated by the European Parliament and voted upon October 27, 1993, with most of the amendments being adopted. An amended Commission proposal for the regulation of Novel Foods and Novel Food Ingredients was published in the *Official Journal of the European Communities* in January 1994 (Anon 1994g). On June 16, 1994 the Internal Market Council failed to adopt a common position on the regulation, and at the time of this writing the future of the European regulation of foods derived from transgenic plants is still uncertain.

Faced with the imminent appearance of food products produced from transgenic plants, and no prospects of a rapid resolution of the debate on the EU Novel Food regulation, the Ministry of Welfare, Health and Cultural Affairs in The Netherlands introduced its national Novel Foods Regulation in July 1993 (Anon 1993k). The regulation was introduced as an interim measure until an EU food regulation is established. The scope of The Netherlands' regulation includes all foodstuffs "produced with the help of genetically modified organisms," but also extends to other categories of new foods. A decision tree approach to the safety evaluation of such foods was proposed previously to the Ministry by the Health Council of The Netherlands (Anon 1992e). Other proposals for evaluating the safety of foods derived from transgenic plants have been published elsewhere by institutes represented on the newly formed Advisory Committee on Novel Foods (Kok 1993; Kok et al. 1993). The Netherlands' regulation contains no specific requirements or guidelines on labeling; any labeling requirements will be based on the opinion of the existing Advisory Committee on the Commodities Act.

In the U.K. the Advisory Committee on Novel Foods and Processes (ACNFP) advises the Ministries of Health and Agriculture on matters relating to novel foods. Clearances for the marketing of novel foods are given under a voluntary scheme, outlined in the ACNFP 1991 publication *Guidelines on the Assessment of Novel Foods and Processes* (Anon 1991c). The scheme is based on a decision-tree approach, which defines the categories of novel foods and the types of information which should be provided for a thorough assessment. The ACNFP considers labeling on a case-by-case basis, but labeling guidelines are provided in an appendix to the Guidelines. The labeling guidelines were reviewed recently by the U.K. Food Advisory Committee (Anon 1993j), following publication of the *Report of the Committee on the Ethics of Genetic Modification and Food Use* (Anon 1993l). The Food Advisory Committee concluded that there is no justification for labeling every food produced with the aid of genetic engineering, but in cases in which there may be ethical or religous concerns, labeling may be warranted.

4.3. VARIETY CERTIFICATION

Various EU directives specifying the requirements for the marketing of varieties of agricultural plants have been in existence since the 1960s. Under these directives plant variety approval is given at the national level, and the variety is then eligible for the Community variety list. The national approval process involves at least 2 years of variety trials, conducted under the auspices of national seed certification authorities, such as the National Institute of Agricultural Botany in the U.K., the Comité Technique Permanent de Sélection in France, and the Bundessortenamt in Germany. The criteria that are considered during the variety trials are distinctness, uniformity, and stability, as well as agronomic value (including yield, disease resistance, quality, and other characteristics) under both intensive and extensive management conditions. The best current varieties serve as performance references. Specific requirements have not been defined for transgenic varieties, but as yet, varietal authorities have had little direct experience with transgenic plants. In the case of transgenic HR varieties demonstration of selectivity will also be a requirement of the herbicide registration process, as it is currently for any selective herbicide, and it remains to be seen whether this will be an additional criterion for varietal approval.

A proposal for a new Community directive amending the existing directives on the marketing of seeds and varieties was drafted and published by the Commission (Anon 1993g). One of the objectives of the new directive is to ensure that the Commission policy of "one door, one key" is implemented for new plant varieties. In essence this means that applications for authorization to market a transgenic variety should contain all of the information necessary to conduct an environmental risk assessment (according to Directive 90/220/EEC) and, for those varieties intended for foods use, all of the information necessary to evaluate food safety (according to the pending Novel Food regulation). While the "one door, one key" approach appears attractive in principle, it remains to be seen how this will work in practice in the early years of transgenic varieties. Breeders experienced in predicting the success of their new varieties scale-up seed production in anticipation of the result. Undoubtedly, they will be more reluctant to scale-up seed production prior to knowing the outcome of the environmental and food safety reviews.

4.4. HERBICIDE SAFETY AND EFFICACY

The EU directive EC 91/414, Placing of Plant Protection Products on the Market (Anon 1991b), was enacted on July 25, 1993. The directive concerns the use of existing and new plant protection products; all products will be reviewed for compliance with the new directive before being listed in Annex I. While there are no specific requirements defined for herbicides used in conjunction with transgenic HR varieties, it is clear that the herbicide must be listed in Annex I of the directive, and must be approved for this particular use. For new herbicides the registration requirements are likely to be identical to the requirements for any other new herbicide. In most cases, however, the herbicides used with transgenic HRCs have been previously approved for other uses (e.g., glyphosate, glufosinate). In such cases the regulatory data packages must be updated to confirm safety and efficacy. The types of additional studies required are likely to include plant metabolism studies, field residue studies, and bioefficacy studies, including selectivity and weed control trials.

Considering the uncertainties of the numerous regulatory approval processes, one of the challenges of companies aiming to introduce transgenic HR varieties will undoubtedly be to ensure that all of the regulatory authorizations are in place in a similar time frame. While the "one door, one key" approach may be one means of ensuring coordinated variety, food and environmental assessments, ultimately the objective must be to avoid penalizing breeders developing transgenic varieties by time delays and excessive regulatory costs.

5. FUTURE REGULATION OF TRANSGENIC HERBICIDE-RESISTANT PLANTS IN EUROPE

It is difficult to see what lies ahead because of many current uncertainties. Clearly there has been a change of focus over time from environmental safety questions to food safety questions.

Understandably, there has been relatively little discussion about the registration of transgenic varieties or about the registration of herbicides for use on HR crops, because the regulatory requirements in these areas are well defined on the basis of the experience over the past several decades.

5.1. SIMPLIFIED PROCEDURES FOR RESEARCH AND DEVELOPMENT WORK WITH TRANSGENIC PLANTS

The Commission has made it clear that in its view there will be no major modifications of the 90/219/EEC and 90/220/EEC directives until all Member States have implemented the directives, and until sufficient experience has been gained in their application. Any short-term simplifications, therefore, must be introduced within the framework of the existing legislation.

In 1993, however, the first steps were taken toward simplifying the procedures for conducting research and development tests with transgenic plants (Anon 1993m). While not compromising environmental safety, the introduction of simplified procedures was justified on the basis of the experience developed in Europe and elsewhere since 1986, when the first field tests with transgenic plants were conducted. A number of other factors have been significant in advancing the debate on reducing the regulatory burden. These factors have no doubt included the EUs competitive position in the area of biotechnology *vís-á-vís* the U.S. and Japan, the introduction in the U.S. of the U.S. Department of Agriculture (USDA) notification system for selected crops meeting specific criteria (Anon 1993n), the findings of the U.K. House of Lords inquiry into biotechnology (Anon 1993c), and publication of the Commission's White Paper on Growth, Competitiveness and Employment, which highlighted the importance of biotechnology for future growth in the EU (Anon 1993o).

Whereas the publication of the USDA notification system outlined both the qualifying criteria and the nature of the simplification, the EU directive 90/220/EEC stipulates that "simplified procedures" must be introduced in a two-step process, starting with the establishment of EU criteria to define eligible crops and traits. This step was completed in October 1993 (Anon 1993m), following consultation among the Commission, Member States, industrial representatives, and academic experts. The initiative for the subsequent step of proposing simplified procedures must come from the Member States, so it is possible that different simplified procedures will be implemented in different Member States, rather than having a community wide simplification. Apart from "simplified procedures" the committee of NCAs (Committee 21) is also able to adapt the Annexes of the directive to technical progress, and a plant-specific Annex II has now been implemented to replace the more general Annex II of the directive 90/220/EEC (Anon 1994a). At the same time an amended SNIF was adopted for reseach and development purposes (Anon 1994i). In the U.K. national fast-track procedures were published in early 1994, outside of the EU simplification process. These procedures provide for 30-d approval for low-hazard crop–trait combinations (Anon 1994b). Such simplifications will significantly reduce the burden of preparing field test applications, particularly if adopted by a majority of Member States.

The report of the U.K. House of Lords Select Committee (Anon 1993c) concluded that the regulation of the new biotechnology is "excessively precautionary," "obsolescent," and "unscientific." The Committee's proposals included product-specific information requirements for research and development, a shorter review period, focus on product-based rather than process-based regulation, and no process-based labeling. These recommendations will not only impact the regulation of transgenic plants in the U.K., but are also likely to prompt a review of the EU directives 90/219/EEC and 90/220/EEC and their implementation throughout the rest of the Community.

In Germany, too, there have been significant moves toward relaxation of the stringent controls on work with GMOs. The German Gene Law, adopted in July 1990 (Anon 1990c), encouraged major German companies, including BASF, Boehringer Mannheim, and Hoechst, to preferentially expand existing facilities outside Germany or to relocate their research or production facilities elsewhere. In May 1993 the German Government announced that it would amend the Gene Law to simplify the approval process (Anon 1993p, q) for research work with GMOs; these amendments were enacted in December 1993 (Anon 1993d). In 1994 the first trials with HR (glufosinate resistance) oilseed rape and corn were approved in Germany (Anon 1994j).

6. REGULATORY AND CONSUMER ACCEPTANCE OF PRODUCTS OF BIOTECHNOLOGY

Products produced by genetic engineering processes have been available in Europe for many years, initially in the pharmaceutical industry (e.g., insulin, growth hormone), and more recently in the food industry (e.g., chymosin for cheese production) and in the animal feed industry (e.g., feed additives such as phytase). Despite having received regulatory approval on the basis of their product characteristics, some of these products have not been used widely because of industry fears of possible consumer rejection of established branded products. Other products of biotechnology have not been able to obtain final regulatory approval, despite clearances on the grounds of safety, efficacy, and quality. In such cases socio-economic need has been introduced as a "fourth hurdle." It is against this background of regulatory and commercial uncertainty that transgenic crops, including HRCs, are being developed in Europe. In addition to carrying the stigma of biotechnology, transgenic HRCs are associated with the generally poor image of pesticides. The benefits of pesticides, and of transgenic HRCs, are quickly overlooked by consumers, who are becoming more critical about the wholesomeness of the food they eat, and about the intensive management practices which are used to ensure an inexpensive and abundant food supply. Politicians, legislators, and regulators are well tuned to the sensitivities of consumers or of consumer associations, and legislation and regulation is strongly influenced by perceived public opinion. Examples of the inseparability of public opinion and regulation are the debates about the use of antibiotic resistance markers (Anon 1993r, s) and the specific labeling of foods derived from transgenic plants (Roth-Behrendt 1993; Anon 1993i). Calls for the labeling of products produced with the aid of genetic engineering have been based on perceptions of public opinion and the consumer's right to know. Recent surveys, however, suggest that consumers may be less resistant to the introduction of transgenic plants and their products than is generally believed. Food surveys in the U.K. indicate low public awareness of foods derived from GMOs, and low interest in the labeling of such products (Anon 1993i). The Dutch consumer research organisation, SWOKA, studied the acceptance of biotechnological developments and applications in the food sector (Hamstra 1993). While products made with the assistance of genetic engineering were less favored than products of traditional biotechnology, consumers were still likely to accept the application of genetic engineering in food production, provided that it did not involve animals, that no adverse effects were expected, and that there were benefits other than those for the producer. On the face of it, transgenic HRCs that provide benefits for farmers and for the environment appear to meet these criteria.

REFERENCES

Anon 1987. *Introduction of Recombinant DNA-Engineered Organisms Into the Environment: Key Issues*, National Academy of Sciences Press, Washington, D.C.

Anon 1990a. Council Directive of 23 April 1990 on the contained use of genetically modified micro-organisms (90/219/EEC). *Off. J. Eur. Communities.* No. L 117: 1.

Anon 1990b. Council Directive of 23 April 1990 on the deliberate release into the environment of genetically modified organisms (90/220/EEC). *Off. J. Eur. Communities.* No. L 117: 15.

Anon 1990c. Gesetz zur Regelung von Fragen der Gentechnik, Bundesgesetzblatt. Jahrgang 1990. Teil I(No. 28): 1080.

Anon 1991a. Council Decision of 4 November 1991 concerning the summary notification information format referred to in article 9 of Directive 90/220/EEC on the deliberate release into the environment of genetically modified organisms (91/596/EEC). *Off. J. Eur. Communities.* No. L 322: 1.

Anon 1991b. Council directive of 15 July 1991 concerning the placing of plant protection products on the market. *Off. J. Eur. Communities.* No. L 230: 1.
Anon 1991c. Advisory Committee on Novel Foods and Processes, Dept. of Health Report on Health and Social Subjects No. 38. *Guidelines on the Assessment of Novel Foods and Processes.* Her Majesty's Stationery Office, London.
Anon 1992a. Scheme of Fees and Charges in Respect of Applications and Consents for the Release and Marketing of GMOs under Part VI of the Environmental Protection Act, 1990, United Kingdom Department of Environment, London.
Anon 1992b. Proposal for a Council Regulation (EEC) on novel foods and novel food ingredients (92/C 190/04), COM(92) 295. *Off. J. Eur. Communities.* No. C 190: 3.
Anon 1992c. Commission Decision of 11 February 1992 concerning the summary notification information format referred to in article 12 of Directive 90/220/EEC (92/146/EEC). *Off. J. Eur. Communities.* No. L 60: 19.
Anon 1992d. Safety Evaluation of Foods Derived by Modern Biotechnology: Concepts and Principles, OECD Publications, Paris.
Anon 1992e. Health Council of the Netherlands: Safety of Food Produced by New Biotechnology. Publ. No. 92/03E. HCN, The Hague.
Anon 1993a. Amended proposal for a Council directive on the legal protection of biotechnological inventions. *Off. J. Eur. Communities.* No. C 44: 36.
Anon 1993b. Amended proposal for a Council regulation (EEC) on community plant variety rights. *Off. J. Eur. Communities.* No. C 113: 7.
Anon 1993c. *Regulation of the United Kingdom Biotechnology Industry and Global Competitiveness,* 7th Rep. House of Lords Select Committee on Science and Technology, Session 1992-93, London.
Anon 1993d. Erstes Gesetz zur Anderung des Gentechnikgesetzes, Bundesgesetzblatt. Jahrgang 1993. Teil I(No. 67): 2059.
Anon 1993e. DOE/ACRE Guidance Note No. 1: The Rgulation and Control of the Deliberate Release of Genetically Modified Organisms. Her Majesty's Stationery Office, London.
Anon 1993f. Gentechnik-Gegner campieren bei Wetze, *Lokales Einbeck*, April 8.
Anon 1993g. Proposal for a Council Directive Amending Directives 66/400/EEC, 66/401/EEC, 66/402/EEC, 66/403/EEC, 69/208/EEC, 70/457/EEC and 70/458/EEC on the Marketing of Beet Seed, Fodder Plant Seed, Cereal Seed, Seed Potatoes, Seed of Oil and Fibre Plants and Vegetable Seed and on the Common Catalogue of Varieties of Agricultural Plant Species. Commission of the European Communities, COM(93) 598final, Brussels, 26 November.
Anon 1993h. Confederation of the Food and Drink Industries of the EEC, Labelling of Foodstuffs. *CIAA Bull.* 3.
Anon 1993i. United Kingdom Food Advisory Committee Consultation Paper, The Labelling of Foods Sourced from Genetically Modified Organisms: Revised Guidelines.
Anon 1993j. Ministry of Agriculture, Fisheries and Food. *Nicholas Soames Announces GM Food Labelling Proposals.* News Release 378/93.
Anon 1993k. Dutch Novel Foods Regulation under the Commodities Act (WVC Warenwetregeling Nieuwe voedingsmiddelen). *Staatscourant* 139(No. DGVgz/VVP/L 931313): 6.
Anon 1993l. Report of the Committee on the Ethics of Genetic Modification and Food Use. Ministry of Agriculture, Fisheries and Food. Her Majesty's Stationery Office, London.
Anon 1993m. Commission decision of 22 October 1993 establishing the criteria for simplified procedures concerning the deliberate release into the environment of genetically modified plants pursuant to Article 6(5) of Council Directive 90/220/EEC. *Off. J. Eur. Communities.* No. L 279: 42.
Anon 1993n. Genetically engineered organisms and products, notification procedures for the introduction of certain regulated articles; and petition for nonregulated status; final rule. *Fed. Reg. 58(No. 60).*
Anon 1993o. Growth, Competitiveness, Employment. The Challenges and Ways Forward into the 21st Century. White paper of the Commission of the European Communities. *Bull. Eur. Communities.* Suppl.: 100.
Anon 1993p. Amendment of the German Genetic Engineering Law (GenTG). *Eur. Biotechnol. Info. Serv. (EBIS) Newsl.* 3(2): 26.
Anon 1993q. Green light for German genetic engineering law changes. *Asgrow* 186: 18.
Anon 1993r. Advisory Committee on Novel Foods and Processes. Consultation Paper on the Use of Antibiotic Resistance Markers in Genetically Modified Food Organisms. Ministry of Agriculture, Fisheries and Food, London.
Anon 1993s. Health Aspects of Marker Genes in Genetically Modified Plants. Report of a WHO Workshop. WHO Food Safety Unit, Geneva.
Anon 1994a. Commission Directive 94/15/EC of 15 April 1994 adapting to technical progress for the first time Council Directive 90/220/EEC on the deliberate release into the environment of genetically modified organisms. *Off. J. Eur. Communities.* No. L 103: 20.
Anon 1994b. DOE/ACRE Guidance Note No. 2: Fast Track Procedures for Certain GMO Releases. Her Majesty's Stationery Office, London.
Anon 1994c. Transgenic tobacco. *Eur. Biotechnol. Newsl.* 175: 6.
Anon 1994d. EC authorizes genetically engineered tobacco plant. *Eur. Chem. News* 61(1615): 39.
Anon 1994e. First genetic oilseed is raising protests on marketing ethics. *Farm. News* April 22.
Anon 1994f. Designer crop is given go-ahead. *Daily Telegraph* April 20.

Anon 1994g. Amended proposal for a European Parliament and Council Regulation (EC) on novel foods and novel food ingredients (94/C16/07), COM(93)631. *Off. J. Eur. Communities.* No. C 16: 10.

Anon 1994h. Statement on the proposal for a regulation on novel foods and novel food ingredients. COMASSO Association of Plant Breeders of the European Economic Community,

Anon 1994i. Commission Decision of 15 April 1994 amending Council Decision 91/596/EEC concerning the summary notification information format referred to in article 9 of Council Directive 90/220/EEC (94/211/EC). *Off. J. Eur. Communities.* No. L 105: 26.

Anon 1994j. Germany backs genetic crop trials. *Agric. Supply Ind.* May 27.

Beversdorf, W.D., Weiss-Lerman, J., Erickson, L.R., and Souza Machado, V. 1980. Transfer of cytoplasmically-inherited triazine resistance from bird's rape to cultivated *Brassica campestris* and *Brassica napus. Can. J. Genet. Cytol.* 22: 167.

Brauer, D. and Schlumberger, H.D. 1993. The U.S. System for the Regulation of Recombinant DNA Operations in Research, Development and Production. Unpublished.

Crawley, M.J., Hails, R.S., Rees, M., Kohn, D., and Buxton, J. 1993. Ecology of transgenic oilseed rape in natural habitats. *Nature* 363: 620.

Cremers, H.C. 1992. Sabotage sets back Dutch biotechnology. *New Sci.* 22 August: 4.

Dietz, A., Niemann, P., Wenzel, G., Heidler, G., and Eggers, T. 1993. *Aspekte des Anbaus herbizidresistenter Kulturpflanzen.* Mitteilungen aus der Biologischen Bundesanstalt fuer Land- und Forstwirtschaft, Berlin Dahlem, 286, Parey, Berlin.

Hamstra, A.M. 1993. *Consumer Acceptance of Food Biotechnology: The Relation between Product Evaluation and Acceptance.* SWOKA Institute for Consumer Research, Sinteur, Leiden.

Hodgson, J. 1992. Europe, Maastricht, and biotechnology. *Bio/Technology* 10: 1421.

Howard, J. and Baszczynski, C. 1992. Herbicide resistance, *Agro-Food-Ind. Hi-Tech* September/October: 3.

Kahn, A. 1993. Commission du Genie Biomoleculaire. Position Statement on the Labelling of Foods Obtained through Genetic Engineering. Publication of the French Ministry of Agriculture and Forestry, Directorate General of Foodstuffs, Paris.

Kiper, M. 1993. Genrisiken sind Spekulation. *Sueddtsch. Ztg.* June 17.

Kok, E.J. 1993. Evaluation of Strategies to Assess the Food Safety of Genetically Modified Agricultural Products. RIKILT-DLO Rep. 92.48.

Kok, E.J., Reynaerts, A., and Kuiper, H.A. 1993. Novel food products from genetically modified plants: do they need additional food safety regulations?, *Trends Food Sci. and Technol.* 4: 42.

Loehr, W. 1993. Umweltgruppen brechen Gentech-Dialog ab. *Tagestagung* June 15.

Newhouse, K., Singh, B., Shaner, D., and Stidham, M. 1991. Mutations in corn (*Zea mays* L.) conferring resistance to imidazolinone herbicides. *Theor. Appl. Gen.* 83: 65.

Ronzheimer, M. 1993. Neues Verfahren — alte Positionen. *Tagesspiegel* June 21.

Roth-Behrendt, D. 1993. Report of the Committee on the Environment, Public Health and Consumer Protection on the Commission Proposal for a Council Regulation on Novel Foods and Food Ingredients. European Parliament Session Documents, July 26.

Van den Daele, W., Pühler, A., Sukopp, H., Eds. 1994. *Verfahren zur Technikfolgenabschätzung des Anbaus von Kulturpflanzen mit Gentechnisch Erzeugter Herbizidresistenz.* Heft 1–19. Wissenschaftszentrum, Berlin.

Chapter 24

THE REGULATION OF HERBICIDE-RESISTANT CROPS IN AUSTRALIA

George D. McLean
Graeme Evans

CONTENTS

1-56670-045-0/96/$0.00+$.50
© 1996 by CRC Press Inc.

1. INTRODUCTION

The commercial development and use of herbicide-resistant crops (HRCs) raises a number of issues of concern within the Australian community which invite regulation by governments. However, regulations that are too restrictive could constrain significant commercial opportunities and economic development, while regulations that are too lax may alienate large sections of the community. In moving to regulate the release and growing of genetically modified plants, the Australian Commonwealth Government will need to consider the issues of public concern. These must be balanced against the potential benefits from the use of HRCs.

Public concern about the hazards of releasing genetically modified organisms (GMOs) into the environment pose a suite of issues that cannot be assessed on purely technical grounds. While the concerns that GMOs hold for some people may recede as the safety of these is assessed, the issues of public concern are politically sensitive and will have to be addressed in legislation, given the inevitable demands from the cropping industries for access to genetically manipulated crop plants.

An increasing number of both preemergent and postemergent herbicides came into the marketplace from the early 1970s, and allowed farmers to intensify cropping while reducing the hazard of erosion in cultivated soils. However, the stability of some farming systems is now being threatened by a growing problem of herbicide resistance in weeds. The strategic use of HRCs could provide farmers with additional options for controlling these weeds.

Australia is heavily dependent on agriculture, which contributes about 30% to our export earnings, more than one third of which is derived from plant products. Industries such as wheat, sugar, and cotton export up to 80% of annual production. To the extent that exports may be vulnerable to the sensitivities of importers to food products derived from transgenic plants, regulations will need to be framed and implemented to ensure that access to overseas markets is maintained.

Recognizing that transgenic plants, including HRCs, will soon be a commercial reality, the Commonwealth Government has acknowledged the need for legislation to regulate their release and use. Given the constitutional responsibility of the states for agricultural matters, a national approach will require the support of all governments. The Commonwealth Government and the state governments are currently discussing the form that the legislation should take.

Like any new technology, the development of crops incorporating herbicide resistance by means of genetic engineering technologies is generating considerable public concern. As such, it is presenting governments with a panoply of issues that need to be worked through before introducing regulations to control field testing and, ultimately, release for commercial production. Government regulation is about steering an even-handed course between the demands of various interest groups to provide the best possible outcome for the community.

Specific public concerns include the safety of releasing GMOs into the environment and a view that the use of HRCs will lead to an increase in herbicide use. Protagonists for the new technology argue that HRCs offer a new approach to weed control that will provide flexibility to farmers and opportunities to increase productivity. Major issues for legislators in Australia are the need to maintain access for our agricultural commodities to international markets, the need to gain consumer acceptance for our products, and, where possible, to differentiate our commodities from those of competitors to gain a marketing advantage. Agricultural products represent about 30% of export earnings, the importance of which is emphasized by the fact that exports account for about 14% of Australia's gross domestic product, a high figure by comparison to the U.S. and Japan, where exports currently represent about 7 and 9% of gross domestic product, respectively (Phillips et al. 1993). For the wheat, cotton, and sugarcane industries, exports represent about 80% or more of annual production.

There are no regulations specifically controlling the growing of HRCs in Australia, reflecting the newness of the technological development. There is, however, a large body of legislation, both state and Commonwealth, that may have application to matters arising from the emergence of recombinant DNA technology, including the development of plants with resistance to herbicides. An analysis of the legislation is provided by Barker (1984). Most of the legislation, if not all of it, predates the commercial reality of transgenic plants. The fragmentary nature of the legislation, falling as it does under different jurisdictions, has resulted in agreement between the Commonwealth Government and the state governments of Australia to develop new legislation specifically regulating recombinant DNA technology, including the release of GMOs into the environment. Developments leading to this agreement are discussed later in this chapter.

Briefly, this chapter describes the framework in which regulations relating to the growing of HRCs need to be developed in Australia. Matters of particular concern are the role of agriculture in the national economy, the fragility of soils available for cropping, and the need to maintain access to international markets, where governments are under pressure to regulate foods derived from GMOs and to enhance the safety of food by reducing the use of pesticides.

2. CROPPING IN AUSTRALIA

The area of Australia is comparable to that of the U.S., Canada, or Europe. However, despite its large size, Australia does not have an abundance of arable land and <4% is used annually for cropping. Low and highly variable rainfall, together with its seasonality, further constrain opportunities for dryland cropping. Wheat is the most valuable crop, followed by cotton and sugarcane. While the wheat crop is almost entirely dryland, >80% of the cotton is irrigated, but a shortage of water and buoyant prices have encouraged expansion in dryland cotton in recent years. About 35% of the sugarcane crop is irrigated. Contraction of sugarcane from some traditional areas of production and expansion into the Burdekin irrigation area in northern Queensland will result in a larger proportion of the crop being irrigated. Horticulture tends to occupy the better classes of land where there is adequate water for irrigation. Potatoes and grapes are the most important crops in terms of value. The former are largely grown for domestic consumption, while the latter provide the basis for a large and growing wine industry that has been buoyed by export sales.

The cropping zone (Figure 1) encompasses a diverse range of climates. The mediterranean climate in southern Australia is characterized by mild temperatures in the winter and hot, dry summers. The influence is particularly strong in the southwestern and south-central regions of the country. Dryland cropping in these regions is dominated by wheat and other winter cereals, barley, oats, and triticale; various grain legumes, particularly lupins; and oilseed rape. Average rainfall throughout the cropping zone of southern Australia is generally <425 mm annually and rainfall during the growing season is about 200 mm (Ridge 1986). Moving north through the eastern cropping zone, the seasonality of the winter-dominant rainfall weakens and precipitation is spread throughout the year. At latitudes below about 32°S, in eastern Australia, the mix of crops is strongly influenced by the pattern of summer-dominant rainfall and includes sorghum, sunflowers, soybean, cotton, and a number of grain legumes. Winter cropping in these areas relies on moisture stored in the soil profile (see, for example, Holland et al. 1987).

Australia is an ancient land mass in which the soils have not been renewed through recent glaciations. As a consequence, soils over vast areas are extremely weathered, shallow, and infertile. Many soils are poorly structured and highly susceptible to erosion. The fragility of the soils in southern Australia was apparent from the early 1900s, when farmers began incorporating long fallows of 10 to 14 months duration into rotations. This long fallowing aimed to conserve moisture and enhance the level of mineralized nitrogen for winter crops, a practice claimed to have been modeled on experience in the Great Plains of North America (Ridge 1986). In Australia, as elsewhere, the system resulted in severe erosion. While the practice was continued by some farmers until quite recent times, fallows began to shorten in the 1950s, when agriculture in southern Australia

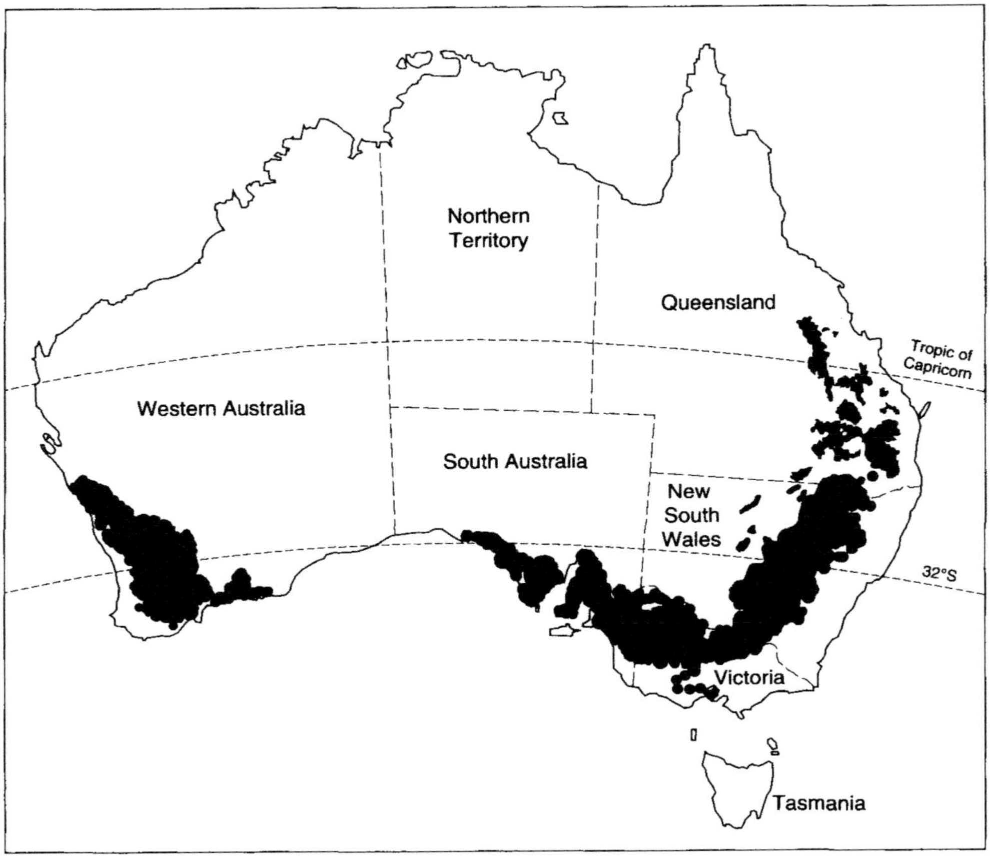

FIGURE 1 Cereal cropping zones of Australia.

was revolutionized by the introduction of annual legumes into pasture leys fertilized with superphosphate and trace elements where required. The revolution was described by Donald and Williams (1954) in the following terms:

> Although subterranean clover first attracted attention because of its qualities as a fodder plant, it has also proved particularly effective in its capacity to convert poor and even worthless land to a moderate or high level of soil fertility and production. There is perhaps no other example of so dramatic an improvement in fertility and production over a vast region as that now being achieved through the use of subterranean clover in southern Australia.

Greatly enhanced levels of soil nitrogen contributed by legumes and the enhanced value of the pasture were the catalysts for farmers to change their practices and adjust the mix of enterprises to improve economic productivity.

Bare fallowing in areas of summer-dominant rainfall remained a common practice for much longer than in the south, partly from the need to conserve moisture and partly because the variability of rainfall results in crops not being planted, leaving open the opportunity for planting in the next growing season, be it winter or summer. Annual pasture legumes play little part in the agricultural systems developed for these areas and the frequency of cropping is higher than in southern situations. Concern about the level of erosion in these areas led to research into reduced tillage practices that has resulted in farmers shortening the fallow period, double cropping, and where possible, adopting other measures to reduce erosion. Reduced tillage technologies have also been adopted in the southern cropping zone, partly in response to the need to control soil erosion and partly in response to the need to shorten rotations when the optimum mix of enterprises favors cropping over meat and wool.

The area planted to winter cereals almost doubled between the early 1960s and the late 1980s, with a peak of 18 million ha in 1983 and 1984. Opportunities for further increases in the area of winter crop production in Australia are very limited (Table 1). Western Australia, South Australia, and probably New South Wales may have reached the sustainable limits for winter cropping in 1983 and 1984 (Nix 1975). This view is supported in a recent study by Hamblin and Kyneur (1993), who examined wheat yield trends and soil fertility in Australia. The study exposed the potential vulnerability of cereal cropping in Australia to declining soil fertility, resulting from a trend to increase cropping at the expense of pasture leys. Rotations incorporating short periods of pasture were not considered to be adequate to maintain soil fertility as determined by stability in soil structure, organic matter content, and nutrient status. Thus, the notion that sustainable limits for winter cropping have been reached appears to have some validity.

The growth in winter cereal production between 1960 and the 1980s was achieved by expansion into new areas and by shortening the pasture phase of the rotations. It was a response to a buoyant world market for grains over much of the period and declining terms of trade for most agricultural commodities, necessitating continued growth in economic productivity as farmers struggled to maintain real incomes. Given that increases in biological productivity, i.e., yield per hectare, were low over the period (Hamblin and Kyneur 1993), improvements in economic productivity were achieved from gains in other areas, such as by improved flexibility of farmers to adjust the mix of farm enterprises, by reducing the amount of land in fallow, and by increasing the efficiency of farm inputs, as occurred with the substitution of capital in the form of farm machinery for labor (Ward et al. 1987). Improvements in weed control technology, particularly the new and growing range of herbicides in the marketplace, and the engineering of machinery able to work in stubble and plant into unprepared seed beds, played an important part in expanding the production of dryland crops.

3. A NICHE FOR HERBICIDE-RESISTANT CROPS IN AUSTRALIA

As a measure of the importance of weeds to crop production in Australia, herbicides currently represent about 60%, by value, of sales of agricultural chemicals, excluding animal health care products (ABARE 1992). The value of herbicide sales almost quadrupled between 1980 and 1990, representing a real increase of about 250%. The growth in sales reflected the increasing dependence on herbicides over the period. To give an example, in Western Australia the area cropped remained more or less constant throughout the 1980s at about 5 million ha, while the area treated with

TABLE 1 Estimated Potential Area (million hectares) for Winter Cereals in Australia

State	Area Available for Cultivation	Crop Area if Rotation of 1 Year in 4 to Cereal	Crop Area in 1983-84 (maximum to date)	Column 3/2 (%)
Queensland	25.7	6.4	1.30	20.3
New South Wales	23.7	5.9	5.44	92.2
Victoria	11.9	3.0	2.35	78.3
Tasmania	0.9	0.2	0.03	15.0
South Australia	11.3	2.8	2.8	100.0
Western Australia	23.1	5.98	5.98	100.0
Australia	96.6	24.1	17.93	74.4

Adapted from Nix 1975.

herbicide increased from 3.9 to 9.5 million ha (Gill 1993). On average, each field received two applications of a herbicide in the period preceding sowing. Herbicide use may not expand further or may even decline in dryland crops because of difficulties being experienced in controlling annual ryegrass (*Lolium rigidum*) which has developed resistance to various herbicides. Anecdotal evidence suggests that in some parts of the cropping zone difficulties of controlling herbicide-resistant weeds have tipped the mix of farm enterprises in favor of meat and wool at the expense of crops. So widespread is the problem that herbicide-resistant ryegrass is seen by some to be the worst weed problem in Australia (Powles, personal communication).

Ryegrass is a component of pastures throughout much of southern Australia. Failure to control it in dryland crops leads to reduced yields and opportunities for ergot in ryegrass to contaminate cereal grains in some years. Ryegrass is also a host for the nematode *Anguina funesta* which replaces seeds with larval galls. The nematode may carry the bacterium *Clavibacter toxicum*, which, when present with the galls, produces a toxin which can kill sheep, cattle, and horses.

As a consequence of the growing dependence on herbicides in dryland crop production, resistant ryegrass biotypes were soon selected, some with multiple resistance to various herbicide groups (Powles and Howat 1990; Powles and Matthews 1992). Factors that contributed to the emergence of ryegrass as a serious weed in dryland crops include:

- The existence of preadaptive genes for resistance to different herbicide groups
- An obligate requirement for cross-pollination
- The development of reduced tillage technologies, which gained considerable impetus with the introduction of diclofop-methyl (Hoegrass®*) in the 1970s
- The dependence of reduced tillage technologies on herbicides
- The trend to increased intensification of cropping, particularly in Western Australia, where lupins are grown in very short rotations with winter cereals

The idea that growing HRCs will lead to an increase in herbicide use may not withstand scrutiny. Most crop land in Australia is already treated with herbicides. Existing products applied for pre- and postemergent weed control may affect the crop to which these are applied, and some products pose problems in the environment. Nonselective herbicides applied to crops incorporating resistance to the chemical of choice hold promise of being less hazardous to the crop and less hazardous to the environment than many of the current range of products. It is also argued that HRCs provide an opportunity to improve the efficiency of herbicide use, as a decision to spray can be delayed until the threat of economic loss from weeds has been assessed. These arguments notwithstanding, there are opportunities to reduce herbicide use and improve efficiency through integrated weed management with strategic use of existing herbicides, as evidenced by experience in Sweden and Denmark, where legislation to reduce pesticide use has focused the attention of agriculturalists on the issue. In any case, HRCs should not be assessed as a substitute for integrated weed management, but rather as another technology to complement existing weed control strategies.

Given the problems of pesticide resistance in insects, helminth parasites of livestock, and microfungi causing diseases in plants, scientists might have foreseen the emergence of herbicide resistance in weeds. Had they done so, they might have developed weed management strategies to lessen the threat, while retaining some depth in the arsenal of herbicides available for weed control. Be that as it may, the approaching reality that HRCs will be available for commercial release introduces new options for weed control that should not be wasted. In crops such as cotton, where manual labor is still used for weed control, the development of herbicide-resistant cultivars holds the promise of significantly lowering production costs.

---

* Registered Trademark of Hoechst A.G., Frankfurt, Germany.

4. INTERNATIONAL TRENDS IN PESTICIDE USE: IMPLICATIONS FOR GROWING HERBICIDE-RESISTANT CROPS

The dependence of modern agricultural systems on pesticides is gaining the attention of many governments around the world. Specific concerns include contamination of the environment, especially water; human health and safety; control of pests and diseases, where resistance to pesticides has occurred, and the need to minimize impediments to international trade. Because of the level of public concern, governments in Sweden, Denmark, and The Netherlands have been able to legislate the reduction of pesticide use by quantifiable amounts over a specified period. Italy, Norway, and Finland and the province of Ontario in Canada have established programs with similar objectives, without enacting legislation. In June 1993 the U.S. announced a commitment to reduce pesticide use and has since indicated an intention that the U.S. Department of Agriculture and the U.S. Environmental Protection Agency will be required to establish commodity-specific use reduction goals for pesticides that are to be met by the year 2000.

Early progress toward meeting the targeted reductions was made relatively easy for farmers in Sweden and Denmark for several reasons:

- The reductions were confined to agricultural pesticides used in crops
- The targets adopted were a reduction in aggregate use of active ingredient in pesticide formulations
- Herbicides were predominant in the mix of pesticides used
- Many of the gains came from an ability to shift to herbicides, such as the sulfonylureas, which have much higher biological activity than the other herbicides in general use

Because of this last point, the term "phantom reduction" was coined to describe the early achievements (Hurst 1992), provoking skepticism among some scientists about the programs and doubts about the need for other countries, such as Australia, to adopt similar measures. The doubts are quickly dispelled, however, by examining other aspects of the programs and by careful analysis of The Netherlands' initiative.

In terms of the value of its agricultural exports, The Netherlands is one of the top three world exporters. It has achieved this position by developing highly intensive and sophisticated systems of primary production, supported by large inputs of pesticides. Like Sweden and Denmark, The Netherlands established a target for reducing pesticide use by 50%, but different targets were established for each of ten industries. Consistent with its policy to reduce pesticide use, The Netherlands government has indicated an intention to apply a very restrictive policy on growing of HRCs (Anon. 1991a).

The pesticide reduction program adopted by The Netherlands is an enormously powerful instrument to manipulate structural change in Dutch agriculture, a move that is essential if the country is to meet its obligations to reduce production under the European Economic Community's (EEC) Common Agricultural Policy. Noting the experience in The Netherlands, it is possible that other countries may adopt similar measures, albeit with modifications to meet individual objectives.

The process of registering pesticides in Sweden, Denmark, and The Netherlands is forcing the pace of change by imposing new and more demanding environmental criteria relating to persistence, mobility, and toxicity to aquatic organisms and to soil biota. Chemicals failing to meet standards set for these criteria at the time of re-registration are withdrawn or use is curtailed severely. A shortened period between registration and re-registration reinforces the need for chemicals to meet the environmental standards. In many cases, as registration lapses, industry is not applying to have chemicals re-registered.

Analysis of the pesticide reduction program in Sweden indicates that the targets set for industry are both technically and economically feasible (Pettersson 1993). Scientists in The Netherlands have demonstrated that the 50% reduction target is possible in integrated arable farming systems (Wijnands and Vereijken 1992), and suggest that with the introduction of fully integrated systems of production, it would be possible to reduce pesticide use by up to 90% (Anon. 1991b). Clearly, technological advances in the management and control of pests, diseases, and weeds are needed if countries are to achieve the legislated targets. These advances will provide those at the forefront of change with significant advantages stemming from increased efficiency and from an ability to sell products with very low levels of pesticide residues or no residues at all.

A risk for countries that do not seek out strategies to reduce the use of pesticides and other agricultural and veterinary chemicals is that their commodities may increasingly be denied access to some markets. This has already occurred: the EEC has refused entry to beef produced using hormonal growth promotants. Recognizing the threat, the former Australian Agricultural and Veterinary Chemicals Council refused to clear the performance-enhancing drug bovine somatotrophin for registration, out of concern that its use could jeopardize the export of dairy products from Australia.

These developments pose a dilemma for the Commonwealth Government and for other countries in regulating the growing of transgenic plants that will incorporate resistance to herbicides. On the one hand, there may be a niche for HRCs in weed management strategies to complement existing control measures. On the other hand, regulation of pesticide use and the growing of transgenic plants in some countries may pose a threat to our exports, either because of problems of market access or because competitors stand to gain a marketing advantage in third countries. The outlook will continue to be uncertain, while some countries remain ambivalent to the emerging technology.

5. THE REGULATORY ENVIRONMENT

The need to regulate the growing of transgenic HRCs reflects the level of concern within the community and industry about:

- Unforeseen consequences resulting from the release of GMOs into the environment
- The possible transfer of alien germplasm in HRCs to other plant species, in particular, the transfer of herbicide resistance to weedy relatives of crops
- Increased selection pressure on weeds to develop resistance to herbicides
- The safety of foods derived from GMOs
- The environmental impact of herbicides arising from an increased dependence on herbicides for weed control
- The opportunity for large, transnational corporations to extract monopoly rents from the sale of seed of HRCs and herbicides
- The need for economic development unimpeded by government regulation
- The acceptance in overseas markets of Australian commodities and products developed from HRCs.

Australia is a federation of six states and two territories. The Commonwealth Government is responsible for such matters as trade and international quarantine, and the state governments are responsible for such matters as health and agriculture, including matters of inter- and intrastate quarantine. Increasingly, the division of responsibility is blurred in an ever more complex world, where the need for a single regulatory authority is greater than it was at the time of federation in 1901.

While the growing of transgenic crops with resistance to herbicides could be regulated in various ways under some existing state and Commonwealth legislation in Australia, it is unlikely that this approach would address the concerns of all interest groups. The deficiencies highlight the need for new legislation, which should be addressed in national legislation now being developed to control the release of GMOs, including crops resistant to herbicides. An analysis of past initiatives to monitor the activities of scientists working on new gene technologies shows that governments are sensitive to public opinion on these matters. The challenge for the Commonwealth Government in developing the new legislation is to steer an even-handed course between the demands of competing interests, recognizing that regulations which are too restrictive will constrain economic opportunities, while a system of regulation that is too open will alienate large sections of the public. If it is to be effective, the legislation must reflect the diverse interests within the community.

5.1. CURRENT LEGISLATION

To the extent that HRCs may pose a threat to humans, agriculture, and the environment, the introduction, release, and consumption of these may be controlled by at least five Acts of the Commonwealth of Australia. A brief description of these follows.

5.1.1. Quarantine Act (1908)

The principal objective of the Act is the exclusion of pests, weeds, and diseases affecting humans, animals, and plants. The Act is administered by the Australian Quarantine and Inspection Service (AQIS), which may use the provisions of the Act to exclude plants, including crop species incorporating genes for resistance to herbicides, if these were assessed as having weed propensities or other pest properties. Powers under the Act to exclude plants that do not carry pests and diseases, or that themselves pose no threat, are limited. The AQIS is of the view, however, that it has a responsibility to cooperate with other Commonwealth instrumentalities regarding the importation of plants, where this is seen to be necessary. Accordingly, AQIS would seek advice from the government's Genetic Manipulation Advisory Committee (GMAC, see Section 5.2.3) on the importation of transgenic plants.

5.1.2. Wildlife Protection (Regulation of Exports and Imports) Act (1982)

The objectives of this Act are to comply with Australia's obligations under the Convention on International Trade in Endangered Species of Wild Fauna and Flora and to further the protection and conservation of the wild fauna and flora of Australia. It does so, in part, by regulating the import of animals and plants of a kind, the establishment of which in Australia could have an adverse effect on the habitats of native Australian animals and plants. The Act is administered by the Australian Nature Conservation Authority. The AQIS has principal responsibility to ensure that the regulations under the Act are complied with.

5.1.3. Agricultural and Veterinary Chemicals Act (1988)

Under the Australian Constitution, the states are responsible for registering agricultural and veterinary chemicals, and it is illegal to use a chemical in a manner contrary to that specified on the registered label. Thus, to legally use a herbicide on a crop incorporating resistance to that chemical would require an extension to the approved uses for the product. Label extensions are more than a mere formality; they require a detailed assessment of the proposed use, the establishment of limits on residues, consideration of health and safety issues, and matters relating to environmental contamination.

For many years clearance of agricultural chemicals on the grounds of health and safety, efficacy, and need was done by the Commonwealth under an arrangement with the states. Once chemicals were cleared for use, matters of registration were handled within the states. This arrangement was changed by the introduction of the Agricultural and Veterinary Chemicals Act, giving the Commonwealth responsibility for clearing chemicals for use. Transitional arrangements now in place give the Commonwealth, through the National Registration Authority for Agricultural and Veterinary Chemicals, responsibility for a uniform system of registration, although the states retain responsibility for control-of-use.

5.1.4. Environment Protection (Impact of Proposals) Act (1974)

The object of this Act is to assess the environmental impact of a proposal about which the Commonwealth must make a decision. The Act is administered by the Commonwealth Environmental Protection Agency. If it were envisaged that the release of any organism could significantly alter the environment, then either an Environmental Impact Statement or an Environment Report could be requested. A recommendation advising release under the Act would depend on an assessment of whether the potential benefits outweighed the perceived costs. No Statements or Reports have yet been requested under the Act to examine the risks arising from the release of GMOs, and it is likely that only the initial proposed releases of genetically modified crops might proceed to this stage.

5.1.5. National Food Authority Act (1991)

The object of the Act is to establish the National Food Authority (NFA) to develop national food standards. The NFA has released a public discussion paper entitled, Assessing the Safety and Consumer Information Requirements of Genetically Manipulated Organisms in Food (NFA 1993). This paper canvasses issues relating to foods from genetically manipulated plants and animals, foods from genetically manipulated microorganisms, food additives, processing aids produced from genetically manipulated organisms, and the labeling of food produced from genetically modified organisms. It is envisaged that the Authority will develop policies on these issues, taking into account the views of the public. At present, there are no specific restrictions on the importation and sale of genetically modified food.

5.2. PROGRESS TOWARD THE DEVELOPMENT OF NEW LEGISLATION

The expectation that genetically modified plants and animals will soon be a commercial reality has been the catalyst for the Commonwealth Government, in consultation with the state governments, to begin framing legislation to regulate the release of transgenic organisms into the environment. The initiative follows extended debate about recombinant DNA technology, monitoring by several nonstatutory bodies and at least two inquiries, the more recent conducted by the House of Representatives Standing Committee on Industry, Science and Technology (*Genetic Manipulation: The Threat or the Glory?* 1992).

In this section we trace the events that have brought the Commonwealth Government to recommend the development of legislative control over the release of GMOs into the environment. The events show that the government's response has been a measured one. Whereas precipitate action to legislate in the early years may have stifled research and commercial initiatives, the delay in framing legislation has bought time that should result in legislation capable of containing the risks, while enabling Australia to capitalize on the opportunities offered by the technology.

5.2.1. Academy of Science Committee on Recombinant DNA Molecules

The hyperbole that accompanied early developments with recombinant DNA technology was matched by concern expressed by scientists, as well as by environmental and consumer interests and by the public at large. At a meeting at the Cold Spring Harbor Laboratory (Cold Spring Harbor, NY, U.S.) in 1973, concerned scientists called for an investigation into the safety of the technology. This led to a meeting of molecular biologists in California in 1975 to discuss the risks associated with the release of GMOs. The Australian Academy of Science sent two representatives to the conference and shortly thereafter the Academy of Science Committee on Recombinant DNA Molecules was appointed to monitor work, to advise on containment procedures, to organize training, and to establish a set of guidelines for contained work. The first guidelines were published in 1975 and revisions were published in 1978, 1979, and 1980.

5.2.2. Recombinant DNA Monitoring Committee

Concern about developments in recombinant DNA technology led the Commonwealth Government in 1980 to review the surveillance of biotechnology in Australia and establish the Recombinant DNA Monitoring Committee (RDMC) in 1981. This Committee was responsible for the surveillance of recombinant DNA work in Australia and reported to the Minister for Science. The RDMC produced guidelines for small-scale work in 1982 and for large-scale work in 1984. Procedures for the assessment of the planned release of recombinant DNA organisms were published in 1987 (RDMC 1987).

5.2.3. Genetic Manipulation Advisory Committee

In responding to a review of RDMC's work (RDMC, 1986) the Government established the GMAC in 1987 to replace the RDMC, to extend monitoring so that innovative genetic manipulation technology was under surveillance, in addition to research on breaking and recombining DNA.

The GMAC established four Sub-Committees, including the Planned Release Sub-Committee to review proposals for releasing into the environment all genetically manipulated live organisms. This Sub-Committee produced guidelines for the planned release of GMOs (GMAC 1993). As a nonstatutory body, GMAC has no power to apply sanctions or otherwise penalize organizations that breach its guidelines on the release of GMOs into the environment. However, it may, if appropriate, recommend the withdrawal of government funding for research and development.

5.2.4. Report of the Victorian Law Reform Commission

In January 1985 the Attorney General of Victoria gave the Law Reform Commission of Victoria a standing reference seeking advice on complex ethical and moral issues arising from new developments in medicine and science. One issue that the Commission studied was genetic manipulation. In June 1988 the Law Reform Commission of Victoria released a discussion paper for public comment, and in 1989 the Report was published (Genetic Manipulation 1989). Among its recommendations, the Commission advised on the need for legislation to control the release of GMOs and on the need for legislation to be enacted either by the Commonwealth or by complementary legislation in the Commonwealth and all states.

5.2.5. Inquiry by the House of Representatives' Standing Committee on Industry, Science and Technology Into the Development, Use and Release of Genetically Modified Organisms

The Commonwealth Government Standing Committee on Industry, Science and Technology launched an inquiry into the development, use, and release into the environment of GMOs in 1992 (*Genetic Manipulation: The Threat or the Glory?* 1992). Among its recommendations the Committee proposed a two-tiered approach for regulating GMO technologies in Australia: (1) retaining GMAC to grant approval for contained work and (2) establishing a GMO Release Authority. In responding to the Report the Government announced an intention to regulate, by legislation, GMO technologies with a single Authority (the Genetic Manipulation Authority, GMA) to control recombinant DNA research and release of GMOs into the environment. Further, it proposed that GMAC be replaced by the new Genetic Manipulation Research Committee (GMRC), with legislatively defined powers to approve contained research and development of GMOs, taking into account wider issues, such as ethical issues proposed by the GMA. The GMRC would also provide specialist scientific advice to the GMA with respect to release proposals. Legislation would be limited to GMOs; "natural" or "unmodified" exotic plants and animals or the products of traditional breeding would remain outside the scope of the legislation. The quarantine service (AQIS) would continue to approve the introduction of plants from overseas. In general, all proposals for release would be publicly notified and submissions invited. The GMA would be able to waive this requirement where the proposal was substantially the same as one considered previously. The report received a favorable response from interest groups.

To progress the development of national legislation, a Commonwealth–State Consultative Group on Genetic Manipulation was convened in November 1992 to prepare a paper for the respective governments. It is envisaged that drafting instructions for legislation will be prepared in 1995.

REFERENCES

ABARE 1992. Commodity Statistical Bulletin, Australian Bureau of Agricultural and Resource Economics, Australian Government Publishing Service, Canberra.

Anon. 1991a. The Multi-Year Crop Protection Plan. Ministry of Agriculture, Nature Management and Fisheries, The Hague, The Netherlands.

Anon. 1991b. Meerjarenplan Gewasbescherming, Tweede Kamer der Staten Generaal, 1990/91, 21 677, no 3-4, Sdu Uitgeverij Plantijnstraat, 's-Gravenhage.

Barker, M.L. 1984. The Recombinant DNA Technique and the Law: A Review of Australian Law Which May Be Relevant to the Regulation of Recombinant DNA Research and Applications. Department of Science and Technology, Canberra.

Donald, C.M. and Williams, C.H. 1954. Fertility and productivity of a podsolic soil as influenced by subterranean clover (*Trifolium subterraneum* L.) and superphosphate. *Aust. J. Agric. Res.* 5: 664.

Genetic Manipulation. 1989. Rep. No. 26. Law Reform Commission of Victoria, Melbourne.

Genetic Manipulation: The Threat or the Glory? 1992. Report by the House of Representatives Standing Committee on Industry, Science and Technology. Australian Government Publishing Service, Canberra.

GMAC. 1993. Guidelines for the Planned Release of Genetically Manipulated Organisms, Genetic Manipulation Advisory Committee, Canberra.

Gill, G.S. 1993. Development of herbicide resistance in annual ryegrass in the cropping belt of Western Australia. In *Proc. I, 10th Australian Weeds Conf. and 14th Asian Pacific Weed Sci. Soc. Conf.* Brisbane, September 6 to 10, p. 282.

Hamblin, A. and Kyneur, G. 1993. *Trends in Wheat Yields and Soil Fertility in Australia*. Bureau of Resource Sciences, Australian Government Publishing Service, Canberra.

Holland, J.F., Doyle, A.D., and Marley, J.M. 1987. Tillage practices for crop production in summer rainfall areas. In *Tillage, New Directions in Australian Agriculture*. Ed. by Cornish, P.S. and Pratley, J.E. Inkata Press, Melbourne, chap. 3.

Hurst, P. 1992. Pesticide reduction programs in Denmark, The Netherlands, and Sweden. *Int. Environ. Affairs* 4: 234.

NFA. 1993. Assessing the Safety and Consumer Information Requirements of Genetically Manipulated Organisms in Food. Discussion Paper. National Food Authority, Canberra.

Nix, H.A. 1975. The Australian climate and its effects on grain quality and yield. In *Australian Field Crops, Volume 1, Wheat and Other Temperate Cereals*. Ed. by Lazenby, A. and Matheson, E.M. Angus & Robertson, Sydney, chap. 7.

Pettersson, O. 1993. Swedish pesticide policy in a changing environment. In *The Pesticide Question; Environment, Economics and Ethics*. Ed. by Pimentel, D. and Lehman, H. Chapman & Hall, New York, chap. 8.

Phillips, B., Winton, J.N., and Gunasekera, H.D.B.H. 1993. Liberalising primary commodity trade in the APEC region. *Agric. Res. Q.* 5: 198.

Powles, S.B. and Howat, P.D. 1990. Herbicide-resistant weeds in Australia. *Weed Technol.* 4: 178.

Powles, S.B. and Matthews, J.M. 1992. Multiple herbicide resistance in annual ryegrass (*Lolium rigidum*): a driving force for the adoption of integrated weed management. In *Resistance '91: Achievements and Developments in Combating Pesticide Resistance, Proc. SCI Symp. Held at Rothamsted Experimental Station, Harpenden, UK, 15–17 July 1991*. Ed. by Denholm, I. and Devonshire, A.L. Elsevier Applied Science, London.

RDMC. 1986. *Monitoring Recombinant DNA Technology: A Five Year Review*. Department of Industry, Technology and Commerce, Australian Government Publishing Service, Canberra.

RDMC. 1987. *Procedures for Assessment of the Planned Release of Recombinant DNA Organisms*. Recombinant DNA Monitoring Committee, Department of Industry, Technology and Commerce, Australian Government Publishing Service, Canberra.

Ridge, P.E. 1986. A review of long fallows for dryland wheat production in southern Australia. *J. Aust. Inst. Agric. Sci.* 52: 37.

Ward, L.D., Brennan, J.P., and Cook, I.C. 1987. Effects of tillage systems on farm management and profitability. In *Tillage, New Directions in Australian Agriculture*. Ed. by Cornish, R.F. and Pratley, J.E. Inkata Press, Melbourne, chap. 16.

Wijnands, F.G. and Vereijken, P. 1992. Region-wise development of prototypes of integrated arable farming and outdoor horticulture. *Netherlands J. Agric. Sci.* 40: 225.

PART VI

RISK ASSESSMENT

Chapter 25

ADDRESSING ENVIRONMENTAL RISKS OF COMMERCIALIZING HERBICIDE-TOLERANT PLANTS: EPA OPTIONS FOR REGULATING HERBICIDE-TOLERANT PLANTS UNDER THE FEDERAL INSECTICIDE, FUNGICIDE, AND RODENTICIDE ACT

Rebecca J. Goldburg
D. Douglas Hopkins

CONTENTS

1-56670-045-0/96/$0.00+$.50
© 1996 by CRC Press Inc.

1. INTRODUCTION

The development of herbicide-tolerant plants (HTPs) is the most frequent and arguably the most controversial application of genetic engineering to plants. About 58% of the approvals for field tests of genetically engineered crops in industrialized countries, and about 34% of such approvals in the U.S., are for HTPs (Beck and Ulrich 1993, Krimsky and Wrubel 1993). Environmentalists and sustainable agriculture advocates have voiced considerable concern about the effects of these crops on herbicide use patterns and the development of sustainable agriculture (Doyle 1985; Goldburg 1992; Goldburg et al. 1990; Gore 1992; Keeney 1989; McAnnelly 1989).

Despite the large number of HTPs under development, federal regulatory agencies have done little to address the various concerns raised about these plants. However, the U.S. Environmental Protection Agency (EPA) Office of Pesticide Programs is currently developing a policy and rule concerning how the agency will address under the Federal Insecticide Fungicide and Rodenticide Act (FIFRA) the environmental and food safety risks of certain pesticidal plants (e.g., plants genetically modified to produce substances that protect the plants from insects and pathogens) (EPA 1993a, b). The EPA could at the same time specifically consider how the agency will address the risks of HTPs, plants genetically modified to tolerate particular herbicides.

This chapter examines three issues: first, a brief review of environmental concerns about herbicides and HTPs; second, an explanation as to why EPAs current regulations and policies under FIFRA will likely fail to adequately protect human health and the environment because they leave significant risks associated with HTPs unregulated; third, a presentation of several proposals for regulating HTPs under FIFRA.

2. ENVIRONMENTAL AND FOOD SAFETY RISKS OF HERBICIDE-TOLERANT PLANTS

2.1. CONCERNS ABOUT PLANTS GENETICALLY MODIFIED TO REDUCE PEST DAMAGE

Researchers are developing for commercial purposes three principal types of transgenic plants designed specifically to reduce pest damage to the plants: insect-resistant plants, virus-resistant plants, and HTPs. These three types of plants all present federal agencies with at least two significant environmental concerns. First, these genetically engineered plants could transfer via pollination their acquired traits to related wild plants, including related weeds (Dale 1992; Doebley 1990; Ellstrand 1988; Ellstrand and Hoffman 1990; Keeler and Turner 1991; Rissler and Mellon 1993; Till-Boutrand et al. 1992). If the transferred trait conferred a selective advantage, it would likely spread, potentially exacerbating weediness. Second, the safety of foods derived from transgenic plants, particularly those engineered to express pesticidal substances, must be confirmed (Goldburg and Tjaden 1990; Hopkins et al. 1991; International Food Biotechnology Council 1990). The EPA is currently evaluating ways to address these concerns under FIFRA (EPA 1993a, b), a Federal Statute which charges EPA with protecting the environment from unreasonable adverse effects of pesticides.

Of the types of transgenic plants now being developed to reduce pest damage, HTPs arguably have the greatest potential to cause environmental degradation. Use of a chemical herbicide to control agricultural weeds is often limited by the sensitivity of a crop to the herbicide. Herbicide-tolerant plants are designed to remove this limitation, and thus are designed specifically to encourage application of their "companion" herbicides. Thus, HTPs present the EPA with an additional environmental concern not presented by other types of plants genetically engineered to reduce pest damage — the potential for increased use of certain herbicides.

The three concerns discussed above are not entirely new, but the advent of modern biotechnology heightens them considerably. Researchers can now transfer to plants genes that impart properties that researchers could never have given the plants using traditional plant breeding techniques. In the case of HTPs researchers have for some time been breeding crops with the goal of increasing tolerance to certain herbicides by exploiting existing genetic variability within a crop (Dyer et al. 1993). In a few cases wide crosses with related weeds have been used to introduce new tolerances to crops. In either case modifying herbicide tolerance through traditional breeding is a relatively slow process that is limited to introducing genetic material from the same or sometimes closely related plant species.

In contrast, with modern gene transfer techniques, researchers can now transfer to crops relatively rapidly any of a panoply of genes from different taxa that confer tolerance to particular herbicides (Dyer et al. 1993; Mazur and Falco 1989; Oxtoby and Hughes 1990). Modern cell culture techniques also allow researchers to exploit somaclonal variation, and thus develop newly tolerant varieties (Dyer et al. 1993; Mazur and Falco 1989). Thus, modern genetic techniques have far greater potential than traditional breeding techniques to encourage the use of particular herbicides by modifying crops so that they tolerate these chemicals.

2.2. ENVIRONMENTAL DEGRADATION CAUSED BY HERBICIDES

Particularly because of their widespread use, herbicides cause substantial environmental degradation in the U.S. Herbicide use has increased sharply over the last few decades, both in total amount and as a percentage of total pesticide use. Herbicides now constitute roughly 60% by weight of active ingredients of all pesticides applied in the U.S. and 84% of pesticides applied to field crops (USDA/ERS 1992). Widespread application of herbicides has resulted in the contamination of groundwater, surface water, and even rainwater by certain herbicides (Goolsby and Battaglin 1993; Goolsby et al. 1993; Hallberg 1986; Kolpin et al. 1993; Nielson and Lee 1987). A number of studies have linked exposure to herbicides to a variety of ailments, including birth defects, central nervous system disorders, and cancer (Burin and Clevenger 1992; Murphy 1986; National Research Council 1987; Nielson and Lee 1987; EPA, Office of Drinking Water 1985; Wigle et al. 1990; Zahm et al. 1990). The National Research Council (1987) estimates that herbicides account for 31% of the oncogenic risk of pesticide residues on fresh food.

2.3. ENVIRONMENTAL EFFECTS OF HERBICIDE-TOLERANT PLANTS

Some individuals argue that the development of HTPs may be environmentally beneficial (Benbrook and Moses 1986; Burnside 1992; Duke et al. 1991; Dyer et al. 1993; ICI Seeds Public Affairs 1991; Mazur and Falco 1989; Oxtoby and Hughes 1990), in large part because HTPs will allow farmers to substitute newer, less environmentally damaging chemicals for older, more damaging chemicals which did not have to meet modern safety standards when they were developed. By allowing farmers to replace certain especially hazardous herbicides with safer ones, they argue, HTPs will provide an immediate net environmental benefit.

Critics of HTPs argue, however, that development of HTPs does not represent an environmentally desirable long term weed control strategy (Goldburg et al. 1990; Goldburg 1992; Gore 1992; Keeney 1989; McAnnelly 1989) in part because HTPs will further entrench farmers' reliance on herbicides, while diverting resources from the development of other weed control strategies. Another major concern is that not all research to develop HTPs is directed toward newer, less environmentally damaging herbicides. As detailed below, some older herbicides that pose or are thought to pose significant environmental hazards are among those targeted by HTP researchers.

Bromoxynil

Toxicological studies indicate that bromoxynil, which originated in 1963 (Thomson 1989), is dermally absorbed and is a developmental toxicant in laboratory animals (Burin and Clevenger 1992; Chitlik 1989). Thus, children of farm workers exposed to bromoxynil may suffer birth defects.

The EPA now requires that mixers and handlers of bromoxynil take a number of measures to protect themselves from exposure to this chemical (Rhône-Poulenc 1994). Bromoxynil is also highly toxic to fish, and thus in farm runoff this chemical can threaten fish populations (Humm et al. 1989; Muir et al. 1991).

Calgene, Inc. (Davis, CA) intends to commercialize cotton genetically engineered to tolerate bromoxynil. Calgene recently successfully petitioned USDA to exempt Calgene's cotton from further regulation by the agency (USDA/APHIS 1994a). Bromoxynil's manufacturer, Rhône-Poulenc, is now testing a large-scale application of bromoxynil to Calgene's cotton (EPA 1994b). Other bromoxynil-tolerant crops are also being developed: R.J. Reynolds (Winston-Salem, NC) has field tested tobacco and USDA and University of Idaho researchers have field tested potatoes genetically engineered to tolerate bromoxynil (USDA/APHIS 1990a, 1992a, 1993b).

2,4-Dichlorophenoxy-Acetic Acid

Considerable evidence indicates that 2,4-dichlorophenoxy-acetic acid (2,4-D), which originated in 1942 (Thomson 1989), causes cancer. A number of (but not all) epidemiological studies indicate that farmers exposed to 2,4-D suffer high rates of non-Hodgkin's lymphoma (Harvard School of Public Health 1990; Wigle et al. 1990; Zahm et al. 1990). A recent study by National Cancer Institute researchers even indicated that pet dogs exposed to 2,4-D-treated lawns develop cancer more often than other dogs (Hayes et al. 1991).

The German chemical firm, Schering, the USDA, and Australia's Commonwealth Scientific and Industrial Research Organization have all sponsored projects to genetically engineer 2,4-D-tolerant crops, including potatoes, cotton, and tobacco (Bayley et al. 1992; Lyon et al. 1993, Streber 1990, Streber and Wilmitzer 1989; USDA/APHIS 1991). In at least some instances these crops are intended to tolerate 2,4-D drift from other crops, rather than direct applications of this chemical. None of these 2,4-D-tolerant crops appear close to commercialization.

Atrazine

Atrazine, which originated in 1959 (Thomson 1989), is classified by the EPA as a possible carcinogen. Recent research indicates that atrazine affects estrogen metabolism, and thus atrazine exposure might increase breast cancer risk (Bradlow 1994; Hileman 1994). In some areas of the U.S., atrazine is the most frequent groundwater and surface water contaminant of any pesticide (Goolsby and Battaglin 1993; Kolpin et al. 1993).

At least two firms, Ciba-Geigy (Basel) and Biotechnica (Cambridge, MA) about 5 years ago abandoned projects to genetically engineer atrazine-tolerant crops, largely in response to environmental concerns (Krimsky and Wrubel 1993). Research to develop atrazine-tolerant crops has continued elsewhere, however. Chinese scientists recently developed atrazine-tolerant soybeans (Ying-Shan 1991). The USDA has in recent years funded university researchers to genetically modify crops to tolerate atrazine (Duke et al. 1991; Goldburg et al. 1990).

One atrazine-tolerant crop is on the market. Seed from canola bred with a closely related weed that tolerates atrazine has been sold since the mid-1980s to Canadian farmers (Duke 1991). The market for this canola seed has been limited by the variety's mediocre yield, however (Dyer et al. 1993).

2.4. EPA OVERSIGHT NEEDED

Without the imposition of appropriate restrictions by the EPA, commercialization of HTPs could well result in increased use of these and other hazardous herbicides, as well as potentially pose gene transfer and food safety risks similar to those associated with other plants genetically modified to reduce pest damage. Of note, some members of the U.S. Congress have expressed similar concerns. Senate Bill S-1916 and a companion House bill, both introduced in November 1991, called on the EPA to:

prepare and submit to the appropriate committees of Congress a report containing recommendations for regulations under the Federal Insecticide, Fungicide, and Rodenticide Act [cite omitted] concerning— (A) the use of herbicides in combination with herbicide resistant plants; and (B) the herbicide resistant plants.

3. REASONS CURRENT FIFRA REGULATIONS WILL NOT ADEQUATELY ADDRESS RISKS POSED BY HERBICIDE-TOLERANT PLANTS

Most HTPs are being designed for use with herbicides that ordinarily would severely injure these crops and therefore have not been registered for use on them. Therefore, under FIFRA, the proposed use of an herbicide on an HTP will in many cases be a "new use" of the herbicide, triggering the requirement for a change in the herbicide's registration to specifically list the new permitted use.* Existing FIFRA regulations governing evaluation of proposed new uses of herbicides do not, however, appear to provide comprehensive protection from the potential environmental problems likely to be caused by HTPs and their companion herbicides. There are at least four reasons for the regulations' inadequacy.

3.1. HERBICIDE USE PATTERNS

The EPA is unlikely under current regulations to consider changes in herbicide use patterns resulting from the development of plants that tolerate these chemicals, because the EPA does not now evaluate the effect of proposed new uses of herbicides on overall herbicide use patterns (Rispin 1991). For example, if a herbicide manufacturer applies to the EPA for a proposed new use for 2,4-D on a crop that has been genetically modified to tolerate 2,4-D, the EPA will not generally consider the effects of the increased total load of 2,4-D in the environment. Only certain limited environmental and food safety matters specific to the crop–herbicide combination are likely to be considered, for example, the safety of 2,4-D residues on the crop and whether 2,4-D poses a particular hazard to aquatic organisms if applied to a wetland crop such as rice.

The issue of whether EPA should consider changes in broad pesticide use patterns when new uses are proposed is not unique to HTPs; any proposed new use of a pesticide could hypothetically result in a significant change in pesticide use. However, HTPs deserve special consideration because they are being developed specifically to alter herbicide use patterns. Just about any crop can now be genetically engineered to tolerate particular herbicides, and accelerating development of HTPs may soon lead to dramatic changes in herbicide use patterns. Increases in the use of relatively old, hazardous herbicides resulting from the development of plants that tolerate chemicals such as 2,4-D, bromoxynil, or atrazine, should be of obvious concern to the EPA. Increases in the use of newer, less hazardous herbicides resulting from the development of plants that tolerate such herbicides should also be of concern to the agency, if such increases were not accompanied by a concomitant decline in the use of other, more hazardous chemicals.

3.2. GENE TRANSFER

Under current regulations for new uses, the EPA probably will not address, and certainly will not address fully, the gene transfer risks associated with particular HTPs. In some cases HTPs should not be grown in geographic areas where they have wild weedy relatives with which they can crossbreed. If the herbicide tolerance trait can be transferred via cross-pollination to related weedy

* 40CFR§152.3(p).

plants, and herbicide use selects for spread of the tolerance trait, the companion herbicide may become ineffective for controlling these weeds (Hoffman 1990; Keeler and Turner 1991; Rissler and Mellon 1993; USDA/APHIS 1990b).

The USDA/APHIS regulates such gene transfer risks by requiring permits under the Federal Plant Pest Act for field tests of genetically engineered plants (Goldburg 1989; Wrubel et al. 1992). The Federal Plant Pest Act does not, however, provide USDA/APHIS with an adequate framework to regulate commercial uses of genetically engineered plants (Goldburg and Hopkins 1992; U.S. Congress Office of Technology Assessment 1993). USDA/APHIS could not possibly require permits for every field use or interstate transport of a genetically engineered crop in commercial use. Therefore, USDA/APHIS has announced that it will exempt from Federal Plant Pest Act permit requirements for genetically engineered plants intended for commercial sale, if USDA/APHIS deems that the plants do not pose a plant pest risk (USDA/APHIS 1993a). It appears that USDA/APHIS may not consider significant gene transfer risks to be of sufficient concern to preclude exemption of a genetically engineered crop from Federal Plant Pest Act permit requirements. The agency recently exempted from Federal Plant Pest Act requirements genetically engineered squash that may pose significant gene transfer risks (USDA/APHIS 1994b; Wilson 1993).

The EPA could choose to address gene transfer risks of HTPs by requiring a companion herbicide to be registered so that its use on the HTP is restricted to geographic regions where there is little risk of gene transfer. As the following two points illustrate, however, even if such labels were widely obeyed, they might not protect adequately against the risks of gene transfer.

First, companies may develop versions of plants that use significantly different mechanisms to tolerate the same herbicide and that therefore pose significantly different gene transfer risks. For example, according to USDA, genetically engineered oilseed crucifers present gene transfer risks in the northern U.S. where potentially they can cross-pollinate with related wild and weedy plants (USDA/APHIS 1990b). It is likely that the EPA would decide not to restrict geographically the use of an herbicide on a "first-generation" herbicide-tolerant canola plant, if the HTP had been engineered to express a male sterility gene that prevented gene transfer. "Second generation" herbicide-tolerant canola plants, which may be engineered for a different mechanism of tolerance, would not necessarily be engineered to express a sterility gene, however, and thus might pose gene transfer risks. Yet, development of the newer HTP would not result in a "new use" of the herbicide that exceeded existing crop type restrictions, and therefore would not trigger an EPA review of the absence of any geographic restriction on use of the herbicide.

Second, seed companies may eventually engineer herbicide tolerance into all their seeds routinely. At the 1991 annual meeting of the Weed Science Society of America, one speaker asserted that "herbicide tolerance will become like heaters in cars. You don't consider the heater when you buy a car, but you wouldn't buy a car without a heater." Under this scenario, limiting herbicide use on particular crops by region would not stop growers from planting HTPs in geographic regions where the plants presented gene transfer risk because growers there (a) may have difficulty purchasing non-herbicide-tolerant seed, and (b) would be under no obligation to do so. Such widespread planting of an HTP would be of concern if use of the companion herbicide on other crops selected for spread of the herbicide tolerance trait, if it was transferred to weeds.

3.3. FOOD SAFETY

Existing FIFRA regulations may prevent the EPA from assuring the food safety of certain second generation HTPs. A principal food safety issue associated with HTPs likely will be the presence of companion herbicides or their degradation products in edible tissues of HTPs. Transgenic 2,4-D-tolerant plants, for example, degrade 2,4-D to 2,4-dichlorophenol (Streber and Wilmitzer 1989), a chemical that the EPA and the Agency for Toxic Substances Disease Registry list as a hazardous substance under the Superfund Amendments and Reauthorization Act of 1986 (DHHS,

Agency for Toxic Substances and Disease Registry 1990). Thus, 2,4-D-tolerant plants may not be acceptable for human consumption following treatment with 2,4-D, depending on whether 2,4-dichlorophenol is rapidly degraded in plant tissue (Goldburg 1992).

Ordinarily, the EPA establishes tolerances for residues of pesticides and degradation products when the agency considers proposed new uses of pesticides. However, the development of a second generation HTP may not result in a "new use," and therefore might not automatically trigger EPAs review of previously established tolerances for the companion herbicide and its degradation products, even if the residue risks presented by the second generation HTP are significantly different.

Mechanisms of herbicide tolerance can vary widely, and companies may develop significantly different versions of HTPs that tolerate the same herbicide. For example, there are several ways plants may be genetically engineered to tolerate glyphosate (Mazur and Falco 1989; Oxtoby and Hughes 1990). As a hypothetical scenario, a first generation glyphosate-tolerant crop, engineered specifically to detoxify this herbicide, might not accumulate glyphosate in its tissues. Based on this herbicide tolerance mechanism, the EPA might allow glyphosate to be labeled for high application rates on the crop. A second generation glyphosate-tolerant crop commercialized later may, instead be engineered to have an altered binding site for this herbicide, rather than to have the detoxification mechanism. Because studies of glyphosate uptake by plants suggest that certain glyphosate-tolerant crops may accumulate this herbicide in their tissues (Duke 1988) rather than metabolize it, glyphosate residues in edible portions of these hypothetical second generation crop plants could exceed established tolerance limits (Goldburg 1992). Commercial production of this second generation crop would not, however, automatically trigger EPAs review of the high application rates specified on the initial glyphosate label, because glyphosate would already have been labeled and registered for use on the crop.

The U.S. Food and Drug Administration (FDA) also may not be aware when new HTPs are commercialized. The FDAs policy for genetically engineered foods (DHHS/FDA 1992) requires no agency notice before most genetically engineered plants are marketed. Moreover, the FDA, which is responsible for enforcing pesticide tolerance levels established by the EPA, cannot be counted on to detect an increase in herbicide residues as in the above scenario. Inspections by the FDA for pesticide residues are infrequent and the agency does not test for residues of all pesticides applied to edible crops (Jacobson et al. 1991; GAO 1986).

The EPA has the authority under existing FIFRA regulations to review second generation HTPs that pose significant food safety risks due to applications of companion herbicides. These regulations define "new use" to include "any proposed use pattern that would require the establishment of, the increase in, or the exemption from the requirement of, a tolerance or food additive regulation under Section 408 or 409 of the Federal Food Drug and Cosmetic Act" and "any additional use pattern that would result in a significant increase in the level of exposure, to the active ingredient [,] of man or other organisms."*

Nevertheless, the EPA will not generally be notified when second generation HTPs are commercialized. The EPA will only be able to use these regulations to protect the food supply if the agency happens to know *a priori* that a particular second generation HTP, and resulting changes in herbicide "use patterns," will lead to significantly increased levels of exposure to a particular active ingredient.

3.4. FAILURE TO TRIGGER NEW USE REVIEW

Potential adverse effects associated with even some first generation HTPs may on occasion escape EPA scrutiny, because the HTPs fail to trigger new use review of companion herbicides. For example, new use reviews for companion herbicides would not necessarily be required for HTPs developed to prevent crop injury from soil carryover of herbicides applied to previous crops. For

* 40CFR§152.3(p)(1) and (3).

example, if made commercially available in the U.S., atrazine-tolerant soybeans (Ying-Shan 1991) would allow increased amounts of atrazine to be applied to corn planted in rotation with soybeans. Labels for atrazine formulations recommend that less than the maximum rate of atrazine be applied to corn planted in rotation with soybeans in some areas (*Crop Protection Chemicals Reference* 1988). Thus, atrazine-tolerant soybeans could encourage an increase in average atrazine application rates and in the number of corn acres treated with atrazine, without any new use review being required.

Similarly, new use reviews for companion herbicides would not necessarily be required for herbicide-tolerant trees developed for use in forestry. Foresters sometimes treat plantations of conifer saplings with nonselective herbicides to "release" the conifers from competition with other vegetation (Nelson 1984). Selectivity can be gained by applying herbicide when conifers are not growing rapidly, and are thus less susceptible to herbicide injury than unwanted vegetation (USDA Forestry Service 1989). The development of even partially herbicide-tolerant conifers (Manders 1992; USDA Forestry Service 1989) would allow foresters more flexibility as to exactly when they applied nonselective herbicides, and how much herbicide they applied (Nelson and Haissig 1986). Foresters are traditionally conservative about applying nonselective herbicides, in part because they do not want to injure their planted stock. Thus, the development of herbicide-tolerant conifers could encourage significant increases in herbicide use (Goldburg et al. 1990; Goldburg 1992), even if the EPA required no new use review for the companion herbicide.

4. PROPOSALS FOR ENVIRONMENTAL PROTECTION AGENCY REGULATION OF THE RISKS OF HERBICIDE-TOLERANT PLANTS AND THEIR COMPANION HERBICIDES

Under FIFRA there are at least two alternative mechanisms whereby the EPA could regulate the gene transfer and food safety risks of HTPs and their companion herbicides. Under either mechanism, the EPA could implement a system of conditional registration that would benefit the EPA, companies, and the public by establishing incentives for honest and thorough appraisal of the effects of HTPs on herbicide use patterns.

4.1. ALTERNATIVE ONE

The EPA could restrict new use registrations of companion herbicides to specific HTPs, thus helping to prevent the kinds of problems resulting from overbroad herbicide registrations discussed above. For example, consider a proposed registration change to allow postemergence use of glyphosate on a hypothetical canola plant genetically engineered to detoxify glyphosate and express a male sterility gene. The EPA could approve a label change approving the use of glyphosate on canola, but restricting it to use only on genetically engineered varieties of canola that detoxify glyphosate and contain a male sterility gene.

Each subsequent "generation" of herbicide-tolerant canola plant, i.e., later developed herbicide-tolerant varieties that use different mechanisms to achieve herbicide tolerance or to control environmental risks such as gene transfer, would require a new use application to further amend the registration of the companion herbicide. Thus, the EPA would be able to evaluate the full range of risks of each new HTP, including risks of food residues and gene transfer, as the new HTP is developed. The EPA would not have to anticipate, at the time a first generation HTP is developed, all the possible unreasonable adverse environmental effects that might be associated with use of a companion herbicide on later generation HTPs.

Under this alternative, the EPA would review only new HTPs for which the registration of the companion herbicide must be amended. The EPA would likely end up evaluating the environmental impacts of most, but not all, new HTPs and their companion herbicides. Agency review of the environmental impacts of commercial planting of herbicide-tolerant conifers developed for forestry and atrazine-tolerant soybeans (discussed above) would likely not be triggered.

FIFRA clearly provides the EPA with the authority to implement this first alternative. FIFRA expressly requires the EPA to consider the potential for a pesticide to cause "unreasonable adverse effects on the environment" when the pesticide is used "to perform its intended function . . . in accordance with widespread and commonly recognized practice." * The intended function of an herbicide obviously includes applying it on an HTP developed specifically to enable use of the herbicide.

If the EPA were to adopt this first alternative, the agency could create special procedures requiring the developer of any new HTP to notify the EPA. The agency could then determine whether the registration of the companion herbicide must be amended to allow its use on the HTP. As a second possibility, the EPA could simply make clear that each time a new HTP is developed, its developer must carefully analyze whether the registration of the companion herbicide must be amended.

4.2. ALTERNATIVE TWO

The EPA could regulate HTPs in the same fashion as the agency intends to regulate the risks of certain plants genetically modified to resist insects, pathogens, and other plant pests. The agency now plans to require registration as pesticides under FIFRA of genetic constructs and their expression products that confer pesticidal properties to the plants (EPA 1993a, b). The EPA could also require registration as pesticides of genetic constructs and their expression products (e.g., bacterial genes, the enzymes they encode) that confer herbicide tolerance when engineered into crop plants. Such "herbicide-tolerant constructs" fit readily into FIFRAs broad definition of pesticides, which includes "plant growth regulator[s]" — "any substance or mixture of substances intended, through physiological action, for accelerating or retarding the rate of growth or rate of maturation or for otherwise altering the behavior of plants." ** When the EPA considered registration of a herbicide-tolerant construct, the agency could also consider registration of any associated new use of a companion herbicide.

If the EPA were to require the registration of herbicide-tolerant constructs as pesticides, the EPA would have authority to review the gene transfer risks and the food safety implications of a herbicide-tolerant construct every time it were inserted in a new crop plant. (Inserting a herbicide-tolerant construct in a new plant would constitute a new use of the construct.) Thus, adoption of alternative two would provide the EPA more comprehensive oversight than alternative one. The EPA would have the authority to review combinations of herbicide-tolerant constructs, crop plants, and companion herbicides, regardless of whether the use of a new HTP as intended would result in a use of a companion herbicide that exceeded the herbicide's existing label restrictions.

The EPA would have to determine whether naturally HTPs and all genetically modified HTPs require registration under this second alternative. Because many types of plants naturally tolerate sizeable applications of certain selective herbicides, registration requirements clearly should apply only to newly developed HTPs. It would make little sense for the EPA to require registration of grasses, for example, that naturally tolerate applications of 2,4-D that kill broadleaved plants. The EPA would also have to decide whether regulations should cover HTPs developed through conventional breeding methods as well as just transgenic HTPs. Genetic engineering has the potential to produce many more new HTPs, and at a considerably faster rate, than conventional breeding. Nevertheless, both transgenic and new HTPs developed by more traditional genetic techniques could

* 7USC§136a(c)(5)(C)(D).

** 7USC§136(u)(v).

have similar environmental effects. Thus, regulations arguably should encompass all genetically modified HTPs.

This second alternative appears to conflict with EPAs draft December 1993 policy statement for plant-pesticides, which states that "substances expressed in plants that would not be considered plant regulators and therefore are not under EPA's authority" include "substances intended to enhance the plant's resistance to chemical herbicides" (EPA, 1993a). It would be shortsighted, however, for the EPA to preclude regulating herbicide-tolerant constructs as pesticides before the agency has fully thought through options for the regulation of the risks of HTPs and their companion herbicides.

4.3. CONDITIONAL REGISTRATION

The EPA could use the agency's March 1994 conditional registration of the herbicide acetochlor (EPA 1994a) as a model for evaluating changes in herbicide use patterns resulting from the development of HTPs. The EPA registered acetochlor largely on the basis of arguments by the herbicide's manufacturers, Zeneca AG and Monsanto Company, that the availability of acetochlor would result in dramatic decreases in overall herbicide use on corn crops. The EPA weighed the manufacturers' arguments for environmental benefits against other evidence that acetochlor is carcinogenic and has the potential to contaminate groundwater.

Taking an unprecedented step, the EPA established a number of conditions for the registration of acetochlor. Most notable, Monsanto and Zeneca must document promised decreases in use of various herbicides during the 5 years following acetochlor's registration. In addition, the companies must fund groundwater and surface water monitoring for contamination by the herbicide. If overall herbicide use patterns do not change as the companies promised, or if acetochlor exceeds specified limits in well water or in community water supplies that depend on surface water, the EPA will cancel acetochlor's registration.

Whether one agrees specifically with the EPAs registration of acetochlor, the conditional nature of the registration provides a potential model for the EPAs registrations of new uses of herbicides with HTPs. The EPA could condition registrations on subsequent confirmations of the applicants' predictions of the effects of HTPs on herbicide use. As in the case of acetochlor, proponents of HTPs commonly argue that use of these plants and their companion herbicides will result in a net environmental benefit by causing a decrease in the use of more hazardous herbicides. Projected beneficial changes in herbicide use patterns are easy to exaggerate, and whether projections are exaggerated, they may be viewed with some skepticism by regulators and the public.

If the EPA implemented a system of conditional registrations for new use of companion herbicides, companies would have a strong incentive to carefully and honestly evaluate the effects of the introduction of HTPs on herbicide use. If a company could not document the promised changes in herbicide use within a specific time period following registration, the new use registration would be canceled. The fact that companies would be held to their claims would give regulators considerable confidence in the information companies provided and would boost public confidence in regulatory decisions concerning HTPs.

Herbicide and HTP manufacturers would also gain from a conditional registration system. The EPA would formally consider the predicted effects of HTP use that these companies typically champion: the potential of HTPs to lead to reductions in the use of older, more hazardous herbicides. Conditional registration would also protect the environment because the EPA would have a clear mechanism to prevent the sale of HTPs which promote increased herbicide use, and particularly increased use of older, more dangerous compounds.

A conditional registration system could function under either alternative one or two, proposed above. In either case the EPA would conditionally register companion herbicides for use with first generation HTPs. The EPA could decide on a case-by-case basis whether to make new use registrations conditional when companion herbicides are proposed for use with later generation

HTPs. In general, conditional registrations of later generation HTPs would be appropriate if (a) the conditional registration of the companion herbicide for use with a first generation HTP had not yet expired, (b) the later generation HTP tolerated significantly higher application rates of the companion herbicide than the first generation HTP, or (c) the applicant company claimed that the later generation HTP would result in further changes in herbicide use patterns.

5. CONCLUSION

Herbicide-tolerant plants are being developed to encourage use of particular companion herbicides. Herbicide-tolerant plants may spread herbicide tolerance genes to wild weedy plants, increase food safety risks from herbicide residues, and lead to increased use of hazardous herbicides.

We propose two alternative mechanisms for the EPA to address these environmental risks, which the EPA and other federal agencies are unable to address fully under current federal regulatory programs. Under alternative one, the EPA would create special FIFRA procedures for reviewing and registering new uses of herbicides on new HTPs. The agency would be able to evaluate the environmental and food safety risks of most new HTPs and their companion HTPs because the EPA would restrict approved new uses of companion herbicides to particular HTPs. Under alternative two, the EPA would interpret FIFRAs broad definition of "pesticide" to require that the genetic constructs and their expression products that confer herbicide tolerance be registered as pesticides.

We also propose a conditional registration system, modeled on the EPAs recent conditional registration of acetochlor, which would supplement either of the above alternatives. In some cases HTPs and accompanying changes in herbicide use may exacerbate environmental degradation caused by particular herbicides. The proposed conditional registration system would provide clear incentives for realistic appraisals of the effects of HTPs on herbicide use patterns.

Whether the EPA chooses to implement any of these proposals or some variation of them, the agency needs to establish a comprehensive regulatory program in order to fully protect the public and natural ecosystems from the potential effects of HTPs and their companion herbicides.

ACKNOWLEDGMENT

The Joyce Foundation provided financial support for the preparation of this chapter.

REFERENCES

Bayley, C., Trolinder, N., Ray C., Morgan, M., Quisenberry, J.E., and Ow, D.W. 1992. Engineering 2,4-D resistance into cotton. *Theor. Appl. Genet.* 83: 645.

Beck, C.I. and Ulrich, T.H. 1993. Environmental release permits. *Bio/Technology* 11: 1524.

Benbrook, C. and Moses, P. 1986. Engineering crops to resist herbicides. *Technol. Rev.*, 54: November/December.

Bradlow, H.L. 1994. Personal communication.

Burin J.G. and Clevenger, A. 1992. *Fifth Developmental Toxicity Review of Bromoxynil.* U.S. Environmental Protection Agency, Washington, D.C.

Burnside, O.C. 1992. Rationale for developing herbicide-resistant crops. *Weed Technol.* 6: 621.

Chitlik, L.D. 1989. *Peer Review Meeting of 12/19/88 on Bromoxynil Developmental Toxicity Issues,* Tox. Chem. No. 119. U.S. Environmental Protection Agency, Washington, D.C.

Crop Protection Chemicals Reference, 4th ed. 1988. John Wiley & Sons and Chemical and Pharmaceutical Publishing, New York, p. 400.

Dale, P.J. 1992. Spread of engineered genes to wild relatives. *Plant Physiol.* 100: 13.

Doebley, J. 1990. Molecular evidence for gene flow among *Zea* species. *BioScience* 40: 443.

Doyle, J. 1985. *Altered Harvest*. Viking Penguin, New York, p. 208.

Duke, S.O., Christy, A.L., Hess, F.D., and Holt, J.S. 1991. *Herbicide-Resistant Crops*. Council on Agricultural Science and Technology, Ames, IA.

Duke, S.O. 1988. Glyphosate. In *Herbicides — Chemistry, Degradation, and Mode of Action*. Ed. by Kearney, P.C. and Kaufman, D.D. Marcel Dekker, New York, chap. 1.

Dyer, W.E., Hess, F.D., Holt, J.S., and Duke, S.O. 1993. Potential benefits and risks of herbicide-resistant crops produced by biotechnology. In *Horticultural Reviews*. Ed. by Janick, J. John Wiley & Sons, New York, chap. 9.

Ellstrand, N. and Hoffman, C. 1990. Hybridization as an avenue of escape for engineered genes. *BioScience* 40: 438.

Ellstrand, N. 1988. Pollen as a vehicle for the escape of engineered genes? *Trends Ecol. Evol.* 3: S30; *Trends Biotechnol.* 6: S30 (combined issue).

Goldburg, R.J. and Hopkins, D.D. 1992. Comments to USDA Concerning the Proposed Interpretive Ruling in Connection with Calgene, Inc. Petition for Determination of Regulatory Status of FLAVR SAVR™ Tomato, August 28.

Goldburg, R J. and Tjaden, G. 1990. Are B.T.K. plants really safe to eat? *Bio/Technology* 8: 1011.

Goldburg, R.J. 1992. Environmental concerns with the development of herbicide-tolerant plants. *Weed Technol.* 6: 647.

Goldburg, R.J., Rissler, J., Shand, H., and Hassebrook, C. 1990. *Biotechnology's Bitter Harvest*. Biotechnology Working Group, Washington, D.C.

Goldburg, R J. 1989. USDA's oversight of biotechnology. In *Reform and Innovation of Science and Education: Planning for the 1990 Farm Bill*. Committee on Agriculture, Nutrition, and Forestry, U.S. Senate, U.S. Government Printing Office, Washington, D.C., p. 137.

Goolsby, D.A. and Battaglin, W.A. 1993. Occurrence, Distribution and Transport of Agricultural Chemicals in Surface Waters of the Midwestern United States. Selected Papers on Agricultural Chemicals in Water Resources of the Midcontinental United States. U.S. Geological Survey, Denver, p. 1.

Goolsby, D.A., Thurman, E.M., Pomes, M.L., Meyer, M., and Battaglin, W.A. 1993. Occurrence, Deposition, and Long Range Transport of Herbicides in Precipitation in the Midwestern and Northeastern United States. Selected Papers on Agricultural Chemicals in Water Resources of the Midcontinental United States. U.S. Geological Survey, Denver, p. 75.

Gore, A. 1992. *Earth in the Balance*. Houghton Mifflin, Boston, p. 140.

Hallberg, G.R. 1986. From hoes to herbicides: agriculture and groundwater quality. *J. Soil Water Conserv.* 41: 357.

Harvard School of Public Health, Program on Risk Analysis and Environmental Health. 1990. *The Weight of Evidence on the Human Carcinogenicity of 2,4-D*. Harvard University Press, Cambridge, MA.

Hayes, H.M., Tarone, T.E., Cantor, K.P., Jessen, C.R., McCurnin, D.M., and Richardson, R.C. 1991. Case-control study of canine malignant lymphoma: positive association with dog-owner's use of 2,4-dichlorophenoxyacetic acid herbicides. *JNCI* 83: 1226.

Hileman, B. 1994. Environmental estrogens linked to reproductive abnormalities. Cancer. *Chem. Eng. News* 19: January 31.

Hoffman, C.A. 1990. Ecological risks of genetic engineering of crop plants. *BioScience* 40: 434.

Hopkins, D.D., Goldburg, R.J., and Hirsch, S.A. 1991. *A Mutable Feast: Assuring Food Safety in the Era of Genetic Engineering*. Environmental Defense Fund, New York.

Humm, J.B., Multer, E.P., and DeFelice, M.S. 1989. Fish and Agricultural Chemicals: Safeguarding Your Pond, Agricultural Guide. University of Missouri Extension Service, Columbia.

ICI Seeds Public Affairs. 1991. *Herbicide Tolerant Plants: Weed Control with the Environment in Mind*. Fernhurst, Surrey, U.K.

International Food Biotechnology Council. 1990. Biotechnologies and food: assuring the safety of foods produced by genetic modification. *Regul. Toxicol. Pharmacol.* 12: S1.

Jacobson, M.F., Lefferts, L.Y., and Garland, A.W. 1991. *Safe Food: Eating Wisely in a Risky World*. Center for Science in the Public Interest and Living Planet Press, Los Angeles, p. 54.

Keeler, K. and Turner, C. 1991. Management of transgenic plants in the environment. In *Risk Assessment in Genetic Engineering: Environmental Releases of Organisms*. Ed. by Levin, M. and Strauss, H. McGraw-Hill, New York, p. 189.

Keeney, D. 1989. Guest editorial: biotechnology and sustainable agriculture: the issues of "misguided resistance." Iowa *Groundwater Assoc. Newsl.* 2: 1.

Kolpin, D.W., Goolsby, D.A., Aga, D.S., Iverson, J.L., and Thurman, E.M. 1993. Pesticides in Near-Surface Aquifers: Results of the Midcontinental United States Ground-Water Reconnaisance, 1991-1992. Selected Papers on Agricultural Chemicals in Water Resources of the Midcontinental United States. U.S. Geological Survey, Denver, CO, p. 64.

Krimsky, S. and Wrubel, R. 1993. *Agricultural Biotechnology: An Environmental Outlook*. Department of Urban of Environmental Policy, Tufts University, Medford, MA.

Kross, B.C. and Hallberg, G.R. 1990. Iowa State-Wide Rural Well-Water Survey. Iowa Department of Natural Resources, Energy and Geological Resources Division, Iowa City, IA, p. 9.

Lyon, B.R., Cousins, Y.L., Llewellyn, D.J., and Dennis, E.S. 1993. Cotton plants transformed with a bacterial degradation gene are protected from accidental spray drift damage by the herbicide 2,4-dichlorophenoxyacetic acid. *Transgenic Res.* 2: 162.

Manders, G., Davey, M.R., and Power, J.B. 1992. New genes for old trees. *J. Exp. Bot.* 43: 1181.

Mazur, B.J. and Falco, S.C. 1989. The development of herbicide-resistant crops. *Annu. Rev. Plant Physiol. Plant Mol. Biol.* 40: 441.

McAnnelly, L. 1989. Chemically dependent crops? *Texas Department of Agriculture, Grassroots* Summer: 1.

Muir, D.C.G., Kenny, D.F., Grift, N.P., Robinson, R.D., Titman, R.D., and Murkin, H.R. 1991. Fate and acute toxicity of bromoxynil esters in an experimental prairie wetland. *Environ. Toxicol. Chem.* 10: 395.

Murphy, S.D. 1986. Toxic effects of pesticides. In *Casarett and Doull's Toxicology*, 3rd ed. Ed. by Klaassen, C., Amdur, M., and Doull, J. Macmillan, New York, chap. 18.

National Research Council. 1987. *Regulating Pesticides in Food: the Delaney Paradox.* National Academy Press, Washington, D.C. p. 69.

Nelson, L.R., Gjerstad, D.H., and Minogue, P.J. 1984. Use of herbicides for industrial forest vegetation management in the southern United States, in *Chemical and Biological Controls in Forestry*, ACS Symp. Ser. No. 238. Ed. by Garner, W.Y. and Harvey, J., Jr. American Chemical Society, Washington, D.C., p. 11.

Nelson, L.R. and Haissig, B.E. 1986. Herbicide stress: use of biotechnology to confer herbicide resistance to selected woody plants. In *Stress Physiology and Forest Productivity.* Ed. by Hennessey, T.C., Dougherty, P.M., Kossuth, S.V., and Johnson, J.D. Martinus Nijhoff, Boston, p. 191.

Nielson, E.G. and Lee, L.K. 1987. The Magnitude and Costs of Groundwater Contamination from Agricultural Chemicals: *A National Perspective.* U.S. Department of Agriculture, Economic Research Service, Washington, D.C.

Oxtoby, E. and Hughes, M.A. 1990. Engineering herbicide tolerance into crops. *Trends Biotechnol.* 8: 61.

Rhône-Poulenc Ag Company. 1994. Label for Buctril Brand Herbicide. Research Triangle Park, NC.

Rispin, A. 1991. Personal communication.

Rissler, J. and Mellon, M. 1993. *Perils Amidst the Promise.* Union of Concerned Scientists, Washington, D.C.

Streber, W.R. and Wilmitzer, L. 1989. Transgenic tobacco plants expressing a bacterial detoxifying enzyme are resistant to 2,4-D. *Bio/Technology* 7: 811.

Streber, W.R. 1990. Additional comments on transgenic plants resistant to 2,4-D by Streber. *Rice Biotech. Q.* 1: 36.

Thomson, W.T. 1989. *Agricultural Chemicals.* Thomson Publications, Fresno, CA.

Till-Bouttraud, I., Reboud, X., Brabant, P., Lefranc, M., Rherissi, B., Vedel, F., and Darmency, H. 1992. Outcrossing and hybridization in wild and cultivated foxtail millets: consequences for the release of transgenic crops. *Theor. Appl. Genet.* 83: 940.

U.S. Congress, Office of Technology Assessment. 1993. Harmful Non-Indigenous Species in the United States, OTA-F-565. U.S. Government Printing Office, Washington, D.C. p. 278.

U.S. Department of Agriculture, Animal and Plant Health Inspection Service. 1990a. Application 90-360-01 under 7 C.F.R. 340, from the U.S. Department of Agriculture Agricultural Research Service, Washington, D.C.

U.S. Department of Agriculture, Animal and Plant Health Inspection Service. 1990b. Workshop on Safeguards for Planned Introduction of Transgenic Oilseed Crucifers. Ed. by McCammon, S.L. and Dwyer, S.G. Cornell University, Ithaca, NY, p. 8.

U.S. Department of Agriculture, Animal and Plant Health Inspection Service. 1991. Application No. 91-024-04 under 7 C.F.R. 340, from the U.S. Department of Agriculture Agricultural Research Service, Washington, D.C.

U.S. Department of Agriculture, Animal and Plant Health Inspection Service. 1992a. Applications 92-057-01 and 92-080-01 under 7 C.F.R. 340, from the University of Idaho. USDA/APHIS, Washington, D.C.

U.S. Department of Agriculture, Animal and Plant Health Inspection Service. 1993a. Genetically engineered organisms and products; notification procedures for the introduction of certain regulated articles; and petition for nonregulated status; final rule. *Fed. Reg.* 58: 17044.

U.S. Department of Agriculture, Animal and Plant Health Inspection Service,. 1993b. Notification 93-120-06n under 7 C.F.R. 340, from R.J. Reynolds. USDA/APHIS, Washington, D.C.

U.S. Department of Agriculture, Animal and Plant Health Inspection Service. 1994a. Availablility of Determination of Nonregulated Status of Calgene, Inc., Genetically Engineered Cotton Lines. *Fed. Reg.* 59: 8452.

U.S. Department of Agriculture, Animal and Plant Health Inspection Service. 1994b. Availability of determination of nonregulated status for virus resistant squash. *Fed. Reg.* 59: 64187.

U.S. Department of Agriculture, Economic Research Service. 1992. Agricultural Resources: Inputs, Outlook and Situation Report, AR-25. USDA/ERS, Washington, D.C.

U.S. Department of Agriculture, Forest Service. 1989. Report of the Forest Service, Fiscal Year 1988. USDA, Washington, D.C. p. 68.

U.S. Department of Health and Human Services, Agency for Toxic Substances and Disease Registry. 1990. Toxicological Profile for 2,4-Dichlorophenol. DHHS, Atlanta, p. iii.

U.S. Department of Health and Human Services, Food and Drug Administration. 1992. Statement of policy: foods derived from new plant varieties; notice. *Fed. Reg.* 57: 22984.

U.S. Environmental Protection Agency. 1993a. Proposed Rule; Plant-Pesticides Subject to the Federal Insecticide, Fungicide, and Rodenticide Act. Draft. December 20.

U.S. Environmental Protection Agency. 1993b. Statement of Policy; Plant-Pesticides Subject to the Federal Insecticide, Fungicide, and Rodenticide Act and the Federal, Food Drug and Cosmetic Act. Draft, December 20.

U.S. Environmental Protection Agency. 1994a. Questions and Answers: Conditional Registration of Acetochlor, EPA, Washington, D.C.

U.S. Environmental Protection Agency. 1994b. Receipt of an application for an experimental use permit for use of bromoxynil on herbicide tolerant cotton plants. *Fed. Reg.* 59: 17536.

U.S. Environmental Protection Agency, Office of Drinking Water. 1985. National primary water regulations: synthetic organic chemicals, inorganic chemicals and microorganisms. *Fed. Reg.* 50: 46936.

U.S. Government Accounting Office. 1986. Pesticides: Need to Enhance FDA's Ability to Protect the Public From Illegal Residues. GAO, Washington, D.C. 58.

Wigle, D.T., Semenciw, R.M., Wilkins, K., Reidel, D., Ritter, L., Morrison, H.I., and Mao, Y. 1990. Mortality study of Canadian male farm operators: non-Hodgkin's lymphoma mortality and agricultural practices in Saskatchewan. *JNCI* 82: 575.

Wilson, H.D. 1993. Free-living Cucurbita pepo in the United States,. Report Prepared for U.S. Department of Agriculture, Animal and Plant Health Inspection Service, Biotechnology, Biologics, and Environmental Protection, Hyattsville, MD, p. 1.

Wrubel, R.P., Krimsky, S., and Wetzler, R.E. 1992. Field testing transgenic plants: an analysis of the U.S. Department of Agriculture environmental assessments. *BioScience* 42: 280.

Ying-Shan, H. 1991. Plant improvements through biotechnology in China. *Rice Biotech. Q.* September/October: 25.

Zahm, S.H., Weisenburger, D.D., Babbitt, P.A., Saal, R.C., Vaught, J.B., Cantor, K.P., and Blair, A. 1990. A case-control study of non-Hodgkin's lymphoma and the herbicide 2,4-diclorophenoxyacetic acid (2,4-D) in eastern Nebraska. *Epidemiology* 1: 349.

Chapter 26

AN AGRICULTURALIST'S VIEWPOINT OF RISKS AND BENEFITS OF HERBICIDE-RESISTANT CULTIVARS

Orvin C. Burnside

CONTENTS

1-56670-045-0/96/$0.00+$.50
© 1996 by CRC Press Inc.

1. INTRODUCTION

Herbicide-resistant cultivars (HRC) do not represent any great change in the technology now being utilized for weed control in crop production worldwide in developed countries or to a lesser extent in developing countries (Burnside 1992; Giaquinta 1992). In fact, crops tolerant to herbicides have been used extensively during the past 4 decades, because crop tolerance is a necessary prerequisite for a selective herbicide. Generally, crop tolerance to herbicides was discovered by screening a large number of organic compounds against a limited number of major crops and weeds of the world to ascertain herbicide selectivity. Once a selective herbicide was identified, the environmental impact of that herbicide was then assessed. Sometimes environmentally "marginal" herbicides have been developed when nothing else was available.

Thus, it would be more logical to identify herbicides that are environmentally "benign" and then look for tolerant crop genotypes or HRC. Using this approach a single herbicide can be evaluated for crop selectivity against a great number of biotypes of a single crop in tissue culture, by selecting from the natural population of a crop, or sometimes gene transfer can be used to produce resistant genotypes. The two major methods of plant gene transfer depend upon *Agrobacterium tumefaciens* (Klee and Rogers 1989) or the physical methods of DNA delivery such as with the particle gun (Christou et al. 1988). This approach to identifying selective herbicides is the reverse of most past screening procedures, but properly utilized it has the potential of reducing the use of environmentally "harsh" herbicides, facilitating greater use of other weed control procedures, solving many present or new weed control problems expeditiously, eliminating the need for chemical "safeners" to protect crops from herbicides, and utilizing herbicides more judiciously.

Herbicide-resistant cultivar technology is being actively pursued, as 347 permits were granted by the Department of Agriculture in Washington, D.C. for field tests on genetically modified plants between 1987 and 1993, and 118, or just under one third, were for herbicide resistance (Hoyle 1993). Scientists in over 18 companies are searching for HRC.

Some environmentalists are concerned about the extent of herbicide use in the U.S., and they believe that HRC will further entrench and extend the pesticide era (Goldburg et al. 1990; Goldburg 1992). Many of us involved with various aspects of agriculture, weed science, and selective herbicide research do not view the development of HRC as anti-environmental (Burnside 1992; Duvick 1992; Knake 1992; Kishore et al. 1992; Giaquinta 1992). In fact, HRC could enable producers to use less herbicide and it should reduce markedly the use of those herbicides that are showing up as persistent in the environment.

It is appropriate to assess the importance of pesticides and specifically herbicides as an aid to crop production in the U.S. In 1991 pesticide sales were a $26.8 billion industry worldwide, with retail sales totaling $6.4 billion in the U.S. or 24% of worldwide pesticide expenditures (McDougall and Phillips 1992). Herbicides represented about 65% of total pesticides used in the U.S. in 1990 and 85% of pesticides used on cropland (Delvo 1990). Herbicide use on crops in the U.S. grew from 52 million kg in 1966 to a peak of 227 million kg in 1982, when over 95% of the hectarage of major crops was treated with herbicides (Delvo 1990; Gianessi 1992). Since 1982 total kilograms of herbicides used has decreased about 10% in the U.S. because of the introduction of compounds that are active at lower rates, reductions in use rates of older herbicides, and hectarage diversion programs. Thus, U.S. farmers rely heavily on herbicides as the major method of weed control in crop production (Duke 1992; Zimdahl 1991). Herbicide use has resulted in substantial decreases in the number of farmers and farm laborers required for crop production in the U.S. (Nalewaja 1974). Adequate alternative methods of weed control are not now available in the event of future restrictions on herbicide usage that may result from public concern about herbicides in ground- and surface water, toxicity to humans, chronic health effects, food safety, and wildlife mortality (Burnside 1993; Duke 1992).

The best way to present an agriculturalist's viewpoint about HRC would be to discuss the pros and cons of different risks and benefits that have been expressed by various groups involved in the development and use of this technology as well as those concerned with implementing of HRC. An understanding of the relative importance of stated risks and benefits with the use of HRC is required to make intelligent decisions regarding this technology. What is needed is to come to some consensus in our assessment of this technology rather than polarize our efforts and frustrate and confuse legislators at the state and federal level who will deal with funding research, education, and regulation of HRC. We must join together in this worthwhile endeavor of keeping our agriculture productive and making sure it is profitable and competitive while protecting our environment and conserving our natural resources.

2. CONCERNS ABOUT HERBICIDE-RESISTANT CULTIVARS

Concern about the use of HRC is common among the general public (Goldburg et al. 1990) because genetically engineered organisms or biotechnology developments have received considerable adverse publicity in the popular press. Herbicide-resistant cultivars have been viewed anywhere from creating an environmental time bomb to more rapidly solving specific weed problems in crop production. Concerns about HRC by one group in society may actually be thought of as a real benefit by another group such as crop producers, e.g., clearance of a new herbicide on a minor acreage crop such as onions (*Allium cepa* L.). However, one must also recognize that grower use of herbicides may have an impact far beyond the farm. Examples are ground- and surface water quality, plant damage from drift or soil residues, soil erosion, farm labor needs, size of farms, wildlife mortality, food and feed safety, and human health (Dyer et al. 1993; Zimdahl 1991). Thus, some oversight and regulation of farmer production practices by state and federal agencies has been the natural outcome of concerns from various interest groups in society (Horne 1992). There will be increasing regulation of pesticide use as agricultural production becomes more intensive in order to accommodate food and fiber demands for an expanding world population. The end result of this conflict over the need for increased food will be either stringent human population control or reoccurring famine and starvation somewhere in the world.

The concerns about HRC are covered here by discussing ten risks often mentioned by various interest groups. They are not listed in any priority order. With this approach, one can present the major concerns with the use of HRC and then discuss the pros and cons of various aspects troubling different interest groups.

2.1. HERBICIDE-RESISTANT CULTIVARS MAY INCREASE OUR DEPENDENCE ON HERBICIDES

We are already heavily dependent upon herbicides for weed control in the U.S. because over 95% of our major crops are treated with herbicides and 85% of pesticides used in crop production are herbicides (Delvo 1990). Critics contend that these HRC will further exacerbate our dependence on herbicides and continue to reduce our use of nonherbicidal weed management technology (Goldburg et al. 1990). Because we are already heavily dependent on herbicides, the primary impact of HRC may be to replace herbicides presently being used on these crops. Environmentally, replacement of present herbicides may be positive or negative, depending on the chemical and physical characteristics of the herbicides in question. Properly used, this technology could substitute for a more toxicologically or environmentally benign herbicide in many crop production systems. Benign herbicides may be effective at lower use rates and show postemergence selectivity vs. prophylatic soil activity, low animal toxicity and persistence, and a broader weed control spectrum that can replace commonly used herbicide combinations.

Minor acreage crops also may receive more attention from herbicide manufacturers because modern technology may make the development of HRC feasible, and herbicides cleared on major crops may be cleared economically on minor crops if much of the efficacy, residue work, and toxicology has already been done. However, minor crops are not major market opportunities because of limited sales potential due to low acreage, and they are high liability risks to the chemical company because of the greater value of these crops. History has shown that herbicide manufacturers have limited interest in accommodating the minor crop market because of regulatory expenses, liability risks, and low profit potential.

Herbicide use in forests has decreased over the past 2 decades because conservation and recreation interests have lobbied U.S. Forest Service personnel and others to reduce herbicide and other pesticide use in these areas. Herbicide-resistant trees have been researched by U.S. Forest Service personnel, but this research also has been questioned (Goldburg et al. 1990). Herbicide applications on publicly owned forests will be a difficult proposition in the future because of disagreements over the importance of timber production when judged against potential hazards to wildlife, fish, and the environment. It would appear that herbicide use in forests will continue to decline unless timber production is given a much higher priority in forest management.

Herbicides are used sparingly in developing countries where labor is inexpensive and readily available because weeds can be economically removed manually (Holm 1975). However, in developed countries the necessary labor for manually removing weeds on major crops is generally not cost effective or available. It has been estimated that 17.7 million people working 40 h/week would be needed over a 6-week period just to weed corn in the U.S. (Nalewaja 1974). This number of workers would be over six times as many people as were employed (unpaid or hired) on all U.S. farms in 1991 (Anon 1992). Thus, our dependence on herbicides will continue until we have major advances in nonchemical weed control technology.

2.2. HERBICIDE-RESISTANT CULTIVARS COULD INCREASE THE USE OF ENVIRONMENTALLY "HARSH VS. BENIGN" HERBICIDES

The technology to develop HRC could be used to develop new uses for environmentally benign herbicides or it could be misused to facilitate the use of environmentally harsh herbicides. Therefore, the technology is neither good nor bad; it all depends upon how it will be utilized (Wyse 1992). However, this technology has a greater chance of being properly utilized because environmental characteristics of the herbicide will be known before the search begins for HRC. An environmentally benign herbicide would mean different things to different individuals, but some general characteristics would be as follows.

2.2.1. Shorter Environmental Half-Life

Herbicide-resistant cultivars could be developed for herbicides that last only hours, or at most days, in soil or water, which would eliminate their accumulation or buildup in the environment. However, in some situations the more persistent herbicides have the advantage of providing weed control over an entire growing season and thus eliminating repeat applications.

2.2.2. Reduced Leaching in Soils

Herbicides that do not leach in soil are effectively prevented from moving into the subsoil or groundwater. Nonpoint source contamination of groundwater by agricultural pesticides is emerging as one of the major environmental issues of the 1990s (Hallberg 1986). Reasons for this concern are intensified use of water by agriculture, industry, and municipalities and documented problems of deterioration of groundwater quality (Smolen et al. 1984). Half of the U.S. population uses groundwater as its primary source of drinking water and 97% of all rural, domestic drinking water is supplied by groundwater wells (EPA 1987). Thus, we have been alerted to a potential problem that is receiving increasing public attention and concern, and the quickest and surest way of reducing or solving the problem is utilizing alternative weed control methods and more judicious use of environmentally benign herbicides, other pesticides, and fertilizers.

2.2.3. Low Volatility

Contamination of air with herbicides can occur by spray drift or subsequent volatilization. Herbicide drift during application is an important problem in herbicide use that must be avoided. Herbicides with low vapor pressure should not move from the target area by volatilization once they are applied. Also, herbicides have a "built-in" detection device that does not allow them to drift or volatilize without visibly damaging adjacent crops or other plants. Therefore, herbicide movement does not go undetected and, for this reason, applicators have learned to avoid herbicide movement onto nontarget species or face litigation. Herbicide formulation or application technology that prevents drift and volatility will become increasingly important to safe and continued use of herbicides.

2.2.4. Toxic to a Broad Spectrum of Weeds

A broad-spectrum herbicide has a distinct advantage, as in essentially all field situations the farmer is dealing with a diverse array of weed species rather than a single species. Thus, a single, broad-spectrum herbicide could provide adequate weed control rather than the present practice of relying on a combination of herbicides. For example, glyphosate [*N*-(phosphonomethyl)glycine] could be used on HRC of soybean [*Glycine max* (L.) Merr.] rather than using one of the multitude of herbicide combinations presently having Environmental Protection Agency (EPA) clearance on this crop (Anon. 1993). Another advantage of a broad-spectrum herbicide would be to avoid controlling only one or a few weed species because the missed species would fill in the void left by the controlled species, and the weed control advantage would be lost.

2.2.5. Nontoxic to Animals

Herbicides vary widely in their toxicity to animals, and they should be selected for use with HRC based on low acute and chronic effects, especially to humans. In general, herbicides are not very toxic to animals, but all should be handled with the utmost precaution because there are obvious exceptions to any generalization and individuals vary in their response to pesticides.

2.2.6. Utilize Selective, Postemergence Herbicides

This approach to weed management would encourage greater use of cultural and mechanical weed control methods (Burnside 1992). A postemergence herbicide would be used only if cultural and mechanical weed control methods failed. Thus, fewer acres would need to be treated with herbicides in contrast to the over 95% of major crop acreage now being treated (Delvo 1990). Postemergence herbicides could be applied in bands rather than broadcast in row crops and this would further reduce herbicide use up to 66%. Weeds between the rows would be removed by cultivation rather than by herbicides. Postemergence vs. soil-applied herbicides are generally applied at lower rates because absorption by soil colloids is avoided (Anon 1993). Also, soil conservation production systems (e.g., reduced tillage, no-tillage, cover crops, etc.) require significant weed control inputs, and selective postemergence herbicides could provide that weed control when it is needed.

2.3. HERBICIDE-RESISTANT CULTIVARS COULD ACCELERATE SELECTION FOR HERBICIDE RESISTANT WEEDS

Resistant weeds will occur following repeated use of the same herbicide or herbicides with the same mode of action, repeated use of any weed control method, growing of crops whose seed can survive in soil (e.g., shattercane [*Sorghum bicolor* (L.) Moench]), outcrossing to weedy relatives of HRC (e.g., jointed goatgrass [*Aegilops cylindrica* Host]), or other methods of biotype selection (Holt 1992; Radosevich et al. 1992). The greater the potential for gene flow or cross-fertilization of HRC with weedy relatives, the greater the chances for introducing herbicide resistance genes into weed species. Many crops have weedy relatives with which they potentially could cross-pollinate (Harlan 1982). The transfer of herbicide resistance via pollen from HRC to a formerly susceptible weed is a risk that warrants attention by researchers. Cross resistance to chemically dissimilar herbicides and to herbicides with different modes of action was first reported in rigid ryegrass (*Lolium rigidum* L.) (Heap and Knight 1982). Investigation is warranted on population dynamics of resistant weeds, resistance management, and whether HRC are more or less competitive with weeds.

2.4. HERBICIDE-RESISTANT CULTIVARS COULD CAUSE A SHIFT IN THE WEED POPULATION IN SOIL

Weed species do shift with different weed management methods, and if the same herbicide was used on the crop(s) grown in an area because of the use of HRC, then use of HRC would exacerbate weed population shifts (Radosevich et al. 1992). Others indicate that HRC would provide better weed control, so ecological diversity would be reduced in treated fields (Burnside 1992; Giaquinta 1992). The truth is, farmers are not seeking ecological plant diversity in their crop production fields because the highest yields occur where competition from other plant species has been eliminated. The principle here is that a given area of land is able to support a certain amount of plant growth depending on climatic, edaphic, and biotic limitations, and if 20% of that plant biomass is weeds, crop yield is reduced by 20%.

A basic principle of crop production is that there are no permanent solutions to weed management. Weeds are a diverse group of species that exhibit considerable plasticity, and thus are able to adapt to any weed control practice because selected species from the great variety available will find a niche for establishment and growth. Thus, the composition of weed species will vary markedly in cultivated fields, nontilled areas, pastures, rangeland, lawns, or gardens. Weed species composition will even vary with different times of seedbed preparation in an individual field, as weed composition will vary from cool season to warm season species, depending upon the planting date of the crop. Thus, one need not become overly concerned about weed population shifts because this is a natural ecological phenomenon to changes in land and crop management.

2.5. HERBICIDE-RESISTANT CULTIVARS WOULD COMPLICATE BUSINESS FOR FARMERS AND SEED SUPPLIERS

Farmers will have more varieties or hybrids to select from because each cultivar must be maintained in several genetic forms to satisfy demands for specific resistance to various herbicides. The farmer then selects HRC that best fits yield goals and weed control needs, but this does complicate and restrict one's crop production options. Also, it increases the chance of a mismatch between HRC planted and herbicide used. Seed company officials are concerned about whether HRC can be profitable, considering that a given herbicide may lose appeal or acceptability while respective HRC are being developed (Duvick 1992). Also, they are concerned that breeding for HRC may distract from breeding for yield and important agronomic traits, alternative systems of weed control will reduce the need for HRC, HRC may accelerate the development of herbicide-resistant weeds (HRW), HRC can become weeds, patenting of HRC genes may affect profitability in the seed trade, profitability may be lost due to the complexity of HRC seed inventory control, and seed price may not justify the added expenses of developing, producing, and selling HRC. One must recognize that these added production costs for HRC will be passed on to the farmer, and on-farm profitability is a concern because the farmer does not set the price for what is produced. The only way the farmer has been able to survive similar production trends in the past was to become more efficient, expand acreage, supplement income with off-farm income, or go out of business.

Use of herbicides has become more complicated over the years, so HRC just become another factor in these changing production practices. A few examples of developments of herbicides and their use which shows this increasing complexity (Wyse 1992) would be as follows: herbicide used alone as a spray, multiple herbicide formulations for different application methods, tank mixtures of several herbicides to increase the spectrum of weed control, prepackaged mixtures of herbicides to also reduce carryover in soil, multiple spray additives for herbicides to change weed control and crop tolerance, combinations of herbicides and fertilizers, antidrift and low volatility herbicide technology, herbicides interactions with other pesticides, compatibility problems of herbicide and other chemical mixtures, differential herbicide tolerance of crop cultivars, weed population shifts and herbicide resistance, and environmental concerns with herbicide use. Thus, weed control technology continues to evolve because weeds escape or evade any weed management system. Weed scientists must continue to reduce their dependence upon a few weed control methods and utilize systems of weed management that include preventive, cultural, mechanical, biological, chemical, and integrated weed control technology (Burnside 1993). No one weed control method is a panacea and weed management requires considerable time and energy. These weed control requirements and challenges need to be recognized by everyone from the farmer to the policymakers.

2.6. HERBICIDE-RESISTANT CULTIVARS COULD LIMIT THE PRODUCTION OPTIONS OF FARMERS

Critics of HRC believe that their use will lead to an abandonment of alternative weed management methods and encourage farmers to continue to repeatedly use the same weed control options (Goldburg 1992). Such a scenario is possible but it depends on whether the technology is used properly or misused. Herbicide-resistant cultivars will be one more weed management option that farmers can add to their weed control arsenal. Farmers need to annually evaluate their weed management systems and initiate changes as they observe weed escapes early in the growing season when corrective control methods can be implemented, and again while harvesting the crop. Herbicide-resistant cultivars may open up new production options for the farmer such as narrow-row crop production, reduced or no-tillage production systems, and growing of new or less weed-competitive crops.

2.7. HERBICIDE-RESISTANT CULTIVARS MAY AFFECT COMPETITIVENESS AMONG AGRICULTURAL COMPANIES

Herbicide company personnel have considered HRC as new opportunities for wider use and increased sales of their most profitable herbicides (Duvick 1992). The primary aim of a seed or chemical company is to develop products that farmers need and sell them at a profit. There has been a major trend by chemical companies to acquire seed companies and develop HRC that tolerate the herbicides they sell. This approach increases the company's market share and tends to increase competition within the industry. Thus, we are seeing fewer but larger pesticide/seed companies, with fewer independent research-based seed companies remaining. This trend of vertical and horizontal integration of pesticide companies certainly increases the competitiveness of the pesticide/seed industry. Also, the increasing cost of getting a pesticide registered with the EPA is rapidly eliminating the small pesticide companies. Competitiveness has been increasing as the companies are now competing for market share in a mature market rather than exploiting new market opportunities in developed countries. However, there is still considerable opportunity for market expansion in the developing countries of the world, so most major pesticide companies have become international companies.

2.8 HERBICIDE-RESISTANT CULTIVARS MAY NOT BE NEEDED FOR OUR MAJOR CROPS

The market potential of major crops is so large that there are 10 to 20 selective herbicides cleared for each crop, and the need for HRC in these crops has been questioned by pesticide critics (Anon. 1993). These critics certainly have a point concerning the need for HRC in large acreage crops. However, competition for a share of these larger markets and observing present HRC developmental activities indicates that there will be HRC available in major crops (Kishore et al. 1992). Nevertheless, the greatest need for HRC will be in minor crops where low acreages do not economically justify standard herbicide clearance procedures. Whether HRC development for minor crops will even be economically feasible is questionable, but if the herbicide has already been cleared on a major crop then many of the clearance costs may have been expended. Thus, HRC may be the most feasible method of getting herbicides cleared in the future for use on many minor crops.

2.9. HERBICIDE-RESISTANT CULTIVARS MAY NOT BE ACCEPTABLE TO THE GENERAL PUBLIC

The national Biotechnology Working Group, representing a number of pro-environmental citizen organizations, came together because of their mutual concern over the research direction of biotechnology (Goldburg et al. 1990). They publicized their objections to the development of HRC by releasing their report entitled, *Biotechnology's Bitter Harvest: Herbicide-Tolerant Crops and the Threat to Sustainable Agriculture,* during three press conferences for legislators March 19, 1990, in Washington, D.C., Davis, CA, and St. Paul, MN. The group indicated that citizen activists concerned with environmental affairs had initially supported increased funding of biotechnology because spokespersons for biotechnology had promised that "... with new genetically engineered crops, and biopesticides to control pests ... chemical pesticides would no longer be needed." The Biotechnology Working Group noted that some of the first commercial products of agricultural biotechnology were herbicide-resistant crops, and thus they complained that "... the direction of agricultural biotechnology is clear: the first major products will not be used to end dependence on toxic chemicals in agriculture. Rather, they will further entrench and extend the pesticide era."

Most individuals involved with various aspects of weed science and HRC research do not view the development of HRC as anti-environmental. Clearly, a dialogue between representatives of each side of the issue are needed to advance mutual understanding of the technology by all concerned.

The EPA authorizes the use of herbicides and other pesticides, and it is a difficult task to please both the pro-environmentalists and chemical company representatives. Animal and Plant Health Inspection Service (APHIS) personnel develop the criteria for evaluation and they issue permits for field testing and release of HRC. In general, personnel of biotechnology and chemical/seed companies want the clearance procedures to be administered at the federal level as are pesticides rather than having to respond to a number of specific laws at the state level. At present, a few states have developed criteria over and above those in the federal guidelines. Interest in genetic engineering does show, however, that the general public is concerned about HRC and other genetically engineered organisms or developments.

2.10. HERBICIDE-RESISTANT CULTIVARS MAY PROMOTE HERBICIDE USE IN DEVELOPING COUNTRIES

Weed scientists are being drawn into an ethical debate over the technology and methods they have developed to mitigate weeds in food and fiber production (Burnside 1993; Duke 1992; Radosevich et al. 1992; Wyse 1992; Zimdahl 1991). The debate has even centered on evolving weed control technology in developing countries, where they now depend largely on human labor for controlling weeds (Tshibaka 1992; Holm 1975). Pro-environmentalists fear, with some justification, that weed control technology will evolve in developing countries just as it has in developed countries (Figure 1). Thus, they believe, the introduction of HRC and increasing use of herbicides will occur on a worldwide basis.

Agriculturalists argue that manual removal of weeds, which is largely the responsibility of females in developing countries, is a menial task that often interferes with the education and personal development of these individuals (Tshibaka 1992; Holm 1975). Also, weed management becomes

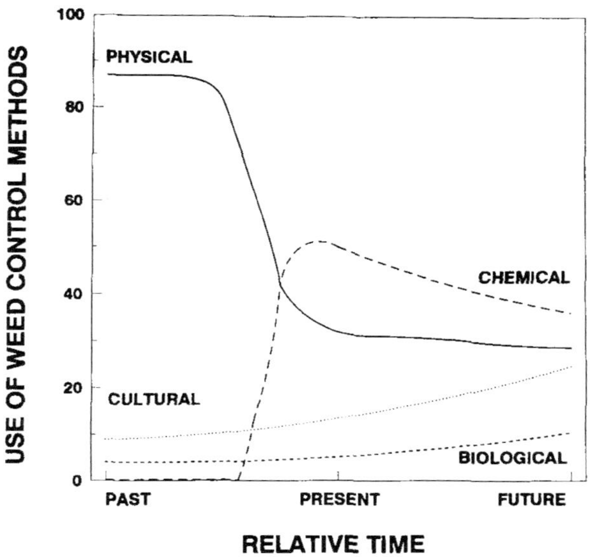

FIGURE 1 Evolving use of the major weed management methods in various systems of crop production in developed countries of the world during ±100 years.

the major time requirement of subsistence agriculture (up to 40% in some countries; Matthews 1981), and farmers are reluctant to try something new or unproven because the food they produce is needed for family survival and they do not wish to risk losing yield. Thus, there is a need for economically feasible technological breakthroughs in order for some of these farmers to evolve out of their subsistence levels of crop productivity while others seek off-farm employment.

3. BENEFITS OF HERBICIDE RESISTANT CULTIVARS

We have already witnessed the introduction of developed HRC with the incorporation of cytoplasmically inherited *s*-triazine resistance into canola (*Brassica napus* L.) hybrids (Beaversdorf et al. 1980). Furthermore, development and introduction of other HRC are increasing at an accelerating pace (Dyer et al. 1993; Gressel 1989; Kishore et al. 1992). Weed scientists also recognize that crop yields, profitability, and competitiveness of U.S. agriculture would drop markedly if herbicide use were curtailed before suitable alternative weed control methods are available (Burnside 1992; Duke 1992; Zimdahl 1991). In fact, a crash program on alternative weed control methods is needed to meet the demands for weed management technology in future agricultural production systems. Thus, it would not be advisable to curtail HRC technology prior to concerted research on alternative weed management technology because HRC have such potential to more selectively manage weeds with more judicious use of herbicides in many crop production systems.

Furthermore, the important role that herbicides play in modern crop production must not be forgotten. In the U.S. 85% of all pesticides used in crop production are herbicides because they not only control the number one crop pest, weeds, but they also reduce horsepower and manpower requirements in crop production (Burnside 1993; Nalewaja 1974). Benefits resulting from the use of herbicides include increased crop yields and quality; reduced soil erosion, compaction, plant residue incorporation, and moisture loss because herbicides substitute for tillage; savings of fuel, machinery, and manpower; more efficient use of water, nutrients, and sunlight in crop production; a marked increase in the hectarage that a farmer can effectively manage; and herbicides have enhanced the profitability and competitiveness of U.S. agriculture. In summary, herbicides have contributed significantly to the most abundant, safe, dependable, and affordable food and fiber supply the world has ever known.

An understanding of not only the risks but also the benefits of HRC is required to make intelligent decisions regarding this technology (Dyer et al. 1993). Thus, a discussion of ten benefits of HRC, not in priority order, follows.

3.1. HERBICIDE-RESISTANT CULTIVARS MAY SOLVE MANY DIFFICULT WEED MANAGEMENT PROBLEMS

Herbicide-resistant cultivars have provided and will increasingly provide new solutions to difficult weed management problems. These are exciting developments for agriculturalists, as the efficacy of weed management has and will continue to advance significantly. However, weed scientists also need to be conversant with and aware of the concerns of other scientists and the general public as they evaluate and promote these developments. This is necessary because of the tremendous change in crop production methods and the environmental impact that past and present weed management technology has had on worldwide agricultural and societal issues. Herbicide-resistant cultivars are just another method of weed management, but they have the potential to control many missed weed species that have escaped prior weed management methods.

One should not become alarmed when one or more weed species escape some cultural, mechanical, biological, or chemical weed control method because that is the rule rather than the exception. One principle that 35 years of weed management research has taught the author is that there is no panacea or permanent weed management solution. No weed management method lasts forever, because some weed species will fill the voids left unoccupied in any crop production system. When we try to divert the maximum amount of plant growth resources or energy requirements into a single crop species we violate nature's law of diversity which favors a mixture of plant species. However, agriculturalists continue to strive for monoculture because it provides greater harvestable food and fiber production per unit area, and it is more economical and manageable.

Herbicide-resistant cultivars have numerous applications in weed management; weed control in reduced or no-tillage situations serves as one example. Farmers are reluctant to give up tillage as a weed control method even though it does increase plant residue incorporation, soil erosion, soil compaction, and soil water loss. Herbicide-resistant cultivars would allow greater crop selectivity while broadening the weed control spectrum and thus compensate for the loss of tillage as a weed control method. The reduced tillage system known as ecofallow and HRC could move crop production systems further west in the Great Plains without exposing those fragile soils to erosion because they provide adequate weed control without excessive tillage so soil moisture is conserved (Burnside 1985; Wicks and Burnside 1981).

3.2. HERBICIDE-RESISTANT CULTIVARS WILL FOSTER THE USE OF ALTERNATIVE WEED CONTROL METHODS

Herbicide-resistant cultivars will allow a grower to use cultural, mechanical, and biological weed control methods, for example, and back them up with a selective, postemergence herbicide if the initial control methods fail. Farmers will be more willing to utilize nonchemical weed control technology if they know that a "back up" method exists to prevent the crop from being lost to weeds. Recent assessments of weed management technology has shown that agriculturists place an overwhelming emphasis on chemical weed control technology (Burnside 1993; Duke 1992; Wyse 1992; Zimdahl 1991). The challenge to public weed scientists, therefore, is to initiate and/or expand preventive, mechanical, cultural, biological, and integrated weed management research or it will not be done. We cannot expect to entice large numbers of chemical company weed scientists into researching alternative weed control technology. They must research marketable products that return a profit or there will be no chemical industry (Burnside 1993). Most of the current alternative weed management technologies do not have the profit potential to merit private-sector funded research.

Herbicide-resistant cultivars are needed to augment weed control technology as weed management evolves from its heavy dependence upon herbicides to a more integrated weed management system (Figure 1). Herbicide-resistant cultivars and the use of postemergence herbicides should increase grower use and confidence in alternative weed control methods.

3.3. HERBICIDE-RESISTANT CULTIVARS WILL FACILITATE THE USE OF ENVIRONMENTALLY "BENIGN" HERBICIDES

If HRC are properly utilized there will be a conscious attempt to utilize environmentally benign rather than harsh herbicides. Herbicide-resistant cultivar technology can be used properly or misused, but it behooves anyone that uses a herbicide to select one with the least adverse effects on the environment, providing it is effective for the weed control needed (Wyse 1992). In general, one should seek HRC for selective, postemergence herbicides that exhibit no toxicity to humans and other animals, broad spectrum weed control at low use rates, some residual weed control with reasonable degradation rates in the environment, low soil leaching so as not to contaminate groundwater, low volatility, compatibility with other chemicals being applied and application equipment, negligible residues in food or feed, new modes of action, effectiveness under various crop production systems, cost effectiveness, and compatibility with crop rotations and integrated weed management systems. Properly used, HRC could eliminate some of the problems we now observe with herbicide use in rural and urban environments.

3.4. HERBICIDE-RESISTANT CULTIVARS WILL FACILITATE HERBICIDE CLEARANCES FOR MINOR ACREAGE CROPS

Herbicide-resistant cultivars can be identified by screening a large population of crop biotypes or millions of crop cells in tissue culture to a selected herbicide. In addition, HRC can be obtained by plant gene transfer techniques, where a gene for herbicide resistance is transferred into a desired crop (Kishore et al. 1992). Tissue culture or gene transfer methods appear to be an economical method of obtaining HRC in minor crops. If the herbicide selected has the needed weed control spectrum, environmental safety, and prior clearance on another crop, then much of the cost of EPA clearance will have already been expended. Thus, it may be economical for a chemical company to obtain EPA clearances for an established herbicide on a number of minor crops. With this approach, agriculturalists may be able to provide selective herbicides for minor crops that are both safe and cost effective. Without HRC for minor crops or some other initiative, growers of these crops will see a continued loss of herbicides available for use on these low acreage crops (Burnside 1993).

3.5. HERBICIDE-RESISTANT CULTIVARS WILL ALLOW FARMERS TO BETTER MANAGE HERBICIDE-RESISTANT WEEDS

Herbicide-resistant weeds were first reported in the 1950s to 1970 (Harper 1956, Radosevich and Appleby 1973, Ryan 1970). Since then there has been a flurry of research activity in this area, both because of an increasing threat to agricultural production from HRW (Holt 1992) and the potential benefits to be gained from developing HRC (Burnside 1992). There are now 40 dicot and 15 monocot plant species known to have biotype resistance to the *s*-triazine herbicides (Holt and LeBaron 1990). Also, at least 44 weed species have been reported to have biotypes resistant to one or more of 15 other herbicides or herbicide families. This list of HRW will continue to grow, especially where we have had repeated and prolonged exposure of weeds to herbicides with the same mode of action. However, agriculturalists have not always recognized the evolution of weeds resistant to herbicides because of the common crop production practices of rotating crops, cultivation methods, and herbicides. Herbicide failure in the past often was attributed routinely to such causes as varying climatic conditions, soil variations, relative weed population pressure, wrong plant growth stage for treatment, and sprayer calibration and application errors. However, it is natural for weed species populations to change in response to repeated herbicide use or to any other weed management method used on a crop.

Herbicide-resistant weeds are best controlled by employing a variety of preventive, cultural, mechanical, biological, chemical, and integrated methods. Thus, HRC represent another control method for HRW that the farmer would have available. Also, HRC technology should markedly improve the ability of growers to control weeds postemergence when needed while initially employing a variety of other weed control methods.

3.6. HERBICIDE-RESISTANT CULTIVARS WILL IMPROVE WEED MANAGEMENT IN CROP PRODUCTION

Herbicide-resistant cultivars most certainly will improve the ability of weed scientists and farmers to control more weeds in more of our crops. Examples would be the control of escaped weed species or groups of species in numerous crops, selective postemergence control of missed grass weeds in corn (*Zea mays* L.) and sorghum (*Sorghum bicolor* [L.] Moench) when fields are too wet to cultivate, complete control of weeds in sugar beet (*Beta vulgaris* L.) production, and control of weeds in minor acreage crops without excessive reliance on hand weeding.

Critics may say that these are the short-term effects of HRC, and that needs to consider the long-term effects. Some long-term impacts of HRC are discussed in other sections of this chapter, but most weed control methods have only a short-term effect, considering the plasticity of weeds. The best approach to weed management would be to use HRC in conjunction with integrated systems of cultural, mechanical, biological, and chemical weed control methods.

3.7. HERBICIDE-RESISTANT CULTIVAR WILL INCREASE THE PROFITABILITY AND COMPETITIVENESS OF CROP PRODUCTION

Herbicide-resistant cultivars will reduce crop injury because of improved crop tolerance to the herbicide, in addition to improved weed control (Burnside 1992; Gressel 1989). Thus, crop yields should be significantly increased with the use of HRC. Farmers will use HRC if they are convinced that this technology will control weed species that are being missed, reduce crop injury, or increase crop yields while being cost effective. No HRC will achieve much of a market share if it does not provide added performance or value over what is presently available. The value-added concept means that farmers expect to gain a given amount of profit when they use HRC plus the herbicide, and thus would be willing to spend some fraction of that expected gain to use the new technology (Duvick 1992). The amount the farmer is willing to spend is a major factor that company personnel will use in setting the price of this technology.

Chemical/seed companies have considered HRC as a means of expanding market share and thus profitability of their crop cultivars and herbicides. Pesticide company personnel hope to gain a competitive edge in the industry by developing HRC that utilize their herbicides. Also, this new technology will give company personnel and farmers greater flexibility in moving away from environmentally harsh to environmentally benign herbicides.

The countries that most rapidly exploit this new technology should provide a competitive edge to the farmers utilizing HRC. However, this type of technology will be exported rapidly to other countries that want to utilize HRC, so the competitive advantage will be relatively short-lived.

3.8. HERBICIDE-RESISTANT CULTIVARS WILL HELP RELEASE PEOPLE FROM THE DRUDGERY OF HAND WEEDING

Most manual weed control occurs in developing countries, where hand weeding is the standard weed control practice, and the weeds removed are used for livestock feed (Holms 1975). In the "slash-and-burn" agriculture practiced in Zaire in central Africa male and female farmers spend about 2 and 10%, respectively, of their crop production time hand weeding their crops (Tshibaka 1992). In developed countries the major weed control methods are herbicide and tillages, with little hand weeding (Figure 1). Hand weeding in the U.S. is still done to remove weed escapes in low growing crops, in sugarbeet production, in high-value minor crops, and in gardens and lawns.

Sugarbeet production generally involves two or three hand weedings, in addition to utilizing herbicides and tillages for weed control, so this crop will be used as an example of how HRC could reduce the menial task of hand weeding. One of the advantages of HRC is that they possess greater crop selectivity or resistance of the cultivar to the herbicide in question. Thus, the herbicide selected to develop a sugarbeet HRC should possess a weed control spectrum broad enough to control weeds presently requiring hand removal in sugarbeet production. This same scenario could be followed for each high value or minor acreage crop for which the elimination of hand weeding would be desirable.

3.9. HERBICIDE-RESISTANT CULTIVARS WILL REDUCE CROP LOSSES FROM HERBICIDE INJURY

All herbicides cause some temporary crop injury. Therefore, weed scientists use the term "selective herbicide," which means "a chemical that is more toxic to some plant species than to others." Common groundsel (*Senecio vulgaris* L.) was the first weed species to be reported as showing resistance to the *s*-triazine herbicides (Ryan 1970). The extent of this resistance was that several *s*-triazine herbicides did not control resistant common groundsel at rates of 15- to over 30-fold above the rate required to kill susceptible biotypes. Similar herbicide resistance can also be introduced into crop cultivars in order to reduce herbicide injury to the crop and increase weed control (Giaquinta 1992). Thus, there is considerable opportunity to reduce herbicide injury to crops by the development of HRC and also to improve weed control in the crop. Nonselective herbicides such as glyphosphate can even be made selective by developing HRC of soybean and other crops (Kishore et al. 1992). Therefore, HRC can markedly improve herbicide selectivity and efficacy in crop production.

3.10. HERBICIDE-RESISTANT CULTIVAR RESEARCH MAY LEAD TO OTHER USEFUL DISCOVERIES

Research with HRC will also lead to other discoveries and developments that will be important to science and possibly to farmers and the general public as well (Burnside 1992). For example, research with herbicide-resistant corn may provide methods for modifying storage lipid biosynthesis in corn seed, and such discoveries may be more important than the sethoxydim (2-[1-(ethoxyimino)butyl]-5-[2-(ethylthio)propyl]-3-hydroxy-2-cyclohexen-1-one] -resistant corn inbred (Dotray et al. 1993; Somers et al. 1993).

Scientists work best when given an area of responsibility or series of needs or problems and are then allowed to exercise their best judgment as to how to solve that problem. Such a research environment forces the scientist to be a self-starter, accept full responsibility for the research challenge, blame oneself rather than others in case of failure, keep oneself thinking about possible solutions, and to be more productive. Thus, we should not impose unnecessary regulations and restrictions that would inhibit creativity, productivity, and discovery. However, we must first and foremost remember that we have to protect the world's inhabitants, environment, and ecosystem.

4. SUMMARY

The technology is now available and development is underway to produce HRC for many of our crops. These are exciting times because HRC can represent a significant step forward in weed management technology but the technology, can also be misused and misunderstood.

Using crops tolerant to herbicides is not new technology, as industry personnel have been using it during the past 5 decades — crop tolerance is necessary in order to have a selective herbicide. In the past chemical company personnel screened a large number of organic compounds against a limited number of major crop and weed species to discover HRC. Now a single herbicide can be evaluated for crop selectivity against a great number of biotypes of a single crop in tissue culture, or sometimes gene transfer can be used to produce tolerant genotypes. Thus, one can now choose an environmentally benign herbicide and search for or develop HRC to match that herbicide.

Herbicide-resistant cultivar technology can be used properly or misused. Those decisions will be up to the scientists, business personnel, and administrators that move this technology from the laboratory to the farmers' fields.

There has been enough adverse publicity about HRC that the general public has mixed feelings about their utilization and impact. Certainly the extensive publicity over the bovine somatotropic hormone as a stimulant to milk production in dairy animals has left many with some mistrust and concern over new developments in biotechnology.

It behooves us to examine both the risks and benefits of HRC and to communicate this information to the general public. Such information is required to make intelligent decisions regarding HRC technology. We need to reach some consensus in our assessment of this technology rather than to polarize our efforts and frustrate and confuse legislators at the state and federal levels who must deal with funding research, education, and regulation of agricultural biotechnology.

When one considers both the pros and cons of HRC, it is obvious that HRC can contribute to the productivity and profitability of agriculture if used wisely in a sound, integrated weed management program. Development of HRC should be limited to those that utilize environmentally benign herbicides and to crop production situations where they can reduce crop injury, improve weed management, increase crop yields or quality, and favorably or at least not adversely affect the environment. It is the responsibility of the Animal and Plant Inspection Service (APHIS) to evaluate the risks and benefits of each HRC and approve of their use on a case-by-case basis rather than issue a blanket authorization or condemnation of this technology.

Adoption and success of HRC will depend on several technical, environmental, business, economic, and societal issues. Let us join together in the worthwhile endeavor of keeping our agriculture productive and making sure it is profitable and competitive while protecting the environment and conserving natural resources.

REFERENCES

Anon. 1992. *Agricultural Statistics*. U.S. Department of Agriculture U.S. Government Printing Office, Washington, D.C.

Anon. 1993. *Crop Protection Chemicals Reference*, 9th ed. John Wiley & Sons, New York.

Beversdorf, W.D., Weiss-Lerman, J., Erickson, L.R., and Machado, V.S. 1980. Transfer of cytoplasmically inherited triazine resistance from birdsrape to cultivated oilseed (*Brassica campestris* and *B. napus*). *Can. J. Genet. Cytol.* 22: 167.

Burnside, O C. 1985. Weed control challenges with conservation tillage in the Great Plains. In *Agricultural Chemicals of the Future*, BARC Symp. Agric. Res. 8. Ed. by Hilton, J.L. Rowman & Allanheld, Totowa, NJ, p. 199.

Burnside, O.C. 1992. Rationale for developing herbicide-resistant crops. *Weed Technol.* 6: 621.

Burnside, O.C. 1993. Weed science — the step child. *Weed Technol.* 7: 515.

Christou, P., McCabe, D., and Swain, W. 1988. Stable transformation of soybean callus by DNA-coated particles. *Plant Physiol.* 87: 671.

Delvo, H.W. 1990. Agricultural Resources — Inputs Situation and Outlook Report. AR-17. Resource Technol. Div., Economic Research Service, U.S. Department of Agriculture, Washington, D.C.

Dotray, P.A., Marshall, L.C., Parker, W.B., Wyse, D.L., Somers, D.A., and Gengenbach, B.G. 1993. Herbicide tolerance and weed control in sethoxydim-tolerant corn (*Zea mays*). *Weed Sci.* 41: 213.

Duke, S.O. 1992. Weed science — the need and the reality. *Phytoparasitica* 20: 183.

Duvick, D.N. 1992. Concerns of seed company officials with herbicide-tolerant cultivars. *Weed Technol.* 6: 640.

Dyer, W.E., Hess, F.D., Holt, J.S., and Duke, S.O. 1993. Potential benefits and risks of herbicide-resistant crops produced by biotechnology. *Hort. Rev.* 15: 367.

Environmental Protection Agency. 1987. Agricultural Chemicals in Groundwater: Proposed Pesticide Strategy. Proposed Strategy Document, Office of Pesticides and Toxic Substances, U.S. EPA, Washington, D.C.

Gianessi, L.P. 1992. U.S. Pesticide Use Trends: 1966-1989. Quality of the Environment Division, *Resources for the Future*, Washington, D.C.

Giaquinta, R.T. 1992. An industry perspective on herbicide-tolerant crops. *Weed Technol.* 6: 653.

Goldburg, R., Rissler, J., Shand, H., and Hasselbrook, C. 1990. *Biotechnology's Bitter Harvest: Herbicide-Tolerant Crops and the Threat to Sustainable Agriculture*. Biotechnology Working Group, Washington, D.C.

Goldburg, R.J. 1992. Environmental concerns with the development of herbicide-tolerant plants. Wee*d Technol.* 6: 647.

Gressel, J. 1989. Conferring herbicide resistance on susceptible crops. p. 237-259. In *Herbicides and Plant Metabolism*. Ed. by Dodge, A.D. Cambridge University Press, Cambridge, U.K.

Hallberg, G.R. 1986. From hoes to herbicides: agriculture and groundwater quality. *J. Soil Water Conserv.* 41: 357.

Harlan, J.R. 1982. Relationships between weeds and crops. In *Biology and Ecology of Weeds*. Ed. by Holzner, W. and Numata, N. W. Junk, The Hague, p. 91.

Harper, J.L. 1956. The evolution of weeds in relation to the resistance to herbicides. *Proc. 3rd Br. Weed Control Conf.* 1: 179.

Heap, I. and Knight, R. 1982. The occurrence of herbicide cross-resistance in a population of annual ryegrass, *Lolium rigidum*, resistant to diclofop-methyl. *Aust. J. Agric. Res.* 37: 149.

Holm, L. 1975. The role of weeds in world food production. *Proc. North Centr. Weed Control Conf.* 30: 18.

Holt, J.S. and LeBaron, H.M. 1990. Significance and distribution of herbicide resistance. *Weed Technol.* 4: 141.

Holt, J.S. 1992. History of identification of herbicide-resistant weeds. *Weed Technol.* 6: 615.

Hoyle, R. 1993. Herbicide-resistant crops are no conspiracy. *Bio/Technology* 11: 783.

Horne, D.M. 1992. EPA's response to resistance management and herbicide-tolerant crop issues. *Weed Technol. 6:* 657.

Kishore, G.M., Padgette, S.R., and Fraley, R.T. 1992. History of herbicide-tolerant crops, methods of development and current state of the art — emphasis on glyphosate tolerance. *Weed Technol.* 6: 626.

Klee, H. and Rogers, S. 1989. Plant gene vectors and genetic transformation: plant transformation systems based on the use of *Agrobacterium tumefaciens*. In *Cell Culture and Somatic Cell Genetics of Plants,* Vol. 6. Ed. by Schell, J. and Vasil, I.K. Academic Press, New York, p. 20.

Knake, E.L. 1992. Technology transfer for herbicide-tolerant weeds and herbicide-tolerant crops. *Weed Technol.* 6: 662.

Matthews, L.J. 1981. The Status of Weed Science Development at County Level. Honorary Membership Address at the Weed Science Society of America Meeting, Las Vegas.

McDougall, J. and Phillips, M. 1992. *Agrochemical Overview 1991. County Natwest Woodmac*, Kintore House, Edinburgh.

Nalewaja, J.D. 1974. Energy requirements for various weed control practices. *Proc. North Centr. Weed Control Conf. 29:* 19.

Radosevich, S.R. and Appleby, A.P. 1973. Relative susceptibility of two common groundsel (*Senecio vulgaris* L.) biotypes to six *s*-triazines. *Agron. J. 65:* 553.

Radosevich, S.R., Ghersa, C.M., and Comstock, G. 1992. Concerns a weed scientist might have about herbicide-tolerant crops. *Weed Technol. 6:* 635.

Ryan, G.F. 1970. Resistance of common groundsel to simazine and atrazine. *Weed Sci. 18:* 614.

Smolen, M.D., Humenik, S.J., Dressing, S.A., and Maas, R.P. 1984. Best Management Practices for Agricultural Nonpoint Source Control. IV. Pesticides. North Carolina Agric. Exp. Stn., Raleigh.

Somers, D.A., Keith, R.A., Egli, M.A., Marshall, L.C., Gengenbach, B.G., Gronwald, J.W., and Wyse, D.L. 1993. Expression of the Acc1 gene-encoded acetyl-coenzyme A carboxylase in developing maize (*Zea mays*) kernels. *Plant Physiol.* 101: 1097.

Tshibaka, T.B. 1992. Labor in the Rural Household Economy of the Zairian Basin. Rep. No. 90. International Food Policy Research Institute, Washington, D.C.

Wicks, G.A. and Burnside, O.C. 1981. Effect of cultivation and herbicides on ecofallow sorghum. *Proc. North Centr. Weed Control Conf.* 36: 124.

Wyse, D.L. 1992. Future impact of crops with modified herbicide resistance. *Weed Technol.* 6: 665.

Zimdahl, R.L. 1991. Weed Science — A Plea for Thought. Cooperative State Research Service, U.S. Department of Agriculture, Washington, D.C.

INDEX

C

D

F

G

K

L

M

N

T

U

V

W

X

Z

For Product Safety Concerns and Information please contact our EU representative GPSR@taylorandfrancis.com Taylor & Francis Verlag GmbH, Kaufingerstraße 24, 80331 München, Germany